T0257455

Concepts and Applications of Polysaccharides

Concepts and Applications of Polysaccharides

Edited by **Erica Young**

New York

Published by Callisto Reference,
106 Park Avenue, Suite 200,
New York, NY 10016, USA
www.callistoreference.com

Concepts and Applications of Polysaccharides
Edited by Erica Young

© 2015 Callisto Reference

International Standard Book Number: 978-1-63239-121-6 (Hardback)

Contents

Preface

Over the recent decade, advancements and applications have progressed exponentially. This has led to the increased interest in this field and projects are being conducted to enhance knowledge. The main objective of this book is to present some of the critical challenges and provide insights into possible solutions. This book will answer the varied questions that arise in the field and also provide an increased scope for furthering studies.

This book is a collection of the varied aspects of polysaccharides in microorganisms, plants and animals. The diversity of these polysaccharides emerges from the structural alterations and the monosaccharide content which is under genetic control. Their physical as well as chemical properties have made them useful in a variety of food, pharmaceutical and industrial applications. These properties determine their biological activity and their functions in different applications. The use of polysaccharides as therapeutics, the role played by them in protection and preservation of food, as carriers of drugs, nutrients and their ability to communicate with molecules for improving textures of food colloids, as well as for efficient delivery are a few important functions. The book is organized under two sections: Applications in Food Industry and Applications in Pharmaceutical Industry.

I hope that this book, with its visionary approach, will be a valuable addition and will promote interest among readers. Each of the authors has provided their extraordinary competence in their specific fields by providing different perspectives as they come from diverse nations and regions. I thank them for their contributions.

Editor

Applications in the Food Industry

Polysaccharides as Carriers and Protectors of Additives and Bioactive Compounds in Foods

Rosa M. Raybaudi-Massilia and Jonathan Mosqueda-Melgar

Additional information is available at the end of the chapter

1. Introduction

Overall quality and shelf-life of fresh foods post-harvest or -slaughter is reduced by several factors including microbial growth, water loss, enzymatic browning, lipid oxidation, off-flavor, texture deterioration, rise in respiration rate and senescence processes, among others. In fresh-cut fruits and vegetables, these events are accelerated due to lesions of tissues during peeling, slicing and cutting [1]; whereas, in meat products these events are accelerated due to lesions of tissues during cutting and grounding.

Fresh-cut fruits and vegetables during mechanical operations are exposed to spoil because the natural protection of fruit (the peel or skin) is generally removed and hence, they become highly susceptible to microbial growth due to the leakage of juices and sugar from damaged tissues which allow the growth and fermentation of some microorganism. Likewise, during processing, enzymes such as polyphenol oxidase, polygalacturonase and pectin methylesterase are released, thus causing, browning and softening from the tissues, respectively. These enzymes come in contact with phenolic compounds for forming brown pigments, and hydrolyze the α-1,4-glucosidic bonds to degrade the tissues [2].

Appearance/color, texture, and flavor are the main quality attributes that affect consumer acceptance of meat, and lipid oxidation is one of the primary causes by quality deterioration in meat and meat products [3]. The meat once cut or sliced is exposed to the surrounding environment, and cell compounds released during mechanical operations react with the environment and cause quality deterioration of tissues. Lipid oxidation occur when oxygen come in contact with lipid present in pieces of meats, being the iron the major catalyst in lipid oxidation processes. This process is associated with the presence of free radicals that lead to the production of aldehydes, which are responsible for the development of rancid flavors and changes in the color of meat [4].

On the other hand, dairy products such as fresh and semi-hard cheeses are complex food products consisting mainly of casein, fat, and water. Such products are highly perishable due to the high content of moisture (only in fresh cheeses) and microorganisms, and some cases high fat-content [5]; therefore, off-flavor, lipid oxidation and microbial spoilage are the major quality deteriorations.

Because of happened issues in the past with fresh or fresh-cut products, new technologies have been applied to counteract these negatives effects. Among them, polysaccharide-based edible films and coatings have emerged like good alternative for enhancing the quality and safety of such foods. Edible films and coatings have been used to reduce the deleterious effect caused by minimal processing. The semipermeable barrier provided by edible coatings is focused to extend shelf-life by reducing moisture and solutes migration, gas exchange, respiration and oxidative reaction rates, as well as suppress physiological disorders on fresh and/or fresh-cut foods [6]. However, the use of edible films and coatings for a wide range of food products, including fresh and minimally processed vegetables and fruits, has received an increasing interest because films and coatings can serve as carriers for a wide range of food additives including: antimicrobials, antioxidants and antisoftening compounds into edible films or coatings to provide a novel way for enhancing the safety and shelf-life of fresh, fresh-cut or ready-to-eat foods [7-10].

The new generation of edible films and coatings is being especially designed to increase their functionalities by incorporating natural or nutraceutical/functional ingredients such as probiotics, minerals and vitamins [10-11]. In addition, the sensory quality of coated products with edible materials can be also improved [2,7-9]. On the other hand, encapsulation (microencapsulation or nanoencapsulation) are being currently applied to foods to preserve and protect the additive or bioactive compounds from the surrounding environment [12-14].

In this chapter will discuss the use of polyssacharide-based edible films and coatings as polymeric matrix to carrier additive or bioactive compounds such as antimicrobial, antioxidant, antisoftening and nutraceutical for enhancing the shelf-life, safety and sensory attributes of fresh food products, as well as methodologies of forming and application of edible films and coatings and futures trends using microencapsulation or nanotechnology.

2. Use of polysaccharide-based edible films and coatings as carriers of additives and bioactive compounds on foods

The overall quality of food products decreases from harvest or slaughter until they are consumed. Quality loss may be due to microbiological, enzymatic, chemical, or physical changes. Therefore, food additives should be added to prevent the quality loss and extend the shelf-life of foods. The use of films and coatings have been a good alternative for carrier different additives and bioactive compounds to the food, as well as, to protect them of the water loss, volatile compounds loss, discoloration, gas permeability and microbial spoilage; since, these can to guarantee the controlled supply of antimicrobial, antioxidant, antisoftening and nutraceutical compounds. Tables 1, 2, 3 and 4 show the additives (antimicrobial, antioxidant, antisoftening and nutraceutical compounds) added on foods through polysaccharide-based edible films and coatings for improving the food quality and safety. Each of these additives is studied in the following sections of this chapter.

2.1. Carriers of antimicrobial compounds

Foods may be contaminated with pathogenic and spoilage microorganisms if bad manufacture practices are carried out during handling, processing, distribution and commercialization [15]. Therefore, antimicrobial compounds should be used during processing and packaging for controlling the microbiological safety and quality, and prolonging the shelf-life of foods. Food antimicrobials are chemical compounds added or naturally occurring in foods to inhibit or inactivate populations of pathogenic and spoilage microorganisms.

Several studies have demonstrated that antimicrobials such as organic acids, enzymes, essential oils, spices and bacteriocin incorporated into polysaccharide-based edible films and coatings have been effective for controlling pathogenic and spoilage microorganisms in different foods (Table 1). In this context, different researchers have demonstrated that the incorporation of bacteriocins into alginate-based film have been effective to inactivate or delay the growth of some pathogenic microorganisms. The alginates are anionic polysaccharides from the cell walls of brown algae that can serve to prepare carriers solution of antimicrobial substances. Hence, Cutter and Siragusa [16], Natrajan and Sheldon [17] and Milette et al. [18] achieved to reduce populations of *Brochothrix thermosphacta* (> 3.0 log CFU/g), *Salmonella enterica* ser. Typhimurium (> 4.0 log CFU/g) and *Staphylococcus aureus* (> 2.5 log CFU/g) on ground beef, poultry skin and beef fillets, respectively, using calcium alginate-based film and coating, and palmitoylated alginate incorporated with nisin (from 5 to 100,000 mg/mL) during storage refrigerated. Likewise, Datta et al. [19] indicated that the growth of *Listeria monocytogenes* and *Salmonella enterica* ser. Anatum was suppressed in the range of 2.2 to 2.8 log CFU/g in smoked salmon coated with an alginate coating containing oyster lysozyme at 160,000 mg/g plus nisin at 10 mg/g during storage at 4ºC by 35 days. Neetoo et al. [20] delayed the growth of *L. monocytogenes* on cold-smoked salmon slices and fillets during the 30 days storage at 4ºC using alginate-based edible coating with sodium lactate (2.4%) and diacetate (0.25%). Marcos et al. [21] reported a bacteriostatic effect against *L. monocytogenes* inoculated in sliced cooked ham during 60 days of storage at 1ºC, when enterocins A and B (2,000 AU/cm^2) were incorporated into an alginate film.

On the other hand, essential oils and their active compounds have been also incorporated into the alginate-based films and coatings to control the growth of pathogenic and spoilage microorganisms in several foods [22-26]. Hence, Oussalah et al. [22] evaluated the effect of an alginate-based film containing essential oils of Spanish oregano, Chinese cinnamon or winter savory at 1% w/v against populations of *S. enterica* ser. Typhimurium or *E. coli* O157:H7 inoculated in beef muscle slices stored at 4 ºC by 5 days. These authors reported that films including essential oils of oregano or cinnamon were more effectives against *S enterica* ser. Typhimurium (>1 log cycle); whereas, films including essential oils of oregano

Type of polysaccharide matrix		Food	Antimicrobial compounds	Reference
Film	Coating			
Alginate	-	Ground beef	Nisin	[16]
Alginate	-	Poultry skin	Nisin	[17]
Alginate	-	Beef fillets	Nisin	[18]
Alginate	-	Smoked salmon	lysozyme / nisin	[19]
Alginate	-	Cooked ham sliced	Enterocin A and B	[21]
Alginate	-	Beef fillets	EOs of oregano, cinnamon, savory	[22]
Alginate	-	Cooked ham and bologna sliced	EOs of oregano, cinnamon, savory	[23]
-	Alginate	Cold-smoked salmon slices and fillets	Sodium lactate and diacetate	[20]
-	Alginate/ apple puree	Fresh-cut apple	Vanillin and EOs of lemongrass, oregano	[24]
-	Alginate	Fresh-cut melon	Malic acid and EOs of lemongrass, cinnamon, plamarose, eugenol, citral, geraniol	[25]
-	Alginate	Fresh-cut apple	Malic acid and EOs of lemongrass, cinnamon, clove, cinnamaldehyde, eugenol, citral	[26]
-	Alginate	Fresh-cut apple	Potassium sorbate	[27]
-	Alginate	Roasted turkey	Sodium lactate and diacetate	[29]
Carrageenan	-	Fresh chicken breast	Ovotransferrin	[28]
Chitosan	-	Cooked ham, bologna and pastrami	Cinnamaldehyde, acetic, propionic and lauric acids	[32]
Chitosa	Chitosan	Mozzarella cheese	Lysozyme	[74]
Chitosan	-	Cheese	Natamycin	[38]
-	Chitosan	Whole strawberry	Potassium sorbate	[33]
-	Chitosan	Rainbow trout	Cinnamon oil	[37]
-	Chitosan	Roasted turkey	Sodium lactate and diacetate	[29]
Chitosa	Chitosan	Cold-smoked salmon	Potassium sorbate, sodium lactate and diacetate	[30]
-	Chitosan / Plastic	Ham steaks	Sodium lactate, diacetate and benzoate, potassium sorbate, or nisin	[36]
Chitosan	-	Pork sausages	Green tea extract	[39]
-	Chitosan / MC	Fresh-cut Pineapple and melon	Vanillin	[35]
CMC	-	Fresh pistachios	Potassium sorbate	[43]
Cellulose	-	Cooked ham sliced	Pediocin	[42]
Cellulose	-	Frankfurter sausages	Nisin	[41]
MC / HPMC	-	Hot Dog Sausage	Nisin	[40]
-	HPMC	Whole oranges	Potassium sorbate, sodium benzoate, sodium propionate	[44]
-	HPMC	Whole strawberry	Potassium sorbate	[33]
-	Pectin	Pork patties	Powder of green tea	[47]
-	Pectin	Roasted turkey	Sodium lactate and diacetate	[29]
-	Starch	Minimally processed carrots	Chitosan	[45]
-	Starch	Roasted turkey	Sodium lactate and diacetate	[29]
-	Starch / Gum	Fruit-based salad, romaine hearts and pork slices	Green tea extract	[46]

MC: methyl cellulose; CMC: carboxy methyl cellulose; HPMC: hydroxyl propyl methyl cellulose; EOs: essential oils

Table 1. Major antimicrobial compounds applied on foods through polysaccharide-based edible films and coatings

Type of polysaccharide matrix		Food	Antioxidant compounds	Reference
Film	Coating			
-	Alginate	Fresh-cut apple	Glutathione and N-Acetyl-cysteine	[61-62]
-	Alginate	Fresh-cut pears	Glutathione and N-Acetyl-cysteine	[63]
-	Alginate	Fresh-cut apples	Calcium chloride	[27]
-	Alginate	Bream (freshwater fish)	Vitamin C and tea polyphenols	[53]
-	Carrageenan	Fresh-cut apples	Ascorbic, citric and oxalic acids	[59]
-	Carrageenan	Fresh-cut banana	Ascorbic acid and cysteine	[60]
-	CMC	Fresh-cut apples and potatoes	Ascorbic acid and TBHQ	[57]
-	Gellan	Fresh-cut apple	Glutathione and N-Acetyl-cysteine	[61-62]
-	Gellan	Fresh-cut pears	Glutathione and N-Acetyl-cysteine	[63]
-	HPMC	Toasted almonds	Ascorbic, citric and EO ginger	[52]
-	MC	Fresh-cut apples	Ascorbic acid	[58]
-	Maltodextrin	Fresh-cut apples	Ascorbic acid	[58]
-	Pectin	Fresh-cut pears	Glutathione and N-Acetyl-cysteine	[63]
-	Pectin	Pork patties	Powder of green tea	[47]

MC: methyl cellulose; CMC: carboxy methyl cellulose; HPMC: hydroxyl propyl methyl cellulose; EOs: essential oils

Table 2. Major antioxidant compounds applied on foods through polysaccharide-based edible films and coatings

Type of polysaccharide matrix		Food	Antisoftening compounds	Reference
Film	Coating			
-	Alginate	Fresh-cut apples	Calcium lactate	[26]
-	Alginate	Fresh-cut melons	Calcium lactate	[27]
-	Alginate	Fresh-cut apples	Calcium chloride	[61-62]
-	Alginate	Fresh-cut melons	Calcium chloride	[67]
-	Alginate	Fresh-cut pears	Calcium chloride	[63]
-	Alginate	Fresh-cut apples	Calcium chloride	[27]
-	Alginate	Fresh-cut papayas	Calcium chloride	[68]
-	Alginate / Apple puree	Fresh-cut apples	Calcium chloride	[24]
-	Carrageenan	Fresh-cut apples	Calcium chloride	[59]
-	Carrageenan	Fresh-cut banana	Calcium chloride	[60]
-	Gellan	Fresh-cut papayas	Calcium chloride	[68]
-	Gellan	Fresh-cut apples	Calcium chloride	[61-62]
-	Gellan	Fresh-cut melons	Calcium chloride	[67]
-	Gellan	Fresh-cut pears	Calcium chloride	[63]
-	Pectin	Fresh-cut melons	Calcium chloride	[67]
-	Pectin	Fresh-cut pears	Calcium chloride	[63]

Table 3. Major antisoftening compounds applied on foods through polysaccharide-based edible films and coatings

Type of polysaccharide matrix		Food	Nutraceutical compounds	Reference
Film	Coating			
-	Alginate	Fresh-cut papayas	Probiotics	[11]
-	Gellan	Fresh-cut papayas	Probiotics	[11]
-	Chitosan	Lingcod fillets	Omega 3 and Vitamin E	[34]
-	Chitosan	Whole strawberry	Calcium	[72]
-	Chitosan	Whole strawberry and red raspberry	Calcium and Vitamin E	[73]
-	Xanthan gum	Peeled baby carrots	Calcium and Vitamin E	[71]

Table 4. Major nutraceutical compounds applied on foods through polysaccharide-based edible films and coatings

was more effective against *E. coli* O157:H7 (> 2 log cycles). Similarly, Oussalah et al. [23] studied the effect of alginate-based edible film containing essential oils of Spanish oregano, Chinese cinnamon, or winter savory at 1% (w/v) against *S. enterica* ser. Typhimurium or *L. monocytogenes* inoculated onto bologna and ham slices. These authors concluded that alginate-based films containing essential oil of cinnamon was the most effective in reducing the populations of both pathogenic microorganisms by more than 2 logs CFU/g on bologna and ham sliced. In the same way, Rojas-Grau et al. [24] studied the antimicrobial effect of essential oils of lemongrass (1and 1.5%) and oregano (0.1 and 0.5%), and vanillin (0.3 and 0.6%) incorporated into coating forming solutions based on alginate and apple puree against the naturally occurring microorganisms and *Listeria innocua* inoculated on fresh-cut apples. These authors found that all the essential oils used significantly inhibited the native flora during 21 days of storage at 4°C, being lemongrass and oregano oils more effective against *L. innocua* than vanillin. Likewise, Raybaudi-Massilia et al. [25] reported significant reduction (3–5 log cycles) of the inoculated *Salmonella enterica* var. Enteritidis population in pieces of melon when an alginate-based edible coating containing malic acid (2.5%), alone or in combination with essential oils of cinnamon, palmarose or lemongrass at 0.3 and 0.7% or their actives compounds eugenol, geraniol and citral at 0.5%, were applied. In addition, inhibition of the native flora by more than 21 days of storage was also observed at 5°C. Similar results were found by Raybaudi-Massilia et al. [26], who evaluated the antimicrobial effect of an alginate-based edible coating with malic acid (2.5%) incorporated, alone or in combination with essential oils of cinnamon bark, clove or lemongrass at 0.3 and 0.7% or their actives compounds cynnamaldehyde, eugenol or citral at 0.5% on fresh-cut apples. They reached to reduce population of *Escherichia coli* O157:H7 (4 log cycles) after 30 days of refrigerated storage (5°C), as well as to inhibit the native flora by more than 30 days.

Other antimicrobial compounds such as potassium sorbate, ovotransferrin, sodium lactate and sodium acetate have been also applied to fresh-cut apples, fresh chicken breast and ready-to-eat roasted turkey through an alginate-based edible coating to inhibit the native flora growth. In such sense, Olivas et al. [27] inhibited the microbial growth of mesophilic and psychrotropic bacteria, moulds and yeasts in apple slices coated with an edible alginate coating containing 0.05% potassium sorbate during 8 days of storage at 5°C. Likewise, Seol et al. [28] reduced populations of total microorganisms (about 2 log cycles) and E. coli (about 3 log cycles) on fresh chicken breast stored at 5°C after 7 days, when a κ-carrageenan-based edible film containing ovotransferrin (25 mg) and EDTA (5 mM) was applied on its surface. Jiang et al. [29-30] showed that potassium sorbate (0.15%), sodium lactate (1.2-2.4%) and sodium diacetate (0.25-0.50%) incorporated into chitosan- or alginate-based edible coating and film were able to inactivate L. monocytogenes (about 1-3 log CFU/g) on ready-to-eat cold-smoked salmon and roasted turkey stored at 4°C. All these results have demonstrated that alginate- κ-carrageenan-based film and coating are excellent carriers of antimicrobial substances on meat, poultry and fruits and vegetables products for reducing populations of pathogenic microorganisms.

In the same way, chitosan which is a linear polysaccharide consisting of (1,4)-linked 2-amino-deoxy-b-D-glucan, and a deacetylated derivative of chitin, and the second most abundant polysaccharide found in nature after cellulose [31] has been used as carrier of antimicrobial compounds in other foods. In such sense, Ouattara et al. [32] evaluated the effectiveness of chitosan films incorporated with acetic or propionic acid, with or without addition of lauric acid or cinnamaldehyde to preserve vacuum-packaged bologna, cooked ham and pastrami during refrigerated storage. The efficacies of the films to inhibit the microbial growth were tested against native lactic acid bacteria, Enterobacteriaceae, and against Lactobacillus sakei or Serratia liquefaciens inoculated on the surface of products. The authors indicated that the growth of lactic acid bacteria were not affected by the antimicrobial films, but the growth of Enterobacteriaceae and S. liquefaciens was delayed or completely inhibited after application. Park et al. [33] showed the antifungal effect of a chitosan-based edible coating containing potassium sorbate (0.3%) to inhibit the Cladosporium sp. and Rhizopus sp, total aerobic count and coliforms growth, and in fresh strawberries stored at 5°C and 50% RH by 23 days. Coating treatment also reduced total aerobic count, coliforms, and weight loss of strawberries during storage. Duan et al. [34] reduced about 1 log cycle the populations of L. monocytogenes, E. coli, or Pseudomonas fluorescens inoculated on the surface of Mozzarella cheese using chitosan composite films and coatings incorporated with lysozyme and storage at 10 °C. Sangsuwan et al. [35] studied the antimicrobial effect of a chitosan/MC film incorporated with vanillin against E. coli and Saccharomyces cerevisiae inoculated on fresh-cut cantaloupe and pineapple. They found that antimicrobial film inactivated populations of E. coli and S. cerevisae on fresh-cut cantaloupe by more than 5 and 0.6 log CFU/g during 8 and 20 days of storage, respectively, at 10°C. Whereas, this antimicrobial film inactivated S. cerevisiae on fresh-cut pineapple by more than 4 log CFU/g during 12 days of storage at 10°C, but against E. coli there was not significant reductions. Ye et al. [36] used a plastic film coated with chitosan and Sodium lactate (1%), diacetate (0.25%), and benzoate (0.1%), potassium sorbate (0.3%) or nisin (5 mg/cm²) for inhibiting the growth of L. monocytogenes on strawberries during 10 days of storage at room temperature (20°C). Ojagh et al. [37] extended the shelf-life of Rainbow trout (a fish native of North America) during 16 days at 4°C incorporating cinnamon oil (at 1.5%) into a matrix of chitosan-based edible coating. Fajardo et al. [38] evaluated the antifungal activity of chitosan-based edible coating containing 0.5mg/mL natamycin on semi-hard "Saloio" cheese; and demonstrated that populations of moulds and yeasts were reduced by about 1.1 log CFU/g compared to control samples after 27 days of refrigerated storage. Jiang et al. [29] showed that a combination of sodium lactate and sodium diacetate incorporated into chitosan edible coating was able to inactivate L. monocytogenes on ready-to-eat roasted turkey stored at 4°C. Siripatrawan and Noipha [39] used a chitosan film containing green tea extract as active packaging for extending shelf-life of pork sausages. These authors completely inhibited the microbial growth in pork sausages refrigerated (4 °C). Hence, chitosan can be used as a natural antimicrobial coating on fresh strawberries to control the growth of microorganisms, thus extending shelf-life of the products

Films and coatings based on cellulose or derivatives such as methyl cellulose (MC), carboxy methyl cellulose (CMC) or hydroxy propyl methyl cellulose (HPMC) containing antimicrobial compounds have been used to control microbial growth and extend the shelf-life of several foods. In such sense, Franklin et al. [40] determined the effectiveness of packaging films coated with a MC/HPMC–based solution containing 100, 75, 25 or 1.563 mg/ml nisin for controlling L. monocytogenes on the surfaces of vacuum-packaged hot dogs. They found that packaging films coated with a cellulose-based solution containing 100 and 75 mg/ml nisin significantly decreased ($P \leq 0.05$) L. monocytogenes populations on the surface of hot dogs by greater than 2 logs CFU/g throughout the 60 days of storage. Nguyen et al. [41] developed and used cellulose films produced by bacteria containing nisin to control L. monocytogenes and total aerobic bacteria on the surface of vacuum-packaged frankfurters. Bacterial cellulose films were produced by Gluconacetobacter xylinus K3 in corn steep liquor-mannitol medium and were subsequently purified before nisin was incorporated into them. Cellulose films with nisin at 25 mg/ml significantly reduced (P<0.05) L. monocytogenes (approximately 2 log CFU/g) and total aerobic bacteria (approximately 3.3 log CFU/g) counts on frankfurters after 14 days of storage as compared to the control samples. Whereas, Santiago-Silva et al. [42] developed and evaluated the antimicrobial efficiency of cellulose films with pediocin (antimicrobial peptide produced by Pediococcus sp.) incorporated at 25% and 50% of cellulose weight on sliced ham. They found that antimicrobial films were more effective against L. innocua than Salmonella sp., since the 50% pediocin-film showed a reduction of 2 log CFU/g in relation to control treatment after 15 days of storage; whereas, the 25% and 50% pediocin-films had similar performance on Salmonella sp. about 0.5 log CFU/g reductions in relation to control, after 12 days of storage at 12ºC. On the other hand, Park et al. [33] achieved to inhibit the growth of Cladosporium sp., Rhizopus sp, total aerobic count and coliforms on fresh strawberry through a HPMC-based edible coating containing potassium sorbate (0.3%) stored at 5°C and 50% RH by 23 days. Sayanjali et al. [43] evaluated the antimicrobial properties of edible films based on CMC containing potassium sorbate (at 0.25, 0.5 and 1.0%) applied on fresh pistachios, and reported that all concentrations of potassium sorbate used inhibited the growth of molds. Valencia-Chamorro et al. [44] studied the antifungal effect of HPMC based coatings with potassium sorbate (2%), sodium benzoate (2.5%), sodium propionate (0.5%) and their combinations on the postharvest conservation of "Valencia" oranges. These authors reported that the application of HPMC coatings reduce significantly the effects caused by *Penicillum digitatum* and *Penicilllum italicum* inoculated in the surface of the oranges, resulting more effectives those coatings with potassium sorbate and sodium propionate combined.

Others polysaccharides-based films and coatings such as pectins and starches have been used also as carriers of antimicrobials compounds in foods. Durango et al. [45] controlled the growth of mesophilic aerobes, yeasts and moulds and psychrotrophics populations in processed minimally carrots during the first 5 days of storage at 15°C using yam starch-based edible coatings containing chitosan (0.5 and 1.5%). In the same way, Chiu and Lai [46] studied the antimicrobial properties of edible coatings based on a tapioca starch/decolorized hsian-tsao leaf gum matrix with incorporated green tea extracts on fruit-based salads, romaine hearts and pork slices. The authors indicated that when green tea extracts at 1% were added into edible coating formulations, the aerobic count successfully decreased and growth of yeasts/molds decreases by 1 to 2 logs CFU/g in fruit-based salads. In addition, they reported that romaine hearts and pork slices coated with these antimicrobial edible coatings reduced populations of Gram positive bacteria from 4 to 6 logs CFU/g during 48 h of refrigerated storage. On the other hand, Kang et al. [47] evaluated the microbiological quality of pork hamburger coated with a pectin-based edible coating with incorporated green tea powder (0.5%), and packed in air or vacuum during 14 days at 10°C. These authors reported that initial population of total aerobic microorganisms (10^4 CFU/mg) decreased until undetectable levels by more than 7 days under vacuum conditions; whereas, in normal conditions of atmosphere (air) a level of 10^5 CFU/mg was reached at the same time. Jiang et al. [29] showed that a combination of sodium lactate and sodium diacetate incorporated into pectin-based edible coating was able to inactivate populations of *L. monocytogenes* on ready-to-eat roasted turkey stored at 4°C.

Previous results have showed that several polysaccharides-based films and coatings (alginate, carrageenan, chitosan, cellulose derivatives, pectin, starch and apple puree) could be used as outstanding carriers of antimicrobial substances for ensuring the quality and safety of foods in the meat, poultry, seafood, dairy, fruits and vegetables industries. In addition, the incorporation of essential oils into films and coatings formulations may contribute to prevent the water vapor permeability and decreases the solubility of films and coatings in foods with high content of humidity.

3. Carriers of antioxidant compounds

Antioxidant compounds can also be incorporated into edible films and coatings to avoid the food oxidation and browning. In such sense, rosemary oleoresin, an extract of spice with antioxidant activity, has been added into starch-alginate coatings to inhibit the lipid oxidation and warmed-over flavor (WOF) development in precooked pork chops [48] and beef patties [49]. In the same way, tocopherols have been incorporated into starch-alginate coatings to retard the formation of WOF in precooked pork chops [50]. Wu et al. [51] studied the effect of starch-alginate (SA), SA-stearic acid (SAS), SA-tocopherol (SAT), SAS-tocopherol (SAST), SAT-coated (SATC), and SAST-coated (SASTC) films on moisture loss and lipid oxidation in precooked ground-beef patties. These authors reported that tocopherol-treated films were more effective ($P < 0.05$) in inhibiting lipid oxidation than those tocopherol-untreated films on ground-beef patties. However, SAS-based films were more effective ($P < 0.05$) in controlling moisture loss than lipid oxidation. Atarés et al. [52] evaluated the antioxidant efficiency of HPMC coatings with ascorbic acid, citric acid or ginger essential oil incorporated on toasted almonds to avoid the lipid oxidation. They concluded that films with ascorbic and citric acid showed a cross-linking effect, and were the most effective protectors against oxidation of almonds, due to both their antioxidant effect and the tighter structure which leads to lower oxygen permeability. Khang et al. [47] found that lipid oxidation decreased and radical scavenging increased in the pork patties coated with a pectin-based edible coating containing green tea leaf extract (0.5%) during 14 days at 10°C. These authors indicated that coated patties held higher moisture contents than the controls in both air- and vacuum packaging. Song et al. [53] indicated that sodium alginate-based edible coating containing vitamin C (5%) or tea polyphenols (0.3%) were able to delay the chemical spoilage and water loss of bream (Megalobrama amblycephala), in addition to enhancing the overall sensory attributes, in comparison with uncoated bream during 21 days storage at 4 ± 1°C.

On the other hand, the color in fresh-cut fruits and vegetables is of great importance, since oxidation and enzymatic browning take place quickly upon contact with oxygen during processing, leading to discoloration [54]. Browning phenomena in fresh-cut products are caused when, after mechanical operations (cutting, slicing, coring, shredding, etc) during processing, enzymes, which are released from wounded tissues, come in contact with phenolic components to give dark colored pigments [55]. Such phenomenon is caused by the action of a group of enzymes called polyphenol oxidases (PPOs), which can oxidize the phenolic substrates to o-quinones in presence of oxygen [56]. Therefore, the application of antioxidant agents incorporated into edible coatings would be a good alternative to ensure the inhibition of browning, to prevent ascorbic acid or vitamin C loss, and extend the shelf-life of fresh-cut fruits and vegetables [9]. In such sense, Baldwin et al. [57] reported that ascorbic acid (0.5%) and ter-butyl-hydro-quinone (0.2%) had a better effect on the inhibition of browning in fresh-cut apples and potatoes throughout storage when these antioxidants were incorporated into an edible coating based on CMC than when these were used in an aqueous dipping solution after 14 days at 4°C. Both methods were effective during the first day of storage, but samples coated with the edible coating prevented browning for a longer time than those samples dipped in an aqueous solution alone. Brancoli and Barbosa-Cánovas [58] achieved a decreasing browning in surface of apple slices during 21 days of storage at 4°C using maltodextrin and MC-based coatings containing ascorbic acid (1%). Likewise, Lee et al. [59] delayed the browning of fresh-cut apples using antibrowning agents such as ascorbic (1%), citric (1%), oxalic (0.05%) acid or their combinations incorporated into edible coatings based on carrageenan. These authors observed an inhibition of the enzymatic browning in fresh-cut apples during 14 days storage at 3°C. In addition, edible coating with antioxidants obtained higher sensory scores (positive effect) during sensory evaluation than non-coated apples. In the same way, Bico et al. [60] reached to retard the browning of fresh-cut bananas using ascorbic acid and cysteine at 0.75% incorporated into an edible coating based on carrageenan during 5 days of storage at 5°C. Rojas-Grau et al. [61-62] inhibited the browning in fresh-cut apples using edible coatings based on alginate or gellan with the addition of glutathione (up to 2%) or N-acetyl-cysteine (up to 2%), or their combination. These authors indicated that a concentration of 1% each of the antibrowning agents was needed to maintain the color of cut apples. Similar results were also obtained by Oms-Oliu et al. [63], who achieved browning inhibition of fresh-cut "Flor de invierno" pears for 14 days at 4°C using N-acetyl- cysteine (0.75%) and glutathione (0.75%) incorporated into edible coatings based on alginate, gellan or pectin. Olivas et al. [27] delayed the development of browning in apple slices during 8 days of storage at 5°C after applying alginate coatings containing calcium chloride (10%). Calcium chloride is an anti-browning agent known to inhibit PPO by interaction of the chloride ion with copper at the PPO active site [64].

Based on the different works reported in the bibliography is possible indicates that several polysaccharides-based films and coatings (alginate, carrageenan, cellulose derivatives, pectin, gellan and maltodextrin) could be used as excellent carriers of antioxidant substances for avoiding the lipid oxidation and enzymatic browning of meat and fruits products.

4. Carriers of antisoftening compounds

Foods more susceptible to the texture loss are fresh-cut fruits and vegetables. This fact is due to that during mechanical operations (peeling, cutting, sliced, shredded) plant tissues are breakdown and enzymes such as pectinolytic and proteolytic are released, thus causing softening [1]. In addition, these enzymes could also affect the morphology, cell wall middle lamella structure, cell turgor, water content, and biochemical components [65]. Pectinase enzymes such as polygalacturonase and pectin methylesterase are responsible for texture losses in plant tissues. Polygalacturonase hydrolyses the α-1,4-glucosidic bond among anhydrogalacturonic acid units, whereas, pectin methylesterase hydrolyses the methyl-ester bonds of pectin to give pectic acid and methanol, thus resulting in texture degradation because of hydrolysis of the pectin polymers [1]. Nonetheless, treatments with calcium can helping to counteract this problem improving the firmness of fruit tissues by reacting with pectic acid present in the cell wall to form calcium pectate, which reinforces the molecular bonding among constituents of the cell wall, thus delaying the senescence and controlling physiological disorders in fruits and vegetables [8,66]. Different studies have demonstrated that the use of polysaccharide based films and coatings (alginate, carrageenan, pectin, gellan and apple puree) as carriers of calcium chloride or lactate have resulted be a good alternative to prevent the firmness or texture loss of the fresh-cut fruits, which could be beneficial to the fresh-cut fruits industry.

In this sense, Oms-Oliu et al. [63,67] and RojasGraü et al. [24,61-62] observed that fresh-cut melons, pears, and apples coated with alginate-, gellan-, pectin- or apple-puree edible coatings containing calcium chloride (2%) maintain in excellent conditions their initial firmness during refrigerated storage (4°C) from 14 to 21 days. The authors indicated that polysaccharide matrices with substances increased the water vapor resistance, thus preventing dehydration, and they had an inhibitory effect on ethylene production, but O_2 and CO_2 production was not affected. Similar effects were achieved by Olivas et al. [27], who preserved the firmness of apple slices stored at 5ºC for 10 days by using an alginate edible coating containing calcium chloride (10%). Raybaudi-Massilia et al. [25,26] showed that the incorporation of calcium lactate (2%) into an alginate-edible coating maintained the firmness of fresh-cut apples and melons during 21 days at 5ºC. Similarly, Tapia et al. [68] improved the firmness of fresh-cut papaya with the addition of calcium chloride (2%) into alginate- and gellan edible coating during the period studied (8 days at 4ºC). Likewise, Lee et al. [59] and Bico et al. [60] kept the firmness of fresh-cut apple and banana slices storage at refrigerated temperature using a carrageenan-based edible coating containing calcium chloride (1%).

5. Carriers of nutraceutical compounds

Nutraceuticals are chemical compounds found as natural components of foods or other ingestible forms that have been determined to be beneficial to the human body in preventing or treating one or more diseases or improving physiological performance [69]. Calcium and vitamin E are the most important nutraceutical compounds and, they can play significant roles in the human body in preventing certain diseases [70]. Nonetheless, probiotics are being used currently as a functional compound in foods, since potential health benefits and biological functions of bifidobacteria in humans like the intestinal production of lactic and acetic acids, pathogens inhibition, reduction of colon cancer risks, cholesterol reduction in serum, improved calcium absorption, and activation of the immune system, among others [11]. Thus, nutraceutical compounds carried into edible coatings and films to strengthen and increase the nutritional value of foods have been researched.

Edible coatings can provide an excellent vehicle to further enhance the health benefit of products like berry fruits where the lack of some important nutraceuticals, such as vitamin E and calcium may be compensated by incorporating them into the coatings [10]. In this way, Mei et al. [71] used xanthan gum coating as a carrier of calcium (as calcium lactate at 5%) and vitamin E (as α-tocopheryl acetate at 0.2%) for covering peeled baby carrots. The authors indicated that calcium and vitamin E contents of the coated samples (85g per serving), increased from 2.6 to 6.6% and from 0 to about 67% of the Dietary Reference Intakes values, respectively. In addition, they found that edible coatings improved the desirable surface color of carrots without significant effects on the taste, texture and fresh aroma. Hernández-Muñoz et al. [72] coated strawberries (*Fragaria* x *ananassa* Duch.) with a chitosan-based edible coating containing 1% calcium gluconate and stored during 4 days at 20°C. These authors found that strawberries coated with chitosan-based edible coating with incorporated calcium were better retained in coating (3,079 g/kg dry matter) than in strawberries dipped in calcium solutions alone (2,340 g/kg), thus resulting in increased nutritional value. Likewise, Han et al. [73] used chitosan-based edible coatings containing 5% Gluconal® CAL or 0.2% DL-α-tocopheryl acetate to enhance the nutritional value of strawberries (*Fragaria* × *ananassa*) and red raspberries (*Rubus ideaus*) stored at 2°C and 88% relative humidity (RH) for 3 weeks or at 23°C up to 6 months. They concluded that chitosan-based coatings containing calcium or Vitamin E significantly increased the content of these nutrients in both fresh and frozen fruits. These researchers also indicated that adding high concentrations of calcium or Vitamin E into chitosan-based coatings did not alter their anti-fungal and moisture barrier functions. Moreover, the coatings significantly decreased decay incidence and weight loss, drip loss and delayed the change in color, pH and titratable acidity of strawberries and red raspberries during cold storage. Duan et al. [74] increased total lipid and omega-3 fatty acid contents of fresh and frozen lingcod by about 3-fold and reduced TBARS (Thio-barbituric acid reactive substances) values in both fresh and frozen samples, incorporating 10% fish oil (containing 91.2% EPA (eicosapentaenoic acid) and DHA (docosahexaenoic acid)) plus 0.8% vitamin E into chitosan-based edible coating.

Developing edible coatings to carry high concentrations of nutraceuticals for nutritionally fortified foods can also be considered as an important way to afford functional characteristics to coated foods. In this context, Tapia et al. [11] managed to incorporate viable *Bifidobacterium lactis* Bb12 strains into alginate and gellan film-forming solutions to coat fresh-cut apples and papayas, and evaluated the effectiveness of such edible coatings to carry and support the probiotic culture. The authors reported that populations > 10^6 CFU/g of the microorganism were kept during 10 days of refrigerated storage. A viable bifidobacteria population of 5 logs CFU/g in the final product has been pointed out as the therapeutic minimum to attain health benefits [75].

In general, fruits, vegetables and seafood industries could apply different polysaccharides-based coatings (alginate, gellan, chitosan and gum) as excellent carriers of nutraceutical compounds for adding nutritive value and functional properties to the products.

6. Methodology for film and coating formation, incorporation of additives/bioactive compounds and ways of applications

6.1. Film and coating formation, and incorporation of additives/bioactives compounds

An edible film is essentially an interacting polymer network of three-dimensional gel structure. Despite the film-forming process, whether it is wet casting or dry casting, film-forming materials should form a spatially rearranged gel structure with all incorporated film-forming agents, such as biopolymers, plasticizers, other additives, and solvents in the case of wet casting. Biopolymers film-forming materials are generally gelatinized to produce film-forming solutions. Sometimes drying of the hydrogels is necessary to eliminate excess solvents from the gel structure. This does not mean that the film-forming mechanism during the drying process is only the extension of the wet-gelation mechanism. The film-forming mechanism during the drying process may differ from the wet-gelation mechanism, though wet gelation is initial stage of the film-forming process. There could be a critical stage of a transition from a wet gel to a dry film, which relates to a phase transition from a polymer-in-water (or other solvents) system to a water-in-polymer system [76].

Two processes can be used for film-production: dry and wet. The dry process of edible film production does not use liquid solvent, such as water or alcohol. Molten casting, extrusion, and heat pressing are good examples of dry process. For the dry process, heat is applied to the film-forming materials to increase the temperature to above the melting point of the film-forming materials, to cause them to flow. The wet process uses solvents for the dispersion of film-forming materials, followed by drying to remove the solvent and form a film structure. For the wet process the selection of solvents is the one of the most important factors. Since the film-forming solution should be edible and biodegradable, only water, ethanol and their mixtures are appropriated as solvents. To produce a homogeneous film structure avoiding phase separation, various emulsifiers can be added to the film forming solution. This solvent compatibility of ingredients is very important to develop homogeneous edible film and coating systems carrying active agents. All ingredients, including active agents as well as biopolymers and plasticizers should be homogeneously dissolved in solvent to produce film-forming solutions [76].

7. Ways of application of films and coatings

Different ways for film and coating application have been reported in the literature; being dipping, spraying, brushing, casting and wrapping the more commons methods [7,76-79]:

- *Dipping:* This method lends to food products that require several applications of coating materials or require a uniform coating on an irregular surface. After dipping, excess coating material is allowed to drain from the product and it is then dried or allowed to solidify. This method has been generally used to apply coating of alginate, gellan, chitosan, MC and pectin to fresh-cut fruit.
- *Spraying:* Film applied by spraying can be formed in a more uniform manner and thinner than those applied by dipping. Spraying, unlike dipping, is more suitable for applying a film to only one side of a food to be covered. This is desirable when protection is needed on only one surface, e.g., when a pizza crust is exposed to a moist sauce. Spraying can also be used to apply a thin second coating, such as the cation solution needed to cross-link alginate or pectin coatings.
- *Brushing:* This method consists in the direct application and distribution of the coating material in a liquid form using a hand brush.
- *Casting:* This technique, useful for forming free-standing films, is borrowed from methods developed for not edible films. For formation of a film the film-forming biopolymers are first dissolved in the solvent. If heating or pH adjustment enhances film formation and/or properties, this is done nest. If a composite film or coating based on an emulsion is desired, a lipid material, and possibly a surfactant, is added. Next the mixture is heated to above the lipid melting point and then homogenized. Degassing is an important step to eliminate bubble formation in the final film or coating. Finally, the edible film or coating is formed by applying the prepared formulation to the desired coating or product surface and allowing the solvent to evaporate

- *Wrapping:* this method is obtained from cast films depending on firmness and flexibility for wrapping surface. It allows films to be cut to any size, and serves as an innovative and easy method for carrying and delivering a wide variety of ingredients such as flavoring, spices and seasoning that can later be used to cover foods. This method is especially useful when applied to highly spicy materials that need to be separated from the food products.

8. Preserving and protecting bioactive compounds through microencapsulation

Microencapsulation is a technique by which solid, liquid or gaseous active ingredients are packaged within a second material for the purpose of protecting or shielding the active compound from the surrounding environment. Thus the active compound is designated as the core material, whereas the surrounding material forms the shell. This technique can be employed in a diverse range of fields such as agricultural, chemical, pharmaceutical, cosmetics, printing and food industry [13].

Microcapsules can be classified on the basis of their size or morphology. Thus, microcapsules range in size from one micron; whereas, some microcapsules whose diameter is in the nanometer range are referred as nanocapsules to emphasize their smaller size. On the other hand, morphology microcapsules can be classified into three basic categories as mono-core (also called single-core or reservoir type), poly-core (also called multiple-core) and matrix types (Figure 1). Mono-core are microcapsules that have a single hollow chamber within the capsule; Poly-core are microcapsules that have a number of different sized chambers within the shell; and matrix type are microparticles that has the active compounds integrated within the matrix of the shell material. However, the morphology of the internal structure of a microparticle depends mostly on the selected shell materials and the microencapsulation methods that are employed [12-13].

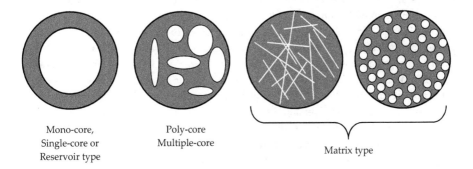

| Mono-core,
Single-core or
Reservoir type | Poly-core
Multiple-core | Matrix type |

Figure 1. Morphology of microcapsules

Current trends of the consumers for eating healthy foods that preventing illness and to be low calories but rich in vitamins, minerals and other bioactive component have conduced to the researches and industrials to develop foods called "functional", where some ingredients to promote health are added. However simply adding ingredients to food products to improve nutritional value can compromise their taste, color, texture and aroma. Sometimes, they are slowly degraded and in consequence, lose their activity, or become hazardous by oxidation reactions. Active compounds can also react with components present in the food system, which may limit bioavailability. Microencapsulation is used to overcome all these challenges by providing viable texture blending, appealing aroma release, and taste, aroma and color masking. This technology enables to the food industries to incorporate minerals, vitamins, flavors and essential oils. In addition, microencapsulation can simplify the food manufacturing process by converting liquids to solid powder, decreasing production costs by allowing batch processing using low cost, powder handling equipment. Microcapsules also help at fragile and sensitive materials survive processing and packaging conditions and stabilize the shelf-life of the active compound.

On the other hand, applications of microencapsulations to foods have been increasing due to the protection of encapsulate materials of factors such as heat and humidity, allowing to maintain its stability and viability. The microcapsules help to food materials to withstand the conditions of processing and packing to improve taste, aroma, stability, nutritional value and appearance of their products. Some of the substances encapsulated have been fertilizers, oil of lemon, lipids, volatile flavors, probiotics, nutraceuticals, seeds of fruits like banana, grapes, guava, papaya, apple, blackberry, granadilla and citrus seeds. In this regard, the encapsulation offers great scope for conservation, germination and exchange of several fruit species, resulting in promising technique for the conservation, transport of transgenic plants and not seed-producing plants, lactase, colorants, enzymes, phytosterols, lutein, fatty acids, plant pigments, antioxidants, aromas and oleoresins, vitamins and mineral [14].

9. Pigments

Pigments are compounds very sensitive due to their instability in the presence of light, air, humidity and high temperatures therefore their use requires a chemical knowledge of their molecules and stability, in order to adapt them to the conditions of use during processing, packaging and distribution. One alternative for their use in the food industry is microencapsulation technology [80]. Carotenoids are used as dyes in food, beverages, cosmetics and animal feed, mainly poultry and fish. During the processing and storage, carotenoids can easily change in different isomers geometric and rust, this result in the reduction or loss of the dye and its biological properties. The main alternatives of applications to increase the stability of carotenoids and, allow its incorporation in hydrophilic environments, is the technique of microencapsulation by the method of spray called spray drying. In the same way, other pigments such as licopeno, lutein, enocianin, astaxantin, antocianins and pigments of nogal and urucú have also been encapsulated [14].

10. Vitamins

Both lipid-soluble (e.g. vitamin A, β-carotene, vitamins D, E and K) and water-soluble (e.g. ascorbic acid) vitamins can be encapsulated using various technologies. The most common reason for encapsulating these ingredients is to extend the shelf-life, either by protecting them against oxidation or by preventing reactions with components in the food system in which they are present. A good example is ascorbic acid (vitamin C), which is added extensively to a variety of food products as either an antioxidant or a vitamin supplement. Its application as a vitamin supplement is impaired by its high reactivity and, hence, poor stability in solution. It can degrade by a variety of mechanisms. For vitamin C encapsulation, both spray-cooling or spray-chilling and fluidized-bed coating can be used when the vitamins are added to solid foods, such as cereal bars, biscuits or bread. For application in liquid food systems, the best way to protect water-soluble ingredients is by encapsulation in liposomes. Liposomes are single or multilayered vesicles of phospholipids containing either aqueous-based or lipophilic compounds. Lipid-soluble vitamins such as vitamin A, β-carotene and vitamins D, E or K are much easier to encapsulate than water-soluble ingredients. A commonly-used procedure is spray-drying of emulsions [81].

11. Minerals

From a nutritional point of view, the iron is one of the most important elements, and its deficiency affects about one-third of the world's population. The best way to prevent this problem is through the iron fortification of foods. However, the bioavailability of iron is negatively influenced by interactions with food ingredients such as tannins, phytates and polyphenols. Moreover, the iron catalyses oxidative processes in fatty acids, vitamins and amino acids, and consequently alters sensory characteristics and decreases the nutritional value of the food. Microencapsulation can be used to prevent these reactions, although bioavailability should be rechecked carefully. The bioavailability of readily water-soluble iron salts such as FeSO4 or ferrous lactate is higher than that of poorly water-soluble (e.g. ferrous fumarate) or water-insoluble (e.g. FePO4) iron. Suitable encapsulation techniques depend on the water solubility of the compound. Liposome technology is the method of choice for iron fortification of fluid food products [81].

12. Polyunsaturated fatty acid

In recent years, have been developing a high consumer preference towards products that possess functional properties. Thus the addition of beneficial substances such as polyunsaturated fatty acids omega 3 and omega 6, to the daily diet of humans has increased significantly since that is associated with the prevention and treatment of heart disease, because they have antithrombotic effects. Also has been associated with inflammatory diseases, autoimmune arthritis and even cancer. However, rich in polyunsaturated fatty acids oils are susceptible to oxidative deterioration and acquire easily bad tastes and odors, also environmental factors as moisture, light and oxygen accelerate its degradation, what he has done to the food and pharmaceutical industry to seek alternatives to prevent their deterioration. The microencapsulation has been an alternative to avoid the deterioration of oils because it can increase the oxidative stability of these and avoid the formation of products of oxidation of high molecular weight, in addition to mask unwanted flavors and aromas. It also gives some properties such as ease of handling and mixing, dispersion, and improvement of the consistency of the product during and after processing [82]. Advances in technologies of microencapsulation and the strategies used in its production have resulted in an increasing number of successful products fortified with omega-3 in the market [83], such as: dietary supplements, dairy, snacks, infant formulas and foods for babies, bakery products and beverages [84].

13. Probiotic bacteria

The microencapsulation has been successfully used to improve the survival of probiotic bacteria in dairy products such as cheese. Ozer et al. [85] demonstrated that colonies of *Lactobacillus ramnosus* microencapsulated in a matrix of alginate maintained their viability over 48 h at pH 2, in comparison with free (without encapsulate) cells that were inactivated completely under the same conditions; Another example related to dairy products is the yogurt, in which Bifidobacteria are encapsulated to increase viability in this fermented beverage;   Also the whey, the liquid product obtained during the preparation of the cheese can be dried by spraying for the production of whey powder and/or whey protein concentrates [14]. Bio-functional foods offer physiological health benefits and disease prevention over and above their nutritional contribution. Microencapsulation has become the recent tool used for protecting and delivering bio-actives in the development of bio-functional foods. Probiotic foods are by far the largest functional food market. They provide several health benefits including immune-stimulation. Viability, physiological and metabolic activity of these bio actives in a food product at the point of sale are important consideration for their efficacy, as they have to survive during shelf life of a food, transit through high acidic and alkaline conditions in the gastro-intestinal tract. Microencapsulation is an inclusion technique for entrapping a bio-functional nutrient or bio-active compound such as probiotic bacteria, folic acid and protease enzymes into a polymeric (gelled) matrix that may be coated by one or more semi-permeable polymers, by virtue of which the encapsulated substance become more stable than the free one [86].

14. Microencapsulation methods

Several encapsulation processes are based on making first droplets of the active compound (in gas, liquid or powder form) and these droplets are subsequently surrounded by the carrier material in a gas or liquid phase via different methods. Excellent reviews on the encapsulation processes have been published in the last years [12-14, 87-88]. For this reason, this section will only show the most commonly used methods in microencapsulation and the steps involved through a list (see Table 5).

Methods	Process steps	Morphology	Load (%)	Particle size (μm)
Spray-drying	Disperse or dissolve active in aqueous coating solution Atomize Dehydrate	Matrix	5-50	10-400
Fluid bed coating	Fluidize active powder Spray coating Dehydrate or cool	Reservoir	5-50	5-5,000
Spray-chilling/cooling	Disperse or dissolve active in heated lipid solution Atomize Cool	Matrix	10-20	20-200
Melt injection	Melt the coating Disperse or dissolve active in the coating Extrude through filter Cooling and dehydrating	Matrix	5-20	200-2,000
Emulsification	Dissolve active and emulsifiers in water or oil phase Mix oil and water phases under shear	Matrix	1-100	0.2-5,000
Preparation of emulsions with multilayers	Prepare o/w emulsions with lipophilic active in oil phase and ionic emulsifiers Mix with aqueous solution containing oppositely charged poly-electrolytes Remove excess of free poly-electrolytes (option) Repeat steps 2 and 3	Reservoir	1-90	0.2-5,000
Coacervation	Prepare o/w emulsions with lipophilic active in oil phase Mix under turbulent conditions Induce three immiscible phases Cool Crosslink (optionally)	Reservoir	40-90	10-800
Preparation of microspheres via extrusion or dropping	Dissolve or disperse active in alginate solution Drop into gelling bath	Matrix	20-50	200-5,000
Preparation of microspheres via emulsification	Emulsify water with biopolymer in oil phase	Matrix	20-50	10-1,000
Co-extrusion	Dissolve or disperse active in oil Prepare aqueous or fat coating Use an concentric nozzle, and press simultaneously the oil phase through the inner nozzle and the water phase through the outer one Drop into gelling or cooling bath	Reservoir	70-90	150-8,000
Encapsulation by rapid expansion of supercritical solution (RESS)	Create a dispersion of active and dissolved or swollen shell material in supercritical fluid Release the fluid to precipitate the shell onto the active	Matrix	20-50	10-400
Freeze- or vacuum drying	Dissolve or disperse active agent and carrier material in water Freeze the sample Drying under low pressure Grinding (option)	Matrix	Various	20-5,000
Preparation of nanoparticles	Various methods	Various	Various	0.1-1

Adapted from Zuidam and Shimoni [12]

Table 5. Overview of the most common microencapsulation processes

15. Advantages and disadvantages of the microencapsulation

According to Zuidam and Shimoni [12] the possible advantages and disadvantages of microencapsulated active compounds in the food industry could be:

Advantages

1. Superior handling of the active agent (e.g., conversion of liquid active agent into a powder, which might be dust free, free flowing, and might have a more neutral smell).
2. Immobility of active agent in food processing systems.
3. Improved stability in final product and during processing (i.e., less evaporation of volatile active agent and/or no degradation or reaction with other components in the food product such as oxygen or water).
4. Improved safety (i.e., reduced flammability of volatiles like aroma, no concentrated volatile oil handling).
5. Creation of visible and textural effects (visual cues).
6. Adjustable properties of active components (particle size, structure, oil- or water-soluble, color).
7. Off-taste masking.
8. Controlled release (differentiation, release by the right stimulus).

Disadvantages

1. Additional costs.
2. Increased complexity of production process and/or supply chain.
3. Undesirable consumer notice (visual or touch) of encapsulates in food products.
4. Stability challenges of encapsulates during processing and storage of the food product.

16. Conclusions and perspectives

Since, very good results have been obtained on enhancing overall quality, browning, oxidation and softening, shelf-life extension, control of decay, and nutraceutical benefits with a number of diverse additives and bioactive compounds incorporated into edible films and coatings designed for different foods, could be useful to consider that the use the edible films and coating as carrier of food additives and bioactive compounds represent a good alternative for food industry to improve the quality and safety of the food, as well as to offer new product to consumers.

Many companies and research institutes are looking for new ingredients with possible health benefits. Ingredients such as phytochemicals, wood-derived ingredients such as phytosterols, pro- and prebiotics, new types of carotenoids, trace minerals and polyphenols will be available in next the years. Microencapsulation will certainly play an important role in this process, although it will always make an ingredient more expensive to use and bioavailability should always be considered carefully.

Author details

Rosa M. Raybaudi-Massilia and Jonathan Mosqueda-Melgar
Institute of Food Science and Technology, Central University of Venezuela, Lomas de Bello Monte, Calle Suapure, Caracas, Venezuela

17. References

[1] Martín-Belloso O, Soliva-Fortuny R, Oms-Oliu G (2006) Fresh-cut fruits. In: Hui Y.H, editor. Handbook of fruits and fruit processing. Iowa, USA: Blackwell Publishing. pp. 129-144.

[2] Olivas G.I, Barbosa-Cánovas G.V (2005) Edible coatings for fresh-cut fruits. Crit. Rev. Food Sci. Nutr. 45(7-8): 657-670.

[3] Min B, Ahn D.U (2005) Mechanism of lipid peroxidation in meat and meat products – A review. Food Sci. Biotechnol. 14(1): 152-163.

[4] Fasseas, M.K, Mountzouris K.C, Tarantilis P.A, Polissiou M, Zervas G (2007) Antioxidant activity in meat treated with oregano and sage essential oils. Food Chem. 106: 1188-1194.

[5] Cerqueira M.A, Lima A.M, Souza B.W.S, Texeira J.A, Moreira R.A, Vicente A.A, (2009) Functional polysaccharides as edible coating for cheese. J. Agric. Food Chem. 57: 1456-1462.

[6] Gennadios A, Hanna M.A, Kurth L.B (1997) Application of edible coatings on meats, poultry and seafoods: A review. LWT. 30: 337-350.

[7] Martín-Belloso O, Rojas-Graü M.A, Soliva-Fortuny R (2009) Delivery of flavor and active ingredients using edible films and coatings. In: Embuscado M.E, Huber K.C, editors. Edible films and coating for food applications. Chapter 10. New York: Springer Science + Business Media, LLC. pp. 295-313.

[8] Raybaudi-Massilia R.M, Mosqueda-Melgar J, Tapia M.S (2010) Edible coatings as carriers of food additives on fresh-cut fruits and vegetables. Stewart Postharvest Rev. 3: 1-7.

[9] Rojas-Graü M.A, Soliva-Fortuny R.C, Martín-Belloso O (2009) Edible coatings to incorporate active ingredients to fresh-cut fruits: a review. Trends Food Sci. Technol. 20:438-447.

[10] Zhao Y (2010) Edible Coatings for enhancing quality and health benefits of berry fruits. In: Qian M.C, Rimando A.M, editors. Flavor and Health Benefits of Small Fruits. Chapter 18, Volume 1035. Washington DC, USA: ACS Symposium Series. pp 281–292.

[11] Tapia M.S, Rojas-Graü M.A, Rodríguez F.J, Ramírez J, Carmona A, Martín-Belloso O (2007) Alginate- and gellan-based edible films for probiotic coatings on fresh-cut fruits. J. Food Sci. 72: E190-E196.

[12] Zuidam N.J, Shimoni E (2010) Overview of microencapsulates for use in food products or processes and methods to make them. In: Zuidam N.J, Nedovic V.A, editors. Encapsulation technologies for active food ingredients and food processing. Chapter 2. New York: Springer Science + Business Media, LLC. pp. 3-29.

[13] Dubey R, Shami T.C, Brasker-Rao K.U (2009) Microencapsulation technology and application. Defence Sci. J. 59(1): 82-95.

[14] Parra-Huertas R.A (2011) Revisión: Microencapsulación de alimentos. Rev. Fac. Nal. Agr. Medellín. 63(2): 5669-5684.

[15] Raybaudi-Massilia R.M, Mosqueda-Melgar J (2009) Factors affecting the incidence of pathogenic microorganisms of fresh-cut produce. Stewart Postharvest Rev. 4: 1-8.

[16] Cutter C.N, Siragusa G.R (1997) Growth of Brochothrix thermosphacta in ground beef following treatments with nisin in calcium alginate gels. Food Microbiol. 14: 425-430.

[17] Natrajan N, Sheldon B (2000) Inhibition of Salmonella on poultry skin using protein and polysaccharide-based films containing a nisin formulation. J Food Prot. 63(9): 1268-1272.

[18] Millette M, Le Tien C, Smoragiewicz W, Lacroix M (2007) Inhibition of Staphylococcus aureus on beef by nisin-containing modified alginate films and beads. Food Contr. 18: 878-884.

[19] Datta S, Janes M.E, Xue Q.G, Losso J, La Peyre J.F (2008) Control of Listeria monocytogenes and Salmonella anatum on the Surface of Smoked Salmon Coated with Calcium Alginate Coating containing Oyster Lysozyme and Nisin. J. Food Sci. 73(2): M67-M71.

[20] Neetoo H, Ye M, Chen H (2010) Bioactive alginate coatings to control Listeria monocytogenes on cold-smoked salmon slices and fillets. Int. J. Food Microbiol. 136(3): 326-331.

[21] Marcos B, Aymerich T, Monfort J.M, Garriga M (2008) High-pressure processing and antimicrobial biodegradable packaging to control Listeria monocytogenes during storage of cooked ham. Food Microbiol. 25: 177-182.

[22] Oussalah M, Caillet S, Salmiéri S, Saucier L, Lacroix M (2006). Antimicrobial effects of alginate-based film containing essential oils for the preservation of whole beef muscle. J. Food Prot. 69: 2364-2369.

[23] Oussalah M, Caillet S, Salmieri S, Saucier L, Lacroix M (2007) Antimicrobial effects of alginate- based film containing essential oils o Listeria monocytogenes and Salmonella thyphimurium present in bologna and ham. J. Food Prot. 70(4): 901-908.

[24] Rojas-Graü M.A, Raybaudi-Massilia R.M, Soliva-Fortuny R.C, Avena-Bustillos R.J, McHugh T.H, Martín-Belloso O (2007) Apple puree-alginate edible coating as carrier of antimicrobial agents to prolong shelf life of fresh-cut apples. Postharvest Biol. Technol. 45: 254-264.

[25] Raybaudi-Massilia R.M, Mosqueda-Melgar J, Martín-Belloso O (2008) Edible alginate-based coating as carrier of antimicrobials to improve shelf life and safety of fresh-cut melon. Int. J. Food Microbiol. 121: 313-327.

[26] Raybaudi-Massilia R.M, Rojas-Graü M.A, Mosqueda-Melgar J, Martín-Belloso O (2008) Comparative study on essential oils incorporated into an alginate-based edible coating to assure the safety and quality of fresh-cut Fuji apples. J. Food Prot. 71: 1150-1161.

[27] Olivas G.I, Mattinson D.S, Barbosa-Cánovas G.V (2007) Alginate coatings for preservation of minimally processed 'Gala' apples. Postharvest Biol. Technol. 45: 89-96.

[28] Seol, K.H., Lim, D.G., Jang, A., Jo, C., y Lee, M. (2009). Antimicrobial effect of κ-carrageenan-based edible film containing ovotransferrin in fresh chicken breast stored at 5 °C. Meat Science 83(3): 479-483.

[29] Jiang Z, Neetoo H, Chen H (2011) Efficacy of freezing, frozen storage and edible antimicrobial coating used in combination for control of *Listeria monocytogenes* on roasted turkey stored at chiller temperatures. Food Microbiol. 28(7): 1394-1401.

[30] Jiang Z, Neetoo H, Chen H (2011) Control of *Listeria monocytogenes* on cold-smoked salmon using chitosan-based antimicrobial coatings and films. J. Food Sci. 76(1): M22-M26.

[31] Dutta P.K, Tripathi S, Mehrotra G.K, Dutta J (2009) Perspectives for chitosan based antimicrobial films in food applications. Food Chem. 114(4): 1173-1182.

[32] Ouattara B, Simard R.E, Piette G, Bejín A, Holler R.A (2000) Inhibition of surface spoilage bacteria in processed meats by application of antimicrobial films prepared with chitosan. Int. J. Food Microbiol. 62: 139-148.

[33] Park S.L, Stan S.D, Daeschel M.A, Zhao Y (2006) Antifungal coatings on fresh strawberries (*Fragaria* × *ananassa*) to control mold growth during cold storage. J. Food Sci. 70(4): M202-M207.

[34] Duan J, Park S, Daeschel M, Zhao Y (2007). Antimicrobial chitosan-lysozyme (CL) films and coatings for enhancing microbial safety of mozzarella cheese. J. Food Sci. 72(9): M355-M362.

[35] Sangsuwan J, Rattanapanone N, Rachtanapun P (2008) Effect of chitosan/methyl cellulose films on microbial and quality characteristics of fresh-cut cantaloupe and pineapple. Postharvest Biol. Technol. 49: 403-410.

[36] Ye M, Neetoo H, Chen H (2008) Control of *Listeria monocytogenes* on ham steaks by antimicrobials incorporated into chitosan-coated plastic films. Food Microbiol. 25(2): 260-268.

[37] Ojagh S.M, Rezaei M, Razavi S.H, Hashem-Hosseini S.M (2010) Effect of chitosan coatings enriched with cinnamon oil on the quality of refrigerated rainbow trout. Food Chem. 120(1): 193-198.

[38] Fajardo P, Martins J.T, Fuciños C, Pastrana L, Teixeira J.A, Vicente A.A (2010) Evaluation of a chitosan-based edible film as carrier of natamycin to improve the storability of Saloio cheese. J. Food Eng. 101(4): 349-356.

[39] Siripatrawan U, Noipha S (2012). Active film from chitosan incorporating green tea extract for shelf life extension of pork sausages. Food Hydrocolloids 27(1): 102-108.

[40] Franklin, N.B., Cooksey, K.D., y Getty, K.J.K. (2004). Inhibition of *Listeria monocytogenes* on the Surface of Individually Packaged Hot Dogs with a Packaging Film Coating Containing Nisin. *Journal of Food Protection* 67(3): 480-485

[41] Nguyen T, Gidley M, Dykes G (2008). Potential of a nisin-containing bacterial cellulose film to inhibit *Listeria monocytogenes* on processed meats. Food Microbiol. 25(3): 471-478.

[42] Santiago-Silva P, Soares N.F.F, Nóbrega J.E, Júnior M.A.W, Barbosa K.B.F, Volp A.C.P, Zerdas E.R.M.A, Würlitzer N.J (2009) Antimicrobial efficiency of film incorporated with pediocin (ALTA®2351) on preservation of sliced ham. Food Contr. 20: 85-89.

[43] Sayanjali S, Ghanbarzadeh B, Ghiassifar S. 2011. Evaluation of antimicrobial and physical properties of edible film based on carboxy methyl cellulose containing potassium sorbate on some mycotoxigenic *Aspergillus* species in fresh pistachios. LWT. 44(4): 1133-1138.

[44] Valencia-Chamorro S.A, Pérez-Gago M.B, del Río M.A, Palou L (2009) Effect of antifungal hydroxy propyl methyl cellulose (HPMC)–lipid edible composite coatings on postharvest decay development and quality attributes of cold-stored "Valencia" oranges. Postharvest Biol. Technol. 54(2): 72-79.

[45] Durango A.M, Soares N.F.F, Andrade N.J (2006) Microbiological evaluation of an edible antimicrobial coating on minimally processed carrots. Food Contr. 17(5): 336-341.

[46] Chiu P.E, Lai L.S (2010) Antimicrobial activities of tapioca starch/decolorized hsian-tsao leaf gum coatings containing green tea extracts in fruit-based salads, romaine hearts and pork slices. Int. J. Food Microbiol. 139(1-2): 23-30.

[47] Kang H.J, Jo C, Kwon J.H, Kim J.H, Chung H.J, Byun M.W (2007) Effect of a pectin-based edible coating containing green tea powder on the quality of irradiated pork patty. Food Contr. 18(5). 430-435.

[48] Handley D, Ma-Edmonds M, Hamouz F, Cuppett S, Mandigo R, Schnepf M (1996) Controlling oxidation and warmed-over flavor in precooked pork chops with rosemary oleoresin and edible film. In: Shahidi F, editor. Natural antioxidants chemistry, health effect, and application. Champaign, USA: AOCS Press. pp. 311-318.

[49] Ma-Edmonds M, Hamouz F, Cuppett S, Mandigo R, Schnepf M (1995) Use of rosemary oleoresin and edible film to control warmed-over flavor in precooked beef patties [abstract]. In: IFT Annual Meeting Book of Abstracts. Chicago, Illinois: Institute of Food Technologists. pp. 139. Abstract Nº 50-6.

[50] Hargens-Madsen M, Schnepf M, Hamouz F, Weller C, Roy S (1995) Use of edible films and tocopherols in the control of warmed-over flavor. J. Am. Diet. Assoc. 95: A-41.

[51] Wu Y, Weller C.L, Hamouz F, Cuppett S, Schnepf M (2001) Moisture loss and lipid oxidation for precooked ground-beef patties packaged in edible starch-alginate-based composite films. J. Food Sci. 66(3): 486-493.

[52] Atarés L, Pérez-Masiá R, Chiralt A (2011). The role of some antioxidants in the HPMC film properties and lipid protection in coated toasted almonds. J. Food Eng. 104(4): 649-656.

[53] Song Y, Liu L, Shen H, You J, Luo Y (2011) Effect of sodium alginate-based edible coating containing different anti-oxidants on quality and shelf life of refrigerated bream (*Megalobrama amblycephala*). Food Contr. 22(3-4): 608-615.

[54] Lin D, Zhao Z (2007) Innovations in the development and application of edible coatings for fresh and minimally processed fruits and vegetables. CRFSFS. 6: 60-75.

[55] Ahvenainen R (1996) Improving the shelf life of minimally processed fruit and vegetables. Trends Food Sci. Tech. 7: 179-187.

[56] Friedman, M. Food browning and its prevention: Overview. Journal of Agricultural and Food Chemistry 1996: 44(3):631–652.

[57] Baldwin E.A, Nisperos M.O, Chen X, Hagenmaier R.D (1996) Improving storage life of cut apple and potato with edible coating. Postharvest Biol. Technol. 9: 151-163.

[58] Brancoli N, Barbosa-Cánovas G.V (2000) Quality changes during refrigerated storage of packaged apple slices treated with polysaccharide films. In: Barbosa-Cánovas G.V, Gould G.W, editors. Innovations in food processing. Pennsylvania: Technomic Publishing Co. pp. 243-254.

[59] Lee J.Y, Park H.J, Lee C.Y, Choi W.Y (2003) Extending shelf-life of minimally processed apples with edible coatings and antibrowning agents. LWT. 36: 323-329.

[60] Bico S.L.S, Raposo M.F.J, Morais R.M.S.C, Morais A.M.M.B (2009) Combined effects of chemical dip and/or carrageenan coating and/or controlled atmosphere on quality of fresh-cut banana. Food Contr. 20: 508-514.

[61] Rojas-Graü M.A, Tapia M.S, Rodríguez F.J, Carmona A.J, Martín-Belloso O (2007) Alginate and gellan based edible coatings as support of antibrowning agents applied on fresh-cut Fuji apple. Food Hydrocolloids 21: 118-127.

[62] Rojas-Graü M.A, Tapia M.S, Martín-Belloso O (2008) Using polysaccharide-based edible coatings to maintain quality of fresh-cut Fuji apples. LWT. 41: 139-147.

[63] Oms-Oliu G, Soliva-Fortuny R, Martín-Belloso O (2008) Edible coatings with antibrowning agents to maintain sensory quality and antioxidant properties of fresh-cut pears. Postharvest Biol. Technol. 50: 87-94.

[64] García E, Barret D.M (2002) Preservative treatments for fresh cut fruits and vegetables. In: Lamikanra O, editor. Fresh-Cut Fruits and Vegetables. Florida: CRC Press. pp. 267-304.

[65] Harker F.R, Redgwell R.J, Hallett I.C, Murray S.H, Carter G (1997) Texture of fresh fruit. Horticultural Rev. 20: 121-224.

[66] Soliva-Fortuny R.C, Martín-Belloso O (2003) New advances in extending the shelf life of fresh-cut fruits: a review. Trends Food Sci. Technol. 14: 341–353.

[67] Oms-Oliu G, Soliva-Fortuny R, Martín-Belloso O (2008) Using polysaccharide-based edible coatings to enhance quality and antioxidant properties of fresh-cut melon. LWT. 41: 1862-1870.

[68] Tapia M.S, Rojas-Graü M.A, Carmona A, Rodriguez F.J, Soliva-Fortuny R, Martín-Belloso O (2008) Use of alginate and gellan-based coatings for improving barrier, texture and nutritional properties of fresh-cut papaya. Food Hydrocolloids. 22: 1493-1503.

[69] Mei Y, Zhao Y (2003) Barrier and Mechanical Properties of Milk Protein-Based Edible Films Containing Nutraceuticals. J. Agr. Food Chem. 51(7): 1914-1918.

[70] Elliott J.G (1998) Application of antioxidant vitamins in foods and beverages. Food Technol. 53: 46-48.

[71] Mei Y, Zhao Y, Yang J, Furr H.C (2002) Using edible coating to enhance nutritional and sensory qualities of baby carrots. J. Food Sci. 67(5): 1964-1968.

[72] Hernández-Muñoz P, Almenar E, Ocio M.J, Gavara R (2006) Effect of calcium dips and chitosan coatings on postharvest life of strawberries (Fragaria x ananassa). Postharvest Biol. Technol. 39: 247-253.

[73] Han C, Zhao Y, Leonard S.W, Traber M.G (2004) Edible coatings to improve storability and enhance nutritional value of fresh and frozen strawberries (Fragaria x ananassa) and raspberries (Rubus ideaus). Postharvest Biol. Technol. 33(1): 167-178.

[74] Duan J, Cherian G, Zhao Y (2010). Quality enhancement in fresh and frozen lingcod (Ophiodon elongates) fillets by employment of fish oil incorporated chitosan coatings. Food Chem. 119(2): 524-532.

[75] Naidu A.S, Bidlack W.R, Clemens R.A (1999) Probiotic spectraof lactic acid bacteria (LAB). Crit. Rev. Food Sci. Nutr. 38: 113-126.

[76] Giancone T (2007) Hydrocolloid-based edible films: composition-structure-properties relationship. Thesis Doctoral. Universitá degli Studi di Napoli Federico II. Department of Food Science. Available: http://www.fedoa.unina.it/1669/1/Giancone_Scienze_Tecnologie.pdf. Accessed 2012 Feb 10.

[77] Gontard N, Guilbert S (1999) Bio-packaging: technology and properties of edible and/or biodegradable material of agricultural origin. In: Mathlouthi M, editor. Food packaging and preservation. Chapter 9. Gaithersburg, USA: Aspen Publisher, Inc. pp. 159-181.

[78] Pavlath A.E, Orts W (2009). Edible films and coatings: Why, what, and how?. In: Embuscado M.E, Huber K.C, editors. Edible films and coatings for food applications. Chapter 1. New York: Springer Science + Business Media, LLC. pp. 1-24.

[79] Chien P.J, Sheu F, Yang F.H (2007) Effects of edible chitosan coating on quality and shelf life of sliced mango fruit. J. Food Eng. 78: 225-229.

[80] Parize A, Rozone T, Costa I Fávere V, Laranjeira M, Spinelli A, Longo E (2008) Microencapsulation of the natural urucum pigment with chitosan by spray-drying in different solvents. African J. Biotechnol. 7(17): 3107-3114.

[81] Schrooyen P, Meer R, Kruif C (2001) Microencapsulation: its application in nutrition. Proc. Nutr. Soc. 60(4): 475-479.

[82] Quispe-Arpasi D, Matos-Chamorro A, Quispe-Condori S (2011) Microencapsulación de aceites y su aplicación en la industria de alimentos. I Congreso Nacional de Investigación, Lima, Perú. November 2-4, 2011. Available: http://papiros.upeu.edu.pe/bitstream/handle/123456789/180/CIn33Articulo.pdf?sequenc e=1 Accessed 2012 Abril 10.

[83] Smith J, Charter E (2010) Functional food product development. United Kingdom: Blackwell Publishing Ltd. 505 p.

[84] Sanguansri L, Augustin M (2007) Microencapsulation and delivery of Omega-3 fatty acids. In: Shi J, editor. Functional food ingredients and nutraceuticals: Processing technologies. Florida, USA: Taylor & Francis. pp. 297-327.

[85] Ozer B, Avni H, Senel E, Atamer M, Hayaloglu A (2009) Improving the viability of Bifidobacterium bifidum BB-12 and Lactobacillus acidophilus LA-5 in white-brined cheese by microencapsulation. Int. Dairy J. 19(1): 22-29.

[86] Kailasapathy K, Madziva H, Anjani K, Seneweera S, Phillips M (2006) Recent trends in the role of micro encapsulation in the development of bio-functional foods. XIVth International Workshop on Bioencapsulation, Lausanne, USA. Available: http://impascience.eu/bioencapsulation/340_contribution_texts/2006-10-05_O7-1.pdf. Accessed 2012 April 10.

[87] Gibbs B.F, Kermasha S, Alli I, Mulligan C.N (1999) Encapsulation in the food industry: A review. Int. J. Food Sci. Nutr. 50: 213-224.

[88] Desobry S, Debeaufort F (2011) Encapsulation of flavors, nutraceuticals, and antibacterials. In: Baldwin E.A, Hagenmaier R, Bai J, editors. Edible coatings and films to improve food quality. 2nd Edition. Chapter 11. Boca Raton, USA: CRC Press. pp. 333-372.

Chapter 2

Chitosan: A Bioactive Polysaccharide in Marine-Based Foods

Alireza Alishahi

Additional information is available at the end of the chapter

1. Introduction

Since people are increasingly conscious of the relationship between diet and health, the consumption of marine-based foods has been growing continuously. Consumers identified seafoods as nutritious and complete foods. Hence, they are perceived them as an excellent source of high quality proteins, valuable lipids with high amounts of PUFA. These compounds are well known to contribute to the enhancement of human health by different alternatives such as reducing the risk of cardiovascular disease, coronary disease and hypertension. Additionally, marine-based food products are easily digested and constitute excellent source of essential minerals. Recently, seafoods have been recognized as nutraceuticals or functional foods. Functional foods, first evolved in Japan in 1980, are defined as foods demonstrating beneficial effect on one or more targeted functions on human organism (Ross, 2000). Marine-based functional foods or nutraceuticals, include omega-3 fatty acids, chitin and chitosan, fish protein hydrolysates, algal constituents, carotenoids, antioxidants, fish bone, shark cartilage, taurine and bioactive compounds (Kadam & Prabhasankar, 2010).

Despite the aforementioned desirable properties, seafood products are highly susceptible to quality deterioration, mainly due to the lipid oxidative reactions, particularly involving polyunsaturated fatty acids (PUFAs). These reactions are enhanced (catalyzed) by the presence of high concentrations of heme and non-heme proteins. These proteins are known to contain iron and other metal ions in their structures (Decker & Hultin, 1992). Moreover, marine-based food quality is highly influenced by autolysis, bacterial contamination and loss of protein functionality (Jeon, Kamil & Shahidi, 2002). More recently, pollution of seafood with different hazardous materials such as refinery, industrial wastes and heavy metals has resulted in elevated concern about the consumption of seafood (Kadam & Prabhasankar, 2010). Additionally, aquaculture industry has increasingly attracted much attention for the intensive farming of fish and shellfish, mainly due to the depleting of wild fish and shellfish stocks worldwide. However, this intensive farming entails several difficulties such as stress, which is the most important factor affecting the immunity system of fish (Ledger, Tucker & Walker, 2002). To address the aforementioned problems, the use of chitosan as protective material appears to be a potential alternative.

In nature, chitosan is found in the cell walls of fungi of the class *Zygomycetes*, in the green algae *Chlorella sp.*, yeast and protozoa as well as insect cuticles and especially in the exoskeleton of crustaceans. Chitosan is a deacetylated derivative of chitin, the second abundant polysaccharide in nature after cellulose. In 1811, the French scientist, Henri Braconnot first discovered chitin in mushroom. In 1820, chitin was isolated from insect cuticles (Bhatnagar & Sillanpa, 2009). In 1859, Rouget reported finding chitosan after boiling chitin in potassium hydroxide (KOH). This treatment rendered the material soluble in organic acids. Hoppe-Seyler named it chitosan in 1894 (Khor, 2001). Chemically, chitosan is a high molecular weight, linear, polycationic heteropolysaccharide consisting of two monosaccharides: N-acetyl-glucosamine and D-glucosamine. They are linked by β-$(1\rightarrow4)$ glycosidic bonds. The relative amount of these two monosaccharides in chitosan vary considerably, yielding chitosans of different degrees of deacetylation varying from 75% to 95%, molecular weight in the range of 50-2000 KDa, different viscosities and pKa values (Tharanathan & Kittur, 2003). In addition, chitosan has three functional moieties on its backbone; the amino group on the C2, the primary and secondary hydroxyl groups on the C3 and C6 positions, respectively. These functional groups play important roles in different functionalities of chitosan. The amino group is the most important among the other moieties, especially in acidic conditions, due to the protonation phenomenon, rendering it able to interact with negatively charged molecules (or sites). Additionally, chitosan polymer interacts with the metal cations through the amino groups, hydroxyl ions and coordination bonds.

Commercially, chitosan is produced from chitin by exhaustive alkaline deacetylation, involving boiling chitin in concentrated alkali for several hours. Because this N-deacetylation is almost never complete, chitosan is classified as a partially N-deacetylated derivative of chitin (Kumar, 2000). From a practical point of view, many of commercial interests and applications of chitosan and its derivatives originate from the fact that this polymer combines several features, such as biocompatibility, biodegradability, nontoxicity and bioadhesion, making it as valuable compound for pharmaceutical (Dias, Queiroz, Nascimento & Lima, 2008), cosmetics (Pittermann, Horner & Wachter, 1997), medical (Carlson, Taffs, Davison & Steward, 2008), food (Shahidi, Kamil & Jeon, 1999; No, Meyers, Prinyawiwatkul & Xu, 2007; Kumar, 2000), textile (El Tahlawy, Bendary, El Henhawy & Hudson, 2005), waste water treatment (Che & Cheng, 2006), paper finishing, photographic paper (Kumar, 2000), and agricultural applications (Hirano, 1996).

Although there have been several prior reviews on the use of chitosan in food applications (No et al., 2007; Shahidi et al., 1999), the use of chitosan in seafood applications, especially its novel application in the form of nanocarriers for bioactive compounds for shelf life extention, has not yet been reported. Recently, a study was published on the use of chitosan nanoparticle for stability enhancement of vitamin C in rainbow trout diet (Alishahi, Mirvaghefi, Rafie-Tehrani, Farahmand, Shojaosadati, Dorkoosh & Elsabee, 2011). Hence, this chapter attempts to survey the applications of chitosan in various fields of marine-based products.

2. Antibacterial activity

The modern era of chitosan research was heralded by publications in the 1990s, that described the antimicrobial potentials of chitosan and its derivatives, exhibiting a wide spectrum of activities against human pathogens and food-borne microorganisms (Chen, Xing & Park & Kong, 2010; No, Park, Lee & Meyers, 2002; Rabea, Badway, Stevens, Smagghe & Steurbaut, 2003: Raafat, Bargen, Haas & Sahl, 2008; Raafat & Sahl, 2009). , The first study reporting antibacterial properties was reported by Allan & Hardwiger (1979). They reported that chitosan showed a broad range of activities and a high inactivation rate against both Gram-positive and Gram-negative bacteria, (Allan & Hardwiger, 1979). However, although several studies have been published in this area, the exact mechanism of the antimicrobial activity of chitosan remains ambiguous.

Six major mechanisms have been proposed in the literature, as follows (Kong et al., 2010; Raafat & Sahl, 2009; Rafaat et al., 2008): (1) the interaction between the positively charged chitosan amine groups and the negatively charged microbial cell membranes, leadingto the leakage of proteinaceous and other intracellular constituents; (2) the activation of several defense processes in the host tissue by the chitosan molecule acting as a water-binding agent and inhibiting various enzymes by blocking their active centers; (3) the action of chitosan as a chelating agent, selectively binding metals and then inhibiting the production of toxins and microbial growth; (4) the formation, generally by high molecular weight chitosan, of an impervious polymeric layer on the surface of the cell, thereby altering cell permeability and blocking the entry of nutrients into the cell; (5) the penetration of mainly low-molecular weight chitosan into the cystosol of the microorganism to bind DNA, resulting in interference with the synthesis of mRNA and proteins; and (6) the adsorption and flocculation of electronegative substances in the cell by chitosan, distributing the physiological activities of the microorganisms, causing their death.. However, it is very important to mention that chitosan is soluble only in acidic media and therefore, the effect of pH on microorganisms must be considered together with the effect of chitosan. Thus, the synergetic effect of chitosan/pH together is probably the most evident explanation of the antimicrobial effect of chitosan.

Complicating the issue, a number of studies aimed at determining the antibacterial activities of chitosan on Gram-positive and Gram-negative bacteria have been reported antithetical outcomes, making their interpretation difficult. More recently, Kong et al. (2010) showed that chitosan and its derivatives are more powerful antibacterial agent against Gram-negative bacteria than against Gram-positive microorganism. Conversely, Raafat and Sahl (2009) reported a study in which they demonstrated that Gram-positive bacteria are more sensitive to the antibacterial effect of chitosan than Gram-negative bacteria. Therefore, the interpretation of the sensitivity of bacteria to chitosan is quite difficult.

To address this problem, Kong et al. (2010) proposed that the variation in the bactericidal efficacy of chitosan arises from different parameters that must be considered when evaluating the antibacterial activity of chitosan. These factors can be categorized into four classes as follows: (1) intrinsic microbial factors, including microbial species and cell age, (2) intrinsic factors of chitosan molecules, namely, the positive charge density, protonation level of the amine group, the chitosan molecular weight and concentration, hydrophilic/hydrophobic characteristics and chelating capacity of the chitosan molecule; (3) its physical state, i.e., either water soluble or solid chitosan; and (4) environmental factors including the ionic strength of the testing medium, pH, temperature and contact time between chitosan and bacterial cells. In addition, it is worth noting that, despite the widely reported antimicrobial properties of chitosan in the literature, the results are mainly based on in vitro experiments. In real-world appliactions, it is important to consider that most foods are complex matrices composed of different compounds (proteins, carbohydrates, lipids, minerals, vitamins, salts and others) and many of them may interact with chitosan to varying levels, possibly leading to a loss or enhancement of its antibacterial activity (Devlieghere, Vermeulen, & Debevere, 2004).

Taking into account the above insights about the antibacterial characteristics of chitosan, the following applications of chitosan in seafood products were considered. Due to the high perishability characteristics of marine-based products, there has been an increased interest in the application of chitosan to extend the shelf life of the products. In this context, chitosan has increasingly gained attention as an antibacterial additive in seafood from both seafood processors and consumers, largely due to a desire to reduce the use of synthetic chemicals in seafood preservation. Cao, Xue and Liu (2009) reported that chitosan at 5 g/L extended the shelf life of oyster (*Crossostrea gigas*) from 8-9 days to 14-15 days. They explained that Pseudomonas and Shewanella are the most prolific microorganisms during cold storage of fish and shellfish and these bacteria can easily be reduced or eliminated with the addition of chitosan at this concentration.

Fish balls have a high water activity and are prone to the growth of microorganisms, with a relatively short shelf life of 4-5 days at a storage temperature of approximately 5 °C. Kok and Park (2007) reported that fish ball shelf life has been increased by adding chitosan which maintained both aerobic and yeast counts at < 1 log CFU/g over 21 days of storage. Kok and Park (2007) also reported that physical state of chitosan is an important parameter for its antibacterial activity. In the dissolved state, chitosan showed excellent antibacterial activity and contributed to the extension of the product shelf life. However, in the study reported by Lopez-Caballero et al. (2005a and b), it was demonstrated that the addition of powdered chitosan to fish patties had no effect on bacterial growth. Roller and Corvill (2000) reported that the spoilage flora was inhibited from log 8 CFU/g in the control sample (without chitosan) to log 4 CFU/g over the 4-week study at 5 °C with the use of chitosan combined with acetic acid in shrimp salad. However, it is important to consider the effect of acetic acid on the ability of chitosan to extend shelf life. Fernandez-Saiz, Soler, Lagaron and Ocio (2010) studied bacterial growth in two conditions. The first one was a fish soup (ANETO ® brand, packaged in TetraBrik ® and fabricated by Jamon Aneto, S. L., Barcelona, Spain). The other medium was a model laboratory growth medium, tryptone soy broth. They reported a significant reduction of the growth of *Listeria monocytogenes*, *Staphylococcus aureus* and *Salmonella spp* when the products were stored in the presence of chitosan. In fish soup and under laboratory conditions, the effect of chitosan in the tested medium at 4, 12, and 37 °C depended significantly on the bacteria type, incubation temperature and food matrix (substrate). The antibacterial effectiveness of chitosan was decreased in the fish soup, suggesting that the constituents of the fish soup had high influence on the antimicrobial efficacy of chitosan. The authors reported that chitosan was probably irreversibly bound by microbial cells or negatively charged compound in the soup and therefore rendered it inactive against the remaining unbound microorganisms. In conclusion, the effect of chitosan-to-cell ratio must be considered. Lopez-Caballero et al. (2005a-b) showed that the addition of chitosan to sausage treated at high pressure yielded a 2-log cycle decrease of total bacterial counts of *Pseudomonas* and *Enterobacteria* at 8-11 days storage. Ye, Neetoo and Chen (2008) stated that chitosan has antibacterial activity that is effectively expressed in aqueous system; however, its antibacterial properties against *L. monocytogenes* in cold-smoked salmon were negligible when chitosan was in the form of an insoluble film. The growth of *L. monocytogenes* in salmon samples wrapped in the chitosan-coated film and plain films was similar. The authors demonstrated that it is possible that chitosan is ineffective in films because it is unable to diffuse through a rigid food matrix such as salmon. Regarding microbial counts, Lopez-Caballero et al. (2005b) showed that chitosan in the soluble state had no significant effect in high-pressure treated cod sausages. However, when chitosan was added in a dry form, higher counts of microorganisms were recorded. This is an indication that chitosan in soluble form contributed to maintaining product safety. The microbial counts in their study were for lactic acid bacteria, *Enterobacteria*, *Pseudomonas* and *Staphylococcus*. Duan, Jiang, Cherian and Zhao (2010) reported that the initial total plate count (TPC) of fresh lingcod was 3.67 log CFU/g, which then rapidly increased to 6.16 and 8.36 log CFU/g on day 6 and day 14, respectively. When chitosan coatings were used, the results showed a 0.15-0.64 reduction in TPC. Moreover, the TPCs of chitosan-coated samples stored under vacuum or modified-atmosphere packaging were significantly lower than those of the control sample during the subsequent cold storage. The combination of chitosan coating and vacuum or modified atmosphere packaging (MAP) resulted in 2.22-4.25 reductions in TPC for the first 14 days of cold storage. The TPCs of chitosan-coated and MPA samples were lower than 10^5 CFU/g even after 21 days of cold storage. This result indicated a significant delay of microbial spoilage. Qi, Zhang and Lan-Lan (2010) reported that because non-fermenting Gram-negative bacteria are dominant in the initial microbial flora of fish and shellfish sourced from cold seawater, controlling the growth of these Gram-negative bacteria may be important for the preservation of oysters. They demonstrated that combined treatment with chitosan and ozonated water had better antibacterial effect than either treatment alone. When only aerobic plate count was measured, the authors showed that the product shelf life with the combination of chitosan with ozonated water was at least

20 days, whereas it was only 8 days for the control sample, 10 days for the ozonated samples, and 14 days for the chitosan-treated samples. Duan, Cherian and Zhao (2009) indicated that fish oil incorporated in chitosan coatings lowered significantly the total and psychrotrophic counts in frozen lingcod fillets over three months cold storage. Ojagh, Rezaei, Razavi and Hosseini (2010) also reported that chitosan coatings enriched with cinnamon oil decreased effectively total viable counts and psychrotropic bacteria in rainbow trout (*Oncorhynchus mykiss*) during 16-day cold storage.

3. Antioxidant activity

Seafood is considered as excellent sources of functional foods for balanced nutrition favorable for promoting good health. The beneficial health effects of marine foods are ascribed to their lipids, mainly the long-chain omega-3 PUFA such as eicosapentaenoic acid (EPA) and docosahexaenoic acid (DHA) (Newton, 2001). However, these valuable compounds in seafood are highly sensitive to oxidative reactions and development of off-flavor even during cold storage (Cadwalladar & Shahidi, 2001). It has been proposed that lipid oxidation in fish and shellfish may be initiated and propagated by a number of mechanisms, namely autooxidation, photosentized oxidation, lipoxygenase, peroxidase and microsomal enzymes (Hsieh & Kinsella, 1989). Additionally, fish and shellfish muscles contain protein-bound iron compounds, for example, myoglobin, hemoglobin, ferritin, transferrin and haemosiderin, as well as other metals. This is a factor that plays an important role in initiating lipid oxidation in marine-based products (Decker & Hultin, 1992). Castell, Maclean and Moore (1965) showed that the relative pro-oxidant activity of metal ions in fish and shellfish muscles decreased in the order of $Cu^{+2}> Fe^{+2}> Cd^{+2}> Li^{+2}> Mg^{+2}> Zn^{+2}> Ca^{+2}> Ba^{+2}$. The iron-bound proteins and other metal ions may be released during the storage period and can thus activate and/or initiate lipid oxidative reactions (Decker & Hultin, 1992).

Along with the growing consumer demand for seafood devoid of synthetic antioxidants, chitosan has been a booming antioxidant agent in fish and shellfish. The antioxidant activity of chitosans of different viscosities (360, 50 and 14 cP) in cooked, comminuted flesh of herring (*Clupea harengus*) was investigated (Kamil, Jeon and Shahidi, 2002). The oxidative stability of fish flesh during cold storage at 4 °C with the addition of chitosan at concentrations of 50, 100 and 200 ppm was compared with that of fish treated with conventional antioxidants, such as butylated hydroxyanisole (BHA) and butylated hydroxytoluene (BHT) and *tert*-butylhydoquinone (THQ) (all at 200 ppm). Among the three chitosan samples tested, the 14 cP chitosan was the most effective in preventing lipid oxidation. The formation of thiobarbituric acid test reactive substances (TBARS) in herring samples containing 200 ppm of 14 cP chitosan was reduced by 52% as after 8 days of storage compared with that the control sample without chitosan. At a chitosan concentration of 200 ppm, the 14 cP chitosan exerted an antioxidant effect similar to that of commercial antioxidants in reducing TBARS values in comminuted herring flesh. A study by Kamil, Jeon and Shahidi (2002) indicated that the antioxidant capacity of chitosan added to fish muscle depends on its molecular weight (MW) and concentration in the product. Similarly, Kim and Thomas (2007) observed that the antioxidative effects of chitosan in salmon depended on its molecular weight (tested at MW = 30, 90 and 120 kDa) and concentration (evaluated at 0.2%, 0.5% and 1% w/w). The authors reporteda that the 30 kDa chitosan showed the highest radical-scavenging activity. The scavenging activities of chitosan werer increased by increasing its concentration. However, varying the concentration showed no significant effects when 120 KDa chitosan was used. Ahn and Lee (1992) studied the preservative effect of chitosan film on the quality of highly salted and dried horse mackerel. The product was prepared by soaking the fresh horse mackerel in 15% salt solution for 30 min, coating with or without (control) chitosan, and drying for 3h at 40 °C in a hot air dryer. During cold storage at a temperature of 5 °C for 20 days, the chitosan-coated samples had lower TBARS and peroxide values (PV) than the control samples. Similarly, Lopez-Caballero et al. (2005) also found that coating codfish patties with a chitosan-gelatin blend considerably lowered lipid oxidation. However, being non soluble in powder form at neutral pH, chitosan had no effect on the prevention of lipid oxidation. Shahidi, Kamil and Jeon (2002) reported that chitosans with different molecular weights and viscosities (14, 57 and 360 cP) were effective in controlling the oxidation of lipids in comminuted cod (*Gadus morhua*) following cooking. Both PV and TBARS values were reduced as a result of treatment of the fish prior to cooking with 50, 100 and 200 ppm of 14, 57, and 360 cP chitosans. Inhibition of the oxidation was concentration-dependent and was the highest for the 14 cP chitosan. The authors stated that the antioxidant activity of chitosans of different viscosity in cooked comminuted cod may be attributed to their metal-chelating capacities. Chitosans with different viscosities were found to protect cooked cod from oxidation at various levels. The observed differences were presumably due to differing degrees of deacetylation and molecular weights of the chitosan molecules. In the study reported by Qin (1993), it has been indicated that the ion-chelating ability of chitosan is strongly affected by the degree of deacetylation. The highly acetylated chitosan has very little chelating activity. In addition, high molecular weight chitosan has a compact structure and the effect of intramolecular hydrogen bonding is stronger, which weakens the activities of the hydroxyl and amino groups. As result, the probability of the exposure of these active moieties may be restricted, resulting in less radical scavenging activity. Obviously, low molecular weight chitosan exhibits higher hydroxyl radical scavenging activity, which is partially attributable to its metal chelating ability. The Fe^{+2} chelating ability of chitosan is mainly attributed to the presence of amino groups, which contain free electron pairs that contribute to form chitosan/Fe^{+2} complex. The Fe^{+2} chelating ability of low molecular weight chitosan is more pronounced than that of high molecular weight chitosan. Consequently, the amino groups in chitosan can react with free radicals to form additional stable macroradicals. Therefore, the active hydroxyl and amino groups in the polymer chains are the origin of the scavenging ability of chitosan (Jeon, Shahidi & Kim, 2000; Feng, Du, Li, Hu & Kennedy, 2008; No et al., 2007). Kamil et al. (2002) explained that in the charged state (protonated amino groups), the cationic groups of chitosans impart intramolecular electric repulsive forces. This phenomenon may be responsible for lesser chelating ability of high viscosity (high Mw) chitosans. Jeon et al. (2002) demonstrated that the antioxidant activity of chitosan is also effective when it is applied as a protective film. In this kind of application, it retards lipid oxidation by acting as a barrier against oxygen penetration. Sathivel, Liu, Huang and Prinyawiwatkul (2007) showed that the TBARS value of coated pink salmon (*O. gorbuscha*) fillets glazed with chitosan (1.3 mg MDA(malondialdehyde)/kg sample) was significantly lower than that of fillets glazed with lactic acid (3 mg MDA/kg sample) or distilled water (1.8 mg MDA/kg sample). The results indicated that chitosan glazing was more effective at reducing lipid oxidation among the studied alternatives. Sathivel (2005) also reported that the TBARS value of pink salmon fillets coated with 1% and 2% chitosan was significantly lower than in both the control sample and protein-coated product after 3 months of frozen storage. The author stated that a higher the concentration of chitosan, resulted in a lower TBARS value. The latter implies that the antioxidant effect of chitosan in the coating state is highly correlated to the coating material thickness, thereby hindering the entrance of oxygen into pink salmon fillet and initiating the oxidative process. Moreover, the primary amino groups of chitosan would form a stable fluorosphere with volatile aldehydes such as malondialdehyde which are derived from breakdown of fats during lipid oxidation (Weist & Karel, 1992).

Duan et al. (2009; 2010) showed that the combination of chitosan with modified atmosphere packaging enhanced the lipid stability of lingcod (*O. elongates*) within 21 days of cold storage. When applied on the surface of lingcod fillets, chitosan coatings may act as a barrier between the fillet and the surrounding atmosphere. This is mainly due to the good oxygen barrier properties of chitosan films, which slow down the diffusion of oxygen from the surrounding air to the surface of fillet and retard lipid oxidation (Aider, 2010). Additionally, Ojagh et al. (2010) reported that chitosan coatings enriched with cinnamon oil could suitably delay lipid oxidation in the refrigerated rainbow trout during 16 days of storage and markedly reduced the TBARS and PV values as compared with the control product. Mao and Wu (2007) showed that lipid oxidation of kamabako gel from grass carp (*Ctenopharyngodon idellus*) significantly decreased when a 1% chitosan solution was added.

4. Bioactive coatings

The modern marine-related food industries are encountering challenges and require for specific alternatives to surmount them. Among these, issues related to seafood packaging for products with a short shelf life are of pivotal importance. Although the utilization of conventional packaging materials such as plastics and their derivatives are effective for seafood preservation, they create serious and hazardous environmental problems, a situation which presents the seafood industry as a source of pollution and social concerns. This problem requires that all stakeholders in this industry and especially scientists specializing in the food engineering and packaging field to seek alternatives to address this serious problem related to the packaging material. A non-negligible aspect, which is the total cost of the final product, is also related to the packaging materials because it is well known that the contribution of the packaging to the product total cost is highly significant. So, the search for more economical packaging materials is a very important subject in the seafood industry (Aider, 2010).

Edible bio-based films have been investigated for their abilities to avoid moisture or water absorption by the seafood matrix, oxygen penetration to the food matrix, aroma loss and solute transport out of the product (Dutta, Tripathi, Mehrotra & Dutta, 2009). Based on this consideration, one of the most perspective active bio-film is the one based on chitosan. More recently, two review studies have reported the application of chitosan as bioactive film in the food industry (Aider, 2010; Dutta et al., 2009). Chitosan film, like many other polysaccharide based films, tend to exhibit resistance to fat diffusion and selective permeability to gases. However, they have a serious lack in terms of resistance to water and water vapor transmission. This behavior is mainly due to the strong hydrophilic character of these biopolymers, a property that leads to high interaction with water molecules (Bordenave, Grelier, & Coma, 2007). Owing to this, polymer blending or biocomposites and multilayer systems are potential approachs to prepare chitosan based bioactive coatings or films with desirable characteristics. In this context, Ye et al. (2008) stated that since edible film formed by chitosan is brittle and does not have good mechanical properties, coating chitosan onto a plastic film would overcome this problem. These authors have used chitosan-coated plastic films in which they have incorporated five antibacterial agents, namely nisin, sodium lactate (SL), sodium diacetate (SD), potassium sorbate (PS) and sodium benzoate (SB) as a novel antibacterial edible film against *Listeria monocytogenes* on cold-smoked salmon. This approach solved problems related to food safety since it is well known that *L. monocytogenes* could grow to high levels on cold-smoked salmon, even at normal refrigeration temperature. The risk related to *L. monocytogenes* is particularly high at abusive storage temperatures. Chtiosan-coated films containing 4.5 mg/cm^2 SL, 4.5 mg/cm^2 SL-0.6mg/cm^2 PS and 2.3 mg/cm^2 SL-500 IU/cm^2 nisin were the most effective treatments against *L. monocytogenes* at ambient temperature. These treatments showed long term antilisterial efficacy during refrigerated storage on vacuum-packed cold-smoked salmon. However, it is important to consider the fact that since antibacterial activity of chitosan may be negligible when it is in the form of insoluble films. Under this state, chitosan is ineffective because it is unable to diffuse through a rigid food matrix such as salmon. Sathivel et al. (2007) showed that skinless pink salmon (*Oncorhynchus gorbuscha*) fillets glazed with chitosan at a solution concentration of 1% (w/w) had significantly (p < 0.05) higher yield and thaw yield than the lactic acid–glazed and distilled water-glazed fillets. This behavior was valid although those fillets all had similar moisture content after thawing. In addition, the rheological study showed that chitosan has pseudoplastic and viscoelastic characteristics. The glass transition temperature for the chitosan film was observed at 80.23 °C. The oxygen, carbon dioxide, nitrogen and water vapor permabilities of the chitosan film were 5.34 10^{-2} (cm^3/ m day atm), 0.17 (cm^3/m day atm), 0.03 (cm^3/m day atm) and 2.92 10^{-10} (g water m/m^2 Pas), respectively. The authors demonstrated that despite the good barrier properties of chitosan against oxygen, it maintained low water vapor transmission because of their hydrophilic nature. Likewise, they stated that chitosan film showed shear thinning and viscoelastic characteristics and temperature dependent viscosity, which allowed uniform glazing on the salmon fillets and prevented rupturing of chitosan glazing during solidification when the glazed fillets were frozen. Therefore, chitosan glazing applied on the surface of the pink salmon fillets might have acted as a barrier between the fillets and the air surrounding, thus slowing down the diffusion of oxygen from the surrounding air into the fillets. Kester and Fennema (1986) reported that chitosan coatings might act as moisture-sacrificing agents of moisture barriers. Thus, moisture loss from the product could be delayed till the moisture contained within the chitosan coating had been evaporated. Sathivel (2005) highlighted that pink salmon fillets coated with chitosan resulted in significantly higher yield, thaw yield, similar drip loss and cook yield, higher moisture content after thawing, less moisture loss than the control samples and somewhat less than protein-coated products. Besides, there were no significant (p < 0.05) effects of coating on color parameters (a^*, b^* and L^* values) for cooked fillets after three months frozen storage. Lopez-Caballero et al. (2005) used chitosan as a material to form a chitosan-gelatin coating for cod patties. They showed that the use of chitosan either as a coating or a powdered ingredient did not affect the product lightness at the end of the storage period. However it resulted in an increase of the product yellowness (b-color parameter). The chitosan coating increased the patty elasticity, whereas the addition of powdered chitosan to the patty mixture increased the other rheological parameters such as gumminess, chewiness, cohesiveness and adhesiveness. Moreover, the coating did prevent spoilage of cod patties as reflected by a decrease in total volatile basic nitrogen (TVBN). Conversely, none of these effects on the bacterial spoilage were observed when the chitosan was added to the patty mixture in a powdered form. Ultimately, the authors reported that the coating had good sensory properties, melted away on cooking and hence did not impart any taste to the product. They provided protection by delaying spoilage. Duan et al. (2010) produced chitosan-krill oil coating and used it in modified atmosphere packaging to extend the shelf life of Lingcod fillets. They reported that chitosan-krill oil coating increased total lipid and omega-3 fatty acid contents of the lingcod by about 2-fold. The reduced chemical changes were reflected by the TVBN values and did not change the color of the fresh fillets, did not affect consumer's acceptance of both raw and cooked lingcod fillets. Consumers preferred the overall quality of chitosan-coated, cooked lingcod fillets over the control. The preference was based on the product firm texture and less fishy aroma and flavor. Considering the lower cost of vacuum packaging, it could be applied in combination with chitosan coatings to maintain the omega-3 fatty acid content and extend shelf life of fresh lean fish such as lingcod. Duan et al. (2009) also showed that fish oil incorporated to chitosan coating decreased the drip loss of frozen samples by 14.1-27.6%. This coating also well fortified the omega-3 fatty acids in lean fish. Cao et al. (2009) and Qi et al. (2010) showed that the chitosan coating could surprisingly increase the shelf life of highly perishable pacific oyster (C. gigas) during 21 days storage. This affirmation was based on TVBN, pH values and sensory evaluation of pacific oyster. They stated that the discrepancies between their results and others were derived from the differences in chemical composition of fish and shellfish in which oyster contains significant levels of carbohydrate (glycogen) and a lower total quantity of nitrogen. Ojagh et al. (2010) synthesized chitosan coatings enriched with cinnamon oil to extend the shelf life of refrigerated rainbow trout and showed that sensory characteristics and TVBN of the end product were drastically improved as the coating was employed on rainbow trout fillets within 16 days cold storage. Similarly, Lopez-Caballero et al. (2005) stated that the addition of dry chitosan led to a noticeable increase in elasticity and product yellowness when cod sausages were enriched with chitosan solution. The TVBN remained stable during 25 days storage and the product elasticity was reinforced.

5. Effluent treatment

The use of chitosan as a coagulating agent for removing suspended solids from various processing streams has been widely investigated including cheese whey and dairy wash water, in the processing of poultry and seafood products (Kumar, 2000; Savant, 2001; Savant & Torres, 2000; Savant & Torres, 2003; Shahidi et al., 1999). Chitosan at a concentration of 10 mg/L reduced up to 98% the total suspended solids in shrimp processing wastewater (Bough, 1976). Protein recoveries from surimi wash water (SWW) using 150 mg/L chitosan-alginate complex per liter SWW at mixing ratio of 0.2 resulted in 78-94% adsorption after 24 h (Wibowo, 2003). This result was higher than the one obtained by using 50 mg/L, which yielded 81-90 % protein adsorption in the same treatment time (Savant, 2001). These reported findings suggest that reaction time and chitosan concentration play an important role in reducing total suspended solids and lowering solution turbidity. Moreover, the differences in molecular weights (MW) and degree of deacetylation (DD) between chitosan samples could explain the significant differences in protein recovery capacity. At the lowest concentration (20 mg/L SWW) tested in the study reported by Wibowo (2003), the experimental chitosan gave higher protein recovery than a commercial sample, which required a 5-fold higher concentration for the same effectiveness. This finding has commercial implications as it would reduce processing costs and the chitosan content in the solids recovered by the treatment (Wibowo, Velazquez, Savant & Torres, 2007a). If implemented commercially, the chitosan-alginate complex may be an effective alternative not only for the recovering of soluble proteins that would otherwise be discarded into the environment, but also as an economically viable downstream process over expensive, commercial membrane treatments and their limited use due to fouling (Savant, 2001). Surimi wash water protein (SWWP) was precipitated by using a chitosan-alginate complex. The precipitate had a crude protein content of 73.1 % and a high concentration of essential amino acids (3% histidine, 9.4% lysine, 3.7% methionine, and 5.1% phenylalanine). In a rat-feeding trial, SWWP as a single protein source showed higher modified protein efficiency and net protein rations than the casein control. Blood chemistry analysis did not reveal any deleterious effect from the full protein substitution or the chitosan in SWWP (Wibowo, Savant, Gherian, Velazquez & Torres, 2007b; Wibowo, Velazquez, Savant & Torres, 2005). Moreover, Guerrero, Omil, Mendez & Lema (1998) showed that the utilization of chitosan at a concentration of 10 mg/L and pH 7 in the process of coagulation-flocculation followed by centrifugation in fish-meal factory effluents decreased the total suspended solids up to 85%. The most important mechanisms explaining the chitosan effectiveness in seafood plant effluents treatment was mainly attributed to its positive charge and interaction with negatively charged compounds in the effluents such as protein. Furthermore, the hydroxyl groups on the chitosan molecule contribute to increase the precipitation of proteins and other suspended solids in the seafood plant effluents (Savant, 2001; Wibowo et al., 2007a,b).

6. Gelling enhancer

Surimi is a refined fish protein product prepared by washing mechanically deboned fish to remove blood, lipids, enzymes and sarcoplasmic protein. The myofibrillar proteins are concentrated in the resulting product and form an elastic gel when solubilized with NaCl and heated (Mao & Wu, 2007). Gel forming properties of myofibrillar proteins are quickly lost by degradation by the action of endogenous proteolytic enzymes if fish is not processed into surimi immediately. The utilization of frozen fish flesh for surimi production is unsuitable due to the rapid loss of protein functionality by freeze denaturation. High quality surimi is produced from fresh, unfrozen fish. Thus, processing at sea has been required in order to obtain high quality surimi. However, the cost of the processing at sea is much higher compared to the land-base processing (Lanier, Manning, Zetterling & Macdonald, 1992). In order to prepare a strong and elastic gel from fish species with low commercial value, low quality surimi is produced onshore with the aid of gel-forming biopolymers such as starch. In this way, chitosan is a good option to be incorporated into the products to improve their techno-functional quality (Kataoka, Ishizaki, & Tanaka, 1998; Mao & Wu, 2007; Li & Xia, 2010). Overall, gel-forming ability of surimi depends on both intrinsic and extrinsic factors, namely fish species, physio-chemical properties of muscle proteins, the presence of endogenous enzymes such as proteinase amd transglutaminase, and the conditions used in the product processing (Benjakul, Visessanguan, Phatchrat & Tanaka, 2003). The strength of gels prepared from low quality walleye Pollock (*Theragra chalcogramma*) was almost doubled by the addition of 1.5% chitosan when salted surimi pastes were set below 25 °C. The polymerization of myosin heavy chain accelerated in the presence of 1.5% chitosan (Kataoka et al., 1198). Along with chitosan, endogenous transglutaminase (TGase) played an important role in the formation of gel. The addition of TGase inhibitor to the salted walleye Pollock surimi inhibited the gel enhancement by chitosan. The mechanisms of chitosan effect on enhancing the gel formation in not clear. However, the participation of hydrophobic interactions, hydrogen bondings, and electrostatic interactions during the setting process has been proposed as a possible mechanism by which chitosan can enhance the formation of cross-linked myosin heavy chain components during their polymerization by endogenous enzymes (Benjakul et al., 2003: Kataoka et al., 1998; Li & Xia, 2010; Mao & Wu, 2007). Benjakul et al. (2003) reported that barred garfish (*Hemiramphus far*) surimi gel showed an increase in the breaking force when 1% chitosan was added. However, gel-forming ability of surimi containing chitosan was inhibited in the presence of EDTA due to the chelating of calcium ions that are necessary for TGase activity. Owing to this, the enhancing effect of chitosan was possibly mediated through the action of endogenous TGase during product processing, resulting in the formation of protein-protein and protein-chitosan conjugates. In conjunction with processing and the addition of calcium ions, TGase may play an important role in the cross-linking of protein-protein and protein-chitosan conjugates by means of the amino groups of chitosan as the acyl acceptor. Conversely, chitosan did not substantially modify the rheological and microstructural properties of horse mackerel gels (*Trachurus spp.*). Also, it had a slight reduction in gel elasticity obtained under high-pressure conditions (Gomez-Guillel, Montero, Sole & Perez-Mateos, 2005). Kok and Park (2007) stated that in the threadfin bream (*Nemipterus spp.*) surimi, the balance of protein-chitosan and protein-protein conjugates determined the surimi gel strength. Similarly, Mao and Wu (2007) showed that in the presence of chitosan in kamaboko gel of grass carp (*Ctenopharyngodon idellus*), protein-chitosan conjugates would be formed between the reactive amino groups of glucosamine and the glutaminyl residue of the myofibrillar proteins. The bonds between chitosan and myofibrillar proteins could be associated with the improvement of texture properties in the gels with final structure formed by both covalent and non-covalent interactions. The effect would be also due to some modifications of the endogenous TGase activity. More recently, Lia and Xia (2010) showed that molecular weight and degree of deacetylation (DD) of chitosan have different impacts on gel properties of salt-soluble meat proteins from silver carp. The gel containing chitosan with DD of 77.3% showed the highest penetration force and storage modulus. The penetration forces of gels increased with increasing the amount of molecular weight of chitosan incorporated in the gel. The interaction between chitosan and salt-soluble meat proteins was mainly stabilized by the electrostatic interactions and hydrogen bonds.

7. Encapsulation

Nowadays, the value of functional foods and bioactive compounds are increasing due to the awareness and consciousness of people about it. Despite this fact, many of these compounds are so much sensitive to environmental factors such as oxygen, light, and temperature. In addition, being incorporated into foods and drugs in delivery systems, these bioactive components are hydrolyzed by harsh conditions in the gastrointestinal tracts (Alishahi et al., 2011). Schep, Tucker, Young, Ledger and Butt (1999) stated that many of oral delivery systems of bioactive compounds in aquaculture met the three major barriers through the gastrointestinal tract, involving the enzymatic barriers from the host luminal and membrane bound enzymes, immunological cells present within both the enterocytes and underlying connective tissue and the physical barrier of the epithelial cells. Based on this consideration, the encapsulation of bioactive compounds and functional foods could be a promising way to overcome these problems. Encapsulation is a process in which thin films, generally of polymeric materials, are applied to little solid particles, liquids or gas droplets. This method is used to entrap active components and release them under controlled conditions (Deladino, Anbinder, Navarro & Martino, 2008). Several materials have been encapsulated for the use in the food industry such as vitamins, minerals, antioxidants, colorants, enzymes and sweeteners (Shahidi & Han, 1993). Chitosan can act as an encapsulating agent because of its non-toxicity, biocompatibility, mucus adhesiveness and biodegradability (Alishahi et al., 2011: Kumar, 2000). Recently, Alishahi et al. (2011) showed that chitosan/vitamin C nanoparticle system successfully increased the shelf life and delivery of vitamin C during 20 days storage of rainbow trout. They showed that shelf life of vitamin C significantly (p <

0.05) increased in rainbow trout feed till 20 days at ambient temperature, while the control which was feed by vitamin C alone, drastically lost its vitamin C content during few days at ambient temperature. Moreover, the controlled release behavior of vitamin C, in vitro and in vivo, showed that vitamin C was released in the gastrointestinal tract of rainbow trout in the controlled manner (up to 48 h) and chitosan nanoparticles could well maintain vitamin C against harsh conditions, acidic and enzymatic hydrolysis, in the gastrointestinal tract of rainbow trout. Also, Alishahi et al. (2011) showed that the chitosan nanoparticles containing vitamin C could significantly (p < 0.05) induce the non-specific immunity system of rainbow trout, as compared with the control. RajeshKumar, VenKatesan, Sarathi, Sarathbabu, Thomas and Anver Basha (2009) demonstrated that chitosan nanoparticles are able to encapsulate DNA and then favorably incorporated into shrimp feed to protect them from white spot syndrome virus. Their results showed that these nanoparticles increased the survival rates of shrimp against white spot syndrome during 30 days post-treatment. Likewise, RajeshKumar, Ishaq Ahmed, Parameswaran, Sudhakaran, Sarath Babu and Sahl Hameed (2008) incorporated chitosan nanoparticles containing DNA vaccine into Asian sea bass (Lates calcarifer) feed. Their results indicated that the sea bass orally vaccinated with chitosan-DNA (pVAOMP38) complex showed moderate protection against experimental *Vibrio anguillarum* infection. Similarly, Tian, Yu and Sum (2008) reported that chitosan microspheres loaded with plasmid vaccine was interestingly used to orally immunize Japanese flounder (*Paralichthys olivaceus*). They explained that the release profile of DNA from chitosan microspheres in PBS buffer (pH 7.4) was up to 42 days after intestinal imbibitions. Aydin and Akbuga (1996) showed that salmon calcitonin, available for clinical use, was suitably encapsulated in chitosan beads and the results confirmed that salmon calcitonin-loaded chitosan beads could be prepared by gelling the cationic chitosan with the anionic counterpart providing a controlled release property. Also, shark liver oil could be efficiently encapsulated in calcium alginate beads coated with chitosan in order to mask its unpleasant taste (Peniche, Howland, Carrillo, Zaldivar & Arguelles, 2004). The chitosan coating allowed controlling the permeability of capsules and avoiding leakage. The shark liver oil loaded chitosan/calcium alginate capsules were initially resistant to the acid environment of the stomach. But after 4 h at the intestinal pH (7.4), the capsule wall weakened and thereby was able to be easily deteriorated and disintegrated by the mechanical and peristaltic movements of the gastrointestinal tract. Likewise, Klinkesorn and Mcclements (2009) stated that the encapsulation of tuna oil droplets with chitosan affected their physical stability and digestibility when they were passed through an in vitro digestion model containing pancreatic lipase. The amount of free fatty acids released from the emulsions decreased as the concentration of chitosan increased. However the relesae was independent of chitosan Mw. These results showed that chitosan was able to reduce the amount of free fatty acids released from the emulsion, which may be attributed to a number of different physiological mechanisms, including formation of a protective chitosan coating around the lipid droplets, direct interaction of chitosan with lipase, or fatty acid binding by the chitosan. Also, they showed that pancreatic lipase was able to digest chitosan and release glucosamine, having important implications for the utilization of chitosan coatings for the encapsulation, protection and delivery of Omega-3 fatty acids. They suggested that encapsulation with chitosan could be used to protect emulsified polyunsaturated lipids from oxidation during storage. However, they will release the functional lipids after they are consumed. Industrially, tuna oil encapsulation with chitosan using ultrasonic atomizer was shown to be the promising technique in the near future (Klaypradit & Huang, 2008).

8. Conclusions

Chitosan, a deacetylated derivative of chitin, has attracted a great attention in the seafood industry due to its non-toxicity, biodegradability, biocompatibility and mucus adhesiveness properties. Chitosan has different characteristics such as antibacterial, antioxidant, film-forming ability, gel enhancer, encapsulating capacity, tissue engineering scaffold, wound dressing, and coagulating agent. Upon knowing these, chitosan could successfully be incorporated into seafood products for both seafood quality and human health enhancement. Regarding its outstanding characteristics, chitosan would be used as functional ingredients in marine-based products and it merits further researches in the future.

Author details

Alireza Alishahi
The University of Agriculture Sciences and Natural Resources of Gorgan, Gorgan, Iran

9. References

Aider, M. (2010). Chitosan application for active bio-based films production and potential in the food industry: A review. *Food Science and Technology*, 43, 837-842.

Ahn, C.B., & Lee, E.H. (1992). Utilization of chitin prepared from the shellfish crust.2. Effect of chitosan film packaging on quality of lightly-salted and dried horse mackerel. *Bulletin of Korean Fish Society*, 25, 51-57.

Alishahi, A., Mirvaghefi, A., Rafie-Tehrani, M., Farahmand, H., Shojaosadati, S.A., Dorkoosh, F.A., & Elsabee, M.Z. (2011). Shelf life and delivery enhancement of vitamin C using chitosan nanoparticles. *Food Chemistry*, In press.

Allan, C. R., & Hardwiger, L. A. (1979). The fungicidal effect of chitosan on fungi of varying cell wall composition. *Experimental Mycology*, 3, 285-287.

Aydin, Z., & Akbuga, J. (1996). Chitosan beads for delivery of salmon calcitonin: preparation and release characteristics. *International Journal of Pharmaceutics*, 131, 101-103.

Benjakul, S., Visessanguan, W., Phatchrat, S., & Tanaka, M. (2003). Chitosan affects transglutaminase – induced surimi gelation. *Journal of Food Microbiology*, 27, 53-66.

Benjakul, S., Viscessanguan, W., Tanaka, M., Ishizaki, S., Suthidham, R., & Sungpech, O. (2001). Effect of chitin and chitosan of gelling properties of surimi from barred garfish (*Hemiramphus far*). *Journal of the Science of Food and Agriculture*, 81, 102-108.

Bhatnagar, A., & Sillanpa, M. (2009). Application of chitin and chitosan-derivatives for water and wastewater- a short review. *Advances in Colloid and Interface Science*, 55, 9479-9488.

Bordenave, N., Grelier, S., & Cama, V. (2007). Water and moisture susceptibility of chitosan and paper-based materials: structure-property relationships. *Journal of Agriculture and Food Chemistry*, 40, 1158-1162.

Bough, W.A. (1976). Chitosan a polymer from seafood waste, for use in treatment of food processing wastes and activated sludge. *Process Biochemistry*, 11, 13-16.

Cadwallader, K.R. & Shahidi, F. (2001). Identification of potent odorants in seal blubber oil by direct thermal desorption-gas chromatography-olfactometry. In omega-3 fatty acids: chemistry, nutrition and health effects. F. Shahidi & J.W. Finley (eds) pp.221-234, symposium series 788, American chemical society, Washington, D.C.

Cao, R., Xue, C.H., & Liu, Q. (2009). Change in microbial flora of pacific oysters (*Crassostera gigas*) during refrigerated storage and its shelf life extension by chitosan. *International Journal of Food Microbiology*, 131, 272-276.

Carlson, R.P., Taffs, R., Davison, W.M., & Steward, P.S. (2008). Anti-biofilm properties of chitosan coated surfaces. *Journal of Biomaterial Science Polymer*, 19, 1035-1046.

Castell, C.H., Maclean, J., & Moore, B. (1969). Rancidity in lean fish muscle. IV. Effect of sodium chloride and other salts. *Journal of Fish Resources and Biodiversity of Canada*, 22, 929-944.

Chi, F.H., & Cheng, W.P. (2006). Use of chitosan as coagulant to treat wastewater from milk processing plant. *Journal of Polymer and the Environment*, 14, 411-417.

Decker, E.A., & Haultin, H.O. (1992). Lipid oxidation in muscle foods via redox ion. In: Angelo, A.J. (ed). Lipid oxidation in food. Washington, D.C.: American Chemical Society, pp. 33-54.

Deladino, L., Anbinder, P.S., Navarro, A.S. & Martino, M.N. (2008). Encapsulation of natural antioxidants extracted from *Ilex paraguariensis*. *Carbohydrate Polymer*, 71, 126-134.

Devlieghere, F., Vermeulen, A., & Debevere, J. (2004). Chitosan: antimicrobial activity, interactions with food components and applicability as coating on fruit and vegetables. *Food Chemistry*, 21, 703-714.

Dias, F.S., Querroz, D.C., Nascimento, R.F., & Lima, M.B. (2008). Simple system for preparation of chitosan microspheres. *Quimica Nova*, 31, 160-163.

Duan, J., Cherian, G., & Zhao, Y. (2009). Quality enhancement in fresh and frozen lngcod (*Ophiodon elongates*) fillets by employment of fish oil incorporated chitosan coatings. *Food Chemistry*, 119, 524-532.

Duan, J., Jiang, Y., Cherian, G., & Zhao, Y. (2010). Effect of combined chitosan-krill oil coating and modified atmosphere packaging on the storability of cold-stored lingcod (*Ophiodon elongates*) fillets. *Food Chemistry*, 122, 1035-1042.

Dutta, P.K., Tripathi, S., Mehratra, G.M., & Dutta, J. (2009). Perspectives for chitosan based antimicrobial films in food applications. *Food Chemistry*, 114, 1173-1182.

El Tahlawy, K.F., El Benday, M.A., Elhendawy, A.G., & Hudson, S.M. (2005). The antimicrobial activity of cotton fabrics treated with different crosslinking agents and chitosan. *Carbohydrate Polymer*, 60, 421-430.

Feng, T., Du, Y. Li. J., Hu, Y., & Kennedy, J.F. (2008). Enhancement of antioxidant activity of chitosan by irradiation. *Carbohydrate Polymer*, 73, 126-132.

Hirano, S. (1996). Chitin biotechnology applications. *Biotechnology Annual Review*, 2, 237-258.

Gomez-Guillen, M., Montero, P., Solas, M.T., & Perez-Mateos, A.G. (2005). Effect of chitosan and microbial transglutaminase on the gel forming ability of horse mackerel (*Trachurus spp.*) muscle under high pressure. *Food Research International*, 38, 103-110.

Guerrero, L., Omil, F., Mendez, R., & Lema, J.M. (1998). Protein recovery during the overall treatment of wastewaters from fish-meal factories. *Bioresources Technology*, 63, 221-229.

Hsieh, R.J., & Kinsella, J.E. (1989). Oxidation of polyunsaturated fatty acids: mechanism, products and inhibition with emphasis on fish. *Advances in Food and Nutrition Research*, 33, 233-241.

Jeon, Y.J., Kamil, J.Y.V.A., &Shahidi, F. (2002). Chitosan as an edible invisible film for quality preservation of herring and Atlantic cod. *Journal of Agriculture and Food Chemistry*, 50, 67-78.

Jeon, J.J., Shahidi, F., & Kim, S.K. (2000). Preparation of chitin and chitosan oligomers and their applications in physiological functional foods. *Food Review International*, 16, 159-176.

Kadam, S.U. & Prabhasankar, P. (2010). Marine foods as functional ingredients in bakery and pasta products. *Food Research International*, 43, 1975-1980.

Kamil, J.Y.V.A., Jeon, J.J., & Shahidi, F. (2002). Antioxidative activity of chitosans of different viscosity in cooked comminuted flesh of herring (*Clupea harengus*). *Food Chemistry*, 79, 69-77.

Kataoka, J., Ishizaki, S., & Tanaka, M. (1998). Effects of chitosan on gelling properties of low quality surimi. *Journal of Muscle Foods*, 9, 209-220.

Kester, J.J., & Fenneema, O. (1986). Edible films and coatings: A review. Food Technology, 40, 47-59.

Kim, K.W., & Thomas, R.L. (2007). Antioxidant role of chitosan in a cooked cod (Godus morhua) model system. *Journal of Food Lipids*, 9, 57-64.

Khor, E. (2001). The relevance of chitin. In: chitin: fulfilling a biomaterial promise. Elsevier, pp. 1-8.

Klaypradit, W., & Huang, Y.W. (2008). Fish oil encapsulation with chitosan using ultrasonic atomizer. *Journal of food science and technology*, 41, 1133-1139.

Klinkesorn, U., & McClement, D.J. (2009). Influence of chitosan on stability and lipase digestibility of lecithin-stabilized tuna oil-in-water emulsions. *Food Chemistry*, 114, 1308-1315.

Kok, T.N., & Park, J.W. (2007). Extending the shelf life of set fish ball. *Journal of Food Quality*, 30, 1-27.

Kong, M., Chen, X.G., Xing, K., & Park, H.J. (2010). Antimicrobial activity of chitosan and mode of action: A state of art review. *International Journal of Food Microbiology*, 144, 51-63.

Kumar, M.N.V. (2000). A review of chitin and chitosan applications. *Reactive Functional Polymer*, 46, 1-27.

Lanier, T.C., Manning, P.K., Zetterling, T., & Macdonald, G.A. (1992). Process innovations in surimi manufacture. In T.C. Lanier & C.M. Lee (eds), Surimi technology, pp. 167-179, Marcel Decker, New York.

Ledger, R., Tucker, I.G., & Walker, G.F. (2002). The metabolic barrier of the lower intestinal tract of salmon to the oral delivery of protein and peptide drugs. *Journal of Controlled Release*, 85, 91-103.

Li, X., & Xia, W. (2010). Effect of chitosan on the gel properties of salt-soluble meat proteins from silver carp. *Carbohydrate Polymer*, 82, 958-964.

Lopez-Caballero, M.E., Gomez-Guillen, M.C., Perez-Mateos, M., & Montero, P. (2005a). A chitosan-gelating blend as a coating for fish patties. *Food Hydrocolloids*, 19, 303-311.

Lopez-Caballero, M.E., Gomez-Guillen, M.C., Perez-Mateos, M., & Montero, P. (2005b). A functional chitosan-enriched fish sausage treated by high pressure. *Journal of Food Science*, 70, 166-171.

Mao, L., & Wu, T. (2007). Gelling properties and lipid oxidation of kamabako gels from grass carp (*Ctenopharyngodon idellus*) influenced by chitosan. *Journal of Food Engineering*, 82, 128-134.

Newton, I.S. (2001). Long chain fatty acids in health and nutrition. In omega-3 fatty acids: chemistry, nutrition and health effects, F. Shahidi & J.W. Finley (eds), ACS symposium series 788, American chemical society, Washington, D.C., pp. 14-27.

No, H.K., Meyers, S.P., Prinyawiwatkul, W., & Xu, Z. (2007). Applications of chitosan for improvement of quality and shelf life of foods: A review. *Journal of Food Science*, 72, 87-100.

No, H.K., Park, N.Y., Lee, S.H., & Meyers, S.P. (2002). Antibacterial activity of chitosan and chitosan oligomers with different molecular weights. *International Journal of Food Microbiology*, 74, 65-72.

Ojagh, S.M., Rezaei, M., Razavi, S.H., & Hossieni, S.M.H. (2010). Effect of chitosan coatings enriched with cinnamon oil on the quality of refrigerated rainbow trout. *Food Chemistry*, 120, 193-198.

Pittermann, W., Horner, V., & Wachter, R. (1997). Food applications of high molecular weight chitosan in skin care applications. In: Muzzarelli, R.A.A. & Peter, M.G. (eds). chitin handbook, Grottammare, Italy: European chitin society, pp. 361.

Peniche, C., Howland, I., Carrillo, O., Zaldivar, C., & Arguelles-Monal, W. (2004). Formation and stability of shark liver oil loaded chitosan/calcium alginate capsules. *Food Hydrocolloids*, 18, 865-871.

Qi, C.R., Bang-Zhang, Y., & Lan-Ian, Z. (2010). Combined effect of ozonated water and chitosan on the shelf life of pacific oyster (*Crassostrea gigas*). *Innovativ Food Science and Emerging Technologies*, 11, 108-112.

Qin, Y. (1993). Chelating property of chitosan fibers. *Journal of Applied Polymer*, 49, 727-731.

Raafat, D., Bargen, K., Haas, A. & Sahl, HG. (2008). Insights into the mode of action of chitosan as an antimicrobial compound. *Applied and Environmental Microbiology*, 74, 3764-3773.

Raafat, D., & Sahl, H.G. (2009). Chitosan and its antimicrobial potential- a critical literature survey. *Microbial Biotechnology*, 2, 186-201.

Rajeshkumar, S., Ishaq Ahmed, V.D., Parameswaran, V., Sudhakaran, R., Sarath Babu, V., & Sahl Hameed, A.S. (2008). Potential use of chitosan nanoparticles for oral delivery of DNA vaccine in Asian sea bass (Lates calcarifer) to protect from *Vibrio anguillarum*. *Fish & Shellfish Immunology*, 25, 47-56.

Rajeshkumar, S., Venkatesan, C., Sarathi, M., Sarathbabu, V., Thomas, J., & Anver Basha, K. (2009). Oral delivery of DNA construct using chitosan nanoparticles to protect the shrimp from white spot syndrome virus (WSSV). *Fish & Shellfish Immunology*, 26, 429-437.

Roller, S., & Corvill, N. (2000). The antimicrobial properties of chitosan in mayonnaise and mayonnaise-based shrimp salads. *Journal of Food Protection*, 63, 202-209.

Ross, S. (2000). Functional foods: the food and drug administration perspective. *The American Journal of Clinical Nutrition*, 71, 1735S-1738S.

Sathivel, S. (2005). Chitosan and protein coatings affect yield, moisture loss, lipid oxidation of pink salmon (*Oncorhynchus gorbushcha*) fillets during frozen storage. Journal of Food Science, 70, 755-459.

Sathivel, S., Liu, Q., Huang, J., & Prinyawiwatkul, W. (2007). The influence of chitosan glazing on the quality of skinless pink salmon (*Oncorhynchus gorbuscha*) fillets during frozen storage. *Journal of Food Engineering*, 83, 366-375.

Savant, V.D. (2001). Protein absorption on chitosan-polyanion complexes: application to aqueous food processing wastes. PhD. Thesis, Food Science and Technology, Oregon State University.

Savant, V.D., & Torres, J.A. (2000). Chitosan based coagulating agents for treatment of cheddar cheese whey. *Biotechnological Progress*, 16, 1091-1097.

Savant, V.D., & Torres, J.A. (2003). Fourier transform infrared analysis of chitosan based coagulating agents for treatment of surimi waste water. *Journal of Food Technology*, 1, 23-28.

Schep, L.J., Tucker, I.G., Young, G., Ledger, R., & Butt, A.G. (1999). Controlled release opportunities for oral peptide delivery in aquaculture. *Journal of Controlled Release*, 59, 1-14.

Shahidi, F., & Han, X. (1993). Encapsulation of food ingredients. *Critical Reviews in Food Science and Nutrition*, 33, 501-547.

Shahidi, F., Kamil, J.Y.V.A., & Jeon, Y.J. (1999). Food applications of chitin and chitosan. *Trends in Food Science & Technology*, 10, 37-51.

Shahidi, F., Kamil, J.Y.V.A., & Jeon, J.J. (2002). Antioxidant role of chitosan in a cooked cod (*Gadus morhua*) model system. *Journal of Food Lipids*, 9, 57-64.

Tian, J., Yu, J., & Sun, X. (2008). Chitosan microspheres as candidate plasmid vaccine carrier for oral immunization of Japanese flounder (*Paralichthys olivaceus*). *Veterinary immunology and Immunopathology*, 126, 220-229.

Tharanathan, R.N., & Kittur, F.S. (2003). Chitin- the undisputed biomolecule of great potential. *Critical Review in Food Science and Nutrition*, 43, 61-87.

Weist, J.L., & Karel, M. (1992). Development of a fluorescence sensor to monitor lipid oxidation. 1. Florescence spectra of chitosan powder and polyamide powder affect exposure to volatile lipid oxidation products. *Journal of Agriculture and Food Chemistry*, 40, 1158-1162.

Wibowo, S. (2003). Effect of the molecular weight and degree of deacetylation of chitosan and nutritional evaluation of solid recovered from surimi processing plant. PhD. Thesis, Food Science and Technology, Oregon State University.

Wibowo, S., Savant, V., Cherian, G., Savange, T.F., Velaquez, G., & Torres, J.A. (2007b). A feeding study to assess nutritional quality and safety of surimi wash water proteins recovered by a chitosan-alginate complex. *Journal of Food Scinece*, 72, 179-184.

Wibowo, S., Velazquez, G., Savant, V., & Torres, J.A. (2005). Surimi wash water treatment for protein recovery: effect of chitosan-alginate complex concentration and treatment time on protein adsorption. *Bioresoures Technology*, 96, 665-671.

Wibowo, S., Velazquez, G., Savant, V., & Torres, A. (2007a). Effect of chitosan type on protein and water recovery efficiency from surimi wash water treated with chitosan-alginate complexes. *Bioresources Technology*, 98, 539-545.

Ye, M., Neetao, H., & Chen, H. (2008). Effectiveness of chitosan-coated plastic films incorporating antimicrobials in inhibition of *Listeria monocytogens* on cold-smoked salmon. *International Journal of Food Microbiology*, 127, 235-240.

Chapter 3

Polysaccharide-Protein Interactions and Their Relevance in Food Colloids

Amit K. Ghosh and Prasun Bandyopadhyay

Additional information is available at the end of the chapter

1. Introduction

Polysaccharides and proteins are natural polymers that are widely used as functional ingredients for various food colloids or emulsion formulations. Majority of food emulsions are constituted with polysaccharide and protein combinations. They are the essential ingredients of any food colloid formulation mainly due to their ability to change product shelf life by varying food texture (Schmidt & Smith, 1992; Schorsch, Jones & Norton 1999). Their interaction in the formulation thus finds many applications particularly in new food formulation development. Due to complex formation and creation of nano or micro structures (aggregation and gelation behavior) they generally change the rheological properties of food colloids which may affect the food product texture and colloidal stability (Benichou, 2002; McClements, 2005, 2006, 2007; Dickinson, 2003). Polysaccharide and protein interactions in solution and interfaces have been studied by several groups (Dickinson, 2003, 2008; Bos &Van Vliet 2001; Carrera & Rodríguez Patino 2005; Krägel, Derkatch, & Miller, 2008; Koupantis, & Kiosseoglou, 2009; Mackie, 2009). However, despite the vast advancement made in the recent past, polysaccharide and protein interactions in food hydrocolloids continue to be one of the most challenging topics to understand.

Proteins, being surface active can play major role in the formation and stabilization of emulsions in the presence of polysaccharide, while interacting through electrostatic or hydrophobic-hydrophobic interactions. On the other hand, polysaccharides being hydrophilic in nature generally remain in aqueous phase thus help in controlling the aqueous phase rheology like thickening, gelling and acting as stabilizing agents. The formation and deformation of polysaccharide-protein complexes and their solubility depend on various factors like charge and nature of biopolymers, pH, ionic strength and temperature of the medium and even the presence of surfactant of the medium (Ghosh & Bandyopadhyay, 2011). If pH of the medium is reduced below isoelctric point (pI) of the protein present then net positive charge of the protein will become prominent which will interact with negatively charged polysaccharide to form stable electrostatic complex. Similarly, if solution pH increased more than protein pI, the net negative charge of protein will tend to form complex with positively charged polysaccharides (Xia & Dubin, 1994; Dickinson, 2008; Turgeon, Schmitt & Sanchez, 2007). Generally, chances of weaker complex formation is more when solution pH is almost equal to protein pI, because at that pH range surface charge of protein becomes nearly zero. However, at very high concentration, similarly charged biopolymers repel each other and the net repulsion make the system unstable (separate as two distinct phases) which is known as *thermodynamic incompatibility*. Incompatibility in the system occurs at pH higher than the protein pI and at higher ionic strength (Grinberg & Tolstoguzov, 1997). Thus by varying pH and ionic strength of the medium one can achieve a control on the polysaccharide-protein interactions.

Polysaccharides and proteins both contribute to the structural and textural properties of food by changing rheology of food emulsions through their gelling networking system (Dickinson, 1992). Non-covalent interactions between polysaccharide and protein in any emulsion formulation play a major role to change the interfacial behavior and stability of the food colloids. The driving force for these non-covalent interactions is electrostatic interactions, hydrophobic interactions, H-bonding and Van der Waals interactions. Recent literatures also focus on how protein and polysaccharide molecules can be linked together by covalent bond. At pH close to protein pI this Maillard-type conjugates were used to improve the colloidal stability and interfacial structure of proteins in certain conditions (Jiménez-Castaño, Villamiel, & López-Fandiño, 2007; Benichou, Aserin, Lutz & Garti, 2007)). Recent developments in the field describe interfacial physico-chemical properties of polysaccharide-protein mixed systems (Rodríguez Patino & Pilosof 2011). In this chapter, we would like to focus more on polysaccharide and protein non-covalent interaction studies and their effect towards food colloids stability.

2. Nature of polysaccharide-protein complex

Polysaccharide and protein complex formation is mainly driven by various non-covalent interactions, like electrostatic, H-bonding, hydrophobic, and steric interactions (Kruif et al 2001). Protein carries +ve or –ve zeta potential based on the pH of the medium (+ve at pH lower than pI and vice-versa). This +ve or –ve electrical charge on the protein chain point towards the presence of different amino acids in the protein molecules and their mode of ionization at different pH ranges (Fig. 1). Carboxylate polysaccharides get deprotonated (become anionic) at a pH range higher than its pKa (Fig. 1). This electrical charge on the back bone of protein or polysaccharide chain is responsible for electrostatic attraction or repulsion between them. Again, presence of -COOH group on the polysaccharide and -NH$_3$, -COOH groups on the protein chain are the sources of hydrogen bonding between these two bio-polymers. Extent of both of this hydrogen bonding and electrostatic interaction depends on the solution parameters such as pH, ionic strength, temperature etc. Except these ionic patches on the bio-polymers, few non-polar segments are also present on the bio-polymers, which are responsible for the hydrophobic staking with each other. Even though solution parameters are important factors to control the different mode of interactions between protein and polysaccharide, type of proteins/polysaccharides, molecular weight, charge density, and hydrophobicity of the bio-polymers are also play significant role towards the extent of complexation between two bio-polymers at a fixed condition.

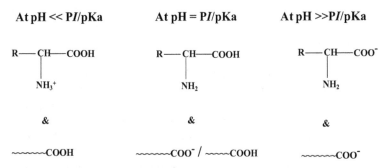

Figure 1. Variation of charge density on the polysaccharide and protein chain at various pH ranges.

In general, interactions between proteins and polysaccharides are quite explored where large numbers of report have been published based on the interactions between oppositely charged "protein-polysaccharide" systems (Dmitrochenko et al 1989; Bengoechea et al 2011, Stone & Nickerson 2012). Although electrostatic attraction is the main driving force for the complexation between protein and polysaccharide, but it is also reported that hydrogen bonding and hydrophobic interaction plays a secondary role for stability of the "protein-polysaccharide" aggregates (McClements, 2006). The extent of hydrogen bonding and hydrophobic interaction also depends on temperature (Weinbreck et al, 2004). In 2009 Nickerson and co-workers(Liu, Low, & Nickerson, 2009) have reported that pea protein and gum acacia complex stabilize at low temperature due to increase in hydrogen bonding interactions and destabilize at high temperature due to decline in hydrogen bonding interactions. Temperature also plays an important role to decide the protein conformations (folded or unfolded). In 2007, Pal (Mitra, Sinha & Pal, 2007) and coworkers have reported that human serum albumin unfolds at higher temperature and undergoes in reversible refolding conformations upon cooling (below 60^0 c). Unfolded conformations of protein expose more reactive sites (amino acids) to the solvent phase, thus more chances of interactions (or binding) with polysaccharide. Binding of anionic polysaccharides (pH~pK_a) to the cationic proteins (at pH<pI) result both soluble and insoluble complexes (Magnusson & Nilsson, 2011). Initial binding of polysaccharides (anionic) to the proteins (cationic) cause charge neutralizations, which lead to the formation of insoluble "protein-polysaccharide" aggregates (Schmitt et al, 1998). Further binding of anionic polysaccharides to those neutral aggregates make it effectively anionic, which leads to formation of soluble complexes. But binding of anionic polysaccharides with anionic proteins (pH>pI) are also known and governed by the interactions between anionic reactive sites of polysaccharide and small cationic reactive sites of protein (Fig. 2). Binding of anionic polysaccharides to the cationic side of proteins (at pH>p*I*) result in formation of anionic "protein-polysaccharide" aggregates, thus soluble complexes. Therefore, concentration of polysaccharides and pH play an important role towards the solubility of "protein-polysaccharide" aggregates.

Two bio-polymers can exist either in a single phase systems or in a phase separated systems depending on the nature of bio-polymers, their concentration, and solution conditions. When two bio-polymers carry opposite charge, then either they agglomerates to form soluble complexes (single phase) or insoluble precipitates (2-phase system). On the other hand, when two non-interacting bio-polymers mixed together, either they exist in a single phase system (where two separate entities distributes uniformly throughout the medium) or exist as two distinct phases (each phases comprise different bio-polymer). Therefore, in the protein-polysaccharide system, phase separation occurs through two different mechanisms which are *associative phase separation* and *segregative phase separation* (Tolstoguzov, 2006). *Associative phase separation* is the aggregation between two oppositely charged bio-polymers (electrostatic attraction driven), leads to the phase separation, where one phase is enriched with two different bio-polymers (coacervation or precipitation) (Fig. 3). *Segregative phase separation* occurs either due to strong electrostatic repulsion (between two similarly charged bio-polymers) or because of very high steric exclusion (between two neutral bio-polymers). In this case, at low concentration, two biopolymers can co-exist in a single phase whereas at higher concentration, it starts phase separation. (Fig. 3).

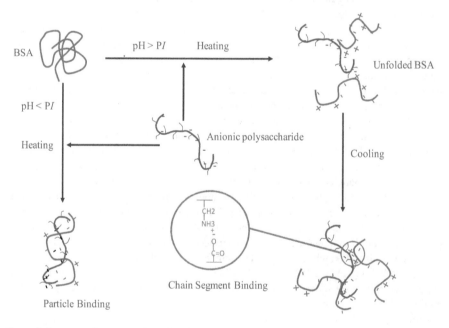

Figure 2. Interaction between polysaccharide and protein at various pH.

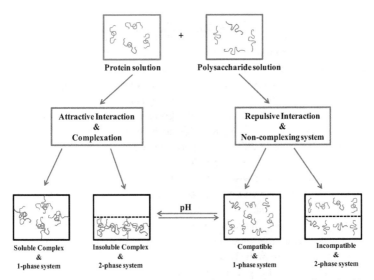

Figure 3. Schematic representation of the possible mode of interaction between polysaccharides and proteins.

3. Functional properties of polysaccharide-protein complexes related to food application

Polysaccharide-protein complexes exhibit a wide range of interesting properties, such as surface activity to stabilize air-water or oil-water interfaces, viscosifying, and gelling properties etc. Viscosifying and gelling ability of polysaccharide-protein complexes help to obtain gel-like processed food products without any thermal treatment in the process. The interfacial properties of these complexes help to impart stability into the emulsion food products. Also, protein–polysaccharide complexes are able to encapsulate several active ingredients; hence they act as delivery systems for many bioactives or sensitive molecules in food formulations. These complexes are also known to vary bulk/interfacial structures, textures and shelf-life stability of the food colloids. In the following section we will discuss this polysaccharide-protein interaction in light of their functional properties.

3.1. Viscosity of polysaccharide-protein complex and air-water foam stability

Viscosity and gelling are the rheological property which depends on the molecular characteristics of the biopolymers, such as their molecular weight, shape, chain flexibility. Other factors are the concentration, interaction between the biopolymers and water, as well as solution parameters like: pH, ionic strength and presence of other components/ligands etc. Interactions between polysaccharide and protein have been proven to widen the functional properties of each individual biopolymer. Rheological properties of polysaccharide-protein complex lead to new rheological behaviors different from each individual biopolymer.

Association of two bio-polymers is expected to increase the bulk viscosity of the system as entities of larger sizes are formed. The rheological behavior of several protein–polysaccharide mixed systems have been studied and ranged from viscous to viscoelastic properties showing elastic behavior is reported. The hydrated polysaccharide-protein complexes increase viscosity and rheology of the system found to be depends on the nature and structure of polysaccharides. Viscous property of gum acacia-protein coacervates attributes to the globular conformation of the polysaccharide, whereas the same protein with linear pectin results in gel-like system. Beside the nature of the individual bio-polymer solution property and concentration of bio-polymers are also known to affect the rheology of the system (Dickinson, 2011). For example, it was found that pH played a major role in the viscosity of the coacervate phase. A maximum viscosity was obtained at pH 4.0, where concentration of whey protein and gum arabic in the coacervate phase was maximum and extent of electrostatic attraction was highest. This suggests that the electrostatic interactions between whey protein and gum arabic were responsible for the highly viscous behavior of the coacervates. Whereas, the same composition of whey protein and gum arabic at pH above protein p*I* (i.e. comparatively lower electrostatic interactions) showed more elastic nature than viscous. Ionic strength and protein/polysaccharide ratio is also known to play an important role towards the rheology of polysaccharide-protein systems. For example, optimal salt concentrations (0.21 M NaCl) favor the coacervation of β-lactoglobulin with pectin at higher concentration and produce much stronger gel strength. For better gelling property, it is necessary to control the parameters which required to form coacervate, because strong associative interaction decrease the solubility of complexes and hence lower the hydration capacity of the complex, which leads to decrease in the viscosity (Schmitt & Turgeon, 2011; Kruif, et al 2004).

Viscoelastic properties of polysaccharide-protein complexes also play an important role towards the foam stability in variety of food products. In case of air-water system foam can be define as the air entrapment by a thin liquid film (water), where this liquid film is stabilized by some surface active molecules. Stability of the foam increases with the increase in the stability of the interfacial liquid film, because lower stability of this interfacial liquid film can lead to the diffusion of air entrapped inside the foam. Viscosity of this liquid film is another parameter by which one can control the diffusion rate of air entrapped inside the foam. Therefore, higher stability and viscosity of the interfacial liquid film leads to lower diffusion of air entrapped inside the foam and increase the foam stability. Schmitt and co-workers have studied the air-water interfacial property of β-lactoglobulin-acacia gum complexes at pH 4.2 [Schmitt et al 2005]. The group has reported that although surface activity of the complex is similar with the pure protein, but complex forms much stronger viscoelastic interfacial film with thickness of about 250 Å. As a result, gas permeability of thin-film stabilized by the complexes was significantly reduced (0.021 cm s^{-1}) compared to pure β-lactoglobulin (0.521 cm s^{-1}). This phenomenon suggests that stability of foam (stabilized by protein-polysaccharide complex) is higher compared to the foam stabilized by protein alone.

The likely explanation of the higher foam stability and different interfacial properties of coacervate is that protein-polysccharide complexes are able to re-organize at the interface by coalescence, forming interfacial microgel. These findings were applied for the ice cream formulation for improved air bubble stability (Schmitt C, Kolodziejczyk E. 2010). Similarly, gelatin has been replaced by whey protein isolate-gum acacia complexes to improve the bubble stability in chilled dairy products (Schmitt C, Kolodziejczyk E. 2010). In case of stabilization by complexes, variation in ratio of biopolymers could be used to control the size of the complexes, hence their surface activity. In addition to that, viscoelastic properties of the air-water interfacial film is possible to tune by either adsorbing two biopolymers simultaneously or by the sequential adsorption of protein followed by polysaccharide. As for example, β-lactoglobulin-pectin complexes are known to stabilize the air-water interface. In this case, thickness of the film obtained from the sequential adsorption of protein and polysaccharide was higher (450 Å) than the adsorption of complexes (250 Å) (Ganzelves et al 2008).

In contrary to air-water foam stability, use of polysaccharide-protein complexes for the stabilization of oil-water emulsion (Martínez et al 2007) has received much more attention. Use of these polysaccharide-protein hydrocolloids as an emulsion stabilizer will be discussed in the next section.

3.2. Oil-water emulsion stability

Emulsion is a uniform dispersion of liquid droplets within a continuous matrix of a second immiscible liquid, stabilized by surface active molecules. These stabilizers are termed as emulsifier. In the context of the present topic, we will limit our discussion to the role of bio-polymers as emulsifier. Generally, emulsifier has the amphiphilic character to adsorb onto the interface of liquid droplets, which can prevent the phase separation of two immiscible liquids. For a fixed emulsifier, stability of the emulsion depends on few factors, such as rate of adsorption of the emulsifier, concentration of emulsifier, etc. At low concentration of emulsifiers, emulsion system fails to retain its initial droplet size. This destabilization can take place through different mechanisms. In case of poor coverage of the interface by liquid droplets, they can coalesce with each other to form a bigger droplet (Fig. 4). Few examples are also reported, where polymer adsorbed onto the interface of liquid droplets thus bridge between two such liquid droplets and initiates bridging flocculation. Interestingly, emulsions at high emulsifier concentration produces stable oil droplets due to better coverage of the interfaces of the liquid droplets (Liu & Zhao, 2011).

Emulsion is possible to achieve by using many surface active agents, such as small surfactant molecules, bio-polymers (proteins or polysaccharides, hydrocolloids (protein-polysaccharide complexes), and inorganic particles. Stability of those emulsion systems mainly governs by the two important factors. First, repulsive force between two closely approaching liquid droplets; second, Ostwald ripening, which involves disappearing of smaller droplets in expense of the formation of larger one. Higher degree of repulsion between the two neighboring droplets results in maximum stability due to least chances of coalescence. Repulsive force between the two liquid droplets govern by the inter droplet distance, i.e. thickness of the thin liquid film between two closely approaching droplets. Thickness of this liquid film depends on the space occupied by the adsorbed molecules (emulsifier) at interface of the droplets. Emulsion generally get stabilized by different emulsifiers present in the formulations such as surfactants, proteins, or hydro-colloids (protein-polysaccharide complexes) and the relative thickness of the liquid film between two closely spaced droplets lies in the order of hydrocolloids (5-10 nm) > proteins (1-5 nm) > surfactants (0.5-1 nm) (Fig. 5). Therefore, stability of the emulsion droplets expected to be higher when they are stabilized by protein-polysaccharide complex compared to the same stabilized by protein or surfactant molecules. In addition to the thickness of the liquid film between two closely spaced droplets, rate of desorption of the emulsifiers from the interface is another important factor. Adsorption of emulsifier molecules (like surfactants, proteins etc.) at the liquid interface is highly reversible. Desorption of the emulsifiers from the liquid interface governs the instability of the system. According to this fact, emulsion stabilized by particles (size ranges from 10 nm to several μm) is likely to have indefinite stability, because of the maximum thickness of thin liquid film in between two closely spaced droplets and maximum desorption energy of the particles from the liquid interface. Despite of this theoretical consideration, experimental evidence by Tcholakova et al. does not support this hypothesis that particle stabilized emulsion are more stable compared to surfactant or protein or hydrocolloids stabilized emulsion (Tcholakova et al. 2008).

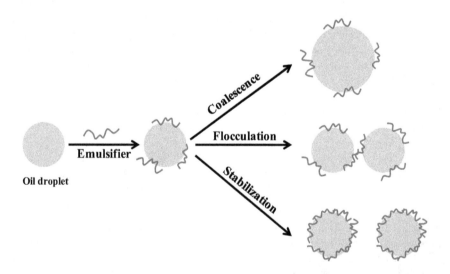

Figure 4. Schematic representation of mode of stabilization and destabilization of oil droplets in an oil-water emulsion.

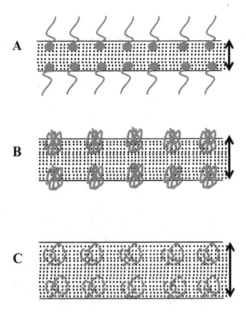

Figure 5. Schematic representation of the relative thickness of the thin liquid film between two closely spaced droplets, stabilized by A) surfactant, B) protein, and C) protein-polysaccharide hydrocolloids.

Another factor guides the emulsion instability is Ostwald ripening, which is disappearance of small size droplets at the expense of the larger droplets formation. Driving force for the Ostwald ripening is the difference in the chemical potential of the smaller and larger droplets. Mass transfer takes place between the droplets by diffusion process. Therefore, Ostwald ripening process requires the solubility of the dispersed phase into the continuous phase to initiate the diffusion process. Type of the emulsifier also plays an important role towards the Ostwald ripening process. Emulsion stabilized by the water soluble surfactant molecules has lower interfacial tension, which reduces the thermodynamic driving force of Ostwald ripening. Adsorption of the surfactant molecules at the droplet interface is a reversible phenomenon. Reversible desorption and adsorption of the surfactant molecules from the interface of the liquid droplet increases the rate of mass transfer between the dispersed droplets, hence increase the Ostwald ripening. The chances of desorption of emulsifier is less in case of the emulsions stabilized by protein molecule because it provides a thicker layer (elastic layer) around the droplets and greater surface coverage of the interfacial area. These factors reduce the ripening process in the protein stabilized emulsion. Ostwald ripening process is possible to avoid completely, only if emulsion is stabilized by insoluble particles (due to very high desorption energy) or thickness of the elastic layer around the dispersed droplets is equal to the droplet radius (Kabalnov, 2001). For this reason particulate emulsions are able to prevent the ripening process completely. Whereas hydrocolloid (protein-polysaccharide complexes) mostly behaves like a soft polymer, more resembles with the protein structures compared to solid particles, which cannot completely avoid the ripening process.

Protein–polysaccharide complex stabilized emulsions are possible to obtain by using two alternative ways. One of them involves addition of charged polysaccharide solution to a primary emulsion which is already stabilized by the protein as single emulsifier, to produce emulsion droplets having a protein–polysaccharide 'bilayer' surface coating (Fig. 6B). Another method involves addition of an aqueous solution containing the protein–polysaccharide complexes as an emulsifying agent following homogenization (Fig. 6A). For convenience, the first method is termed as 'bilayer emulsions' and second one is termed as 'mixed emulsions'. The bilayer approach is also commonly known as 'layer-by-layer' approach. Recently, it has attracted significant importance because of its use in nano-encapsulation and protection of emulsions against severe environmental stresses. The major problem lies in 'layer-by-layer' approach, where emulsion droplet tends to flocculate. Flocculation during the 'layer-by-layer' adsorption takes place because of two different mechanisms: a) bridging flocculation, b) depletion flocculation. *Bridging flocculation* takes place at low polysaccharide concentration when droplet collisions occur faster than the rate of polysaccharide saturation of the protein-coated droplet surfaces. *Depletion flocculation* occurs at higher polysaccharide concentration when unadsorbed polysaccharide exceeds a critical value. For this reason it is convenient to make emulsions with protein and polysaccharide present together before homogenization compared to 'layer-by-layer' approach. Recently direct comparison between the two techniques has been demonstrated experimentally, which shows that the more convenient mixed emulsion approach leads to better stability behavior than the bi-layer approach(Camino et al 2011).

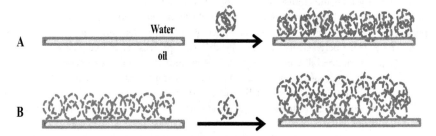

Figure 6. Schematic illustration of two alternative approach of preparing oil-water emulsion using protein-polysaccharide complexes as an emulsifier. A) Mixed emulsions, both protein and polysaccharide present together during emulsification. B) Bilayer emulsion, polysaccharide added to emulsion prior stabilized by protein.

3.3. Encapulation and release of active molecules

Generally, encapsulation includes all aspects of protection or stabilization of active molecules (flavors and bio-actives) against several external drastic conditions (such as heat, redox potential, shear, temperature, light, oxygen, moisture, etc.). Controlled release facilitates the delivery of the encapsulated material to the targeted place with the optimal kinetics. Conditions for the encapsulation of active molecules depends on the sensitivity (thermal and redox stability) and nature (solubility in oil and water) of the active components but release can be controlled by mechanical process, pH variations (acidic conditions in the stomach, neutral in the intestine) or enzymatic actions etc. As for example, Peniche et al. has shown the encapsulation of shark liver oil (rich in poly unsaturated fatty acids) using chitosan-alginate system to mask the unpleasant taste of oil. These capsules disrupts by enzymes, like lipase or pancreatine. Initially it was resistant to the acid environment of the stomach, but after 4 hour in the intestinal pH (pH 7.4), the capsule walls weakened and delivers the active molecules.

Another important application of this aspect is the encapsulation of flavor molecules. Recently, Yeo et al. has shown that gelatin–acacia gum coacervate can encapsulate flavors which can be released during cooking in baked goods (Yeo et al. 2005). Weinbreck et al. (Weinbreck et al. 2004) has shown that Whey protein–acacia gum coacervates can encapsulate lemon and orange flavors and their release under mechanical action like chewing. Encapsulation was one of the first applications of gum arabic-gelatin coacervates (Bungenberg de Jong and Kruyt, 1929). Viscous coacervate was made at a temperature higher than the gel point of gelatin and during cooling, interfacial rigidity increases which lead to a stable gelled shell around the microcapsule. This rigid shell disrupts after consumption, gelatin melts easily in the mouth and therefore releases the encapsulated actives. In addition to gelatin-acacia gum, several other protein-polysaccharide systems have been evolved (whey proteins, plant proteins, pectin, and xanthan gum) to broaden the encapsulation techniques. Beside these polysaccharide-protein combinations, process parameter for encapsulation (pH, ionic strength, macromolecular ratio, and drying/homogenization procedure) also plays an important role to modulate the physical properties (thickness, swelling rate, etc.) of the coacervate layer in the microcapsules (Savary et al 2010). Use of cross-linking agents can further harden the coacervate layer after formation of the microcapsules. As for example, use of trans-glutaminase can introduce covalent linkages between carboxyl group of a glutamine and amino group of lysine in the protein molecules. Alternatively, formaldehydes and glutaraldehydes have also been studied although they are non-food grade reagents. Recently, tannic acid, plant phenolics, citral molecules and glycerin have been studied as food grade alternatives (McClements, 2010).

Contrary to the encapsulation through coacervation, the bi-layer emulsion technique (formed by successive adsorption of biopolymers at the interface) is another way to study the microcapsule properties. Recently, McClements group has described that bi-layer approach of encapsulation has a better control of the interface structure, charge, thickness and permeability with improved stability and controlled release of actives. Group has reviewed this research area and discussed multilayer emulsions in light of bioavailability control and release of actives to the specific site of action depending on layer composition and properties (McClements, 2011). Sagis et al. has used high molecular weight pectins and pre-heated whey proteins (denaturation of protein) for the encapsulation through multilayer approach. (Sagis et al 2011).

4. Summary

Basic understanding of supramolecular chemistry which span among origin and nature of the various non-covalent molecular interactions between polysaccharides and proteins can be widely used to create various desirable nano and macro structures which are quite significant in food colloids/formulations. Food product texture modulation and colloidal stability can be achieved by controlling protein-polysaccharide interactions. Modulation of this interaction by varying medium conditions like pH, ionic strength etc. one can create many possibilities towards rheological properties of food colloids which may affect the emulsion stability. Protein-polysaccharide interactions are well characterized in various pH conditions. Although, the interaction depends on type and nature of biopolymers, no structure-activity correlation has been established until now. Moving forward, there is a huge demand to establish a correlation between biopolymer structures and interaction efficiency. The creative manipulation of polysaccharide-protein interactions can open up a completely new dimension in health and nutrition platform. Food particle travels from mouth to gut in various pH environments (for example pH decreases when it moves from mouth saliva to stomach and increases when partially digested food particle passes from stomach to small intestine), thus one can design a smart polysaccharide-protein complex system which can encapsulate or slow down and trigger or release of nutrients in various stages of digestion process depending on the pH of the system and any particular health demand. Such kind of pH sensitive system design and product development which works in vivo is a real challenge for the food scientists. Unfortunately, general understanding of protein-polysaccharide interactions, their bulk and interfacial properties toward emulsion stability is not completely understood and requires more systematic investigation in future to unveil its full potential.

Author details

Amit K. Ghosh and Prasun Bandyopadhyay*
Unilever R & D Bangalore, Whitefield, Bangalore, India

5. References

Bengoechea, C., Jones, O. G., Guerreo, A., & McClements, D. J. *Food Hydrocolloid*, Vol. 25, 2011, pp. 1227-1232

Benichou, A., Aserin, A. & Garti, N. *Journal of Dispersion Science and Technology*. Vol. 23, 2002, pp. 93-123

Benichou, A., Aserin, A., Lutz, R. & Garti, N. *Food Hydrocolloid*, Vol. 21, 2007, pp. 379-391

Bos, M. A, & Van Vilet, T. *Advances in Colloid & Interface Science*. Vol. 91, 2001, 437-471

Bungenberg, H. G. & Kruyt, H. R. Procedings K. Ned. Akad. Wet. Vol. 32, 1929, pp. 849-856

---

* Corresponding Author

Camino, N. A., Sánchez, C. C., Rodríguez Patino, J. M. & Pilosof, A. M. R. *Food Hydrocolloid,* Vol. 25, 2011, pp. 1-11

Carrera, C. & Rodríguez Patino, J. M. *Food Hydrocolloid.* Vol. 19, 2005, pp. 407-416

Choi, S. J., Decker, E. A., Henson, L., Popplewell, L. M. & McClements, D. J. *Food Chemistry* Vol. 122, 2010, pp. 111-116

Dickinson, E. *An introduction to food colloids.* 1992, Oxford University Press

Dickinson, E. *Colloid & Surfaces B: Biointerfaces.* Vol. 20, 2001, pp. 197-210

Dickinson, E. *Food Hydrocolloid,* Vol. 25, 2011, pp. 1966-1983

Dickinson, E. *Food Hydrocolloid.* Vol. 17, 2003, pp. 25-40

Dickinson, E. *Soft Matter.* Vol. 4, 2008, pp. 932-942

Dmitrochenko, A.P., Antoonov, Yu. A., & Tolseoguzov, V. B. *Applied Biochemistry and Microbiology,* Vol 25, 1989, 353-360

Ganzelves, R. A., Fokkink, R., van Vilet, T.,Cohen Stuart, M. A. de Jongh, H.H.J. *Journal of Colloid and Interface Science* Vol. 317, 2008, 137-140

Ghosh, A, & Bandyopadhyay, P. *Chemical Communications.* Vol. 47, 2011, pp. 8937-8939

Grinberg, V. Y. & Toolstoguzov, V. B. *Food Hydrocolloid.* Vol. 11, 1997, pp. 145-158

Jiménez-Castaño, L., Villamiel, M & López-Fandiño, R. *Food Hydrocolloid,* Vol. 21, 2007, pp. 433-443

Kabalnov, A. *Journal Dispersion Science Technology* Vol. 22, 2002, pp. 1-12

Koupantis, T. & Kiosseoglou, V. *Food Hydrocolloid.* Vol. 23, 2009, pp. 1156-1163

Krägel, J. Derkatch, S. R. & Miller, R. *Advances in Colloid & Interface Science.* Vol. 144, 2008, 38-53

Kruif, de C. G. & Tuinier, R. *Food Hydrocolloid,* Vol. 15, 2001, pp. 555-563

Kruif, de C. G., Weinbreck, F. & Vries, de R. *Current Opinion in Colloid & Interface Science.* Vol. 9, 2004, 340-349

Liu, L., Zhao, Q., Liu, T. & Zhao, M. *Food Hydrocolloid,* Vol. 25, 2011, pp. 921-927

Liu, S., Low, N. H., & Nickerson, M. T. *Journal of Agricultural and Food Chemistry,* Vol. 57, 2009, pp. 1521-1526

Mackie, A. *Current Opinion in Colloid & Interface Science.* Vol. 9, 2009, 357-361

Magnusson, E., & Nilsson, L. *Food Hydrocolloid,* Vol. 25, 2011, pp. 764-772

Martínez, K. D., Sánchez, C. C., Ruíz-Henestrosa, V. P. Rodríguez Patino, J. M. & Pilosof, A. M. R. *Food Hydrocolloid,* Vol. 21, 2007, pp. 813-822

McClements, D. J. *Biotechnology Advances,* Vol. 24, 2006, pp. 621-625

McClements, D. J. *Biotechnology Advances.* Vol. 24, 2006, pp. 621-625

McClements, D. J. *Food Emulsions.* (2nd Ed), 2005, Boca Raton, FL: CRC Press

McClements, D. J. *Soft Matter.* Vol. 7, 2011, pp. 2297-2316

McClements, D. J. *Understanding and controlling the microstructure of complex foods.* Abington: woodhead Publishing 2007

Mei, L. Y., Choi, S. J., Alamed, J., Henson, L., Popplewell, L. M. & McClements, D. J. *Journal of Agricultural and Food Chemistry* Vol. 57, 2009, pp. 11349-11353

Mitra, R. K., Sinha, S. S. & Pal, S. K. *Langmuir,* Vol. 23, 2007, pp. 10224-10229

Peniche, C., Howland, I., Carrillo, O., Zaldı´var, C., Argqelles-Monal, W., *Food Hydrocoll.* 18 2004, 865– 871.

Rodríguez Patino J. M., & Pilosof A. M. R. *Food Hydrocolloid*, Vol. 25, 2011, pp. 1925-1937

Salgam, D., Venema, P., Vries, de R. J., Sagis, L. M. C., Linden, E. van der. *Food Hydrocolloid*, Vol. 25, 2011, pp. 1139-1148

Savary, G., Hucher, N., Bernadi, E., Grisel, M. & Malhiac, C. *Food Hydrocolloid*, Vol. 24, 2010, pp. 178-183

Schmidt, K. A & Smith, D. E. *Journal of Dairy Science*. Vol.5, 1992, pp. 36-42

Schmitt C, Kolodziejczyk E. In: Williams PA, Phillips GO, editors. Gums and stabilizers for the food industry, 15. Royal Society of Chemistry; 2010. p. 211–21

Schmitt, C. & Turgeon, S. L. *Advances in Colloid & Interface Science*. Vol. 167, 2011, 63-70

Schmitt, C., Kolodziejczyk E., Lesser, M. E. *Food Colloids: interactions, microstructure and processing*, RSC, 2005, pp. 284-300

Schmitt, C., Sanchez, C., Desobry-Banon, S., & Hardy, J. *Critical Reviews in Food Science and Nutrition*, Vol. 38, 1998, pp. 689-753

Schorsch, C., Jones, M. G. & Norton, I. T. *Food Hydrocolloid*. Vol. 13, 1999, pp. 89-99

Stone, A. K., Nickerson, M. T. *Food Hydrocolloid*, Vol. 27, 2012, pp. 271-277

Tcholakova, S., Denkov, N. D., & Lips, A. (2008). *Physical Chemistry Chemical Physics*, 10, 1608–1627

Tolstoguzov, V. *Food Polysaccharides and their Application*, Taylor & Francis, 2006, pp. 589-627

Turgeon, S. L., Schmitt, C., & Sanchez, C. *Current Opinion in Colloid & Interface Science*. Vol. 12, 2007, 166-178

Weinbreck, F., Minor, M., de Kruif, C., G., *J. Microencapsul.*, 21, 2004, 667

Weinbreck, F., Nieuwenhuijse, H., Robijn, G. W, de Kruif, C. G. *Agricultural and Food Chemistry*, Vol. 52, 2004, pp. 3550-3555

Xia, J, & Dubin, P. L. *Macromolecular Complexes in Chemistry and Biology*, Springer-Verlag, Berlin, 1994, pp. 247

Yeo, Y., Bellas, E., Firestone, W., Langer, R., Kohane, D. S., *J Agric Food Chem.*, 53, 2005, 7518

Dietary Fiber and Availability of Nutrients: A Case Study on Yoghurt as a Food Model

Marina Dello Staffolo, Alicia E. Bevilacqua,
María Susana Rodríguez and Liliana Albertengo

Additional information is available at the end of the chapter

1. Introduction

Dietary fibers are consumed from cereals, fruit and vegetables, but now are also added in purified form to food preparations since the roles of dietary fibers in preventing and treating some diseases have been well documented. Dietary fiber intake in Western countries is currently estimated to be 16.3-43.4 g per person per day [1]. According to current recommendations (Food and Nutrition Board, Institute of Medicine, 2001), the average daily requirement of dietary fiber is 25 g per day for women younger than 50, 21 g per day for women older than 50; 38 g per day for men younger than 50, and 30 g per day for men older than 50 [2].

The addition of dietary fibers to foods confers three different types of benefits. Their nutritional value motivates consumers to eat increased quantities of dietary fibers, which is advised by nutritionists. Their technological properties are of great interest to food manufacturers. Finally, dietary fibers may also be used to upgrade agricultural products and by-products for use as food ingredients. Consequently, both the nutritional value and technological properties of dietary fibers are important in the potential development of a wide range of fiber-enriched foods for example: bakery products, snacks, sauces, drinks, cereals, cookies, dairy products, meat products [3].

Different types of dietary fibers have different structures and chemical compositions, and correspondingly are of varying nutritional and technological interest. Although many studies have confirmed the nutritional benefits of dietary fibers (preventing diabetes mellitus, cardiovascular diseases, various types of cancer, and improving immune functions), the results depend on the types of dietary fibers studied or on the experimental conditions used [3]. The intake of dietary fiber might influence in different ways the absorption of nutrients [4]. With respect to glucose, an increase in the total fiber content of food can delay the glycemic response

[5]. However, there is fairly consistent evidence that soluble types of fiber reduce blood glucose and purified insoluble fibers have little or no effect on postprandial blood glucose [6]. In the other direction, dietary fiber has been shown to impair the absorption of minerals and trace elements in the small intestine because of their binding and/or sequestering effects [7]. This is associated with negative impacts on mineral bioavailability, particularly in high-risk population groups [8]. Glucose, calcium and iron have gained increasing interest in nutrition fields. Glucose is the key in carbohydrate and lipid metabolism influencing the management of body weight [9]. Calcium is involved in most metabolic processes and the phosphate salts of which provide mechanical rigidity to the bones and teeth. Intake of calcium is related to the prevention of osteoporosis [10,11]. Iron (Fe) deficiency is a leading nutritional concern worldwide, affecting 20–50% of the world's population [12].

There is an unequivocal need for predicting absorption of these nutrients. The aim of most of the investigations in this field is to make evident that fiber may be an important determinant of the utilisation of these nutrients in the diet. Much research has been done to better understand the physicochemical interactions between dietary fiber and these nutrients in the past decades [13-15]. Several of these investigations have applied in vitro digestive models to study their absorption in foods [16-18]. However, few works have been done to study their absorption from fermented milk products [19]. Yogurt is one of the dairy products, which should continue to increase in sales due to acceptance for the consumers and diversification in the range of yogurt-like products, including reduced fat content yogurts, yogurts with dietary fibers, probiotic yogurts, symbiotic yogurts, yogurt ice-cream, etc [20]. For a long time, yogurt has been recognised as a healthy food and as an important nutritional source [21].

The interactions between fibers added to yogurt from different sources (animal and plant fibers) and with different behaviors (soluble, insoluble and viscous fibers) and glucose, calcium and iron, have been studied using chemical experimental models of the human digestive tract to evaluate the availability of these nutrients.

2. Definition and composition of dietary fiber

The term 'dietary fiber' (DF) first appeared in 1953 and referred to hemicelluloses, cellulose and lignin [22]. Since the 1970s it has been recognised as having health benefits. Burkitt [23] recommended that individuals should increase their DF intake in order to increase their stool volume and stool softness. This was based on comparisons between Africa and the UK concerning fiber intakes and disease incidence. Trowell [24] first defined DF as 'the remnants of the plant cell wall that are not hydrolysed by the alimentary enzymes of man'. From those days the definition has undergone numerous revisions that were summarized accurately by Tungland and Meyer [4].

The Codex Alimentarius Commission adopted a new definition of fiber in July 2009, designed to harmonize the use of the term around the world. It describes fiber as elements not hydrolised by endogenous enzymes in the small intestine (indigestibility) as well as having physiological effects beneficial to health. Dietary fibers are carbohydrate polymers with ten or more monomeric units and belonging to one of three categories of carbohydrates polymers: edible carbohydrate polymers naturally occurring in food, carbohydrate polymers which have been obtained from raw food material by physical, enzymatic, or chemical means, and synthetic carbohydrate polymers [25–27].

The chemical nature of fibers is complex; dietary fibers are constituted of a mixture of chemical entities [28]. Dietary fiber is composed of nondigestible carbohydrate, lignin and other associated substances of plant origin, fibers of animal origin and modified or synthetic nondigestible carbohydrate polymers. The nondigestible carbohydrates are composed of the following polysaccharides: cellulose, β-glucan, hemicelluloses, gums, mucilage, pectin, inulin, resistant starch; oligosaccharides: fructo-oligosaccharides, oligofructose, polydextrose, galacto-oligosaccharides; and soybean oligosaccharides raffinose and stachyose [4]. Chitosan is an example of fiber of animal origin, derived from the chitin contained in the exoskeletons of crustaceans and squid pens; its molecular structure is similar to that of plant cellulose [29]. Cereals are the principal source of cellulose, lignin and hemicelluloses, whereas fruits and vegetables are the primary sources of pectin, gums and mucilage [30]. Each polysaccharide is characterised by its sugar residues and by the nature of the bond between them [28]. They are presented in Table 1.

Fibers	Main chain	Branch units
Cellulose	β-(1,4) glucose	
β-glucans	β-(1,4) glucose and β-(1,3) glucose	
Hemicelluloses		
Xylans	β-D-(1,4) xylose	
Arabinoxylans	β-D-(1,4) xylose	Arabinose
Mannans	β-D-(1,4) mannose	
Glucomanns	β-D-(1,4) mannose and β-D-(1,4) glucose	
Galactoglucomannans	β-D-(1,4) mannose, β-D-(1,4) glucose	Galactose
Galactomannans	β-(1,4) mannose	α-D-galactose
Xyloglucans	β-D-(1,4) glucose	α-D-xylose
Pectin		
Homogalacturonan	α-(1,4)-D-galacturonic acid (some of the carboxyl groups are methyl esterified)	
Rhamnogalacturonan-I	(1,4) galacturonic acid, (1,2) rhamnose and 1-, 2-, 4-rhamnose	Galactose, arabinose, xylose, rhamnose, galacturonic acid
Rhamnogalacturonan-II	α-(1-4) galacturonic acid	Unusual sugar such as: apiose, aceric acid, fucose
Arabinans	α-(1-5)-L-arabinofuranose	α-arabinose
Galactans	β-(1-4)-D-galactopyranose	
Arabinogalactanes-I	β-(1-4)-D-galactopyranose	α-arabinose
Arabinogalactanes-II	β-(1-3)-and β-(1-6)-D-galactopyranose	α-arabinose
Xylogalacturonan	α-(1-4) galacturonic acid	xylose
Inulin	β-(2-1)-D-fructosyl-fructose	
Gum		
Carrageenan	Sulfato-galactose	
Alginate	β-(1,4)-D-mannuronic acid or α-(1-4)-L-guluronic acid	

Fibers	Main chain	Branch units
e.g. 1: seed gum from Abutilon indicum	β-(1,4)-D-mannose	D-(1,6) galactose
e.g. 2: seed gum from Lesquerella fendleri	Rhamnose, arabinose, xylose, Mannose, galactose, glucose, galacturonic acid	
Oligofructose (enzymatic hydrolysis of inulin)	β-(2-1)-D-fructosyl-fructose	
Polydextrose (synthetic)	D-Glucose	
Resistant maltodextrins (heat and enzymatic treatment of starch)	α (1-4)-D-Glucose	α (1-6)-D-Glucose
Lignin	Polyphenols: Syringyl alcohol (S), Guaiacyl alcohol (G) and p-coumaryl alcohol (H)	
Chitosan	β-(1-4)-linked D-glucosamine and N-acetyl-D-glucosamine	

Table 1. Chemical composition of dietary fibers [28].

3. Classification of dietary fiber

Several different classification systems have been suggested to classify the components of dietary fiber: based on their role in the plant, on their fiber constituents (Table 2), on the type of polysaccharide, on their simulated gastrointestinal solubility, on site of digestion and on products of digestion and physiological classification. However, none is entirely satisfactory, as the limits cannot be absolutely defined. Two very accepted classifications are those which use the concept of solubility in a buffer at a defined pH, and/or the concept of fermentability in an in vitro system using an aqueous enzyme solution representative of human alimentary enzymes. Generally, well fermented fibers are soluble in water, while partially or poorly fermented fibers are insoluble [4]. However, dietary fiber is conventionally classified in two categories according to their water solubility: insoluble dietary fiber (IDF) such as cellulose, part of hemicellulose, and lignin; and soluble dietary fiber (SDF) such as pentosans, pectin, gums, and mucilage [31].

Taking into account the physiological and physicochemical behavior of fibers it could be necessary to add two subcategories among the group of soluble fibers: viscous soluble fibers (such as guar gum, glucomannans, pectins, oat β-glucan, psyllium, mucilages, etc) and nonviscous soluble fibers (such as lactulose, oligosaccharides, fructo-oligosaccharides, inulin, etc.). Jenkins et al. [6, 32] mentioned the term viscous soluble fiber in their works about fibers and low glycaemic index, blood lipids and coronary heart diseases. Dikeman and Fahey [33] also mentioned the term in their work in which they investigated the viscosity in relation with dietary fiber including definitions and instrumentation, factors affecting viscosity of solutions, and effects on health. Thus, dietary fiber could be classified in soluble (viscous and non viscous) and insoluble fiber. The latter do not form gels due to their water insolubility and fermentation is very limited [34]. However, numerous commercial fibers are available in the market for use in food technology and have both insoluble and soluble fiber components in the same product. This is due to the fact that many of these are powders which come from the extraction, concentration and drying of the fiber contained in cereals, fruits and vegetables. Therefore, it might be more appropriate to classify fibers based on their content of soluble and insoluble fractions (Table 3). It is recognised that the physiological and physicochemical effects of dietary fibers depend on the relative amount of individual fiber components, especially as regards the soluble and insoluble fractions [28].

Fractionation of dietary fibers aims to quantify those constituents, to isolate fractions of interest and to eliminate undesirable compounds. Techniques for fractionation of dietary fibers are limited in number. Several researchers have determined the cellulose, hemicelluloses and lignin contents of dietary fibers from different food sources. Claye, Idouraine, and Weber [35] isolated and fractionated insoluble fibers from five different sources. Using cold and hot water extraction, enzymatic and chemical treatment, they obtained four fractions: cellulose, hemicellulose A and B, and lignin. The fractionation methods are varied and were developed according to the material tested. Therefore, there is no global method used. The existing methods described universal techniques of fractionation. Each analyst must modify previously used approaches to develop a method optimal to the material being tested [28].

Fiber Constituents	Principal groupings	Fiber components/sources
Nonstarch polysaccharides & oligosaccharide	Cellulose	Cellulose-Plants (vegetables, sugar beet, various brans)
	Hemicellulose	Arabinogalactans, _-glucans, arabinoxylans, glucuronoxylans, xyloglucans, galactomannans, pectic substances.
	Polyfructoses	Inulin, oligofructans
	Gums & Mucilages	Seed extracts (galactomannans –guar and locust bean gum), tree exudates (gum acacia, gum karaya, gum tragacanth), algal polysaccharides (alginates, agar, carrageenan), psyllium
	Pectins	Fruits, vegetables, legumes, potato, sugar beets
Carbohydrate analogues	Resistant starches and maltodextrins	Various plants, such as maize, pea, potato
	Chemical synthesis	Polydextrose, lactulose, cellulose derviatives (MC, HPMC)
	Enzymatic synthesis	Neosugar or short chain fructooligosaccharides (FOS), transgalactooligo collagen, chondroitin saccharides (TOS), levan, xanthan gum, oligofructose, xylooligosaccharides (XOS), guar hydrolyzate, curdlan.

Fiber Constituents	Principal groupings	Fiber components/sources
Lignin	Lignin	Woody plants
Substances associated with nonstarch polysaccharides	Waxes, cutin, Suberin	Plant fibers
Animal origin fibers	Chitin, chitosan, collagen, chondroitin	Fungi, yeasts, invertebrates

Table 2. Dietary fiber constituents [4]

Category	Subcategory	Fiber fraction	Main food source
Soluble fiber	Viscous	β-glucans	Grains (oat, barley, rye)
		Pectins	Fruits, vegetables, legumes, sugar beet, potato
		Gums & Mucilages	Leguminous seed plants (guar, locust bean), seaweed extracts (carrageenan, alginates), plant extracts (gum acacia, gum karaya, gum tragacanth), microbial gums (xanthan, gellan), psylluim
	Nonviscous	Sugars	Lactulose
		Oligosaccharides	Various plants and synthetically produced (polydextrose, fructooligosaccharides, galactooligosaccharides, transgalactooligosaccharides)
		Inulin	Chicory, Jerusalem artichoke, sugar beet, onions
Insoluble fiber		Cellulose	Plants (vegetables, various brans)
		Hemicellulose	Cereal grains
		Lignin	Woody plants
		Cutin/suberin/other plant waxes	Plant fibers
		Chitin and chitosan, collagen	Fungi, yeasts, invertebrates
		Resistant starches	Plants (corn, potatoes, grains, legumes, bananas)
		Curdlan (insoluble β-glucan)	Bacterial fermentation

Table 3. Table 3. Classification of dietary fiber based on solubility

4. Analytical methods for studying dietary fibers

The complexity of fibers is given by their chemical nature and polymerisation degree that they possess. This requires various analytical methods for the measurement of dietary fiber, to precisely estimate its composition in food and food by-products. Methods for the determination of dietary fiber may be divided into three categories: non-enzymatic-gravimetric, enzymatic-gravimetric, and enzymatic-chemical methods. The latter includes enzymatic-colorimetric and enzymatic-chromatographic (GLC/ HPLC) methods [28]. Nowadays, the most commonly used methods for dietary fiber measurement are the enzymatic-gravimetric Association of Official Analytical Chemists (AOAC) method [36] and enzymatic-chemical method [37]. The method of Van Soest [38] is generally used in veterinary studies.

5. Dietary fiber and human health

5.1. Fiber, lipid metabolism and cardiovascular disease

The earliest and most widely researched topic related to dietary fiber and human health is reducing the risk factors for coronary heart disease [24]. Total serum cholesterol and low density- lipoprotein (LDL) cholesterol levels are generally accepted as biomarkers indicating of potential risk for developing the disease [5]. In consequence, research has primarily focused on their reduction as a means to diminish the risk of developing cardiovascular disease. Substantial experimental data support that blood cholesterol can be lowered using viscous soluble fibers that produce relatively high viscosity in the intestinal tract [39-41]. It is known that viscous soluble and insoluble dietary fibers can bind bile acids and micelle components, such as monoglycerides, free fatty acids, and cholesterol, which decrease the absorption and increase the fecal excretion of these entities [42,43]. For insoluble dietary fibers such as lignin or citric fiber this reducing effect is rather low compared to viscous soluble dietary fibers and is mainly based on direct binding of bile acids. In the small intestine the bile acids are bound by the insoluble dietary fibers through hydrophobic interactions and excreted from enterohepatic circulation together with the undigested insoluble dietary fibers which results in a lowering of the blood biomarkers levels [44,45]. Furthermore, free fatty acids and cholesterol bound by dietary fibers cannot be absorbed by the body and will be excreted. The biomarkers-lowering effect of viscous soluble dietary fibers such as psyllium, oat β-glucan or pectin is based on different mechanisms. The binding of water in the chyme and the resulting increase in viscosity is regarded as the main effect. This leads to a reduced diffusion rate of bile acids, which cannot be reabsorbed by the body and thus are excreted [46-48]. Besides, some studies indicate that there are also direct binding forces such as hydrophobic interactions between soluble dietary fibers and bile acids [49]. Dietary fiber also modifies lipid metabolism by influencing the expression of key genes. Acetyl-CoA carboxylase is the rate-limiting enzyme in lipogenesis and is regulated by AMP-activated protein kinase (AMPK). In a 10-week study comparing obese and lean rats, adding 5 g of P. ovata to rat chow increased the phosphorylation of AMPK, consequently inhibiting acetyl-CoA carboxylase [50]. Fructooligosaccharide (10g/100g) also has been shown to decrease the hepatic acetyl-CoA carboxylase expression in rats [51]. In view of new research [52,34] the United States Food and Drug Administration has approved health claims supporting the role of dietary fiber in the prevention of coronary heart disease [53].

5.2. Fiber, carbohydrate metabolism, and diabetes mellitus

It is known that exists a link among an elevated body mass index, waist circumference and the risk of type 2 diabetes mellitus [54,55]. The role of DF in weight reduction has been examined in animal and human studies. A reduced risk for type 2 diabetes mellitus (T2DM) appears to depend on the type and dose of dietary fiber and the study population [36]. In mice, 10% psyllium and 10% sugar cane fiber decreased the fasting blood glucose and fasting plasma insulin when added to a high fat diet for 12 weeks, compared to the insoluble fiber cellulose [56]. β-Glucan also improved the glucose tolerance and decreased the serum insulin in mice when added to a high fat diet at a 2% and 4% level [57]. In humans, muffins high in β-glucan and resistant starch lowered the postprandial blood glucose and insulin levels [58]. A prospective cohort design with 252 women was used to measure energy intake, dietary fat intake, fiber intake, body weight, body fat percentage, physical activity, season of assessment, age and time between assessments. They concluded that increasing dietary fiber intake significantly reduced the risk of gaining weight and fat in women, independent of several potential confounders such as: physical activity, dietary fat intake, and others [59]. Another study about the consumption of soluble viscous fiber, that included one hundred and seventy six men and women, reached the same conclusions [60]. Other biomarkers such as Glycemic Index (GI) and Glycemic Load (GL) when they are high were both associated with an increased risk of diabetes in a meta-analysis of observational studies [61,62]. On the other hand, numerous epidemiological studies performed to date relate to a high intake of dietary fiber with low levels of GI and GL [63-65]. Moreover, DF has also shown to be effective improving altered parameters in obesity and T2DM [66,67]. Soluble viscous fiber plays an important role in controlling satiety and postprandial glycemic and insulin responses [68] and some studies showed that insoluble dietary fiber improved the quality of life for these patients [69]

The protective effect of DF on obesity and T2DM has been historically attributed to greater satiety due to an increased mastication, calorie displacement, and decreased absorption of macronutrients [55]. This mechanism is associated with the ability of soluble fibers to form viscous solutions that prolong gastric emptying, consequently inhibiting the transport of glucose, triglycerides and cholesterol across the intestine [70-72]. Recently, it was observed that both soluble and insoluble DFs also modifies carbohydrate metabolism by influencing the expression of hormones such as glucose-dependent insulin tropic polypeptide and glucagon-like peptide-1, that stimulate postprandial insulin release, enhance glucose tolerance, and delay gastric emptying [73-76].

5.3. Fiber and gut microflora

The large intestine plays host to a large and diverse resident microflora. Over the last 10-15 years, 16S ribosomal RNA analyses has allowed a more complete characterization of the diverse bacterial species that make up this population [77]. Around 95% of human colonic microflora (as estimated from faecal sampling) appear to be within Bacteroides and Clostridium phylogenic groups, with less than 2% of the total microflora being made up of

Lactobacilli and bifidobacteria [78]. In general, the colonic microflora is partitioned from the rest of the body by the mucus layer and mucosa. Loss of this partitioning effect is associated with disease processes within the large intestine [79], but it is unsure whether this is a cause or -effect of the disease process. Within the healthy large intestine, the main way the colonic microflora interacts with the host is through its metabolites [80]. Some of these metabolites are putatively damaging to the underlying mucosa, such as indoles, ammonia and amines while others are potentially beneficial to the host, including short chain fatty acids (SCFA) [81] and lignans that the mammalian gut can absorb [82,83]. SCFA are produced by bacterial fermentation of dietary carbohydrate sources, of which dietary fiber is the main type in the large intestinal lumen.

Dietary fiber, plays a profound role on the number and diversity of bacteria that inhabit the large intestine. In the absence of dietary fiber or other luminal energy sources, resident bacteria in the colon will turn to large intestinal mucus as an energy source prior to attacking the underlying mucosa [84]. As bacteria require the necessary enzymes to break down saccharide bonds of the diverse range of dietary fibers, fiber will clearly affect microfloral population dynamics. The presence of any fermentable dietary fiber is likely to lead to an increase in microfloral bifidobacterial and Lactobacillus levels, as these bacteria ferment carbohydrates. Previous studies in humans have suggested dietary fibers like alginate [85], chitosan [86], and inulin [87] lead to a reduction in potentially harmful microfloral metabolites. A range of small human interventions with various fermentable dietary fibers have shown significant, but small, clinical benefit in a number of intestinal diseases and disorders either on their own or in combination with probiotics [88-90].

Catabolism by microbial populations may also be important for decrease the levels of cholesterol and lipids. Bacteria such as Lactobacillus and Bifidobacteria can exert a hypocholesterolemic effect by enhancing bile acid deconjugation [91,92]. Furthermore, Lactobacillus and Bifidobacteria remove cholesterol in vitro by assimilation and precipitation [93,94]. Fermentation products further affect lipid metabolism. Propionate inhibits the incorporation of acetic acid into fats and sterols, resulting in decreased fatty acid and cholesterol synthesis [95].

5.4. Fiber and Immune function

Besides its absorptive functions, the gastrointestinal tract is involved with a range of immune functions. The mucosa effectively partitions the rest of the body from digestive enzymes, large numbers of bacteria and assorted toxins that occur within the gut. The mucosa has two main roles in immunity. Firstly, the mucosa samples luminal contents to assess the threat to the body because the gut comes into contact with a wide range of external antigenic compounds. This is carried out by the gut-associated lymphoid tissue or GALT [96]. In the second place, gut epithelium must also protect itself from the luminal stress of damaging agents and shear forces [97]. To do this, protective mucus is secreted along almost the entirety of the gastrointestinal tract (excluding the oesophagus and possibly Peyer's patches). Within the mouth, mucin is secreted alongside other salivary secretions and acts as a lubricant. In the stomach and intestine, mucus is secreted as a protective barrier [98].

There is a paucity of data regarding intake of DFs and immune function associated with the gut or otherwise in humans [99]. Animal studies within this area are also sparse. Field et al. [100] carried out studies with dogs and they found that fermentable fiber intake resulted in increased intra-epithelial T-cell mitogen response [92]. In a recent study it has been observed that DF may interact directly with immunoregulatory cells. Mucosal macrophages and dendritic cells have receptors with carbohydrate-binding domains that bind β-glucans and cause a decrease in IL-12 and increase in IL-10, which is consistent with an anti-inflammatory phenotype [101]. No previous study has assessed the impact of DFs on the human mucus barrier due to the invasiveness of procedures involved with measuring the mucus barrier directly. However, the effects of different types of DFs on the intestinal mucus barrier have been studied in animal models. Fibers and fiber sources such as alginates, ispaghula husk, wheat bran, ulvan and carrageenan all appear to benefit the protective potential of the colonic mucus barrier [100,102].

5.5. Fiber and prevention of cancer

Cancer continues to be one of the number one health concerns of populations worldwide. Most cancers strike both men and women at about the same rate, with exception of cancers of the reproductive system. Of particular concern is cancer of the colon, ranking among the top 3 forms of cancer in the U.S.A., for both men and women. Colon cancer is also one of the leading causes of cancer morbidity and mortality among both men and women in the Western countries, including the U.S.A. [103]. The European Prospective Investigation of Cancer (EPIC) is a project that includes more than half a million people in 10 European countries and they results indicate that dietary fiber provides strong protective effects against colon and rectal cancer. In one of its papers, the authors clarify that methodological differences in some previous studies (e.g., study design, dietary assessment instruments, definition of fiber) may account for the lack of convincing evidence for the inverse association between fiber intake and colorectal cancer risk [104]. A careful work within the same project was conducted as a prospective case–control study nested within seven UK cohort studies which included 579 patients who developed incident colorectal cancer and 1996 matched control subjects. They used standardized dietary data obtained from 4- to 7-day food diaries that were completed by all participants to calculate the odds ratios for colorectal, colon, and rectal cancers. In this work, the researchers confirmed that the intake of dietary fiber is inversely associated with colorectal cancer risk [105]. Taking into account these studies, the United States Food and Drug Administration has approved health claims in 2010 supporting the role of DF in the prevention of cancer [106].

Human metabolic and animal model studies indicate that beneficial effects of dietary fiber in relation to colon cancer development depend on the composition and physical properties of the fiber [107,108]. The effect of soluble fiber sources is mainly based on their fermentation and on the effects of short-chain fatty acids produced, especially of butyric acid. It has been known since 1982 that the colonic mucosa uses these acids, especially butyrate, as a preferential energy source [109]. Butyric acid stimulates the proliferation of normal cell lines both in vitro and in the normal epithelium, but retards the growth of carcinoma cell lines and induces apoptosis in cultured colonic adenoma and carcinoma cells [110,111]. Insoluble fiber has been found to have a protecting effect by absorbing hydrophobic carcinogens [112-114]. A third potentially effective mechanism is that of the accompanying phenolic compounds. Several phenolic compounds, having antioxidative properties, are present especially in cereal fiber sources. They are released from their bound states by bacterial enzymatic action in the colon, and can act in the intestine locally as anticarcinogens both in preventing cancer initiation and progression [115-117].

6. Digestive and absorptive functions of the gastrointestinal tract and dietary fibers

The gastrointestinal tract (GIT) is the initial site of action from which dietary fibers produce systemic effects presented in the previous section. The physiological effects of dietary fiber depend on a myriad of variables, but generally they depend on the type (soluble or insoluble), the dose of a specific fiber consumed, the composition of the entire fiber-containing meal, and the individual physiological profile of the subject who consume the fiber-containing meal [5]. The GIT serves as an interface between the body and the external environment. The main function of the GIT is to absorb nutrients from ingested foods. The organs of the GIT are connected to the vascular, lymphatic and nervous systems to facilitate regulation of the digestive function [118]. To carry out this function digestive processes are realized by secretion of enzymes and associated co-factors, and through maintenance of the gut lumen at optimal pH for digestion [119]. Gastrointestinal secretion of enzymes and other factors, alongside control of gut motility is governed by a series of complex neurohumoral pathways (mediated by acetylcholine, gastrin, motilin, cholecystokinin, gastric inhibitory peptide (GIP), secretin, etc.) that begin to operate by luminal content. Two main features of luminal content which appear to govern gastrointestinal physiology are luminal chemical profile and luminal bulk. The nutrient/chemical profile of the gut lumen is sensed by specialised chemosensor enteroendocrine cells within the epithelium [120], while mechanoreceptors (stretch activated neural cells) occurring within the myenteric and submucosal plexusues [121] are activated as a result of mechanical pressure from luminal contents. The main absorptive area in the gut is the small intestine, which is involved in the absorption of the subunits of digestible macronutrients, as well as vitamins, minerals and other micronutrients [87]. Ingested foods must be mechanically homogenised with digestive secretions in order to allow better hydrolysis of macronutrients, and, in some cases, to allow micronutrient release. Mastication in the buccal cavity mix food with salivary secretions among them α-amylase starts digestion of starches [122]. Food boluses entering the stomach are maintained there for mixing with gastric secretions. A strongly acidic secretion allows denaturation of proteins and solubilisation of other factors. Gastric proteases (mainly pepsin) cleave bonds in proteins to form a range of shorter peptides and amino acids. Gastric lipase initiates digestion of dietary lipids [123]. By the time the majority of luminal contents leave the stomach, they have been processed into creamy, homogenous slurry, known as chyme. As luminal contents appear in the upper section of the small intestine (the duodenum), they are met with alkali (bicarbonate-rich) secretions from the liver, pancreas and intestinal crypts. Pancreatic exocrine secretions also contain a myriad of enzymes for digestion of all macronutrients [124].

Classically, dietary fiber is cited as reducing whole gut transit time, thereby increasing frequency of defecation. This effect can be explained on the one hand, due to DF increase the intestinal luminal bulk resulting in an increased peristalsis which reduce the whole gut transit time. DFs that increment the luminal bulk are those that have a high water-binding capacity [125]. Furthermore, feed-forward and feedback from other portions of the gut as a result of fiber intake could also affect motility of the different organs of the GIT. Prolongation of nutrient release into the intestinal lumen from the stomach is likely to result in a lengthened phase of hormonal feedback from the duodenum, terminal ileum and colon, leading to a delay in gastric emptying [84]. At the same time, this delay in the gastric emptying towards small intestine are likely to increased motility distally (and therefore decreased transit time). The most researched area of the effects of dietary fibers on gastric motility is linked to gastric emptying. A range of studies have demonstrated that inclusion of viscous fibers in liquid test meals results in delayed gastric emptying, and are particularly consistent in the case of pectins in human studies [126,127]. In a study comparing the physiological effects of a mixed meal containing high levels of natural fibers (fruit, vegetables and whole grains) against one without these fibers (instead containing fruit and vegetable juice and refined grains), concluded that removal of natural fiber decreased gastric emptying mean rate of approximately 45 min in a crossover feeding trial in 8 healthy adult participants [128]. The dietary fibers that raise the bulk of luminal contents of the large bowel are those that are not well fermented by the colonic microflora, and those that have a high water-binding capacity [84].

6.1. Nutrients absorption

To date, evidence has been obtained in different types of studies that dietary fiber can influence the metabolism of carbohydrates and lipids preventing the development of diabetes mellitus and cardiovascular disease. Intake of dietary fiber can influence the absorption of nutrients in different ways. It has been postulated that the presence of any dietary fiber in the upper GI tract will result in a decreased rate of intestinal uptake of a range of nutrients. However, it is necessary to consider what physicochemical factors of dietary fibers are important in these roles [84]. In previous animal studies, Kimura et al. [129], noted higher levels of cholesterol excretion in rats fed diets containing 1000 mg/kg of degraded alginates with molecular weights of 5 and 10 KDa compared to the effect of a diet with a lower molecular weight (1 kDa) alginate or a control (no fiber) diet. While such absorption-lowering effects can be beneficial in reducing energy uptake, it must also be noted that such factors are also likely to reduce the bioavailability of minerals, vitamins and phyto-chemicals. Dietary fiber fractions differ largely in their abilities to affect mineral and trace element availability and this might have negative impacts in high-risk population groups. Small human feeding studies have suggested that inclusion of food hydrocolloids like alginates [130], guar gum [131,132] and β-glucan [58,133-135] into test meals results in a blunting of postprandial glycaemic and insulinaemic responses.

7. *In vitro* chemical experimental models

To study the absorption of nutrients *in vivo*, feeding methods, using animals or humans, usually provide the most accurate results, but they are time consuming and costly, which is why much effort has been devoted to the development of in vitro procedures [136]. The *in vitro* digestive chemical experimental model enabled mimicking, in the laboratory, the in vivo reactions that take place in the stomach and duodenum. In principle, in vitro digestion models provide a useful alternative to animal and human models by rapidly screening food ingredients. The ideal in vitro digestion method would provide accurate results in a short time [137] and could thus serve as a tool for rapid food screening or delivery systems with different compositions and structures [19]. In vitro methods cannot be used alone for important decisions taken by industry or international organizations because human studies are required for such determinations, but are important for screening purposes and to project future studies.

8. Work objective

Considering the fact that dietary fibers are new ingredients widely applied in foods, it is important to know their effect on absorption of nutrients and micronutrients. For this reason, the interaction between nutrients and fibers from different sources (animal and plant fibers) and types (soluble and insoluble fibers) has been studied using chemical experimental models of the human digestive tract to evaluate the availability for absorption of glucose, calcium and iron using yoghurt as a food model.

9. Materials and methods

9.1. Dietary fibers employed

The plant fibers used in this work were: inulin (Frutafit-inulin, Imperial Sensus, The Netherlands), bamboo (Qualicel, CFF, Gehren, Germany), wheat (Vitacel WF 101, JRS, Rosenberg, Germany), apple (Vitacel AF 400-30, JRS, Rosenberg, Germany) and psyllium (Metamucil, Procter and Gamble Co., Cincinnati, OH, USA). Metamucil is a pharmaceutical formula with Plantago ovata seed husk (49.15% w/w) and sucrose (50.85%). Suppliers of wheat and apple fiber indicated that these products are free from phytic acid, and besides, the wheat fiber is gluten free. The inulin utilized in this work has a degree of polymerisation ≥ 9 as declared by suppliers.

The dietary fiber from animal source utilised in these assays was chitosan. It was obtained from crustacean chitin in the Laboratorio de Investigación Básica y Aplicada en Quitina (LIBAQ-INQUISUR-CONICET), Universidad Nacional del Sur, Bahía Blanca, Argentina. Chitin firstly was isolated from shrimp (Pleoticus mülleri) waste by the process that was described in our previous work [138]. Chitosan was prepared directly by heterogeneous deacetylation of chitin with 50% (w/w) NaOH. For the biopolymer characterisation, moisture and ash contents were determined at 100–105 °C and 500–505 °C, respectively.

Deacetylation degree was obtained using FT-IR spectroscopy (Nicolet iS10 FT-IR Spectrometer, Thermo Fisher Scientific, USA) with samples in the form of KBr at a ratio of 1:2. Viscosity of 1% chitosan in 1% acetic acid solution was measured with a Brookfield model DV-IV + viscosimeter (Brookfield, USA) with spindle 21 and a 50 rpm rotational speed at 25 °C [139].

9.2. Analysis for dietary fiber

Total, soluble and insoluble dietary fiber contents of chitosan and plant fibers were analysed according to the enzymatic–gravimetric method of the Association of Official Analytical Chemists (AOAC) Official Method 991. 43 [140]. Apple, bamboo, psyllium and wheat fibers were investigated to obtain contents of main cell wall constituents (lignin, cellulose, hemicellulose). These components were determined by modifications of the method described by Robertson and van Soest [38, 141] using ANKOM200/220 Fiber Analyzer (ANKOM Technology, Macedon, NY, USA). This method measures Acid Detergent Fiber (ADF), Neutral Detergent Fiber (NDF) and Lignin. Cellulose and hemicellulose contents were obtained by calculations. To determine ADF, duplicate samples were agitated under pressure with hot acid detergent solution for 60 min, rinsed in hot water and dried. To determine lignin content, duplicated samples were digested in 72% (v/v) sulfuric acid, following ADF analysis. Cellulose content of samples was calculated from ADF minus the lignin content. To determine NDF, duplicated samples were shaken with neutral detergent solution and heat-stable α-amylase for 60 min, rinsed and dried. Hemicellulose content of samples was calculated as NDF minus ADF [141].

9.3. Yoghurt preparation

Yoghurt was prepared using reconstituted whole milk powder (15% w/w) and 5% sucrose. This mix was homogenized and heated to 85 °C for 30 min., cooled to ambient temperature and inoculated with 0.03% starter culture. Starter was constituted by a 1:1 mixture of Streptococcus thermophilus (CIDCA collection 321) and Lactobacillus delbrueckii subsp. bulgaricus (CIDCA collection 332). Samples were incubated at 43°C to reach a pH of 4.4–4.6 and stored at 4°C, after completion of the fermentation process 1.3% (w/w) of each dietary fiber was added to samples of yoghurt [142]. The amount of fiber was selected following US regulations for fiber-fortified products [143].

To study glucose availability 0.6 g of glucose (Sigma-Aldrich Co., St. Louis, MO, USA) was added for each sample of yoghurt with each type of dietary fiber. In calcium availability studies the digestive mimicking was done without the addition of exogenous calcium because yoghurt is a source of calcium in the diet [138]. To evaluate the interactions between the fibers and iron, 0.8% (w/w) of ferrous sulfate was added to yoghurt samples with each type of fiber [139]. This addition was in accordance with local regulations governing iron supplementation in milk products. Ferrous sulfate ($FeSO_4 \cdot 7\ H_2O$) of 99.9% purity was used as purchased (Sigma-Aldrich Co., St. Louis, MO, USA).

9.4. Digestive chemical experimental model

Two types of digestive simulations were performed to study the interactions between dietary fibers used and the macro and micro nutrients tested. To evaluate the interaction of glucose and calcium with the fibers, gastric and duodenal environments were simulated. To examine the interactions between the fibers and iron was used in addition, a dialysis membrane to imitate the iron passage through the intestinal wall. Digestive enzymes were not utilized in these models because they do not hydrolyze fibers. The importance of duodenal simulation in these studies is because most dietary glucose, calcium and iron are absorbed in the duodenum.

The experiments to study the availability of glucose and calcium were performed in the following steps: a mix of 12.5 g of yogurt with 0.3 g of each fiber was stirrer in 50 mL of 0.1 M HCl (Merck) pH 1.0–2.0, 30 rpm and 37°C to reproduce the gastric environment. After 1 hour simulations were taken from the acidic medium to pH 6.8–7.2 with 15 g/L of $NaHCO_3$ (Sigma Chemical Co., St. Louis, MO, USA). The stirring speed was increased from 30 to 300 rpm and the temperature was maintained at 37°C to reproduce the duodenal environment. Then simulations were allowed to rest for 15 min until two phases separated. Samples to determine glucose and calcium concentration were taken from the supernatant. Glucose and calcium amounts, determined by this way, represent the bioavailability fraction of those nutrients. A control without fibers was made to consider glucose and calcium 100% availability [138].

Experiments to study the interaction of dietary fibers with iron were carried out in the following manner. Yoghurts with ferrous sulfate and each fiber were stirred in 50 mL of 0.1 M HCl (Merck) for 1 h at pH 1.0–2.0, 30 rpm and 37 °C to reproduce the gastric environment. During this first step of simulation pH was checked each 15 min with a pH Meter Hach model EC-30 (USA) and it remained constant (pH 1.0–2.0). To reproduce the chemical duodenal environment pH level was increased to pH 6.8–7.2 with 0.2 M NaHCO3 (Sigma-Aldrich Co., St. Louis, MO, USA), stirring speed was increased from 30 to 300 rpm to imitate the peristaltic movement and temperature was maintained at 37°C. Simulations were immediately transferred into a dialysis tubing cellulose membrane (D9527-100 FT, (Sigma-Aldrich Co., St. Louis, MO, USA). This cellulose membrane (molecular weight cut-off 12,400) was previously prepared, as indicated by suppliers, and it was cut into 28 cm length pieces. The loaded tubes were immersed in 100 mL of distilled water; at 37°C. Iron concentrations were determined from the dialysed medium at 30 and 60 minutes. Control yoghurt with ferrous sulfate without fibers was subjected to the digestive simulation and was considered as 0% iron retention to calculate iron retention percentages for each fiber [139].

9.5. Analytical techniques

To determine glucose concentration an enzymatic method was used. Glucose reacts with 10 kU/L glucose oxidase (GOD), and 1 kU/L peroxidase (POD) in presence of 0.5 mM 4-aminophenazone (4-AP) and 100 mM phosphate buffer (pH 7.0) containing 12 mM hydroxybenzoate (Wiener Lab Glicemia enzymatic AA Kit, Argentina). An amount of digestive simulation solution (10 mL) was mixed with 1.0 mL of reagent, tubes were incubated for 5 min in water bath at 37°C and developed colour were read in spectrophotometer (Spectronic 20 Genesys TM, Spectronic Instrument, USA) at 505 nm. Final reaction colour is stable for 30 min. Glucose calibration curve was carried out. The amounts of glucose used in this study correspond to available carbohydrates in the human mixed diet.

To determine calcium concentration a spectrophotometric method was used. Calcium reacts with 3.7 mmol/L cresolphtalein complexone (Cpx) at pH 11 (buffer 0.2 mol/L aminomethylpropanol (AMP) solution in 35%v/v methanol) (Wiener Lab Ca-color Kit, Argentina). Assays were carried directly in spectrophotometer test tubes: 50 1L Cpx were mixed with a plastic rod and absorbance was read in spectrophotometer (Spectronic 20 Genesys TM, Spectronic Instrument, USA) at 570 nm (internal blank), then 20 mL of each digestive mimicking sample were added, immediately mixed and read after 10 min. A standard curve was developed [138].

To determine iron concentration in the dialysates a spectrophotometric method was used, 500 μL of dialyzates was reduced with 2 mL of mercaptoacetic acid (succinic acid buffer, pH 3.7). Then, iron reacted with one drop of pyridyl bis-phenil triazine sulfonate (PBTS) producing a pink color due to the complex formed (Wiener Lab Fe-colour Kit, Rosario, Argentina). Absorbance was read on a spectrophotometer (Spectronic 20 Genesys Thermo Electron Scientific Instruments Corp., Madison, WI, USA) at 560 nm (internal blank). All glassware used in sample preparation and analysis was rinsed with 10% (v/v) concentrated HCl (37%) and deionised water before using, to avoid mineral contamination. A regression equation ($y = 2.5333x + 0.0042$, $R2 = 0.995$) derived from data generated from standards of $FeSO_4$ was used to calculate iron concentrations in the samples. Iron retention percentages for each studied fibers were calculated as a percentage of the amount of iron measured in the dialysed medium obtained with the control yoghurt without fibers [139].

9.6. Statistical analysis

Experiments were performed at least five times for each dietary fiber using freshly prepared yogurt. For total iron concentration in dialyzates, each individual sample was run in duplicate. Averages and standard deviations were calculated and expressed in each case as the mean ± SD for n replicates. Normality of the data was checked with the Lilliefors test. The influence of different dietary fibers on the retention percentages of glucose, calcium and iron were statistically analyzed by a one-way analysis of variance (ANOVA) ($p < 0.05$) to find significant differences and Tukey's test to compare means.

10. Results and discussion

10.1. Characterisation of fibers

The dietary fibers used in this study have different water solubility characteristics: inulin is a soluble fiber, bamboo and wheat are insoluble fibers, apple is partially insoluble fiber, and psyllium forms a viscous dispersion at concentrations below 1% and a clear gelatinous mass at 2%. Chitosan is a fiber of a different origin, i.e. from animal source and is soluble in an acidic medium and flocculates in an alkaline medium. We used these fibers because they present different physicochemical behaviors that have been described in literature [144,4]. The commercial fiber compositions used in this study, regarding total, soluble and insoluble fractions, are shown in Table 4. Analysis for dietary fiber using the AOAC method 991.43 showed that wheat and bamboo have high amounts of insoluble fraction.

Fiber	Insoluble fiber	% Insoluble fiber	Soluble fiber	% Soluble fiber	Total fiber
Apple	44.8 ± 0.4	77.1	13.3 ± 0.7	22.9	58.1 ± 1.0
Bamboo	91.4 ± 0.5	95.9	3.2 ± 0.8	3.4	95.3 ± 0.9
Chitosan	98.0 ± 1.0	100	nd	nd	98.0 ± 1.0
Inulin	nd	Nd	≥ 85.5	100	≥ 85.5
Psyllium	37.5 ± 0.6	82.9	7.1 ± 0.5	15.7	45.2 ± 0.8
Wheat	92.1 ± 0.6	97.6	2.3 ± 0.6	2.4	94.4 ± 1.1

nd: no detectable

Table 4. Total, soluble and insoluble fiber content (g/100g) of employed fibers

Inulin presents only soluble fraction as expected. Psyllium and apple have both soluble and insoluble fractions. Apple fiber is characterized by a well balanced proportion between soluble and insoluble fraction [145]. The total dietary fiber content is 45.2% for psyllium, which is an acceptable value, taking into account that the supplier declared a 49.15% content for Plantago ovata seed husk in Metamucil preparation. Van Craeyveld et al. [146] reported 3.4% (dm) ash and 7.1% (dm) protein contents for Plantago ovata seed husks. The total dietary fiber content is 58.1% for apple, which is about 10–14% higher than the values reported by Sudha et al. [147]; however, this value was in accordance with suppliers. The chitosan used in this study has 98% of insoluble fraction and no detectable soluble fraction. Furthermore the characteristics of this biopolymer are a deacetylation degree of 89%, a viscosity of 120 mPa.s, 6.7 g% moisture and 0.67 g% ash content.

Plant fiber characterisations were completed with the study of Acid Detergent Fiber (ADF) and Neutral Detergent Fiber (NDF), lignin, cellulose and hemicellulose contents (Table 5). Apple presents the highest lignin content. Wheat fiber mainly has cellulose. Bamboo has proportional amounts of cellulose and hemicellulose, but compared with other fibers, has the highest hemicellulose content. These results are in accordance with their plant fiber origins and previous works [145-149]. Frutafit-Inulin was not analysed because its composition was ≥85.5% (w/w) of inulin, ≤9.5% of mono and disaccharides, ≤0.1% of ash with polymerisation degree ≥9 according to suppliers. Chitosan was not analysed either, because of its animal origin.

Fibre	ADF	NDF	Lignin	Cellulose	Hemicellulose
Apple	38.6 ± 0.9	44.3 ± 0.7	8.4 ± 0.8	30.2 ± 1.7	5.7 ± 1.6
Bamboo	50.2 ± 0.7	90.4 ± 0.6	5.0 ± 0.3	45.2 ± 1.0	40.2 ± 1.7
Psyllium	7.3 ± 0.4	36.8 ± 0.9	0.8 ± 0.1	6.5 ± 0.4	29.5 ± 1.3
Wheat	74.8 ± 0.3	89.7 ± 0.6	2.6 ± 0.4	72.2 ± 0.7	14.9 ± 0.9

Table 5. Acid Detergent Fiber (ADF), Neutral Detergent Fiber (NDF), lignin, cellulose and hemicellulose (%) of fibers

Scientists who deal with animal nutrition usually use Van Soest's method to analyse feed. Scientists working on human nutrition use methods of the AOAC, because of their interest in soluble fiber. It is known that soluble fiber plays an important role in human health and the food industry. However, it could be useful in human nutrition to know the composition of insoluble fiber, as it is possible that insoluble fibers do not all have the same effect on human health. The NDF and insoluble fiber methods were applied to the same samples. Insoluble fiber includes hemicellulose, cellulose, lignin, cutin, suberin, chitin, chitosan, waxes and resistant starch. NDF includes hemicellulose, cellulose and lignin. Escarnot et al. [149] studied three wheat varieties and four spelt genotypes. They analysed three milling fractions from those grains for insoluble and soluble fiber contents, lignin, hemicellulose and cellulose. They found a very high correlation ($r^2 = 0.99$) between the two methods, showing that NDF and insoluble fiber methods cover the same types of fiber. For insoluble fiber analysis, the NDF method is faster and more thorough.

10.2. Digestive chemical model and glucose, calcium and iron retention percentages

Dietary fiber have been found to have the capacity of binding different substances like bile salts and glucose which have implications in cholesterol lipid and carbohydrate metabolism respectively, as presented in the preceding sections. However, the continuous introduction of new ingredients in the food industry requires further studies in order expand knowledge of the impact on nutrient absorption.

Figure 1 shows the behavior of samples during the digestive tract simulation to evaluate glucose and calcium retention percentages and macroscopical differences between them could be observed. Different simulated digestive contents for different fibers before dialysis in assays to determine the iron retention percentages are not shown because they are similar to those presented in Fig. 1. Simulation of gastrointestinal environment during dialysis of different yoghurts can be observed in Figure 2. Changes in pH during gastrointestinal simulation produces different behaviors depending on the type of fiber employed. The apple fiber is a fine powder with brownish color, probably due to the content of phenolics compounds [150]. When apple fiber is added to the yogurt and subjected to the gastrointestinal simulation this color persists (Figure 1). In Figure 2 it can be seen that Psyllium fiber gives a viscous dispersion [151,152]. Due to changing pH values in the

(1) Wheat, (2) psyllium, (3) apple, (4) inulin, (5) chitosan and (6) bamboo fibers.

Figure 1. Photograph of the macroscopic view of different fibers in the in vitro digestive tract simulation.

(1) Yoghurt without fiber, (2) chitosan, (3) psyllium, (4) wheat.

Figure 2. Different fiber behaviors in the dialysis step of digestive simulation.

digestive tract, Chitosan precipitates while passing through the first portion of the small intestine, forming flocculus. Chitosan, that is a positively charged polysaccharide, is insoluble in neutral and alkaline pH. It is only soluble in acidic pH because below pH 6.5 (pKa = 6.5), the amine groups of chitosan are positively charged. When it is solubilised in dilute acid, chitosan has a linear structure [153]. At pH > 6.5, the polymer loses its charges from the amine groups and therefore becomes insoluble in water and precipitate forming flocculus.

Using the model that reproduces *in vitro* gastrointestinal conditions we determined glucose availability reduction and the results are shown in Figure 3. Significant differences (p < 0.05) are observed in glucose availability reduction percentage for the different fiber samples. In the gastrointestinal conditions chitosan formed a flocculus that entrapped glucose so its availability reduction is the highest. Psyllium increases viscosity medium and glucose availability reduction is 15.3 ± 1.8%; wheat has 9.5 ± 2.1% of glucose retention and inulin 5.7 ± 1.8%, apple and bamboo showed no availability reduction. This *in vitro* study supports the view that certain types of dietary fiber reduce the rate of glucose absorption but chitosan has the most pronounced effect. The behavior in delaying absorption could be likely to alter the gut endocrine response both by carrying material further down the small intestine prior to absorption as well as by producing a flatter blood glucose profile.

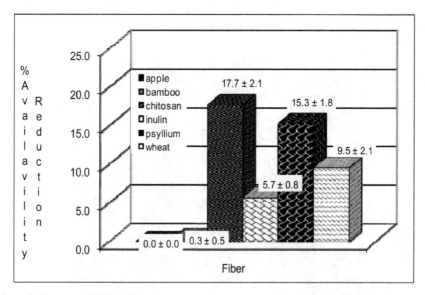

Figure 3. Glucose availability reduction

On the other hand, dietary fiber may influence the availability of minerals, such as calcium, magnesium [154] and iron [155]. Animal studies have found that dietary chitosan possibly arrests the absorption of calcium [156,157].

To study calcium availability the same model for glucose was used but without the addition of exogenous calcium because yogurt is an important source of this mineral in the human diet. Data are shown in Figure 4. Statistical analysis confirmed significant differences (p < 0.05) among the behavior of the different fibers with calcium. It is observed 16.5 ± 1.6% of calcium availability reduction for apple fiber that have significant differences with the others fibers. However availability reduction responses, between insoluble fibers (wheat and bamboo) and soluble ones (inulin and psyllium plantago), have no significant differences (p < 0.05) by Tukey's test. Again, like results obtained with glucose, this study demonstrated that the chitosan effect is more pronounced and higher than for the other studies [138].

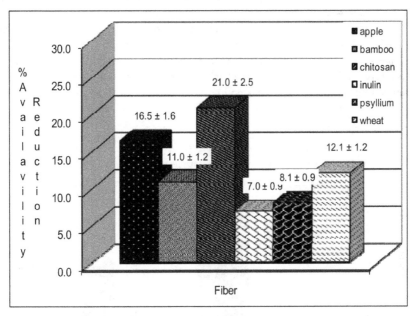

Figure 4. Calcium availability reduction.

To study iron retention percentages by the fibers tested in the present study, the introduction of cellulose dialysis tubes in the digestive chemical experimental model is utilised. The use of a membrane dialysis tube reproduces, in the laboratory, the duodenum wall and its utilisation is presumably a significant factor that determines iron absorption according to Miret et al. [158].

In yoghurt, caseins are modified as a consequence of its production process. Bioactive peptides are formed from caseins during the elaboration of milk products (cheese, yoghurt) under the action of endogenous enzymes of milk (plasmin, cathepsin, among others) or of microorganisms [159]. These peptidic fragments that are already present in yoghurt, could fix iron and calcium according to Bouhallab and Bouglé [159]. Then, these complex matrices (yoghurts with each type of fiber and iron or calcium) are subjected to the gastrointestinal simulation. Control yoghurt with ferrous sulfate without fiber was also subjected to the digestive simulation and considered to be 0% iron retention (100% iron dialyzated) to calculate iron retention percentages for each fiber. Similarly, control yoghurt without fibers was subjected to the digestive simulation to estimate calcium 100% availability. With these control yoghurts, we could consider the interaction of iron or calcium with casein peptidic fragments.

Iron retention percentages of different fibers are presented in Figure 5. Bamboo and wheat fibers, both insoluble, have low iron retention percentages between 2–5% at 30 min with a maximum of 10% at 60 min. There are no significant differences ($p < 0.05$) between them by Tukey's test. Bamboo and wheat are high in cellulose content. Cellulose could retain iron by physical adsorption according to results reported by Torre et al. [15]. They worked with high dietary fiber food materials studying the physicochemical interactions with Fe(II), Fe(III) and Ca(II) without an *in vitro* digestive model. They found that the interaction between Fe(II) and cellulose could be explained better by physical adsorption than complex formation. Inulin, a soluble fiber, has no iron retention at 30 or 60min of simulation. This result is in accordance with studies that confirm that inulin does not interfere with iron absorption [17,20,160,161].

Although psyllium and apple fiber contain both soluble and insoluble fractions, they have significantly different responses ($p < 0.05$). The apple fiber incorporated in yoghurt has no

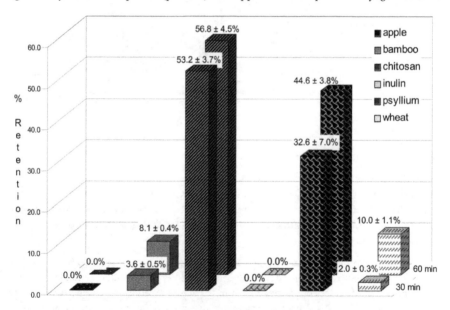

Figure 5. Iron retention percentages of yoghurts with different studied fibers

influence on iron retention. Psyllium shows, on average, 44.6 ± 3.8% iron retention at 60min, which may be mainly attributed to the formation of high viscous dispersion that could be interfering with iron absorption (Figure 2). In addition, the differing behaviors between apple fiber and psyllium could be explained by the different chemical composition of these fibers. Psyllium has high hemicellulose content and apple has the highest lignin content and cellulose. However, bamboo has a low iron retention percentage. The bamboo behavior could be explained due to the composition with equal proportions of soluble and insoluble fractions. More research is needed of this type of fiber.

Chitosan presents the highest iron retention percentages at 30 min (53.2 ± 3.7%) and 60 min (56.8 ± 4.5%), which shows significant differences (p < 0.05) with other fibers. This biopolymer, which has an animal origin, contains 98% insoluble fiber, and flocculates in the first portion of the small intestine. This flocculus (Figure 2), which could entrap iron, clearly decrease iron dialysis. However, certain amount of iron could go through the cellulose membrane and could be measured to calculate the iron retention percentage. Certain amount of casein-peptide-fragments interacting with iron could remain in solution. Nevertheless, their presence does not interfere with the calculation of iron retention percentages as proven by the digestive simulations performed with control yoghurts.

This study shows that the effect of chitosan on iron absorption is more pronounced and higher than those measured for the other studied plant fibers, as dietary fiber is a significant factor that influences iron absorption. The iron retention percentages of different fibers used in this work could be explained mainly as a result of physicochemical phenomena, like adsorption, formation of viscous dispersion and flocculus.

Yoghurt contains peptidic fragment s from caseins. The caseins are amphiphilic phosphoproteins and their isoelectric point (pI) value is 4.6. At pH above the pI, caseins are negatively charged and soluble in water. The caseins have an electronegative domain preferentially located in small peptidic fragments known as α_{s1}-Casein, β-casein and κ-casein. These structural features of the caseins may render these molecules adept at forming complexes with multivalent cationic macromolecules, such as chitosan [153]. In yoghurt (pH = 4.4–4.6) aggregation of the casein-peptide-fragments occur because of a reduction in the electrostatic repulsion at around their pI value. Anal et al. [153] studied the interactions between sodium caseinate and chitosan, under a range of conditions. This study showed that soluble or insoluble chitosan–caseinate complexes can be formed depending on the pH. The characteristics of the complexes are determined by the biopolymer types and their concentration, as well as by environmental conditions. In a certain pH range (5.0–6.0), nanocomplexes of chitosan and sodium caseinate with diameter between 250 and 350 nm were formed. The chitosan and sodium caseinate complexes associated to form larger particles, which resulted in phase separation appear when the pH was either in the range 4.0–4.5 or >6.5. At pH 3.0–3.8, where chitosan and sodium caseinate have similar charges, they may dissociate from each other and become solubilized in solution. According to these authors, yoghurts with chitosan could contain chitosan-casein-peptidic complexes apart from free chitosan molecules in solution.

Besides, calcium existing in yoghurt or the added iron could interact with free chitosan molecules and those complexes. In our work, yoghurts with chitosan are subjected to gastrointestinal simulations. In the first step, our food passes through the simulated stomach (pH = 1.0–2.0) and it could be expected that casein peptidic fragments, chitosan, iron or calcium, all remain in solution. While the food passes through the first portion of the simulated small intestine, changes in pH can lead to formation of chitosan-casein peptidic complexes and iron or calcium could be interacting with them. At pH 6.8–7.0, free chitosan molecules and chitosan-casein-peptidic complexes precipitate forming flocculus. The force of the coagulum formed is high and can be seen in Figures 1 and 2. The results reported by Ausar et al. [162] indicate that hydrophobicity of the casein-chitosan complex is the main mechanism by which the casein-chitosan flocculation is produced.

Chitosan is essentially a positively charged polysaccharide. Iron and calcium are cations. Anal et al. [153] measured zeta potential of chitosan solutions, sodium caseinate solutions and chitosan-caseinate mixtures in a range of pH (3.0–6.5). They found that the pure chitosan solutions were strongly positively charged between pH 3.0 and 6.0. The zeta potential values of chitosan solutions decreased with increasing pH and were slightly negative (approximately – 2.5 mV) at pH 6.5. In our study, in the range of pH 3.0–6.0, isolated molecules of chitosan were probably interacting with iron or calcium by adsorption rather than by electrostatic forces. Besides, Anal et al. [153] observed that the zeta potentials of the chitosan–caseinate solutions were negative at pH > 5.5. In this range of pH, in our work, electrostatic interaction could exist between chitosan-caseinate complexes and iron or calcium. However, when chitosan precipitates, it captures the iron or calcium either by electrostatic forces or by adsorption [138,139].

The behavior of chitosan with calcium and iron in the digestive simulations were similar and can be explained in the same manner. However, the behavior between the other fibers used and the same micronutrients in the digestive simulations were significantly different. The flocculus formation by chitosan is a very strong kind of behavior which is independent of the use of the dialysis membrane. Evidently other types of interactions are brought into play for the other fibers that need further studies to determine them.

11. Conclusion

Results showed that the different plant fibers decreased glucose, calcium and iron availabilities whereas the effect of chitosan (fiber from animal source) was more pronounced. These findings could be positive or negative depending on the nutrient and the nutritional stage or health of the population who would receive the food under study. However, the *in vitro* digestive chemical experimental model may be used to increase the understanding of the interactions between animal and plant fibers with nutrients and micronutrients. This knowledge is very important from the point of view of health and for food industry and technologists.

Author details

Marina Dello Staffolo
CIDCA (Centro de Investigación y Desarrollo en Criotecnología de Alimentos), CONICET–CCT La Plata, Fac. Cs. Exactas, Universidad Nacional de La Plata, La Plata, Argentina

Alicia E. Bevilacqua
CIDCA (Centro de Investigación y Desarrollo en Criotecnología de Alimentos), CONICET–CCT La Plata, Fac. Cs. Exactas, Universidad Nacional de La Plata, La Plata, Argentina
Departamento de Ingeniería Química, Facultad de Ingeniería, UNLP

María Susana Rodríguez and Liliana Albertengo
Instituto de Química del Sur (INQUISUR), CO NICET–UNS Bahía Blanca, Departamento de Química, Universidad Nacional del Sur, Argentina

Acknowledgement

Financial support from CONICET and SeCyT-Universidad Nacional del Sur is gratefully acknowledged. Marina Dello Staffolo and Alicia Bevilacqua express their gratitude to Universidad Nacional de La Plata and bamboo, inulin and wheat suppliers (Imperial Sensus, CFF and JRS).

12. References

[1] Green CJ. Fiber in enteral nutrition. South African Journal of Clinical Nutrition 2000;13(4) 150-160.

[2] Food and Nutrition Board, Institute of Medicine (2001). Dietary reference intakes. Proposed definition of dietary fiber. A report of the panel on the definition of dietary fiber and the standing committee on the scientific evaluation of dietary reference intakes. Washington, DC: National Academy Press.

[3] Thebaudin JY., Lefebvre AC.,Harrington M., Bourgeois CM. Dietary fibres: Nutritional and technological interest. Trends in Food Science & Technology 1997;81 41-48.

[4] Tungland BC., Meyer D. Nondigestible oligo and polysaccharides (dietary fibre): their physiology and role in human health and food. Comprenhensive Reviews in Food Science and Food Safety 2002;1(3) 90-109.

[5] Trout D., Behall K., Osilesi O. Prediction of glycaemic index for starchy foods. American Journal of Clinical Nutrition 1993;58 873-878.

[6] Jenkins DJA., Kendall CWC., Axelsen M., Augustin LSA., Vuksan V. Viscous and nonviscous fibres, nonabsorbable and low glycaemic index carbohydrates, blood lipids and coronary heart disease. Current Opinion in Lipidology 2000;11(1) 49-56.

[7] Luccia BHD., Kunkel ME. In vitro availability of calcium from sources of cellulose, methylcellulose, and psyllium. Food Chemistry 2002;77 139-146.

[8]   Bosscher D., Van Ciaillie-Bertrand M., Deelstra H. Effect of thickening agents, based on soluble dietary fibre on the availability of calcium, iron, and zinc from infants formulas. Nutrition 2001;17 614-618.

[9]   Salas-Salvadó J., Martinez-González MA., Bulló M., Ros E. The role of diet in the prevention of type 2 diabetes. Nutrition, Metabolism & Cardiovascular Diseases 2011;21, B32-B48.

[10] Nordin BEC. Calcium and Osteoporosis. Nutrition 1997;13 664-686.

[11] Fardellonea P., Cottéb F-E., Rouxc C., Lespessaillesd E., Merciere F., Gaudinf A-F. Calcium intake and the risk of osteoporosis and fractures in French women. Joint Bone Spine 2010;77 154-158.

[12] Beard J., Stoltzfus RJ. Iron-deficiency anemia: Reexamining the nature and magnitude of the public health problem. Journal of Nutrition 2001;131 563S-703S.

[13] Thompson SA., Weber CW. Influence of pH on the binding of copper, zinc and iron in six fiber sources. Journal of Food Science 1979;44 752-754.

[14] Laszlo JA. Mineral binding properties of soy hull. Modelling mineral interactions with an insoluble dietary fiber source. Journal of Agriculture and Food Chemistry 1987;35 593-600.

[15] Torre M., Rodriguez AR., Saura-Calixto F. Interactions of Fe(II), Ca(I1) and Fe(III) with high dietary fibre materials: A physicochemical approach. Food Chemistry 1995;54 23-31.

[16] Chau C-F., Chen C-H., Lin C-Yi. Insoluble fibre-rich fractions derived from Averrhoa carambola: Hypoglycemic effects determined by in vitro methods. Lebensmittel-Wissenschaft und-Technologie 2004;37 331-335.

[17] Bosscher D., Van Caillie-Bertrand M., Van Cauwenbergh R., Deelstra H. Availabilities of calcium, iron, and zinc from dairy infant formulas is affected by soluble dietary fibres and modified starch fractions. Nutrition 2003;19 641-645.

[18] Argyri K., Birba A., Miller DD., Komaitis M., Kapsokefalou M. Predicting relative concentrations of bioavailable iron in foods using in vitro digestion: New developments. Food Chemistry 2009;113 602-607.

[19] Hur SJ., Lim BO., Decker EA., McClements DJ. In vitro human digestion models for food applications. Food Chemistry 2011;125 1-12.

[20] Laparra JM., Tako E., Glahn RP., Miller DD. Supplemental inulin does not enhance iron bioavailability to Caco-2 cells from milk- or soy-based, probiotic-containing, yoghurts but incubation at 37°C does. Food Chemistry 2008;109 122-128.

[21] Tamine A., Robinson R. Yoghurt Science and Technology, 3rd ed. Boca Raton: CRC Press; 2007.

[22] Hispley EH. Dietary 'fibre' and pregnancy toxaemia. British Medical Journal 1953;2 420-422.

[23] Burkitt DP. The role of refined carbohydrate in large bowel behaviour and disease. Plant Foods for Man 1973;1 5–9.

[24] Trowell H. Atherosclerosis, Ischemic heart disease and dietary fiber. American Journal of Clinical Nutrition 1972;25 926-932.

[25] Cummings TJ., Mann J., Nishida C., Vorster H. Dietary fibre: An agreed definition. Lancet 2009;373 365-366.

[26] Harris SS., Pijls L. Dietary fibre: Refining a definition. Lancet 2009;374 28.

[27] Phillips GO., Cui SW. An introduction: Evolution and finalisation of the regulatory definition of dietary. Food Hydrocolloid 2011;25 139-143.

[28] Elleuch M., Bedigian D., Roiseux O., Besbes S., Blecker C., Attia H. Dietary fibre and fibre-rich by-products of food processing: Characterization, technological functionality and commercial applications: A review. Food Chemistry 2011;124 411-421.

[29] Borderías AJ., Sánchez-Alonso I., Pérez-Mateos M. New application of fibre in foods: Addition to fishery products. Trends in Food Science and Technology 2005;16 458-465.

[30] Normand FL., Ory RL., Mod RR. Binding of bile acids and trace minerals by soluble hemicelluloses of rice. Food Technology 1987;41 86-99.

[31] Chawla R., Patil GR. Soluble Dietary Fiber. Comprenhensive Reviews in Food Science and Food Safety 2010;9 178-196.

[32] Jenkins DJA., Jenkins MJA., Wolever TMS., Taylor RH. Slow release carbohydrate: Mechanism of action of viscous fibers. Journal of Clinical Nutrition and Gastroenterology 1986;1 237-241.

[33] Dikeman CL., Fahey GCJr. Viscosity as Related to Dietary Fiber: A Review. Critical Reviews in Food Science and Nutrition 2006;46(8) 649-663.

[34] Lattimer J., Haub M. Effects of dietary fiber and its components on metabolic health. Nutrients 2010;2 1266-1289.

[35] Claye SS., Idouraine A., Weber CW. Dietary fibre, what it is and how it is measured. Food Chemistry 1996;57 305-310.

[36] Prosky L., Asp N-G., Scheweizer TF., DeVries JW., Furda I. Determination of insoluble and soluble, and total dietary fibre in foods and food products: Interlaboratory study. Journal of the Association of Official Analytical Chemists1988;71 1017-1023.

[37] Englyst HN., Quigley ME., Hudson GJ. Determination of dietary fiber as non-starch polysaccharides with gas–liquid chromatographic or spectrophotometric measurement of constituent sugars. Analyst 1994;119 1497-1509.

[38] Van Soest PJ. Use of detergents in the analysis of fibrous feeds II. Determination of plant cell-wall constituents. Journal of the Association of Official Analytical Chemistry 1963;48 829-835.

[39] Ripsin CM., Keenan JM., Jacobs DR., Elmer PJ., Welch RR., Van Horn L., Liu K., Turnbull WH., Thye FW., Kestin M., Hegsted M., Davidson DM., Davidson MH., Dugan LD., Demark-Wahnefried W., Beling S. Oat products and lipid lowering: a meta-analysis. Journal of the American Medical Association 1992;267 3317-3325.

[40] Olson BH., Anderson SM., Becker MP., Anderson JW., Hunninghake DB., Jenkins DJA., LaRosa JC., Rippe JM., Roberts DCK., Stoy DB., Summerbell CD., Truswell AS., Wolever TMS., Morris DH., Fulgoni VL. Psyllium-enriched cereals lower blood total cholesterol and LDL-cholesterol, but not HDL cholesterol, in hypercholesterolemic adults: results of a meta-analysis. Journal of Nutrition 1997;127 1973-1980.

[41] Anderson JW., Allgood LD., Lawrence A., Altringer LA., Jerdack GR., Hengehold DA., Morel JG. Cholesterol-lowering effects of psyllium intake adjunctive to diet therapy in

men and women with hypercholesterolemia: meta-analysis of 8 controlled trials. American Journal of Clinical Nutrition 2000;71 472-479.

[42] Chau CF., Huang YL. Effects of the insoluble fiber derived from Passiflora edulis seed on plasma and hepatic lipids and fecal output. Molecular Nutrition & Food Research 2005;49 786-790.

[43] Cho IJ., Lee C., Ha TY. Hypolipidemic effect of soluble fiber isolated from seeds of Cassiatora Linn in rats fed a high-cholesterol diet. Journal of Agriculture and Food Chemistry 2007;55 1592-1596.

[44] Chau CF., Huang YL., Lin CY. Investigation of the cholesterol-lowering action of insoluble fibre derived from the peel of Citrus sinensis L. cv. Liucheng. Food Chemistry 2004;87(3) 361-366.

[45] Phillips MC. Mechanisms of cholesterol-lowering effects of dietary insoluble fibres: Relationships with intestinal and hepatic cholesterol parameters. British Journal of Nutrition 2005;94 331-337.

[46] Gonzalez M., Rivas C., Caride B., Lamas A., Taboada C. Effects of orange and apple pectin on cholesterol concentration in serum, liver and faeces. Journal of Physiology and Biochemistry 1998;54(2) 99-104.

[47] Anttila H., Sontag-Strohm T., Salovaara H. Viscosity of β-glucan in oat products. Agricultural and Food Science 2004;13 80-87.

[48] Wood PJ. Relationships between solution properties of cereal β-glucans and physiological effects – A review. Trends in Food Science and Technology 2004;15 313-320.

[49] Zacherl C., Eisner P., Engel K-H. In vitro model to correlate viscosity and bile acid-binding capacity of digested water-soluble and insoluble dietary fibres. Food Chemistry 2011;126 423-428.

[50] Galisteo M., Moron R., Rivera L., Romero R., Anguera A., Zarzuelo A. Plantago ovata husks-supplemented diet ameliorates metabolic alterations in obese Zucker rats through activation of AMP-activated protein kinase. Comparative study with other dietary fibers. Clinical Nutrition 2010;29 261-267.

[51] Delzenne NM., Kok N. Effects of fructans-type prebiotics on lipid metabolism. American Journal of Clinical Nutrition 2001;73 456S-458S.

[52] Eshak ES., Iso H., Date C., Kikuchi S., Watanabe Y., Wada Y., Tamakoshi A. and the JACC Study Group. Dietary fiber intake is associated with reduced risk of mortality from cardiovascular disease among Japanese men and women. Journal of Nutrition 2010;140 1445-1453.

[53] Health claims: fruits, vegetables, and grain products that contain fiber, particularly soluble fiber, and risk of coronary heart disease. Code of Federal Regulations 2010;Secc. 101, 77.

[54] Kaczmarczyka MM., Michael J., Miller MJ., Freunda GG. The health benefits of dietary fiber: Beyond the usual suspects of type 2 diabetes mellitus, cardiovascular disease and colon cancer. Metabolism 2012, article in press. doi:10.1016/j.metabol.2012.01.017.

[55] Slavin JL. Dietary fiber and body weight. Nutrition 2005;21 411-418.

[56] Wang ZQ., Zuberi AR., Zhang XH., Macgowan J., Qin J., Ye X. Effects of dietary fibers on weight gain, carbohydrate metabolism, and gastric ghrelin gene expression in mice fed a high-fat diet. Metabolism 2007;56 1635-1642.

[57] Choi JS., Kim H., Jung MH., Hong S., Song J. Consumption of barley β-glucan ameliorates fatty liver and insulin resistance in mice fed a high-fat diet. Molecular Nutrition & Food Research 2010;54 1004-1013.

[58] Behall KM., Scholfield DJ., Hallfrisch JG., Liljeberg-Elmstahl HG. Consumption of both resistant starch and β-glucan improves postprandial plasma glucose and insulin in women. Diabetes Care 2006;29 976-81.

[59] Tucker LA., Thomas KS. Increasing total fiber intake reduces risk of weight and fat gains in women. Journal of Nutrition 2009;139 576-581.

[60] Birketvedt GS., Shimshi M., Erling T., Florholmen J. Experiences with three different fiber supplements in weight reduction. Medical Science Monitor 2005; 11 I5-I8.

[61] Barclay AW., Petocz P., McMillan-Price J., Flood VM., Prvan T., Mitchell P. Glycemic index, glycemic load, and chronic disease riskea meta-analysis of observational studies. American Journal of Clinical Nutrition 2008;87 627-637.

[62] Sluijs I., van der Schouw YT., van der A DL., Spijkerman AM., Hu FB., Grobbee DE., Beulens JW. Carbohydrate quantity and quality and risk of type 2 diabetes in the European Prospective Investigation into Cancer and Nutrition-Netherlands (EPIC-NL) study. American Journal of Clinical Nutrition 2010;92 905-911.

[63] Anderson JW., Baird P., Davis RH Jr., Ferreri S., Knudtson M., Koraym A., Waters V., Williams CL. Health benefits of dietary fiber. Nutrition Review 2009;67 188-205.

[64] Salmerón J., Ascherio A., Rimm EB., Colditz GA., Spiegelman D., Jenkins DJ. Dietary fiber, glycemic load, and risk of NIDDM in men. Diabetes Care 1997;20 545-550.

[65] Salmerón J, Manson JE, Stampfer MJ, Colditz GA, Wing AL, Willett WC. Dietary fiber, glycemic load and risk of non-insulin-dependent diabetes mellitus in women. Journal of the American Medical Association 1997;277 472-477.

[66] Sierra M., Garcia JJ., Fernandez N., Diez MJ., Calle AP. Therapeutic effects of psyllium in type 2 diabetic patients. European Journal of Clinical Nutrition 2002;56 830-842.

[67] Chandalia M., Garg A., Lutjohann D., von Bergmann K., Grundy SM., Brinkley LJ. Beneficial effects of high dietary fiber intake in patients with type 2 diabetes mellitus. The New England Journal of Medicine 2000;342 1392-1398.

[68] Salas-Salvadó J., Farrés X., Luque X., Narejos S., Borrell M., Basora J. Fiber in Obesity-Study Group. Effect of two doses of a mixture of soluble fibres on body weight and metabolic variables in overweight or obese patients: a randomised trial. British Journal of Nutrition 2008;99 1380-1387.

[69] Weickert MO., Mohlig M., Koebnick C., Holst JJ., Namsolleck P., Ristow M. Impact of cereal fibre on glucose-regulating factors. Diabetologia 2005;48 2343-2353.

[70] Raninen K., Lappi J., Mykkanen H., Poutanen K. Dietary fiber type reflects physiological functionality: comparison of grain fiber, inulin, and polydextrose. Nutrition Review 2011;69 9-21.

[71] Lairon D., Play B., Jourdheuil-Rahmani D. Digestible and indigestible carbohydrates: interactions with postprandial lipid metabolism. Journal of Nutrition Biochemistry 2007; 18 217-27.

[72] Anderson JW., Chen WJ. Plant fiber. Carbohydrate and lipid metabolism. American Journal of Clinical Nutrition 1979;32 346-363.

[73] Naslund E., Bogefors J., Skogar S., Gryback P., Jacobsson H., Holst JJ. GLP-1 slows solid gastric emptying and inhibits insulin, glucagon, and PYY release in humans. American Journal of Physiology 1999;277 R910-R916.

[74] Reimer RA., Grover GJ., Koetzner L., Gahler RJ., Lyon MR., Wood S. The soluble fiber complex PolyGlycopleX lowers serum triglycerides and reduces hepatic steatosis in high-sucrose fed rats. Nutrition Research 2011;31 296-301.

[75] Massimino SP., McBurney MI., Field CJ., Thomson AB., Keelan M., Hayek MG. Fermentable dietary fiber increases GLP-1 secretion and improves glucose homeostasis despite increased intestinal glucose transport capacity in healthy dogs. Journal of Nutrition 1998;128 1786-93.

[76] Delzenne NM, Cani P. Nutritional modulation of gut microbiota in the context of obesity and insulin resistance: potential interest of prebiotics. International Dairy Journal 2010;20 277-280.

[77] Eckburg PB., Bik EM., Bernstein CN., Purdom E., Dethlefsen L., Sargent M. Microbiology: diversity of the human intestinal microbial flora. Science 2005; 308 1635–1638.

[78] Sghir A., Gramet G., Suau A., Rochet V., Pochart P., Dore J. Quantification of bacterial groups within human fecal flora by oligonucleotide probe hybrid-ization. Applied and Environmental Microbiology 2000; 66(5) 2263–2266.

[79] Muller CA., Autenrieth IB., Peschel A. Innate defenses of the intestinal epithelial barrier. Cellular and Molecular Life Sciences 2005; 62(12) 1297–1307.

[80] Macfarlane, GT., Macfarlane S. Human colonic microbiota: ecology, physiology and metabolic potential of intestinal bacteria. Scandinavian Journal of Gastroenterology 1997; Supplement 32(222) 3-9.

[81] Cook SI., Sellin JH. Review article: short chain fatty acids in health and disease. Alimentary Pharmacology and Therapeutics 1998;12(6) 499–507.

[82] Mazur WM., Duke JA., Wahala K., Rasku S., Adlercreutz H. Isoflavonoids and lignans in legumes: nutritional and health aspects in humans. Journal of Nutritional Biochemistry 1998;9(4) 193-200.

[83] Rowland I., Faughnan M., Hoey L., Wahala K., Williamson G., Cassidy A. Bioavailability of phyto-oestrogens. British Journal of Nutrition 2003, 89(Suppl. 1) 45–58.

[84] Brownlee IA. The physiological roles of dietary fibre. Food Hydrocolloids 2011;25 238–250.

[85] Terada A., Hara H., Mitsuoka T. Effect of dietary alginate on the faecal microbiota and faecal metabolic activity in humans. Microbial Ecology in Health and Disease 1995;8(6) 259–266.

[86] Terada, A., Hara, H., Sato, D., Higashi, T., Nakayama, S., Tsuji, K. Effect of dietary chitosan on faecal microbiota and faecal metabolites of humans. Microbial Ecology in Health and Disease 1995;8(1) 15–21.

[87] Grasten S., Liukkonen KH., Chrevatidis A., El-Nezami H., Poutanen K., Mykkanen H. Effects of wheat pentosan and inulin on the metabolic activity of fecal microbiota and on bowel function in healthy humans. Nutrition Research 2003;23(11) 1503–1514.

[88] Bittner AC., Croffut RM., Stranahan MC., Yokelson TN. Prescriptassist probiotic–prebiotic treatment for irritable bowel syndrome: an open- label, partially controlled, 1-year extension of a previously published controlled clinical trial. Clinical Therapeutics 2007;29 (6) 1153–1160.

[89] Fujimori S., Gudis K., Mitsui K., Seo T., Yonezawa M., Tanaka S. A randomized controlled trial on the efficacy of synbiotic versus probiotic or prebiotic treatment to improve the quality of life in patients with ulcerative colitis. Nutrition 2009;25(5) 520–525.

[90] Molist F., de Segura AG., Gasa J., Hermes RG., Manzanilla EG., Anguita M. Effects of the insoluble and soluble dietary fibre on the physicochemical properties of digesta and the microbial activity in early weaned piglets. Animal Feed Science and Technology 2009;149(3–4) 346-353.

[91] Noriega L., Cuevas I., Margolles A., de los Reyes-Gavilán CG. Deconjugation and bile salts hydrolase activity by Bifidobacterium strains with acquired resistance to bile. International Dairy Journal 2006;16(8) 850-855.

[92] Klaver FA., van der Meer R. The assumed assimilation of cholesterol by Lactobacilli and Bifidobacteriu m bifidum is due to their bile salt-deconju gating activity. Applied and Environmental Microbiology 1993;59 1120-1124.

[93] Tahri K., Grill JP., Schneider F. Bifidobacteri a strain behaviour toward cholestero l: coprecipita tion with bile salts and assimilation. Current Microbiology 1996;33 187-93.

[94] Pereira DI., Gibson GR. Cholesterol assimilat ion by lactic acid bacteria and bifidobacteria isolated from the human gut. Applied and Environmental Microbiology 2002;68 4689-4693.

[95] Demigne C., Morand C., Levrat MA., Besson C., Moundras C., Remesy C. Effect of propionate on fatty acid and cholesterol synthesis and on acetate metabolism in isolated rat hepatocytes. British Journal of Nutrition 1995;74 209-219.

[96] Nagler-Anderson C. Man the barrier! Strategic defences in the intestinal mucosa. Nature Reviews Immunology 2001;1(1) 59-67.

[97] Strugala V., Allen A., Dettmar PW. Pearson JP. Colonic mucin: methods of measuring mucus thickness. Proceedings of the Nutrition Society 2003; 62(1) 237-243.

[98] Taylor C., Allen A., Dettmar PW., Pearson JP. Two rheologically different gastric mucus secretions with different putative functions. Biochimica et Biophysica Acta - General Subjects 2004;1674(2) 131-138.

[99] Watzl B., Girrbach S., Roller M. Inulin, oligofructose and immunomodulation. British Journal of Nutrition 2005;93(Supp) S49-S55.

[100] Field CJ., McBurney MI., Massimino S., Hayek MG., Sunvold GD. The fermentable fiber content of the diet alters the function and composition of canine gut associated lymphoid tissue. Veterinary Immunology and Immunopathology 1999;72(3-4) 325-341.

[101] Wismar R., Brix S., Frokiaer H., Laerke HN. Dietary fibers as immunoregulatory compounds in health and disease. Annals of the New York Academy of Sciences 2010;1190 70-85.

[102] Brownlee IA., Dettmar PW., Strugala V., Pearson JP. The interaction of dietary fibres with the colon. Current Nutrition and Food Science 2006;2(3) 243-264.

[103] Parker SL., Tong T., Bolden S., Wingo PA. Cancer Statistics. CA- A Cancer Journal for Clinicians 1997;47 5-27.

[104] Bingham SA., Day NE., Luben R., Ferrari P., Slimani N., Norat T., Clavel-Chapelon F., Kesse E., Nieters A., Boeing H., Tjønneland A., Overvad K., Martinez C., Dorronsoro M., Gonzalez, C.A., Key TJ., Trichopoulou A., Naska A., Vineis P., Tumino R., Krogh V., Bueno-de-Mesquita HB., Peeters PHM., Berglund G., Hallmans G., Lund E., Skeie G., Kaaks R., Riboli E. Dietary fibre in food and protection against colorectal cancer in the European Prospective Investigation into Cancer and Nutrition (EPIC): an observational study. The Lancet 2003;361 1496-1501.

[105] Dahm CC., Keogh RH., Spencer EA., Greenwood DC., Key TJ., Fentiman IS, Shipley MJ., Brunner EJ., Cade JE., Burley VJ., Mishra G., Stephen AM., Kuh D., White IR., Luben R., Lentjes MAH., Khaw KT, Bingham SA. Dietary Fiber and colorectal cancer risk: A Nested case–control Study Using Food Diaries. Journal of the National Cancer Institute 2010;102(9) 614-626.

[106] Health claims: fruits, vegetables, and grain products that contain fiber, particularly soluble fiber, and risk of coronary heart disease. Code of Federal Regulations 2010;101.77.

[107] Reddy BS., Engle A., Simi B., Goldman M. Effect of dietary fiber on colonic bacterial enzymes and bile acids in relation to colon cancer. Gastroenterology 1992;102 1475-1482.

[108] Reddy BS. Dietary fiber and colon cancer: animal model studies. Pre-Med Journal 1995;6 559-565.

[109] Roediger WEW. Utilization of nutrients by isolated epithelial cells of rat colon. Gastroenterology 1982;83 424-429.

[110] Kim YS., Tsao D., Siddiqui B., Whitehead JS., Arnstein P., Bennett J., Hicks J. Effects of sodium butyrate and dimethylsulfoxide on biochemical properties of human colon cancer cells. Cancer 1980;45 1185-1192.

[111] Hague A., Manning AM., Haneon KA., Huschstscha LL., Hart D., Paraskeva C. Sodium butyrate induces apoptosis in human colonic tumour cell lines in a p53-independent pathway implications for the possible role of dietary fibre in the prevention of large-bowel cancer. International Journal of Cancer 1993;55 498-505.

[112] Harris PJ., Roberton AM., Hollands HJ., Ferguson LR. Adsorption of a hydrophobic mutagen to dietary fibre from the skin and flesh of potato tubers. Mutation Research 1991;260 203-213.

[113] Harris PJ., Triggs CM., Roberton AM., Watson ME., Ferguson LR. The adsorption of heterocyclic aromatic amines by model dietary fibres with contrasting compositions. Chemico-Biological Interactions 1996;100 13-25.

[114] Ferguson LR., Roberton AM., Watso ME., Kestell P., Harris PJ. The adsorption of a range of dietary carcinogens by α-cellulose, a model insoluble dietary fiber. Mutation Research. 1993;319 257-266.

[115] Halliwell B., Zhao K., Whiteman M. The gastrointestinal tract: a major site of antioxidant action? Free Radical Research 2000;33 819-830.

[116] Gee JM., Johnson IT. Polyphenolic compounds: interactions with the gut and implications for human health. Current Medicinal Chemistry 2001;8 1245-1255.

[117] Mälki Y. Trends en dietary fiber research and development: a review. Acta Alimentaria 2004;33(1) 39-62.

[118] Schneeman BO. Gastrointestinal physiology and functions. British Journal of Nutrition 2002;88(2) S159-S163.

[119] Hersey SJ., Sachs G. Gastric acid secretion. Physiological Reviews 1995;75(1) 155-189.

[120] Sternini C., Anselmi L., Rozengurt E. Enteroendocrine cells: a site of 'taste' in gastrointestinal chemosensing. Current Opinion in Endocrinology, Diabetes and Obesity 2008;15(1) 73-78.

[121] Furness JB. Types of neurons in the enteric nervous system. Journal of the Autonomic Nervous System 2000;81(1–3) 87-96.

[122] Pedersen AM., Bardow A., Jensen SB., Nauntofte B. Saliva and gastrointestinal functions of taste, mastication, swallowing and digestion. Oral Diseases 2002;8(3) 117-129.

[123] Embleton JK., Pouton CW. Structure and function of gastro-intestinal lipases. Advanced Drug Delivery Reviews 1997;25(1) 15-32.

[124] Miled N., Canaan S., Dupuis L., Roussel A., Riviere M., Carriere F. Digestive lipases: from three-dimensional structure of physiology. Biochimie 2000;82(11) 973-986.

[125] Chaplin, MF. Fibre and water binding. Proceedings of the Nutrition Society 2003;62(1) 223-227.

[126] Holt S., Heading RC., Carter DC. Effect of gel fibre on gastric emptying and absorption of glucose and paracetamol. Lancet 1979;1(8117) 636-639.

[127] Sanaka M., Yamamoto T., Anjiki H., Nagasawa K., Kuyama Y. Effects of agar and pectin on gastric emptying and post-prandial glycaemic profiles in healthy human volunteers. Clinical and Experimental Pharmacology and Physiology 2007;34(11) 1151-1155.

[128] Benini L., Castellani G., Brightenti F., Heaton KW., Brentegani MT., Casiraghi MC. Gastric emptying of a solid meal is accelerated by the removal of dietary fibre naturally present in food. Gut 1995;36(6) 825-830.

[129] Kimura Y., Watanabe K., Okuda H. Effects of soluble sodium alginate on cholesterol excretion and glucose tolerance in rats. Journal of Ethnopharmacology 1996;54(1) 47-54.

[130] Torsdottir I., Alpsten M., Holm G., Sandberg AS., Tolli J. A small dose of soluble alginate-fiber affects postprandial glycemia and gastric emptying in humans with diabetes. Journal of Nutrition 1991;121(6) 795-799.

[131] Fairchild RM., Ellis PR., Byrne AJ., Luzio SD., Mir MA. A new breakfast cereal containing guar gum reduces postprandial plasma glucose and insulin concentrations in normal-weight human subjects. British Journal of Nutrition 1996;76(1) 63-73.

[132] Sierra M., Garcia JJ., Fernandez N., Diez MJ., Calle AP., Sahagún AM. Effects of ispaghula husk and guar gum on postprandial glucose and insulin concentrations in healthy subjects. European Journal of Clinical Nutrition 2001;55(4) 235-243.

[133] Behall KM., Scholfield DJ., Hallfrisch JG. Barley β-glucan reduces plasma glucose and insulin responses compared with resistant starch in men. Nutrition Research 2006;26(12) 644-650.

[134] Jenkins AL., Jenkins DJA., Zdravkovic U., Wulrsch P., Vuksan V. Depression of the glycemic index by high levels of β-glucan fiber in two functional foods tested in type 2 diabetes. European Journal of Clinical Nutrition 2002;56(7) 622-628.

[135] Tosh SM., Brummer Y., Wolever TMS., Wood PJ. Glycemic response to oat bran muffins treated to vary molecular weight of β-glucan. Cereal Chemistry 2008;85(2) 211-217.

[136] Boisen S., Eggum BO. Critical evaluation of in vitro methods for estimating digestibility in simple-stomach animals. Nutrition Research Reviews 1991;4 141-162.

[137] Coles LT., Moughan PJ., Darragh AJ. In vitro digestion and fermentation methods, including gas production techniques, as applied to nutritive evaluation of foods in the hindgut of humans and other simple stomached animals. Animal Food Science and Technology 2005;123 421-444.

[138] Rodríguez MS., Montero M., Dello Staffolo M., Martino M., Bevilacqua A., Albertengo L. Chitosan influence on glucose and calcium availability from yoghurt: In vitro comparative study with plants fibre. Carbohydrate Polymers 2008;74 797–801.

[139] Dello Staffolo M., Martino M., Bevilacqua A., Montero M., Rodríguez MS., Albertengo L. Chitosan Interaction with Iron from Yoghurt Using an In Vitro Digestive Model: Comparative Study with Plant Dietary Fibers. International Journal of Molecular Science 2011;12 4647-4660.

[140] Total, soluble, and insoluble dietary fibre in foods. AOAC Official Methods 991.43. In: Cunniff P. (ed.) Official Methods of Analysis (16th ed.); Arlington: AOAC International; 1995. p7–9.

[141] Robertson JB., Van Soest PJ. The detergent system of analysis and its applications to human foods. In: James WPT., Theander O., (eds.) The Analysis of Dietary Fiber and Food. New York: Marcel Dekker; 1981. p139-153.

[142] Dello Staffolo M., Bertola N., Martino M., Bevilacqua A. Influence of dietary fibre addition on sensory and rheological properties of yoghurt. International Dairy Journal 2004;14 263-268.

[143] Fernández-García E., McGregor JU. Fortification of sweetened plain yoghurt with insoluble dietary fibre. European Food Research and Technology 1997;204, 433-437.

[144] Kumar MNVR., Muzzarelli RAA., Muzzarelli C., Sashiwa H., Domb AJ. Chitosan chemistry and pharmaceutical perspectives. Chemical Reviews 2004;104 6017-6084.

[145] Gorinstein S., Zachwieja Z., Folta M., Barton H., Piotrowicz J., Zember M., Comparative content of dietary fiber, total phenolics, and minerals in persimmons and apples. Journal of Agricultural and Food Chemistry 2001;49 952-957.

[146] Van Craeyveld V., Delcour JA., Courtin CM. Extractability and chemical and enzymic degradation of psyllium (Plantago ovata Forsk) seed husk arabinoxylans. Food Chemistry 2009;112 812-819.

[147] Sudha ML., Baskaran V., Leelavathi K. Apple pomace as a source of dietary fiber and polyphenols and its effect on the rheological characteristics and cake making. Food Chemistry 2007;104 686-692.

[148] Dierenfeld ES., Hintz HF., Robertson JB., van Soest PJ., Oftedal OT. Utilization of bamboo by the giant panda. Journal of Nutrition 1982;112 636-641.

[149] Escarnot E., Agneessens R., Wathelet B., Paquot M. Quantitative and qualitative study of spelt and wheat fibres in varying milling fractions. Food Chemistry 2010;122 857-863.

[150] Sun-Waterhouse D., Farr J., Wibisono R., Saleh Z. Fruit-based functional foods I: Production of food-grade apple fibre ingredients. International Journal Food Science and Technology 2008;43 2113-2122.

[151] Chan JKC., Wypyszyk VA. Forgotten natural dietary fiber: Psyllium mucilloid. Cereal Food. World 1988;33 919-922.

[152] Fischer MH., Yu N., Gray GR., Ralph J., Andersond L., Marletta JA. The gel-forming polysaccharide of psyllium husk (Plantago ovata Forsk). Carbohydrate Research 2004; 339 2009-2017.

[153] Anal AK., Tobiassen A., Flanagan J., Singh H. Preparation and characterization of nanoparticles formed by chitosan–caseinate interactions. Colloids and Surfaces B: Biointerfaces 2008;64 104-110.

[154] Reinhold JG., Faradji B., Abadi P., Ismail-Beigi F. Decreased absorption of calcium, magnesium, zinc and phosphorus by humans due to increased fibre and phosphorus consumption of wheat bread. Journal Nutrition 1976;106 493-503.

[155] Vitali D., Dragojević IV., Šebečić B., Vujić L. Impact of modifying tea–biscuit composition on phytate levels and iron content and availability. Food Chemistry 2007;102 82-89.

[156] Deuchi K., Kanauchi O., Shizukuishi M., Kobayashi E. Continuous and massive intake of chitosan affects mineral and fat-soluble vitamin status in rats fed on a high-fat diet. Bioscience Biotechnology and Biochemist 1995;59 1211-1216.

[157] Yang CY., Oh TW., Nakajima D., Maeda A., Naka T., Kim CS. Effects of habitual chitosan intake on bone mass, bone-related metabolic markers and duodenum CaBP D9K mRNA in ovariectomized SHRSP rats. Journal of Nutritional Science and Vitaminology 2002;48 371-378.

[158] Miret S., Simpson RJ., McKie AT. Physiology and molecular biology of dietary iron absorption. The Annual Review of Nutrition 2003;23 283-301.

[159] Bouhallab S., Bouglé D. Biopeptides of milk: Caseinophosphopeptides and mineral bioavailability. Reproduction Nutrition Development 2004;44 493-498.

[160] Van den Heuvel EG., Schaafsma G., Muys T., van Dokkum W. Nondigestible oligosaccharides do not interfere with calcium and nonheme-iron absorption in young, healthy men. American Journal of Clinical Nutrition 1998;67 445–451.

[161] Azorín-Ortuño M., Urbán C., Cerón JJ., Tecles F., Allende A., Tomás-Barberán FA., Espín JC. Effect of low inulin doses with different polymerisation degree on lipid metabolism, mineral absorption, and intestinal microbiota in rats with fat-supplemented diet. Food Chemistry 2009; 113 1058-1065.

[162] Ausar SF., Bianco ID., Badini RG., Castagna LF., Modesti NM., Landa CA, Beltramo DM. Characterization of casein micelle precipitation by chitosans. Journal of Dairy Science 2001;84 361-369.

Chapter 5

Plant Biotechnology for the Development of Design Starches

María Victoria Busi, Mariana Martín and Diego F. Gomez-Casati

Additional information is available at the end of the chapter

1. Introduction

Obesity is a major public health problem due to its pronounced increase and prevalence worldwide. The World Health Organization indicated that in 2005 at least 1.6 billion of adult people were overweight and about 400 million of adults were obese. Predictions for 2015 are even more alarming because indications are that more than 700 million of people will be obese. One of the most common problems associated with obesity is the current lifestyle. Overweight is one of the main risk factors in the development of many chronic diseases, such as respiratory and heart diseases, type 2 diabetes, hypertension and some types of cancer. The increased risk of acquiring some of these diseases is associated with small changes in weight but it can be prevented if appropriate changes in lifestyle are introduced [5].

Furthermore, gastrointestinal infections remain a major health problem despite new advances in medicine. The global incidence of deaths caused by this type of disease is about of 3 million deaths per year. Although this problem is more severe in developing countries, it also occurs in industrialized countries where the incidence of intestinal infection affects about 10% of the population. In most people, the enteropathogenic bacteria cause gastroenteritis that can be treated with drugs and an adequate rehydration. However, in populations such as old people, children, people with chronic intestinal inflammation and immunodeficiencies, it could be a serious problem, leading to the production of septicemia and death. The control of intestinal infections with antibiotics has been one of the medical breakthroughs of the twentieth century. However, the misuse and abuse of these compounds, has led to increased bacterial resistance. Thus, it becomes extremely important to look for new strategies to prevent and/or treat infections. One promising approach is based on the modulation and control of the intestinal microflora through the diet [6].

Starch is a substantial component of the human diet, mainly in populations that are fed on agricultural crops, providing about 50% of daily energy uptake, mostly through unrefined cereals. In contrast, in westernized societies, the average consumption of grains is much lower, reaching only 25%. Polysaccharides such as glycogen or starch are some of the polymers which can be digested by the enzymes of the human gut. This digestion occurs in the small intestine, except a portion named resistant starch (RS) which is degraded in the large intestine. RS is defined as the set of starch and products of starch degradation (oligosaccharides and others) that are not absorbed in the small intestine but are fermented in the colon producing short-chain fatty acids (such as butyrate), and promotes the normal function of the colonocytes. Because the good function of the human gut is given by the consumption of foods rich in starch, and the change of dietary habits towards healthier eating is not a simple job, the enrichment of some foods with RS becomes the most promising option for a healthy diet [3,7,8].

RS function as dietary fibers, including pre-biotic effect on colon microflora, altering lipid metabolism, improving cholesterol metabolism, and reducing the risk of ulcerative colitis and colon cancer. Since RS is not digested in the small intestine it also reduces the glycemic index of the food [9] (Table 1).

POTENTIAL PHYSIOLOGICAL EFFECTS	POSSIBLE PROTECTIVE EFFECTS
Prebiotic and improved bowel health	Colonic health; colorectal cancer, inflammatory bowel disease, constipation, ulcerative colitis
Improve insulinaemic and glycaemic responses	Diabetes, the metabolic syndrome, impaired insulin and glucose responses
Improvement of blood lipid profile	Lipid metabolism, cardiovascular disease, the metabolic syndrome
Increased satiety and synergistic interactions with other dietary components	Obesity, improved metabolic control and enhanced bowel health
Adjunct to oral rehydration therapies and increased micronutrient absorption	Treatment of chronic diarrhea and cholera; osteoporosis
Thermogenesis	Diabetes and obesity

Table 1. Physiological effects of resistant starch (adapted from [3])

There are at least four mechanisms by which resistant starches are obtained [7,10]: RS1: physically inaccessible starch, usually encapsulated in indigestible tissues (encapsulated or embedded within a matrix of lipid and/or protein) ; RS2: starch granules resistant to degradation, with two subtypes, RS2a with low amylose (0 - 30%), which generally loses its strength when cooked, and RS2b, starches with high amylose content which retains its granular structure during processing, RS3: starch retrograde which requires cooking to be released from the granules, and the starch retrograde capacity is affected by the intrinsic biosynthetic process; finally, RS4: chemically modified starches; although this mechanism is the most used to produce resistant starch, there are no reports of changes in plant that can mimic those obtained by chemical methods. Because each of these processes is independent, it is possible that in some foods resistant starches are derived from more than one mechanism. In these classes, RS1, RS2 and RS3 can be influenced by genetic manipulation of plants. The high amylose starches have the greatest potential to generate resistant starch through two mechanisms, RS2b and RS3 [11]. To achieve this, three strategies have been proposed: reduction of branching enzyme activity; reduction of the amylopectin synthesis rate without altering the synthesis of amylose or/and the increment of the amylose synthesis without altering the synthesis of amylopectin (Table 2).

TYPE OF RESISTANT STARCH	EXAMPLES OF OCCURRENCE	RESISTANCE REDUCED BY
RS1: Physycally inaccessible	Whole or partly milled grains and seed, legumes, pasta	Chewing, milling
RS2: Resistant granules	Raw potatoes, green bananas, high-amylose starches, some legumes. Ungelatinised resistant granules, hydrolysed slowly by α-amylases	Cooking and food processing
RS3: Retrograded	Cooked and cooled potato, food products with repeated and/or prolonged heat treatment, bread, cornflakes	Processing conditions
RS4: Chemically modified	Modified starches due to cross-bonding with esters, ethers, etc. Some cakes, breads and fibre-drinks that were made with modified starches.	Less susceptible to digestibility *in vitro*

Table 2. Nutritional classification of resistant starches (adapted from [3] and [7]).

Given that each of these mechanisms is independent, it is possible that any food could contain RS derived from more than one mechanism. Moreover, RS, RS1, RS2, and RS3 content in foods can be modified by crop genetics [10]. Examples of major components of dietary RS are retrograded amylose (RS1), such as cooked and cooled starchy foods like pasta salad, and native starch granules (RS2), such as those found in high amylose maize starch and bananas [12]. On the other hand, RS3 preserves its nutritional functionality during the cooking process. Thus, it may be used as a food ingredient. RS3 is produced in two steps: gelatinization, which is a disruption of the granular structure by heating with excess of water [13] and retrogradation, a slow recrystallization of the starch molecules upon cooling or dehydration [14]. The resistant fraction may be then isolated using amylolytic enzymes such as pancreatic amylase [15], or Termamyl—heat stable α-amylase [16]. It has been shown that the later approach leads to formation of very thermally stable RS3, and to yields up to 40% [9,14].

Finally, but not least, is the role of investigation and development conducted by researchers from universities and industry. The incorporation of progress in science and the use of currently existing technology contributes to the production of healthy foods, and in this context, designing plants with biology tools to improve their current molecular nutritional qualities is a challenge [5].

The first use of transgenesis in plants in the 1980s brought the arrival of a powerful tool for the study of metabolic regulation and crop improvement. Of particular interest from a health and commercial viewpoint was the potential for increasing yield making alteration of carbon partitioning between sucrose, starch and amino acids [17]. Since that time, plant biotechnology and its commercialization are in exponential phase. Already In 1998, more than 28 million hectares of transgenic crop plants were grown worldwide. Of these 28 million hectares, the largest area was in the USA (22 million hectares) followed by Canada (1.8 million hectares), Argentina (1.8 million hectares) and China (estimated at 1.1 million hectares). It was also estimated that in the US 40% of the cotton, 24% of corn and 40% of soybean planted was transgenic [18].

Given the large amount of information available from molecular biology studies and from genomic programs about the starch biosynthetic genes from crop plants, it is now relatively simple to identify the changes at the DNA level to generate desired starch phenotypes [19-21]. Transgenic approaches to altering the composition of crop plants involve two general approaches: overexpression of an endogenous or foreign gene in the target tissue, and use of RNAi technology to specifically suppress the activity of a specific plant gene [10]. We propose in this chapter to give an overview of starch synthesis to review the potential target technologies and to summerize the successful work done by numerous research groups in different plant species using different strategies.

2. Overview of the starch biosynthesis and degradation in plants

Polyglucans are the most important and widespread carbohydrate storage compounds found in nature, with glycogen and starch being the most abundant forms. Both polysaccharides are comprised of glucose chains linked by an α-(1,4) bond, and branched at α-(1,6). Glycogen is a homogeneous water-soluble polymer with relatively uniformly distributed branches [22] and is found in organisms such as archaea, bacteria and certain eukaryotes. Starch is made up of amylose (a largely unbranched, minor component) and amylopectin (an asymmetrically branched major component) and is present in the cytoplasm of *Rhodophyceae* (red algae) and *Glaucophyta* [23], but is confined to the plastid stroma (chloroplasts in green tissues and amyloplasts in reserve organs) in green algae and higher plants. In fact, starch synthesis is restricted to the *Archaeplastida*, whose origins are thought to be via a single endosymbiotic event involving ancestors of cyanobacteria and a heterotrophic host [24], rendering the organelle known as the plastid, which is capable of oxygenic photosynthesis. Recent phylogenetic studies indicate that the plastidial starch pathway is complex, and made up of genes with both cyanobacterial and eukaryotic origins [25,26], and is in sharp contrast to the lower-complexity pathway of cytosolic starch synthesis found in the *Rhodophyceae* and *Glaucophyta* [27]. Phylogenetic analysis of the enzymes of the starch biosynthetic pathway strongly suggests that the pathway was originally cytosolic (in the common ancestor of the *Archaeplastida*), and then re-directed to plastids via three discrete steps, leaving some enzymes involved in the metabolism of malto-oligosaccharides (MOS) and amylopectin degradation in the cytoplasm. The three evolutionary steps involved are: (1) plastidial synthesis of unbranched MOS; (2) glycogen synthesis (including priming steps and branching activities); and (3) plastidial starch synthesis, resulting in the eventual loss of cytosolic starch synthesis. Interestingly, the relocation of the starch synthesis pathway to plastids coincides with the evolution of light-harvesting complexes [26,28].

There are four biochemical steps in each tissue that are required for the synthesis of starch, substrate activation, chain elongation, chain branching, and chain debranching [10] and it involves at least three enzymes such as ADP-glucose pyrophosphorylase (ADPGlc PPase, EC 2.7.7.27), starch synthase (SS, EC 2.4.1.21), and branching enzyme (BE, EC 2.4.1.18) [29,30] (Figure 1).

The first step of the starch biosynthesic pathway is the synthesis of the activated monomer ADPglucose (ADPGlc) from glucose-1-phosphate and ATP, synthesized by ADPGlc PPase. This reaction is the key step for the control of carbon flux through the starch biosynthetic pathway [29,30].

The second step of the starch biosynthesis pathway is the reaction catalyzed by starch synthase, in which the glucosyl moiety of ADPGlc is transferred to the non-reducing end of a pre-existing α-1,4 glucan polymer [10]. To date, five SS isoforms have been described based on sequence similarities: granule-bound SS (GBSS), involved mainly in amylose synthesis and the soluble isoforms: SSI (involved in the synthesis of small chains of amylopectin), SSII and SSIII (with a major role in amylopectin synthesis) and SSIV (recently found to be involved in the control of starch granule number and starch granule initiation) [31-34].

To produce an efficient clustering of the branch points and the formation of crystalline lamella, several debranching enzymes (DBE) are required [35,36]. In addition, the degradation of the crystalline granules depends on a recently discovered group of enzymes – the glucan, water dikinases (GWDs) – which phosphorylate crystalline sections of the granules. Such phosphorylation is catalyzed by two GWD types: the GWD1, involved in the tagging of the glucan chains by C-6 phosphorylation, which is a prerequisite for subsequent C-3 phosphorylation by the second isoform, the GWD3/PWD (glucan, water dikinase 3/phosphoglucan, water dikinase) [37-39]. These enzymes seem to have evolved concomitantly with the appearance of starch deposition [36,40].

A fourth obligatory step in starch biosynthesis has been identified through genetic studies but is poorly understood in terms of the biochemical mechanism that mediates the effect. This step is the cleavage of α-1,6 linkages by isoamylase-type DBE [10,41]. The DBE are crucial for the generation of longer, clustered linear segments in the amylopectin molecule that can crystallize and increase the density of the polysaccharide [42]. Plants contain four DBE genes, three of which are classified as isoamylases on the basis of their sequence homologies and substrate specificities, and one pullulanase-type debranching enzyme [10,43].

While the steps leading to the synthesis of starch are common in most cereals, there are differences in the location and engagement of enzymes, depending on whether the synthesis is in leaf or endosperm (Figure 1).

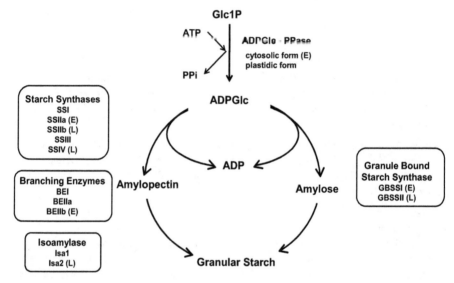

Figure 1. Starch biosynthesis pathway in plants from Glc1P. The scheme indicates the involvement of different isoforms in cereal leaf (L) or endosperm (E). When not specified, the enzymes are dual localized in both compartments (adapted from [10] and [44]).

3. Carbohydrate binding modules

Many of the enzymes involved in the pathway of polysaccharide biosynthesis present a carbohydrate binding domain in its structure. The first carbohydrate binding domain described was a cellulose-binding domain [45-47]; but later it has been found other modules in related enzymes that bind polysaccharides other than cellulose. These findings compelled to redefine the nomenclature of these domains, and now are called CBM (carbohydrate binding module). To date over 300 putative sequences in more than 50 different species have been identified, and binding domains have been classified into 64 families based on amino acid sequence, the substrate binding specificity and structure [48] (see Carbohydrate-Binding Module Family Server, http://afmb.cnrsmrs. fr/~pedro/CAZY/cbm.html).

CBMs have been found in several non-hydrolytic and hydrolytic proteins. Those with hydrolytic activity, such as cellulases, have a complex molecular structure comprised of discrete modules (one catalytic domain and one or more CBMs) that are normally linked by unstructured sequences. The CBMs increase the speed of enzymatic reactions by conducting the catalysis in a close and prolonged physical association with substrates [48] The CBMs present in non-hydrolytic proteins constitute a subunit of the catalytic domain hosts generating cohesive multienzyme complex, which lose enzymatic activity when the CBMs are removed from the structure [48]. Therefore, the CBMs have three general roles with respect to the function of their cognate catalytic modules: (i) a proximity effect, (ii) a targeting function and (iii) a disruptive function [49].

The SBD (starch binding domain) is usually a distinct sequence-structural module that improves the efficiency of an amylolytic enzyme, improving the binding to starch and its hydrolysis. Because this module was first recognised in amylases and thus revealed to cope with raw starch, it was named the raw (granular) starch binding site [50]. At the present, due to the occurrence of SBDs in a wide spectrum of non-amylolytic enzymes, it has become logical to expect a more variable function of these modules. However, there is little evidence that SBD could bind polysaccharides different to starch, although the ability of pure starch binding and degrading seems to be reserved for microorganisms [50,51].

The CBMs have been clasified in ten families based on sequence comparison: (i) CBM20, such as the C-terminal SBD from *Aspergillus niger* glucoamylase; (ii) CBM21, located at the N-terminal domain in amylase proteins; (iii) CBM25, containing one (i.e. β-amylase from *Bacillus circulans*) or two (i.e. *Bacillus sp.* α-amylase) domains; (iv) CBM26, mainly organized in tandem repeats (i.e. C-terminal domains from *Lactobacillus manihotivorans* α-amylase); (v) CBM34, present in the N-terminal domains of neopullulanase, maltogenic amylase and cyclomaltodextrinase; (vi) CBM41, N-terminal SBD, present mostly in bacterial pullulanases; (vii) CBM45, originating from eukaryotic proteins from the plant kingdom (i.e. N-terminal modules of α-glucan water dikinases and α-amylases); (viii) CBM48, which display glycogen-binding properties (including SBD from the GH13 pullulanase and regulatory modules of mammalian AMP-activated protein kinase); (ix) CBM53, SBD modules from SSIII and (x) CBM58, find in α-amylase/neopullulanase of *Bacteroides thetaiotaomicron* showing maltoheptaose binding [52-54] (http://www.cazy.org). This modules becomes important in breaking down the structure of the substrate due to the presence of two polysaccharide-binding sites [55].

Using bioinformatics techniques several SBDs and several sequences exhibiting similarities to SBDs have also been recognised in enzymes and proteins that are not necessarily amylases such as dual-specific phosphatases. These enzymes should deserve special attention because of their participation in various important physiological processes in plants and mammals. It is worth mentioning that in plants these processes concern starch metabolism, whereas in mammals they participate in the metabolism of glycogen [56,57]. The presence of an SBD motif in protein phosphatases reflects their regulatory function since they are involved in polysaccharide metabolism indirectly via modulation of activity of degradative enzymes (i.e. also amylases), such as isoamylase, β-amylase and disproportionating enzyme [85]. In particular, the initial steps of starch degradation at the granule surface are regulated mainly by phosphorylation [50,51]. Furthermore, a starch biosynthetic enzyme, the starch synthase III (SSIII) from *Arabidopsis thaliana* (AtSSIII) has been reported by our group to have a regulatory role in the synthesis of transient starch [33]. This enzyme contains 1025 amino acid residues and has an N-terminal transit peptide for chloroplast localization followed by three in tandem starch-binding domains (SBD D1, D2 and D3, residues 22-591), which bind to raw starch and its individual components, amylose or amylopectin [53,54,58,59]. The adsorption experiments show that the SBD123 region binds preferentially to amylose, and that the D1 domain is mainly responsible for this selective binding. The D2 domain contains two binding sites including amino acid residues Y394 (binding site 1) and W366 (binding site 2) which act in cooperation with the D1 domain in the binding activity while G335 and W340 have a minor role [54]. It is worth mentioning that our work was the first report on the existence of an SBD in a synthesizing enzyme (AtSSIII) and the first experimental evidence of its starch binding capacity.

4. Altering the composition and amount of starch by biotechnological manipulation of enzymes

The alteration of starch quantity and quality can be achieved through the overexpression of some enzymes involved in starch synthesis [60], by mutations or RNAi technology, such as the inhibition of potato SSII, SSIII and GBSS [61], or the decrease in the expression of wheat BEIIa and BEIIb [62,63]. In this way, by affecting the catalytic activity of enzymes involved in the synthesis of amylose or amylopectin, it could be possible to obtain starches for different purposes. Table 3 presents a summary of some of the varieties of plants (transgenic, mutant or silenced by RNAi) that exhibit altered levels of amylose.

The production of high amylose starch is of particular interest because its amount is correlated with the amount of RS in food. Foods with higher content of RS have the potential to improve human health and lower the risk of serious noninfectious diseases. As described above, the amylose content can be increased by the inactivation of the enzymes involved in amylopectin synthesis. In this way, RNAi was used to down-regulate the two different isoforms of starch-branching enzyme (BE) II (BEIIa and BEIIb) in wheat endosperm. Whereas the inhibition of BEIIb expression alone had no effect on amylose content; the decrease of both, BEIIa and BEIIb expression, resulted in the accumulation of starch containing more than 70% of amylose. When this high amylose starch was used to feed rats as a whole meal, it was observed that short-chain fatty acids such as butyrate, propionate and acetate increased with respect to controls. Short chain fatty acids are derived from the anaerobic fermentation of polysaccharides in the large intestine and are important in improving colonic health. These results indicate that this high-amylose wheat has a significant potential to improve human health through its RS content [62].

The decrease of BEIIb enzyme activity in rice is also traditionally associated with elevated amylose content, increased gelatinization temperature, and a decreased proportion of short amylopectin branches. To further elucidate the structural and functional role of this enzyme, the phenotypic effects of down-regulating BEIIb expression in rice endosperm were characterized by Buttardo and coworkers [64] by artificial microRNA (amiRNA) and hairpin RNA (hp-RNA) gene silencing. The results showed that RNA silencing of BEIIb expression in rice grains did not affect the expression of the other major isoforms of BE or SS proteins. The increase in about 2-fold of amylose content was not due to an increase in the relative proportion of amylose chains but instead was due to significantly elevated levels of long and intermediate chains of amylopectin. Rice altered by the amiRNA technique produced a more extreme starch phenotype than those modified using the hp-RNA technique, with a greater increase in the proportion of long and intermediate chains of amylopectin. The major structural modifications of starch produced in the amiRNA lines led to more severe alterations in starch granule morphology and crystallinity as well as digestibility of freshly cooked grains [64].

LINES	AMYLOSE (%)	EVENT	ENZYME INVOLVED
Standard maize	20-30	reference	NA
Sugary-1	37	mutation	Isoamylase
Dull	31	mutation	SSIII
Amylose extender	56	mutation	BEIIb
Indica rice	27	reference	NA
Japonica rice	16	mutation	SSIIa
OsSSIIIa	8	retrotransposon or mutagenesis	SSIIIa
OsSSI	60	mutation	SSI
Standard barley	20-30	reference	NA
Waxy	9	mutation	GBSS
Himalaya 292	71	mutation	SSIIa
BEIIa + BEIIb	75	RNA-silencing	BEIIa/BEIIb
Wheat	18-36	reference	NA
Sgp-1 triple null	31-38	mutation	SSIIa
BEIIa + BEIIb	70	RNAi	BEIIa/BEIIb
Potato	29	reference	NA
BE-II	38	antisense	BEII
BEI + BEII	77-87	antisense	BEI/BEII

Ref: NA, not applicable

Table 3. Amylose content of different lines (wt, mutant and/or transgenic). Adapted from [10].

The roles of BEIIa and BEIIb in defining the structure of amylose and amylopectin were also examined in barley (*Hordeum vulgare*) endosperm. Barley lines with low expression of either BEIIa, BE IIb or both isoforms were generated through RNA-mediated silencing technology. These lines enabled the study of the role of each of these proteins in determining the amylose content, the distribution of chain lengths, and the frequency of branching in both amylose and amylopectin. A high amylose phenotype (> 70%) was observed in lines expressing lower levels of BEIIa and BEIIb, while a reduction in the expression of either of these isoforms alone had minor impact on amylose content. The structure and properties of the barley high amylose starch resulting from the decrease in the expression of both BEII isoforms were found to be similar to those observed in amylose mutants of maize, which result from mutations that decrease the expression of the BEIIb gene. The analysis of amylopectin chain length distribution indicated that both BEIIa and BEIIb isoforms have distinct roles in determining the fine structure of amylopectin. A significant reduction in the frequency of branches in amylopectin was observed only when both BEIIa and BEIIb were reduced, whereas there was a significant increase in the branching frequency of amylose when BEIIb alone was reduced [61,65].

Other way of modifying amylase content is by SS expression. Amylose and amylopectin of rice mutants deficient in endosperm SS isoforms, either SSI (ΔSSI) or SSIIIa (ΔSSIIIa), were found to have an altered structure respect to to their parent (cv. *Nipponbare*, Np). The amylose content was higher in the mutants (Np, 15.5%; ΔSSI, 18.2%; ΔSSIIIa, 23.6%), and the molar ratio of branched amylose and its side chains was increased. In addition, the chain-length distribution of the β-amylase limit dextrins of amylopectin showed high regularity, which is consistent with the reported cluster structure. The mole % of the B(1)-B(3) fractions was changed slightly in ΔSSI, which is consistent with the proposed role of SSI in elongating the external part of clusters. In ΔSSIIIa, it has been observed a significant increase in the B(1) fraction and a decrease in both, the B(2) and B(3) fractions. The internal chain length of the B(2) and B(3) fractions appeared to be slightly altered, suggesting that the deficiency in SS affected the actions of branching enzyme(s) [66].

In another approach, SSIIIa null mutants of rice (*Oryza sativa*) were generated using retrotransposon insertion and chemical mutagenesis. The amylopectin B(2) to B(4) chains with degree of polymerization (DP) >/= 30 and the M(r) of amylopectin were reduced to about 60% and 70% in the mutants, suggesting that SSIIIa plays an important role in the elongation of amylopectin B(2) to B(4) chains. Chains with DP 6 to 9 and DP 16 to 19 decreased while chains with DP 10 to 15 and DP 20 to 25 increased in the amylopectin mutants. These changes in the SSIIIa mutants are almost opposite images of those of SSI-deficient rice mutant and were caused by 1.3- to 1.7-fold increase of the amount of SSI in the mutant endosperm. Furthermore, the amylose content and the extralong chains (DP >/= 500) of amylopectin were increased by 1.3- and 12-fold, respectively. These changes of starch composition of the mentioned mutants are due to the increase in about 1.7-fold of GBSSI activity. The starch granules of the mutants were found to be smaller with round shape and less crystalline. Thus, SSIIIa deficiency, the second major SS isoforrm in developing rice endosperm, affected either the structure of amylopectin, amylase content, and also the physicochemical properties of starch granules in two ways: directly by the SSIIIa deficiency itself and indirectly by up-regulation of both SSI and GBSSI mRNA [67].

By a different approach Safford et al [68] reported no effect on the amylose content of potato starch after the downregulation of the expression of the major branching enzyme isozyme (BE). However, a notable increase (50 – 100%) of the phosphorous content was detected.

Although the almost complete suppression of the branching enzyme activity (less than 5% respect to wt levels) in transgenic potato tubers, no changes in amylose content of the starches derived from these transgenic lines were detected. Differences in the gelatinization properties (an increase of up to 5°C in the peak temperature and viscosity onset temperature) are reported, suggesting that these changes correlated with the branching pattern of the starch that result in changes of the double helix length. It is also possible that the increased phosphate content observed in the transgenic starches resulted in the elevation of the gelatinization temperature [68].

Other strategy to obtain high amylose starches was carried out by Itoh et al [60]. The Waxy (Wx) gene encodes a granule-bound starch synthase (GBSS) that plays a key role in the amylose synthesis of rice and other plant species. In rice, it has been described two functional Wx alleles: Wx(a), which produces higher amounts of amylose, and Wx(b), which produces low amounts of this polymer due to a mutation in the 5' splicing site of intron 1. When the Wx(a) cDNA was introduced into null-mutant Japonica rice (wx) the amylose content were 6-11% higher than that of the original cultivar, Labelle, which carries the Wx(a) allele, although the levels of the Wx protein in the transgenic rice were equal to those of cv. Labelle [60].

Finally, using *A. thaliana* null mutant lines for the SSIII locus, it has been postulated that SSIII has a regulatory role in the starch synthesis process [33]. These mutant lines show a higher accumulation of leaf starch during the day due to an apparent increase in biosynthetic rate. Besides, starch granules show physical alterations and higher phosphate content [33]. These data suggest that SSIII might have a negative regulatory role in starch synthesis. Previously, SSIII had been associated to a starch-excess phenotype, although indirectly through its association with regulatory proteins such as 14-3-3 [69]. In addition other SS isoform, SSIV, has been described to be essential for the initiation process of starch granule synthesis since *A. thaliana* SSIV mutant plants show just one large starch granule per plastid. The role of this isoform in the formation of the starch granule could be replaced in part by the SSIII isoform since the concomitant elimination of both enzymes in Arabidopsis block the starch synthesis. These data suggests that the remaining synthase activities are unable to start the synthesis of the starch granule. Recently, SSIV has been postulated to be also involved in the regulation of starch accumulation since its overexpression increases the starch levels in Arabidopsis leaves by 30%–40%. In addition, SSIV-overexpressing lines display a higher growth rate. The increase in starch content as a consequence of enhanced SSIV expression is also observed in long-term storage starch organs such as potato tubers [70].

5. Use of carbohydrate-binding modules to change amylose - amylopectin ratio and obtaining of modified starches.

In the past few years the search for different strategies in order to produce starches with new properties was intensified. One of these strategies is to evaluate the possibility whether the microbial starch binding domains (SBDs) could be used as a universal tool for starch modification in plant biotechnology.

It has been reported that SBDs are also present in microbial starch degrading enzymes. As mentioned above, one of the functions of SBD is to attach amylolytic enzymes to the insoluble starch granule. The amino acid sequences of these modules are very well conserved among different enzymes (i.e. glucoamylase, α-amylase, β-amylase, etc.), as well as among different species such as *Clostridium thermosulfurogenes, Bacillus circulans, Aspergillus niger, Klebsiella pneumonia, Streptomyces limosus, Pseudomonas stutzeri,* etc. [50,71-73]. Several studies have shown that these enzymes lose (most of) their catalytic activity towards raw starch granules upon removal of the SBD, whereas their activity on soluble substrates remains unaltered. Besides their affinity for starch granules, SBDs can also bind maltodextrins and cyclodextrins [71]. Ji et al [72], explored the possibility of engineering artificial granule-bound proteins, which can be incorporated in the granule during biosynthesis. The SBD-encoding region of cyclodextrin glycosyltransferase from *B. circulans* was fused to the sequence encoding the transit peptide (amyloplast entry) of potato GBSSI. The synthetic gene was expressed in the tubers of two potato cultivars and one amylose-free (amf) potato mutant. The results showed that SBDs are accumulated inside starch granules, not at the granule surface and amylose-free granules contained 8 times more SBD than the amylose-containing ones. However, no consistent differences in physicochemical properties between transgenic SBD starches and their corresponding controls were found, suggesting that SBD can be used as an anchor for effector proteins without having side-effects [72].

On the other hand it was also evaluated whether is it possible to produce an amylose-free potato starch by displacing GBSSI, from the starch granule by engineering multiple-repeat CBM20 SBD (two, three, four and five). The constructs were introduced in wild type potato cultivar, and the starches of the resulting transformants were compared with those expressing amf potato clones. The amount of SBDs accumulated in starch granules was increased progressively from SBD to SBD3 and not when were used SBD4 and SBD5; however, a reduction in amylose content was not achieved in any of the transformants. It was shown that SBDn expression can affect the physical process underlying granule assembly in both potato genetic backgrounds, without altering the primary structure of the constituent starch polymers and the granule melting temperature. Granule size distribution of the starches obtained from transgenic plants was similar to untransformed controls, irrespective of the amount of SBDn accumulated. In the amf background, granule size is severely affected [74].

In the case of starches which require chemical modifications to enhance their properties, such as the improved stability in solution by acetylation, a drawback is generated when pollutant chemicals are used. A biological alternative to the derivatization process was investigated by the expression of an amyloplast-targeted *Escherichia coli* maltose acetyltransferase (MAT) in tubers of wild-type and mutant amf potato plants. MAT was expressed alone, or fused in its N- or C-terminus to a SBD to be target to the starch granule. Starch granules derived from transgenic plants contained acetyl groups in low number. In addition, MAT protein on the starch granules present catalytic activity even after post-harvesting, when supplied with glucose or maltose and acetyl-coenzyme A, but it was not able to acetylate starch polymers in vitro. Starch granules from transformants where MAT was expressed alone also showed MAT catalytic activity, indicating that MAT is accumulated in starch granules, and could bind to the polymer without the presence of any SBD. Furthermore, the fusion of MAT and SBD affects granule morphology: in potato transformants, the percentage of altered granules when the SBD was located at the C-terminal end correlated with the amount of fusion protein accumulated. When SBD was located at the N-terminus of MAT or it is absent, no differences were found respect to the untransformed controls, indicating that not only is the simultaneous presence of SBD and MAT important for altering granule morphology, but also their localization in the fusion protein [75].

Another approach to obtain modified starches involves the bacterial glucansucrases [76]. Certain bacteria possess an array of enzymes, so-called glucansucrases, which can attach (contiguous) 1,6-linked or 1,3-linked glucosyl residues to maltodextrins. This, together with the presence of sucrose inside the potato tuber amyloplast [77], suggests that glucansucrases are of great interest for diversifying starch structure. With few exceptions, glucansucrases are extracellular enzymes, which are produced by lactic acid bacteria such as *Leuconostoc mesenteroides*, oral *Streptococci* and some species of *Lactococcus* and *Lactobacillus* [78]. The glucansucrases catalyze the polymerization of glucose residues from sucrose, which leads to the production of a large variety of α-glucans with different sizes and structures, and composed of diverse linkage types. Most glucansucrases share a common structure composed of four different regions: a signal peptide, a variable region, a catalytic domain, and a glucan-binding domain (GBD) [76].

Production of water-insoluble mutan polymers in wild type potato tubers was investigated by Kok-Jacon et al (2005) after expression of full-length GTFI (mutansucrase) and a truncated version without glucan-binding domain from *Streptococcus downei*. Mutan polymers are bacterial polysaccharides that are secreted by oral microorganisms and have adhesive properties and different degrees of water-solubility [81]. They account for about 70% of the carbohydrates present in dental plaque [79] in addition to dextrans and levans [80]. When the short form of the protein was expressed, low amounts of mutan polymer attached to the starch granules has been detected. Besides, these plants exhibited severely altered tuber phenotype and starch granule morphology in comparison to those expressing the full-length GTFI gene, whereas no changes at the starch level were observed. Finally, the rheological properties of the starch obtained from plants expressing the truncated protein were also altered, showing a higher retrogradation during cooling of the starch paste [80].

Subsequently, the same group of investigators fused the truncated form of a mutansucrase (without glucan binding domain) to an N- or C- terminal SBD. The different enzymes were introduced into two genetically different potato backgrounds (wild type and amf lines), in order to attach the enzyme to the growing starch granules, and to facilitate the incorporation of mutan polymers in starch. Starches from the chimeric transformants seemed to contain less amounts of mutan than those from plants expressing the mutansucrase alone, suggesting that SBD might inhibit the catalytic activity of the enzyme. Scanning electron microscopy showed that expression of SBD-mutansucrase fusion proteins resulted in alterations of granule morphology in both genetic backgrounds. Surprisingly, the amf starches containing the chimeric form had a spongeous appearance, as the granule surface contained many small holes and grooves, indicating that this fusion protein can interfere with the lateral interactions of amylopectin sidechains. No differences in physicochemical properties of the transgenic starches were observed [82].

Finally, all the knowledge gained about the characteristics, structure, function and occurrence of SBD and GBD will support current and future experimental research. Since SBD are domains which retain their structural fold and functional properties independently of the remaining parts of the protein molecule including the catalytic domain, they can be applied in various fields of biotechnology [48,83-86]. It is important to note that most of the applications have involved only the CBM20 SBD. One of the most attractive fields is represented by starch processing in the food industry, especially the hydrolysis of starch into maltodextrins and maltooligosaccharides [87]. Since conventional processes require starch gelatinization at elevated temperature and thus use of thermostable amylolytic enzymes [88], the possibility of carrying out the processes without gelatinization, by utilizing new enzymes with attached SBD is desirable [52,89,90].

6. Conclusions

Food production in terms of quality and quantity, as well as for new plants commodities and products in developed and developing countries, cannot based only on classical agriculture [91]. The metabolic engineering of plants has yielded remarkable results by increasing the production of minor components (essential oils, vitamin A, vitamin E and flavonoids) and, as well as the composition of major components, such as starch or fatty acids [92]. The improvement in the food we eat is necessary and crucial in societies that have bad eating habits. The health benefits provided by the intake of resistant starches have been properly tested and it will be desirable that these kinds of starches could be incorporated into the human diet. Molecular tools available at the present and those likely to be developed in the near future, will enable the development of new strategies to increase the content of resistant starch in grains and other vegetables. Manipulation of the starch synthesis pathway through the modification of enzymes belonging to this route, and the use of CBM (and specifically SBD) of both microbial and plant, are alternatives that are desirable to explore in more detail.

Author details

María Victoria Busi and Diego F. Gomez-Casati
Centro de Estudios Fotosintéticos y Bioquímicos (CEFOBI-CONICET),
Suipacha 570, Rosario, Argentina
IIB-INTECH, Universidad Nacional de General San Martín (UNSAM),
San Martín, Buenos Aires, Argentina

Mariana Martín
Centro de Estudios Fotosintéticos y Bioquímicos (CEFOBI-CONICET),
Universidad Nacional de Rosario, Suipacha 570, Rosario, Argentina

Acknowledgement

Our work is supported in part by grants from the Biotechnology Program from Universidad Nacional de General San Martin (UNSAM) (PROG07F / 2-2007), Consejo Nacional de Investigaciones Científicas y Técnicas (CONICET, PIP 00237) and Agencia Nacional de Promoción Científica y Tecnológica (ANPCyT, PICT 2010 – 0543 and PICT 2010 – 0069). MVB, MM and DGC are research members from CONICET.

7. References

[1] Halpin, C. (2005). Gene stacking in transgenic plants--the challenge for 21st century plant biotechnology. Plant Biotechnol J 3, 141-55.

[2] Cassidy, A. (2004) Serials, Nuts and Pulses. In Plants: Diet and health (Goldberg, G., ed.), pp. 134-146. Blackwell Sciences, Oxford, UK.

[3] Nugent, A.P. (2005). Health properties of resistant starch. Nutrition Bull 30

[4] Zhao, F.J. and Shewry, P.R. (2010). Recent developments in modifying crops and agronomic practice to improve human health Food Policy

[5] Lopez, X. (2009). Obesity: How did resistant starches might contribute? Enfasis Alim 2, 32-38.

[6] Dominguez Vergara, A.M., Vazquez-Moreno, L. and Ramos-Clamont Mon Fort, G. (2009). Review of the role of prebiotic oligosaccharides in the prevention of gastrointestinal infections. Arch Latin Nutr 59, 358-368.

[7] Topping, D.L., Fukushima, M. and Bird, A.R. (2003). Resistant starch as a prebiotic and synbiotic: state of the art. Proc Nutr Soc 62, 171-6.

[8] Robertson, M.D. (2012). Dietary-resistant starch and glucose metabolism. Curr Opin Clin Nutr Metab Care

[9] Shamaia, K., Bianco-Peled, H. and Shimonic, E. (2003). Polymorphism of resistant starch type III Carbohydr Polym 54, 363-369.

[10] Morell, M.K., Konik-Rosse, C., Ahmed, R., Li, Z. and Rahman, S. (2004). Synthesis of resistant starch in plants. J AOAC Int 87, 740-748.

[11] Bird, A.R., Flory, C., Davies, D.A., Usher, S. and Topping, D.L. (2004). A novel barley cultivar (Himalaya 292) with a specific gene mutation in starch synthase IIa raises large bowel starch and short-chain fatty acids in rats. J Nutr 134, 831-5.

[12] Higgins, J.A., Higbee, D.R., Donahoo, W.T., Brown, I.L., Bell, M.L. and Bessesen, D.H. (2004). Resistant starch consumption promotes lipid oxidation. Nutr Metab (Lond) 1, 8.

[13] Farhat, I.A., Protzmann, J., Becker, A., Valles-Pamies, B., Neale, R. and Hill, S.E. (2001). Effect of the extent of conversion and retrogradation on the digestibility of potato starch. Starch/Stärke 53, 431-436.

[14] Sivak, M. and Preiss, J. (1998). Advances in Food Nutrition Research. Academic Press, San Diego, California. 41

[15] Alonso, A.G., Calixto, F.S. and Delcour, J.A. (1998). Influence of botanical source and proccesing on formation of resistant starch type III. Cereal Chem 75, 802-804.

[16] Thompson, D.B. (2000). Strategies for the manufacture of resistant starch. Trends Food Sci Technol 11, 245-253.

[17] Jenner, H.L. (2003). Transgenesis and yield: what are our targets? Trends Biotechnol 21, 190-2.

[18] Willmitzer, L. (1999). Plant Biotechnology: Output traits-the second generation of plant biotechnology products is gaining momentum. Curr Opin Biotechnol 10, 161-162.

[19] Morell, M.K. et al. (2003). Barley sex6 mutants lack starch synthase IIa activity and contain a starch with novel properties. Plant J 34, 173-85.

[20] Blauth, S.L., Kim, K.N., Klucinec, J., Shannon, J.C., Thompson, D. and Guiltinan, M. (2002). Identification of Mutator insertional mutants of starch-branching enzyme 1 (sbe1) in Zea mays L. Plant Mol Biol 48, 287-97.

[21] Gao, M., Fisher, D.K., Kim, K.N., Shannon, J.C. and Guiltinan, M.J. (1997). Independent genetic control of maize starch-branching enzymes IIa and IIb. Isolation and characterization of a Sbe2a cDNA. Plant Physiol 114, 69-78.

[22] Roach, P.J. (2002). Glycogen and its metabolism. Curr Mol Med 2, 101-20.

[23] Dauvillee, D. et al. (2009). Genetic dissection of floridean starch synthesis in the cytosol of the model dinoflagellate Crypthecodinium cohnii. Proc Natl Acad Sci U S A 106, 21126-30.

[24] Cavalier-Smith, T. (2009). Predation and eukaryote cell origins: a coevolutionary perspective. Int J Biochem Cell Biol 41, 307-22.

[25] Patron, M.J. and Keeling, P. (2005). Common evolutionary origin of starch biosynthetic enzymes in green and red algae. J Phycol 41, 1131-1141.

[26] Deschamps, P., Moreau, H., Worden, A.Z., Dauvillee, D. and Ball, S.G. (2008). Early gene duplication within chloroplastida and its correspondence with relocation of starch metabolism to chloroplasts. Genetics 178, 2373-87.

[27] Deschamps, P., Haferkamp, I., d'Hulst, C., Neuhaus, H.E. and Ball, S.G. (2008). The relocation of starch metabolism to chloroplasts: when, why and how. Trends Plant Sci 13, 574-82.

[28] Tetlow, I.J. (2011). Starch biosynthesis in developing seeds. Seed Sci Res 21, 5-32.

[29] Smith, A.M., Denyer, K. and Martin, C. (1997). The Synthesis of the Starch Granule. Annu Rev Plant Physiol Plant Mol Biol 48, 67-87.

[30] Ball, S.G. and Morell, M.K. (2003). From bacterial glycogen to starch: understanding the biogenesis of the plant starch granule. Annu Rev Plant Biol 54, 207-33.

[31] Delvalle, D. et al. (2005). Soluble starch synthase I: a major determinant for the synthesis of amylopectin in Arabidopsis thaliana leaves. Plant J 43, 398-412.

[32] Maddelein, M.L. et al. (1994). Toward an understanding of the biogenesis of the starch granule. Determination of granule-bound and soluble starch synthase functions in amylopectin synthesis. J Biol Chem 269, 25150-7.

[33] Zhang, X., Myers, A.M. and James, M.G. (2005). Mutations affecting starch synthase III in Arabidopsis alter leaf starch structure and increase the rate of starch synthesis. Plant Physiol 138, 663-74.

[34] Delrue, B. et al. (1992). Waxy Chlamydomonas reinhardtii: monocellular algal mutants defective in amylose biosynthesis and granule-bound starch synthase activity accumulate a structurally modified amylopectin. J Bacteriol 174, 3612-20.

[35] Wattebled, F., Planchot, V., Dong, Y., Szydlowski, N., Pontoire, B., Devin, A., Ball, S. and D'Hulst, C. (2008). Further evidence for the mandatory nature of polysaccharide debranching for the aggregation of semicrystalline starch and for overlapping functions of debranching enzymes in Arabidopsis leaves. Plant Physiol 148, 1309-23.

[36] Coppin, A. et al. (2005). Evolution of plant-like crystalline storage polysaccharide in the protozoan parasite Toxoplasma gondii argues for a red alga ancestry. J Mol Evol 60, 257-67.

[37] Ritte, G., Heydenreich, M., Mahlow, S., Haebel, S., Kotting, O. and Steup, M. (2006). Phosphorylation of C6- and C3-positions of glucosyl residues in starch is catalysed by distinct dikinases. FEBS Lett 580, 4872-6.

[38] Baunsgaard, L., Lutken, H., Mikkelsen, R., Glaring, M.A., Pham, T.T. and Blennow, A. (2005). A novel isoform of glucan, water dikinase phosphorylates pre-phosphorylated alpha-glucans and is involved in starch degradation in Arabidopsis. Plant J 41, 595-605.

[39] Kotting, O., Pusch, K., Tiessen, A., Geigenberger, P., Steup, M. and Ritte, G. (2005). Identification of a novel enzyme required for starch metabolism in Arabidopsis leaves. The phosphoglucan, water dikinase. Plant Physiol 137, 242-52.

[40] Kotting, O., Kossmann, J., Zeeman, S.C. and Lloyd, J.R. (2010). Regulation of starch metabolism: the age of enlightenment? Curr Opin Plant Biol 13, 321-9.

[41] Myers, A.M., Morell, M.K., James, M.G. and Ball, S.G. (2000). Recent progress toward understanding biosynthesis of the amylopectin crystal. Plant Physiol 122, 989-97.

[42] Blennow, A. and Svensson, B. (2010). Dynamics of starch granule biogenesis -the role of redox regulated enzymes- and low-affinity carbohydrate binding modules. Biocatal Biotransf 28, 3-9.

[43] Rahman, S. et al. (2003). The sugary-type isoamylase gene from rice and Aegilops tauschii: characterization and comparison with maize and arabidopsis. Genome 46, 496-506.

[44] Morell, M.K. and Myers, A.M. (2005). Towards the rational design of cereal starches. Curr Opin Plant Biol 8, 204-10.

[45] Gilkes, N.R., Warren, R.A., Miller, R.C., Jr. and Kilburn, D.G. (1988). Precise excision of the cellulose binding domains from two Cellulomonas fimi cellulases by a homologous protease and the effect on catalysis. J Biol Chem 263, 10401-7.

[46] Tomme, P., Driver, D.P., Amandoron, E.A., Miller, R.C., Jr., Antony, R., Warren, J. and Kilburn, D.G. (1995). Comparison of a fungal (family I) and bacterial (family II) cellulose-binding domain. J Bacteriol 177, 4356-63.

[47] Tomme, P., Warren, R.A. and Gilkes, N.R. (1995). Cellulose hydrolysis by bacteria and fungi. Adv Microb Physiol 37, 1-81.

[48] Shoseyov, O., Shani, Z. and Levy, I. (2006). Carbohydrate binding modules: biochemical properties and novel applications. Microbiol Mol Biol Rev 70, 283-95.

[49] Boraston, A.B., Bolam, D.N., Gilbert, H.J. and Davies, G.J. (2004). Carbohydrate-binding modules: fine-tuning polysaccharide recognition. Biochem J 382, 769-81.

[50] Machovic, M. and Janecek, S. (2006). Starch-binding domains in the post-genome era. Cell Mol Life Sci

[51] Machovic, M. and Janecek, S. (2006). The evolution of putative starch-binding domains. FEBS Lett 580, 6349-56.

[52] Janecek, S., Svensson, B. and MacGregor, E.A. (2011). Structural and evolutionary aspects of two families of non-catalytic domains present in starch and glycogen binding proteins from microbes, plants and animals. Enzyme Microb Technol 49, 429-440.

[53] Wayllace, N.Z., Valdez, H.A., Ugalde, R.A., Busi, M.V. and Gomez-Casati, D.F. (2010). The starch-binding capacity of the noncatalytic SBD2 region and the interaction between the N- and C-terminal domains are involved in the modulation of the activity of starch synthase III from Arabidopsis thaliana. Febs J 277, 428-40.

[54] Valdez, H.A., Peralta, D.A., Wayllace, N.Z., Grisolía, M.J., Gomez-Casati, D.F. and Busi, M.V. (2011). Preferential binding of SBD from Arabidopsis thaliana SSIII to polysaccharides. Study of amino acid residues involved. Starch/Stärke 63, 451-460.

[55] Southall, S.M., Simpson, P.J., Gilbert, H.J., Williamson, G. and Williamson, M.P. (1999). The starch-binding domain from glucoamylase disrupts the structure of starch. FEBS Lett 447, 58-60.

[56] Kerk, D., Conley, T.R., Rodriguez, F.A., Tran, H.T., Nimick, M., Muench, D.G. and Moorhead, G.B. (2006). A chloroplast-localized dual-specificity protein phosphatase in Arabidopsis contains a phylogenetically dispersed and ancient carbohydrate-binding domain, which binds the polysaccharide starch. Plant J 46, 400-13.

[57] Roma-Mateo, C. et al. (2011). Laforin, a dual specificity protein phosphatase involved in Lafora disease, is phosphorylated at Ser25 by AMP-activated protein kinase. Biochem J

[58] Palopoli, N., Busi, M.V., Fornasari, M.S., Gomez-Casati, D., Ugalde, R. and Parisi, G. (2006). Starch-synthase III family encodes a tandem of three starch-binding domains. Proteins 65, 27-31.

[59] Valdez, H.A., Busi, M.V., Wayllace, N.Z., Parisi, G., Ugalde, R.A. and Gomez-Casati, D.F. (2008). Role of the N-terminal starch-binding domains in the kinetic properties of starch synthase III from Arabidopsis thaliana. Biochemistry 47, 3026-32.

[60] Itoh, K., Ozaki, H., Okada, K., Hori, H., Takeda, Y. and Mitsui, T. (2003). Introduction of Wx transgene into rice wx mutants leads to both high- and low-amylose rice. Plant Cell Physiol 44, 473-80.

[61] Jobling, S.A., Westcott, R.J., Tayal, A., Jeffcoat, R. and Schwall, G.P. (2002). Production of a freeze-thaw-stable potato starch by antisense inhibition of three starch synthase genes. Nat Biotechnol 20, 295-9.

[62] Regina, A. et al. (2006). High-amylose wheat generated by RNA interference improves indices of large-bowel health in rats. Proc Natl Acad Sci U S A 103, 3546-51.

[63] Sestili, F., Janni, M., Doherty, A., Botticella, E., D'Ovidio, R., Masci, S., Jones, H.D. and Lafiandra, D. (2010). Increasing the amylose content of durum wheat through silencing of the SBEIIa genes. BMC Plant Biol 10, 144.

[64] Butardo, V.M. et al. (2011). Impact of down-regulation of starch branching enzyme IIb in rice by artificial microRNA- and hairpin RNA-mediated RNA silencing. J Exp Bot 62, 4927-41.

[65] Regina, A., Kosar-Hashemi, B., Ling, S., Li, Z., Rahman, S. and Morell, M. (2010). Control of starch branching in barley defined through differential RNAi suppression of starch branching enzyme IIa and IIb. J Exp Bot 61, 1469-82.

[66] Hanashiro, I., Higuchi, T., Aihara, S., Nakamura, Y. and Fujita, N. (2011). Structures of starches from rice mutants deficient in the starch synthase isozyme SSI or SSIIIa. Biomacromolecules 12, 1621-8.

[67] Fujita, N. et al. (2007). Characterization of SSIIIa-deficient mutants of rice: the function of SSIIIa and pleiotropic effects by SSIIIa deficiency in the rice endosperm. Plant Physiol 144, 2009-23.

[68] Safford, R.E. et al. (1998). Consequences of antisense RNAi of starch branching enzyme activity on properties of potato starch. Carbohydr Polym 35, 155-160.

[69] Sehnke, P.C., Chung, H.J., Wu, K. and Ferl, R.J. (2001). Regulation of starch accumulation by granule-associated plant 14-3-3 proteins. Proc Natl Acad Sci U S A 98, 765-70.

[70] Szydlowski, N. et al. (2009). Starch granule initiation in Arabidopsis requires the presence of either class IV or class III starch synthases. Plant Cell 21, 2443-57.

[71] Svensson, B., Jespersen, H., Sierks, M.R. and MacGregor, E.A. (1989). Sequence homology between putative raw-starch binding domains from different starch-degrading enzymes. Biochem J 264, 309-11.

[72] Ji, Q., Vincken, J.P., Suurs, L.C. and Visser, R.G. (2003). Microbial starch-binding domains as a tool for targeting proteins to granules during starch biosynthesis. Plant Mol Biol 51, 789-801.

[73] Janecek, S. and Sevcik, J. (1999). The evolution of starch-binding domain. FEBS Lett 456, 119-25.

[74] Firouzabadi, F.N., Vincken, J.P., Ji, Q., Suurs, L.C., Buleon, A. and Visser, R.G. (2007). Accumulation of multiple-repeat starch-binding domains (SBD2-SBD5) does not reduce amylose content of potato starch granules. Planta 225, 919-33.

[75] Nazarian Firouzabadi, F., Vincken, J.P., Ji, Q., Suurs, L.C. and Visser, R.G. (2007). Expression of an engineered granule-bound Escherichia coli maltose acetyltransferase in wild-type and amf potato plants. Plant Biotechnol J 5, 134-45.

[76] Kok-Jacon, G.A., Ji, Q., Vincken, J.P. and Visser, R.G. (2003). Towards a more versatile alpha-glucan biosynthesis in plants. J Plant Physiol 160, 765-77.

[77] Gerrits, N., Turk, S.C., van Dun, K.P., Hulleman, S.H., Visser, R.G., Weisbeek, P.J. and Smeekens, S.C. (2001). Sucrose metabolism in plastids. Plant Physiol 125, 926-34.

[78] Robyt, J.F. (1995). Mechanisms in the glucansucrase synthesis of polysaccharides and oligosaccharides from sucrose. Adv Carbohydr Chem Biochem 51, 133-68.

[79] Loesche, W.J. (1986). Role of Streptococcus mutans in human dental decay. Microbiol Rev 50, 353-80.

[80] Kok-Jacon, G.A., Vincken, J.P., Suurs, L.C. and Visser, R.G. (2005). Mutan produced in potato amyloplasts adheres to starch granules. Plant Biotechnol J 3, 341-51.

[81] Sutherland, I. (2001). Biofilm exopolysaccharides: a strong and sticky framework. Microbiology 147, 3-9.

[82] Nazarian Firouzabadi, F., Kok-Jacon, G.A., Vincken, J.P., Ji, Q., Suurs, L.C. and Visser, R.G. (2007). Fusion proteins comprising the catalytic domain of mutansucrase and a starch-binding domain can alter the morphology of amylose-free potato starch granules during biosynthesis. Transgenic Res 16, 645-56.

[83] Juge, N. et al. (2006). The activity of barley alpha-amylase on starch granules is enhanced by fusion of a starch binding domain from Aspergillus niger glucoamylase. Biochim Biophys Acta 1764, 275-84.

[84] Ohdan, K., Kuriki, T., Takata, H., Kaneko, H. and Okada, S. (2000). Introduction of raw starch-binding domains into Bacillus subtilis alpha-amylase by fusion with the starch-binding domain of Bacillus cyclomaltodextrin glucanotransferase. Appl Environ Microbiol 66, 3058-64.

[85] Latorre-Garcia, L., Adam, A.C., Manzanarco, P. and Polaina, J. (2005). Improving the amylolytic activity of Saccharomyces cerevisiae glucoamylase by the addition of a starch binding domain. J Biotechnol 118, 167-76.

[86] Christiansen, C., Abou Hachem, M., Janecek, S., Vikso-Nielsen, A., Blennow, A. and Svensson, B. (2009). The carbohydrate-binding module family 20--diversity, structure, and function. Febs J 276, 5006-29.

[87] Van Geel-Schutten, G.H. et al. (1999). Biochemical and structural characterization of the glucan and fructan exopolysaccharides synthesized by the lactobacillus reuteri wild-type strain and by mutant strains. Appl Environ Microbiol 65, 3008-14.

[88] Bertoldo, C. and Antranikian, G. (2002). Starch-hydrolyzing enzymes from thermophilic archaea and bacteria. Curr Opin Chem Biol 6, 151-60.

[89] Vikso-Nielsen, A., Andersen, C., Hoff, T. and Pedersen, S. (2006). Development of new alpha-amylases for raw starch hydrolysis. Biocatal Biotransf 24, 121-127.

[90] Wang, P., Singh, V., Xue, H., Johnston, D.B., Rausch, K.D. and Tumbleson, M.E. (2007). Comparisson of raw starch hydrolysis enzyme with conventional liquetfaction and saccharification enzymes in dry-grind corn proccesing. Cereal Chem 84, 10-14.

[91] Altman, A. (1999). Plant Biotechnology in the 21st century: the challenges ahead. Electr J Biotechnol 2, 51-55.

[92] Morandini, P. (2009). Rethinking metabolic control. Plant Sci 176, 441-451.

Applications in the Pharmaceutical Industry

Chapter 6

1,3-β-Glucans: Drug Delivery and Pharmacology

Mohit S. Verma and Frank X. Gu

Additional information is available at the end of the chapter

1. Introduction

Natural polysaccharides are used in a variety of applications due to their unique properties. These applications range from paper manufacturing to wound healing [1]. One interesting class of polysaccharides comprises 1,3-β-glucans, which are glucopyranose polysaccharides with (1,3) glycosidic linkages and varying degree of (1,6) branches [2]. 1,3-β-glucans can form single or triple helical structures, which can be used to synthesize resilient gels by applying heat and humidity [3,4]. The properties of these gels are governed by the structure of the polysaccharide, which is determined by the degree of branching and the molecular weight. Thus, controlling the microscopic structure allows control over the macroscopic function. This is especially advantageous in the field of drug delivery because the encapsulation of different agents can be facilitated by the use of different polysaccharides. The properties of 1,3-β-glucans can also be modified by covalently attaching functional units to the polysaccharide backbone [5].

1,3-β-glucans are derived from microbial [6] and fungal [2] sources and hence have innate immunomodulatory properties. When these 1,3-β-glucans are a component of the foreign pathogens, they can act as recognition sites for macrophages to facilitate the elimination and removal of these pathogens [7]. When extracted 1,3-β-glucans are administered to animals or humans, they recruit macrophages and stimulate the immune system through a similar mechanism [8,9]. This result has been utilized for various pharmacological applications including cancer inhibition [10-17], infection resistance [18-21] and wound healing [22-24]. Current research is focusing on combining the structural properties of 1,3-β-glucans with the pharmacological ones to further enhance the efficacy of hybrid systems thus created.

2. Crystallinity of 1,3-β-glucans

2.1. Structure

1,3-β-glucans are semi-crystalline polysaccharides comprising a combination of single helices, triple helices and random coils. The crystallinity of these polysaccharides has been studied using X-ray diffraction (XRD). In this study, curdlan, which is a linear 1,3-β-glucan, was used as a model polysaccharide [3]. Different forms and states of curdlan demonstrate different crystallinity. One example that was studied in detail is the annealed "dry" state, where the curdlan is dissolved in dimethyl sufoxide, extracted in methanol and annealed in the presence of water at 145 °C. Curdlan is then dried *in vacuo* to obtain the sample for XRD experiments. The results from XRD measurements conclude that six-fold triple-helices are formed with an advance of 2.935 Å per monomer unit. The model of this structure confirms that the three strands of triple helices are held together by hydrogen bonding between O(2) hydroxyls while the helices are brought together by O(4) and O(6) hydrogen bonding [3].

An alternate structure of curdlan helices is presented based on semi-empirical modeling. It is proposed that hydrogen bonding of the strands occurs along the helix axis rather than perpendicular to it. The different structures are illustrated in Figure 1. It is demonstrated that this alternate structure provides a more stable structure of curdlan and hence is likely to have higher population [25].

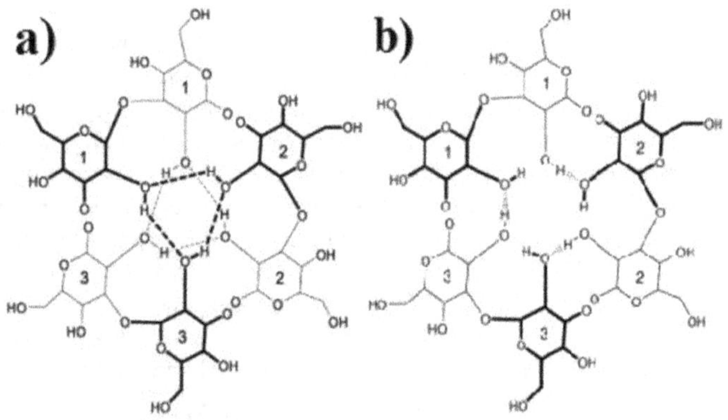

Figure 1. Illustration of possible orientations of curdlan triple helices: a) hydrogen bonding perpendicular to helix axis; b) hydrogen bonding along helix axis. Figure adapted from [25].

1,3-β-glucans have also been complexed with nucleotides to form crystalline structures. In the example of curdlan and poly(cytidylic acid) complex, semi-empirical modeling suggests that two glucose units of different curdlan chains form hydrogen bonds with one base of the nucleotide chain [26]. This property of curdlan complexing with nucleotides has been exploited in forming liquid crystalline gels with deoxyribonucleic acid (DNA). Such structures could be synthesized at varying scales ranging from nanometers to centimeters [27].

2.2. Liquid crystalline gels

Curdlan can be used to form liquid crystalline gels when it is exposed to transition metal salts [28,29]. The crystallinity of these gels depends on the molecular weight of the gels [30]. DNA has also been used to synthesize gel beads [31]. When used together, DNA and curdlan provide control over the size and morphology of the synthesized hybrid structures. Various structures can be obtained by modifying the concentration of curdlan and DNA [27].

Curdlan is insoluble in water but it dissolves in alkaline solutions. Thus, DNA and curdlan are mixed together in a basic solution and then this mixture is added to a solution of calcium chloride salt either directly or through a dialysis membrane. Direct addition leads to formation of structures at the nanometer and millimeter scales. Dialysis allows for the formation of centimeter sized gels. The macroscopic structures are assessed by using crossed nicols (Figure 2), while the nano- and micro-structures are characterized using transmission electron microscopy (TEM, Figure 4). When viewing the centimeter scale gels between two perpendicularly placed polarizers, orthogonal dark lines are observed on the gels. These lines are known as isogyres and indicate the anisotropy in liquid crystalline gels. It is observed that increasing the concentration of DNA decreases the crystallinity of the gel as the isogyres become less defined. This is illustrated in Figure 2 [27].

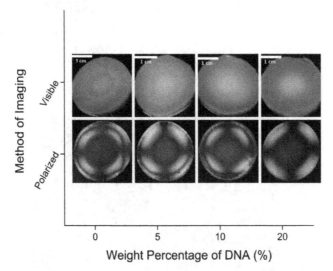

Figure 2. Liquid crystalline gels of curdlan and DNA: 100% curdlan, 5% DNA, 10% DNA and 20 % DNA as seen under visible light (top) and crossed nicols (bottom). Scale bars are 1 cm each. Figure obtained from [27].

A similar phenomenon was observed at the milimeter scale when the structure was observed under the microscope. It was seen that although DNA provided rigidity and well-defined shape to the structure, it reduced the crystallinity. This is likely because DNA forms a less crystalline structure compared to curdlan. It is possible that DNA might not be forming helices with curdlan, but instead forming a gel with microphase separation. The results from millimetre scale are highlighted in Figure 3. The opacity and lack of isogyres in DNA sample implies low crystallinity. Thus, a simple method is presented to determine degree of crystallinity of gels qualitatively [27].

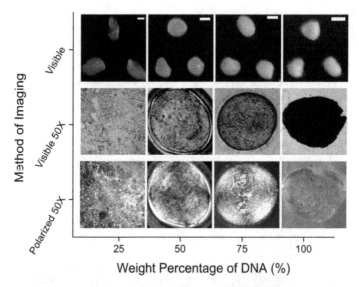

Figure 3. Spherical gels of curdlan and DNA observed under visible light (top and middle) and polarized light (bottom). Scale bars are 1 mm each. Figures obtained from [27].

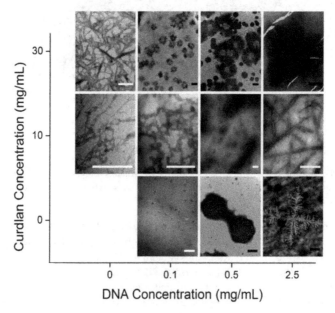

Figure 4. TEM images of micro- and nano-structures of curdlan and DNA. White scale bars are 500 nm and black scale bars are 2000 nm. Figure reproduced from [27].

At the micro- and nano-scale levels, the morphology of the gels could be changed between spheres and fibers by modulating the concentration of the DNA or curdlan. This is summarized in Figure 4. These hybrid liquid crystalline gel systems have the potential of creating advanced drug delivery vehicles where the crystalline regions of the system prevent degradation of the biomolecule and the amorphous regions maintain functionality of the encapsulated moiety. This hybrid system also serves as a tool for further studies of the molecular structure undertaken by 1,3-β-glucan and DNA in various microenvironments [27].

3. Applications in drug delivery

The ability to control the structure of 1,3-β-glucan based carriers has encouraged the application of these glucans in drug delivery. The glucans can be used as gels, nanoparticles, microparticles or complexes.

3.1. Encapsulation within gels

Curdlan is the commonly used 1,3-β-glucan for formation of gels. Curdlan gels can be prepared by heating the suspension of curdlan in aqueous solution and then cooling it down. If the suspension is heated to 60 °C, a low-set thermally reversible gel is formed whereas if the temperature is above 80 °C, a high-set thermally irreversible gel is formed. Drugs such as indomethacin, salbutamol sulfate and prednisolone have been encapsulated in curdlan gels. The gels are prepared by mixing curdlan in drug solutions at 5-10% curdlan concentration, adding the suspension to a glass test tube, heating the test tube in water bath at the desired temperature for 10 minutes and then unmolding the gel. The experiments conducted in [32] have demonstrated that a high-set gel can lower the rate of drug release. The curdlan based gels have been able to provide a sustained release of drugs for compared to commercially available formulations. These gels can be used as drug delivery suppositories for rectal administration, which bypasses hepatic first pass clearance [32].

Curdlan gels have also been used for developing protein delivery devices. Since proteins can denature at high temperatures, the temperature required for forming curdlan gels needs to be lowered. Curdlan can form aqueous gels in the presence of hydrogen bond disrupting agents such as dimethyl sulfoxide, urea and thiocyanates [33]. This property encourages the use of chaotropes for lowering the gelling temperature. It has been demonstrated that the presence of 8 M urea can decrease the gelling temperature of curdlan from 55 °C to 37 °C. This has been used for encapsulating bovine serum albumin (BSA) as a model protein. Although gels synthesized using urea are able to demonstrate sustained release of BSA over 100 hours, the toxicity of urea is a concern. Urea also has the possibility of disrupting the hydrogen bonds of BSA and hence denaturing the protein. Thus, an alternative method of reducing the gelling temperature has been developed by modifying the backbone of curdlan to form a hydroxyethyl derivative. This system was also able to form gels at 37 °C but the BSA release was sustained for a shorter time period of 75 hours [34].

3.2. Microparticles and nanoparticles

Nanoparticles are typically used in drug delivery applications because of high drug encapsulation efficiency [35], controlled drug release and incorporation of diagnostic agents [36,37]. One of the major challenges with drug delivery is achieving specificity with cellular uptake. One method of overcoming this hurdle is by attaching targeting ligands on the surface of nanoparticles to allow enhanced uptake by specific cells [38]. Another strategy is the encapsulation of these drug loaded nanoparticle in glucan based microspheres. Glucan microspheres are derived from the cell walls of *Sacchamoryces cerevisiae* (Baker's yeast) and are 2-4 μm in size. These are porous and hollow microparticles composed mainly of 1,3-β-D-glucan and small quantities of chitin [39]. The β-glucan on the microspheres serves as a specific target for uptake by immune cells such as macrophages and dendritic cells [40]. Glucan microspheres can be used for delivering various payloads such as proteins [19], DNA [40], siRNA [16,41] and small molecules [39]. Typically small molecules are not easily entrapped within these glucan microspheres and hence the use of nanoparticles is necessary. These nanoparticles can then be loaded in glucan microspheres for enhanced uptake. Recently, two types of nanoparticles have been encapsulated within glucan microspheres: fluorescent polystyrene nanoparticles and doxorubicin loaded mesoporous silica nanoparticles [39]. The polystyrene particles are encapsulated using capillary forces from the pores of microspheres, whereas the silica nanoparticles are loaded using electrostatic interactions. These particles are then used *in vitro* to demonstrate enhanced uptake by dectin-1 expressing fibroblast cells (NIH3T3-D1 cell line). It has been observed that doxorubicin loaded silica nanoparticles were more effective when encapsulated within glucan microspheres because of enhanced uptake. These results were consistent with fluorescently tagged polystyrene nanoparticle uptake as well [39].

Aside from being used as a targeting moiety, 1,3-β-glucans have also been used as structural units for encapsulating insoluble drugs. One example is the use of short chain curdlan with a molecular weight of 990 Da, derived from Vietnam medicinal mushroom *Hericium erinaceum*. This curdlan has been used to synthesize a nanoparticle formulation for encapsulating curcumin, which is a water-insoluble compound with promising anti-cancer activity. Curcumin is derived from the rhizomes of the herb *Curcuma longa* and has demonstrated cancer prevention and suppression through various signaling pathways [42]. The delivery of curcumin has been enhanced by encapsulating it within curdlan to form 50 nm nanoparticles. These nanoparticles are prepared by adding curcumin dissolved in ethanol to a solution of curdlan in water and stirring at room temperature. The solvents are then evaporated and nanoparticles are purified by centrifugation to remove excess curdlan and larger aggregates. These curcumin loaded curdlan nanoparticles have been able to inhibit tumor growth in Hep-G2 cell line *in vitro* [10].

Amphiphilic derivatives of 1,3-β-glucans have also been extensively studied for formation of micellar nanoparticles with drug encapsulating capabilities. One example of an amphiphilic system is based on cholesterol-carboxymethylcurdlan. The hydrophilicity of curdlan is increased by modifying the backbone with carboxymethyl groups. The loading of hydrophobic drugs is increased by inclusion of cholesterol moieties in the curdlan backbone [17]. This system uses a remote loading method for encapsulating epirubicin, where a pH gradient between the interior and exterior of the nanoparticle is utilized to achieve high ratio of drug to carrier [43]. Remote loading method is implemented by preparing blank cholesterol-carboxymethylcurdlan nanoparticles using probe sonication, resuspending dried nanoparticles in ammonium sulfate, performing buffer exchange in sodium chloride and finally adding the desired amount of epirubicin to the solution. This method is able to achieve up to 39.6% drug loading, which is remarkable for polymeric systems. This curdlan based delivery device has been able to enhance the cytotoxicity of epirubicin *in vitro* when assessed using human cervical carcinoma (HeLa) cell lines as compared to free epirubicin. The delivery vehicle also increases the circulation half-life *in vivo* by 4.31 times as compared to free epirubicin when tested in Wistar rats. These results provide promising opportunities for utilizing a curdlan based drug delivery vehicle for enhanced efficacy of encapsulated drugs.

Alternatively, curdlan can also be used as the hydrophobic component in an amphiphilic drug delivery vehicle. To achieve this a graft copolymer of curdlan and poly(ethylene glycol) (PEG) has been synthesized [5]. PEG is a commonly used polymer for enhanced biocompatibility [44-48]. Doxorubicin can be incorporated in graft copolymer nanoparticle by nanoprecipitation method. In this technique, the copolymer and drug are dissolved in a common water miscible solvent and then added to magnetically stirred water in a drop-wise manner. The mixing causes self-assembly of polymers where hydrophobic components are at the core of the nanoparticle and hydrophilic components are at the surface. In the case of doxorubicin and curdlan-graft-PEG, dimethyl sulfoxide is used as the common solvent [5]. A schematic of the nanoparticle is presented in Figure 5.

The structure of curdlan-graft-PEG nanoparticles has been confirmed using TEM. The samples were stained using phosphotungstic acid, which acts as a negative stain. This makes the background appear dark and the samples of interest appear bright. Additionally, since phosophotungstic acid is hydrophilic, hydrophobic components will be excluded from the acid and hence appear brighter than hydrophilic components. As observed in Figure 6, the nanoparticles are about 109 nm in size. Doxorubicin is visible as the bright center of the nanoparticles. Figure 6 (B) shows the three distinct layers of the nanoparticle, which correspond well with the schematic presented in Figure 5 [5].

The doxorubicin formulation with curdlan-graft-PEG was tested *in vitro* to determine its drug release profile and it demonstrated sustained release over 24 hours following a Fickian diffusion release model. Since PEG graft is supposed to improve the biocompatibility of the nanoparticles, a hemolysis assay was implemented to assess this property. Sheep red blood cells (RBCs) were used for this assay and nanoparticles of interest were incubated with the

RBCs for one hour at 37 °C. The amount of lysis was compared to a negative control of veronal buffered saline and positive control of deionized water. Curdlan-graft-PEG nanoparticles showed hemolysis below 5% at clinically relevant concentrations [5], which is considered biocompatible [35].

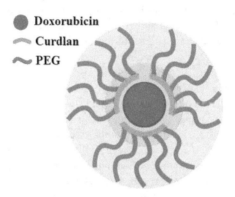

Figure 5. Schematic of nanoparticle synthesized using curdlan-graft-poly(ethylene glycol) and doxorubicin. Doxorubicin is at the core, surrounded by curdlan and PEG forms the shell. Figure adapted from [5].

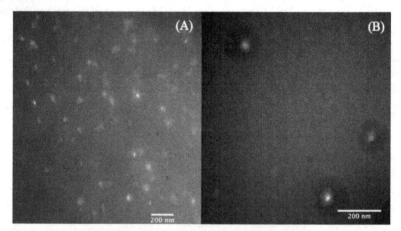

Figure 6. TEM images of doxorubicin encapsulated within curdlan-graft-PEG nanoparticles. (A) Bright center indicates doxorubicin and surroundings are darker because of higher electron density; (B) Three layers of the nanoparticle can be seen: bright doxorubicin core, dense curdlan layer in the middle and sparse PEG layer on the shell. Scale bars are 200 nm. Figure adapted from [5].

Therefore, nanoparticles and microparticles based on 1,3-β-glucans provide promising opportunities for drug delivery as structural units and as targeting ligands. The combination of these properties can be exploited for developing a drug delivery vehicle with enhanced potency.

3.3. Glucan complexes with polynucleotides

Polynucleotides have found several applications in drug delivery as active agents due to their therapeutic effects [16,41,49]. One of the challenges faced by polynucleotide delivery is their rapid degradation *in vivo*. Encapsulation of polynucleotides thus becomes necessary for maintaining their function. 1,3-β-glucans are a preferred choice for forming these complexes because of their helix forming capabilities. In the past, several polysaccharides have been tested for their ability to form complexes with the polynucleotide poly(C) and the complexation was assessed using circular dichorism (CD) where a change in the spectrum indicates complex formation. Only schizophyllan (1,3-β-glucan with one 1,6 branch every three units) and lentinan (1,3-β-glucan with two 1,6 branch every five units) have demonstrated changes in CD spectra, whereas curdlan, amylose (1,4-α-glucan), dextran (1,6-α-glucan) and pullulan (1,4-α-1,6-α-glucan) did not show any changes. This implies that only soluble 1,3-β-glucans are able to form complexes with poly(C). Commercially available curdlan is unable to form a complex because of its low solubility in water, which leads to precipitation [50]. The solubility of curdlan can be improved by reducing the molecular weight of the polymer. At lower molecular weight, curdlan is able to form complexes with poly(C) but these complexes are not as stable as the ones formed with schizophyllan [51]. Complexes have also been formed with poly(A) using schizophyllan but poly(U) showed no complexation, which suggests that polysaccharides can bind to polynucleotides in a specific manner [52]. Curdlan backbone has also been modified with carbohydrate molecules using click chemistry to improve the solubility and hence induce complex formation with poly(C) [53].

Hitherto, only homo-sequence polynucleotides have been discussed. Often therapeutic polynucleotides are composed of heterogeneous base pairs. Drug delivery vehicles for hetero-sequence oligonucleotides have also been developed by synthesizing cationic curdlan chains. CpG DNA has demonstrated immune stimulating effects but it needs to be preserved from degradation [54]. Cationic curdlan, synthesized using click chemistry, has been able to form complexes with CpG DNA and increase the cellular uptake in macrophage-like cell line J774.A1. The complex also induces an increase in cytokine (IL-12) secretion, which suggests activation of the macrophage cells [54]. Another strategy of binding hetero-sequence oligonucleotides is by modification of one terminal to attach a homo-sequence such as poly(A). A schizophyllan derivative has been utilized for binding modified antisense oligonucleotides. Schizophyllan has been modified with galactose and PEG units to enhance cellular uptake. It has been observed that the antisense effect was maximized with the use of schizophyllan derivative when the complex is administered to hepatoblastoma HepG2 and melanoma A375 cell lines [55]. Thus, 1,3-β-glucan based drug delivery devices serve as biocompatible carriers for a variety of nucleotides.

4. Applications in immunotherapy

1,3-β-glucans have been known to generate an immune response including stimulation of cytokine production, oxidative burst, increased phagocyte and lymphocyte proliferation as well as phagocytosis of opsonized tissues. Various mechanisms are responsible for this activation and this response has been utilized for varied applications including cancer resistance, disease immunity and wound healing [56].

4.1. Biological pathways of activation

The complement system is responsible for innate immunity and can be activated by either classical, alternative or lectin pathways. Using lentinan, pachyman and pachymaran polysaccharides, it has been demonstrated that 1,3-β-glucans use the alternative pathway of complement activation for generating an immune response since they show an increased consumption of C3 and C5 proteins from the complement cascade [57]. Additionally, opsonization is an important component of innate immune response. Typically complement protein C3b gets coated on pathogens and is later detected by complement receptor 1 or deactivated to form iC3b for regulation. In the presence of 1,3-β-glucans, iC3b can be detected by complement receptor 3 (CR3) or dectin-1 and this mechanism can be exploited for attacking cancer cells coated with iC3b [8]. Although the effect of CR3 has been negligible in immune activation in murine models [9,58], functional CR3 is essential for phagocytosis in human neutrophils [59].

Most of the existing studies have focused on macrophages and neutrophils but some 1,3-β-glucans such as lentinan are responsible for stimulation of T-cells and natural killer cells and hence affect the acquired immune response [60,61]. Curdlan has also been used to demonstrate increased proliferation of lymphocytes, which can in turn enhance the immune response [62]. Oxidative burst is yet another immunomodulatory effect that has been demonstrated by 1,3-β-glucans. While curdlan has shown the induction of inducible nitric oxide synthase in rat macrophages [63], other 1,3-β-glucans have shown an increase in the production of reactive oxygen species [62]. Some studies have demonstrated that immune activation is only possible by linear 1,3-β-glucans [64] while others emphasize that complex branching is important for most effective stimulation of immune response [65]. Although, several aspects of the biological pathways of activation have been discovered, further studies are necessary to gain a better understanding of the intricate interactions between 1,3-β-glucans and the immune system.

4.2. Tumor suppression

The use of 1,3-β-glucans in cancer therapy has been present in Japan since 1986, where they have been used for gastric, lung and cervical cancers [66]. Lentinan and pachymaran have demonstrated high tumor inhibition ratios of 99.6% and 96%, when tested against subcutaneous implantation of sarcoma 180 in mice [64]. It is hypothesized that the effect of lentinan and pachymaran is highly dependent on the activation of T-lymphocytes because the removal of the thymus from mice caused a suppression of antitumor effects from the 1,3-β-glucans [60]. Additionally, it is also speculated that deactivation of protein helices might be important for antitumor effects because a study has demonstrated that only polysaccharides that deactivated bovine serum albumin showed antitumor activity [67]. Besides lentinan and pachymaran, other 1,3-β-glucans have also exemplified tumor suppression. Some of the prominent examples are scleroglucan with an inhibition ratio of 90.4%, curdlan with inhibition of 99-100% [68], grifolan with inhibition of 97.9% [69] and 1,3-β-glucan from *Agaricus blazei* with inhibition of 99.3% [65]. These results have been encouraging and hence 1,3-β-glucans have been used in combinatorial therapies with antibodies in implanted human tumor xenografts from melanoma, epidermoid carcinoma, breast carcinoma, metastatic lymphoma and daudi lymphoma. The mice had higher survival rates in the presence of 1,3-β-glucans as compared to treatment with antibodies alone [12].

Subsequently, water soluble 1,3-β-glucans have been derived to improve the usability of these polysaccharides. Some examples of these polysaccharides include carboxymethylpachymaran with tumor inhibition ratio of 99.6%, hydroxymethylpachymaran and hydroxypropylpachymaran with up to 100% tumor inhibition when assessed against solid sarcoma 180 at a dose of 5 mg/kg [70]. Several derivatives of curdlan including carboxymethyl, glucosyl, sulfoethyl and sulfopropyl attachments to the backbone have also retained antitumor activity [71,72]. These examples demonstrate the versatility of 1,3-β-glucans in cancer inhibition and thus these polysaccharides can be modified to suit the desired application.

4.3. Infection prevention

Most common infections are caused by bacteria and fungi and since 1,3-β-glucans can induce inflammatory response, these glucans can be used for providing infection resistance. It has been shown that when administering 1,3-β-glucan to mice, their survival against *Staphylococcus aureus* infection increased from 70% to 97% [73]. Other 1,3-β-glucans such as glucan phosphate, laminarin and scleroglucan have been studied in detail for assessing their pharmacokinetic profile following oral administration in rats. It has been observed that these glucans are able to translocate from the gastrointestinal tract to systemic circulation. The glucans were able to increase secretion of interleukins, increase expression of dectin-1 on macrophages and increase expression of toll-like receptor 2 on dendritic cells. Thus, these effects increase the long-term survival of rats from 0% to 40% when challenged with *Candida albicans* fungal infection and from 0% to 50% when challenged with *Staphyloccous aureus* bacterial infection [74].

When considering larger mammals, infection resistance in pigs, dogs and horses has also found applications of 1,3-β-glucans. When piglets were fed with β-glucans after weaning, they demonstrated lower infection from enterotoxigenic *Escherichia coli*, which was highlighted by decreased diarrhoea and decreased content of inoculated *Escherichia coli* in the faeces as compared to control groups. These results present a significant advancement in veterinary medicine for pigs because the immunity of pigs usually suffers severely right after weaning and the current vaccines against *Escherichia coli* take a long period to be effective [20]. In the case of horses, the administration of 1,3-β-glucan to pregnant mares has been able to increase the cellular immune response in foals. This becomes very useful for preventing premature deaths of neonates [75]. While most studies have focused on the effects of 1,3-β-glucans on improving innate immunity, when these polysaccharides are administered to dogs, they showed enhancement in humoral immunity as indicated by changes in serum IgM and IgA levels [18].

1,3-β-glucans have also been used in aquatic animals as exemplified by the use of schizophyllan with 60-80% survival rate and lentinan and scleroglucan with 55-75% survival rate when tested against *Edwardsiella tarda* bacteria attack on *Cyrinus carpio L.* carp. The survival rate was also increased when infected with *Aeromonas hydrophila* with survival of 60% with schizophyllan, 70% with lentinan and 80% with scleroglucan administration. In the absence of glucans, the carp underwent complete mortality upon any infection [76]. These results have been repeatable when tilapia and grass carp is exposed to *Aeromonas hydrophila* [77]. As more studies are conducted on the use of 1,3-β-glucans, the quality of veterinary health care can be improved further and these results can eventually be transferred to human applications.

In addition to bacterial and fungal resistance, 1,3-β-glucans have also demonstrated promising results against malaria [78], herpes simplex virus [79] and human immunodeficiency virus (HIV) [80-82]. Expanding on HIV research, various complexes have been synthesized with curdlan sulfate in order to enhance the efficacy of the polysaccharide. Some prominent examples include covalent conjugation of azidothymidine to curdlan sulfate for drug delivery to the lymph nodes and bone marrow [83] and conjugation of fullerene C_{60} with curdlan for combining their anti-HIV effects [84].

4.4. Wound healing

1,3-β-glucans can have an impact on wound healing by recruiting macrophages to the wound site [85] and by increasing collagen deposition [24]. Beta glucan collagen matrix wound dressings have been used in children suffering from partial thickness burns and the dressings were able to simplify wound care by reducing analgesic requirements, improving cosmetic results and eliminating the need for repetitive dressing changes [86]. Other composites of β-glucan have been created with poly(vinyl alcohol) [23] and chitosan [22]. Poly(vinyl alcohol)/β-glucan composite was able to speed up the wound healing process when tested using rat models and hence decreased the healing time by 48% as compared to cotton gauze [23]. When using a composite of β-glucan and chitosan, a transparent dressing was obtained, which showed better results compared to commercially available chitosan based Beschitin® W. The synthesized chitosan composite did not dissolve during application period and was easy to remove because it did not adhere to wounds [22].

5. Conclusion

An assortment of 1,3-β-glucans have been explored for their structural and pharmacological capabilities. The ability of 1,3-β-glucans to form helical structures and gels has been advantageous for forming complexes with small molecules and macromolecules. The immunomodulatory effects of 1,3-β-glucans have served to fight cancer and infections and to promote wound healing. Research is moving towards combining the ability of 1,3-β-glucans to encapsulate bioactive agents with their own bioactivity for creating potent therapeutic devices against current challenges.

Author details

Mohit S. Verma and Frank X. Gu*

Chemical Engineering, University of Waterloo, Waterloo, ON, Canada
Waterloo Institute for Nanotechnology, University of Waterloo, Waterloo, ON, Canada

Acknowledgement

This work was financially supported by Natural Sciences and Engineering Research Council of Canada (NSERC) and 20/20 NSERC – Ophthalmic Materials Network. Mohit S. Verma is also financially supported by NSERC Vanier Canada Graduate Scholarship. We would also like to acknowledge Benjamin C. Lehtovaara for his contributions to the outline of this book chapter.

6. References

[1] Czaja W, Krystynowicz A, Bielecki S, Brown RM. Microbial cellulose - the natural power to heal wounds. Biomaterials 2006;27(2):145-151.

[2] Wasser SP. Medicinal mushrooms as a source of antitumor and immunomodulating polysaccharides. Applied Microbiology and Biotechnology 2002;60(3):258-274.

[3] Deslandes Y, Marchessault RH, Sarko A. Packing Analysis of Carbohydrates and Polysaccharides .13. Triple-Helical Structure of (1-]3)-Beta-D-Glucan. Macromolecules 1980;13(6):1466-1471.

[4] McIntire TM, Brant DA. Observations of the (1 -> 3)-beta-D-glucan linear triple helix to macrocycle interconversion using noncontact atomic force microscopy. Journal of the American Chemical Society 1998;120(28):6909-6919.

[5] Lehtovaara BC, Verma MS, Gu FX. Synthesis of curdlan-graft-poly(ethylene glycol) and formulation of doxorubicin-loaded core-shell nanoparticles. Journal of Bioactive and Compatible Polymers 2012;27(1):3-17.

[6] Sutherland IW. Microbial polysaccharides from Gram-negative bacteria. International Dairy Journal 2001;11(9):663-674.

[7] Steele C, Rapaka RR, Metz A, Pop SM, Williams DL, Gordon S, et al. The beta-glucan receptor dectin-1 recognizes specific morphologies of Aspergillus fumigatus. Plos Pathogens 2005;1(4):323-334.

[8] Vetvicka V, Thornton BP, Ross GD. Soluble beta-glucan polysaccharide binding to the lectin site of neutrophil or natural killer cell complement receptor type 3 (CD11b/CD18) generates a primed state of the receptor capable of mediating cytotoxicity of iC3b-opsonized target cells. Journal of Clinical Investigation 1996;98(1):50-61.

[9] Brown GD, Herre J, Williams DL, Willment JA, Marshall ASJ, Gordon S. Dectin-1 mediates the biological effects of beta-glucans. Journal of Experimental Medicine 2003;197(9):1119-1124.

* Corresponding Author

[10] Le Mai Huong, Ha Phuong Thu, Nguyen Thi Bich Thuy, Tran Thi Hong Ha, Ha Thi Minh Thi, Mai Thu Trang, et al. Preparation and Antitumor-promoting Activity of Curcumin Encapsulated by 1,3-beta-Glucan Isolated from Vietnam Medicinal Mushroom Hericium erinaceum. Chemistry Letters 2011;40(8):846-848.

[11] Liu J, Gunn L, Hansen R, Yan J. Combined yeast-derived beta-glucan with anti-tumor monoclonal antibody for cancer immunotherapy. Experimental and molecular pathology 2009;86(3):208-214.

[12] Cheung NKV, Modak S, Vickers A, Knuckles B. Orally administered beta-glucans enhance anti-tumor effects of monoclonal antibodies. Cancer Immunology Immunotherapy 2002;51(10):557-564.

[13] Na K, Park KH, Kim SW, Bae YH. Self-assembled hydrogel nanoparticles from curdlan derivatives: characterization, anti-cancer drug release and interaction with a hepatoma cell line (HepG2). Journal of Controlled Release 2000;69(2):225-236.

[14] Ross GD, Vetvicka V, Yan J, Xia Y, Vetvickova J. Therapeutic intervention with complement and beta-glucan in cancer. Immunopharmacology 1999;42(1-3):61-74.

[15] Sasaki T, Abiko N, Nitta K, Takasuka N, Sugino Y. Anti-Tumor Activity of Carboxymethylglucans obtained by Carboxymethylation of (1-]3)-Beta-D-Glucan from Alcaligenes-Faecalis Var Myxogenes Ifo-13140. European Journal of Cancer 1979;15(2):211-215.

[16] Tesz GJ, Aouadi M, Prot M, Nicoloro SM, Boutet E, Amano SU, et al. Glucan particles for selective delivery of siRNA to phagocytic cells in mice. Biochemical Journal 2011;436:351-362.

[17] Li L, Gao F, Tang H, Bai Y, Li R, Li X, et al. Self-assembled nanoparticles of cholesterol-conjugated carboxymethyl curdlan as a novel carrier of epirubicin. Nanotechnology 2010;21(26):265601-265601.

[18] Stuyven E, Verdonck F, Van Hoek I, Daminet S, Duchateau L, Remon JP, et al. Oral Administration of beta-1,3/1,6-Glucan to Dogs Temporally Changes Total and Antigen-Specific IgA and IgM. Clinical and Vaccine Immunology 2010;17(2):281-285.

[19] Huang H, Ostroff GR, Lee CK, Wang JP, Specht CA, Levitz SM. Distinct Patterns of Dendritic Cell Cytokine Release Stimulated by Fungal beta-Glucans and Toll-Like Receptor Agonists. Infection and Immunity 2009;77(5):1774-1781.

[20] Stuyven E, Cox E, Vancaeneghem S, Arnouts S, Deprez P, Goddeeris BM. Effect of beta-glucans on an ETEC infection in piglets. Veterinary Immunology and Immunopathology 2009;128(1-3):60-66.

[21] Murphy EA, Davis JM, Brown AS, Carmichael MD, Carson JA, Van Rooijen N, et al. Benefits of oat beta-glucan on respiratory infection following exercise stress: role of lung macrophages. American Journal of Physiology-Regulatory Integrative and Comparative Physiology 2008;294(5):R1593-R1599.

[22] Kofuji K, Huang Y, Tsubaki K, Kokido F, Nishikawa K, Isobe T, et al. Preparation and evaluation of a novel wound dressing sheet comprised of beta-glucan-chitosan complex. Reactive & Functional Polymers 2010;70(10):784-789.

[23] Huang M, Yang M. Evaluation of glucan/poly(vinyl alcohol) blend wound dressing using rat models. International Journal of Pharmaceutics 2008;346(1-2):38-46.

[24] Portera CA, Love EJ, Memore L, Zhang LY, Muller A, Browder W, et al. Effect of macrophage stimulation on collagen biosynthesis in the healing wound. American Surgeon 1997;63(2):125-130.

[25] Miyoshi K, Uezu K, Sakurai K, Shinkai S. Proposal of a new hydrogen-bonding form to maintain curdlan triple helix. Chemistry & Biodiversity 2004;1(6):916-924.

[26] Miyoshi K, Uezu K, Sakurai K, Shinkai S. Polysaccharide-polynucleotide complexes. Part 32. Structural analysis of the Curdlan/poly(cytidylic acid) complex with semiempirical molecular orbital calculations. Biomacromolecules 2005;6(3):1540-1546.

[27] Lehtovaara BC, Verma MS, Gu FX. Multi-phase ionotropic liquid crystalline gels with controlled architecture by self-assembly of biopolymers. Carbohydrate Polymers 2012;87(2):1881-1885.

[28] Dobashi T, Nobe M, Yoshihara H, Yamamoto T, Konno A. Liquid crystalline gel with refractive index gradient of curdlan. Langmuir 2004;20(16):6530-6534.

[29] Dobashi T, Yoshihara H, Nobe M, Koike M, Yamamoto T, Konno A. Liquid crystalline gel beads of curdlan. Langmuir 2005;21(1):2-4.

[30] Nobe M, Kuroda N, Dobashi T, Yamamoto T, Konno A, Nakata M. Molecular weight effect on liquid crystalline gel formation of curdlan. Biomacromolecules 2005;6(6):3373-3379.

[31] Moran MC, Miguel MG, Lindman B. DNA gel particles: Particle preparation and release characteristics. Langmuir 2007;23(12):6478-6481.

[32] Kanke M, Tanabe E, Katayama H, Koda Y, Yoshitomi H. Application of Curdlan to Controlled Drug-Delivery .3. Drug-Release from Sustained-Release Suppositories In-Vitro. Biological & Pharmaceutical Bulletin 1995;18(8):1154-1158.

[33] Renn DW. Purified curdlan and its hydroxyalkyl derivatives: preparation, properties and applications. Carbohydrate Polymers 1997;33(4):219-225.

[34] Kim BS, Jung ID, Kim JS, Lee J, Lee IY, Lee KB. Curdlan gels as protein drug delivery vehicles. Biotechnology Letters 2000;22(14):1127-1130.

[35] Verma MS, Liu S, Chen YY, Meerasa A, Gu FX. Size-tunable nanoparticles composed of dextran-b-poly(D,L-lactide) for drug delivery applications. Nano Research 2012;5(1):49-61.

[36] Sun D. Nanotheranostics: Integration of Imaging and Targeted Drug Delivery. Molecular Pharmaceutics 2010;7(6):1879-1879.

[37] Rosen JE, Yoffe S, Meerasa A, Verma MS, Gu FX. Nanotechnology and Diagnostic Imaging: New Advances in Contrast Agent Technology. Journal of Nanomedicine & Nanotechnology 2011:1000115.

[38] Cho KJ, Wang X, Nie SM, Chen Z, Shin DM. Therapeutic nanoparticles for drug delivery in cancer. Clinical Cancer Research 2008;14(5):1310-1316.

[39] Soto ER, Caras AC, Kut LC, Castle MK, Ostroff GR. Glucan particles for macrophage targeted delivery of nanoparticles. Journal of Drug Delivery 2012;2012:143524.

[40] Soto ER, Ostroff GR. Characterization of multilayered nanoparticles encapsulated in yeast cell wall particles for DNA delivery. Bioconjugate chemistry 2008;19(4):840-848.

[41] Aouadi M, Tesz GJ, Nicoloro SM, Wang M, Chouinard M, Soto E, et al. Orally delivered siRNA targeting macrophage Map4k4 suppresses systemic inflammation. Nature 2009;458(7242):1180-1184.

[42] Sa G, Das T. Anti cancer effects of curcumin: cycle of life and death. Cell Division 2008;3:14.

[43] Lewrick F, Suss R. Remote loading of anthracyclines into liposomes. Methods in Molecular Biology 2010;605:139-45.

[44] Dhar S, Gu FX, Langer R, Farokhzad OC, Lippard SJ. Targeted delivery of cisplatin to prostate cancer cells by aptamer functionalized Pt(IV) prodrug-PLGA-PEG nanoparticles. Proceedings of the National Academy of Sciences of the United States of America 2008;105(45):17356-17361.

[45] Esmaeili F, Ghahremani MH, Ostad SN, Atyabi F, Seyedabadi M, Malekshahi MR, et al. Folate-receptor-targeted delivery of docetaxel nanoparticles prepared by PLGA-PEG-folate conjugate. Journal of Drug Targeting 2008;16(5):415-423.

[46] Cheng J, Teply BA, Sherifi I, Sung J, Luther G, Gu FX, et al. Formulation of functionalized PLGA-PEG nanoparticles for in vivo targeted drug delivery. Biomaterials 2007;28(5):869-876.

[47] Luu YK, Kim K, Hsiao BS, Chu B, Hadjiargyrou M. Development of a nanostructured DNA delivery scaffold via electrospinning of PLGA and PLA-PEG block copolymers. Journal of Controlled Release 2003;89(2):341-353.

[48] Avgoustakis K, Beletsi A, Panagi Z, Klepetsanis P, Karydas AG, Ithakissios DS. PLGA-mPEG nanoparticles of cisplatin: in vitro nanoparticle degradation, in vitro drug release and in vivo drug residence in blood properties. Journal of Controlled Release 2002;79(1-3):123-135.

[49] Davis HL, Weeratna R, Waldschmidt TJ, Tygrett L, Schorr J, Krieg AM. CpG DNA is a potent enhancer of specific immunity in mice immunized with recombinant hepatitis B surface antigen. Journal of Immunology 1998;160(2):870-876.

[50] Kimura T, Koumoto K, Sakurai K, Shinkai S. Polysaccharide-polynucleotide complexes (III): A novel interaction between the beta-1,3-glucan family and the single-stranded RNA poly(C). Chemistry Letters 2000(11):1242-1243.

[51] Koumoto K, Kimura T, Kobayashi H, Sakurai K, Shinkai S. Chemical modification of curdlan to induce an interaction with poly(C)(1). Chemistry Letters 2001(9):908-909.

[52] Sakurai K, Shinkai S. Molecular recognition of adenine, cytosine, and uracil in a single-stranded RNA by a natural polysaccharide: Schizophyllan. Journal of the American Chemical Society 2000;122(18):4520-4521.

[53] Hasegawa T, Numata M, Okumura S, Kimura T, Sakurai K, Shinkai S. Carbohydrate-appended curdlans as a new family of glycoclusters with binding properties both for a polynucleotide and lectins. Organic & Biomolecular Chemistry 2007;5(15):2404-2412.

[54] Krieg A. CpG motifs: the active ingredient in bacterial extracts? Nature medicine 2003;9(7):831-835.

[55] Karinaga R, Anada T, Minari J, Mizu M, Koumoto K, Fukuda J, et al. Galactose-PEG dual conjugation of beta-(1 -> 3)-D-glucan schizophyllan for antisense oligonucleotides delivery to enhance the cellular uptake. Biomaterials 2006;27(8):1626-1635.

[56] Lehtovaara BC, Gu FX. Pharmacological, structural, and drug delivery properties and applications of 1,3-beta-glucans. Journal of Agricultural and Food Chemistry 2011;59(13):6813-6828.

[57] Hamuro J, Hadding U, Bittersuermann D. Solid-Phase Activation of Alternative Pathway of Complement by Beta-1,3-Glucans and its Possible Role for Tumor Regressing Activity. Immunology 1978;34(4):695-705.

[58] Brown GD, Taylor PR, Reid DM, Willment JA, Williams DL, Martinez-Pomares L, et al. Dectin-1 is a major beta-glucan receptor on macrophages. Journal of Experimental Medicine 2002;196(3):407-412.

[59] van Bruggen R, Drewniak A, Jansen M, van Houdt M, Roos D, Chapel H, et al. Complement receptor 3, not Dectin-1, is the major receptor on human neutrophils for beta-glucan-bearing particles. Molecular Immunology 2009;47(2-3):575-581.

[60] Maeda YY, Chihara G. Lentinan, a New Immuno-Accelerator of Cell-Mediated Responses. Nature 1971;229(5287):634-634.

[61] Fujimoto T, Omote K, Mai M, Natsuumesakai S. Evaluation of Basic Procedures for Adoptive Immunotherapy for Gastric-Cancer. Biotherapy 1992;5(2):153-163.

[62] Sonck E, Stuyven E, Goddeeris B, Cox E. The effect of beta-glucans on porcine leukocytes. Veterinary Immunology and Immunopathology 2010;135(3-4):199-207.

[63] Ljungman AG, Leanderson P, Tagesson C. (1 -> 3)-beta-D-glucan stimulates nitric oxide generation and cytokine mRNA expression in macrophages. Environmental Toxicology and Pharmacology 1998;5(4):273-281.

[64] Chihara G, Hamuro J, Maeda Y, Arai Y, Fukuoka F. Antitumour Polysaccharide Derived Chemically from Natural Glucan (Pachyman). Nature 1970;225(5236):943-944.

[65] Ohno N, Furukawa M, Miura NN, Adachi Y, Motoi M, Yadomae T. Antitumor beta-glucan from the cultured fruit body of Agaricus blazei. Biological & Pharmaceutical Bulletin 2001;24(7):820-828.

[66] Bohn JA, BeMiller JN. (1->3)-beta-D-glucans as biological response modifiers: A review of structure-functional activity relationships. Carbohydrate Polymers 1995;28(1):3-14.

[67] Hamuro J, Chihara G. Effect of Antitumor Polysaccharides on Higher Structure of Serum-Protein. Nature 1973;245(5419):40-41.

[68] Sasaki T, Abiko N, Sugino Y, Nitta K. Dependence on Chain-Length of Anti-Tumor Activity of (1-]3)-Beta-D-Glucan from Alcaligenes-Faecalis Var Myxogenes, Ifo 13140, and its Acid-Degraded Products. Cancer Research 1978;38(2):379-383.

[69] Iino K, Ohno N, Suzuki I, Miyazaki T, Yadomae T. Structural Characterization of a Neutral Antitumour Beta-D-Glucan Extracted with Hot Sodium-Hydroxide from Cultured Fruit Bodies of Grifola-Frondosa. Carbohydrate Research 1985;141(1):111-119.

[70] Hamuro J, Yamashit.Y, Ohsaka Y, Maeda YY, Chihara G. Carboxymethylpachymaran, a New Water Soluble Polysaccharide with Marked Antitumour Activity. Nature 1971;233(5320):486-488.

[71] Demleitner S, Kraus J, Franz G. Synthesis and Antitumor-Activity of Derivatives of Curdlan and Lichenan Branched at C-6. Carbohydrate Research 1992;226(2):239-246.

[72] Demleitner S, Kraus J, Franz G. Synthesis and Antitumor-Activity of Sulfoalkyl Derivatives of Curdlan and Lichenan. Carbohydrate Research 1992;226(2):247-252.

[73] Diluzio NR, Williams DL. Protective Effect of Glucan Against Systemic Staphylococcus-Aureus Septicemia in Normal and Leukemic Mice. Infection and Immunity 1978;20(3):804-810.

[74] Rice PJ, Adams EL, Ozment-Skelton T, Gonzalez AJ, Goldman MP, Lockhart BE, et al. Oral delivery and gastrointestinal absorption of soluble glucans stimulate increased resistance to infectious challenge. Journal of Pharmacology and Experimental Therapeutics 2005;314(3):1079-1086.

[75] Krakowski L, Krzyzanowski J, Wrona Z, Siwicki AK. The effect of nonspecific immunostimulation of pregnant mares with 1,3/1,6 glucan and levamisole on the immunoglobulins levels in colostrum, selected indices of nonspecific cellular and humoral immunity in foals in neonatal and postnatal period. Veterinary Immunology and Immunopathology 1999;68(1):1-11.

[76] Yano T, Matsuyama H, Mangindaan REP. Polysaccharide-Induced Protection of Carp, Cyprinus-Carpio L, Against Bacterial-Infection. Journal of Fish Diseases 1991;14(5):577-582.

[77] Wang WS, Wang DH. Enhancement of the resistance of Tilapia and grass carp to experimental Aeromonas hydrophila and Edwardsiella tarda infections by several polysaccharides. Comparative Immunology Microbiology and Infectious Diseases 1997;20(3):261-270.

[78] Evans SG, Morrison D, Kaneko Y, Havlik I. The effect of curdlan sulphate on development in vitro of Plasmodium falciparum. Transactions of the Royal Society of Tropical Medicine and Hygiene 1998;92(1):87-89.

[79] Zhang M, Cheung PCK, Ooi VEC, Zhang L. Evaluation of sulfated fungal beta-glucans from the sclerotium of Pleurotus tuber-regium as a potential water-soluble anti-viral agent. Carbohydrate Research 2004;339(13):2297-2301.

[80] Yoshida T, Hatanaka K, Uryu T, Kaneko Y, Suzuki E, Miyano H, et al. Synthesis and Structural-Analysis of Curdlan Sulfate with a Potent Inhibitory Effect Invitro of Aids Virus-Infection. Macromolecules 1990;23(16):3717-3722.

[81] Yoshida T, Yasuda Y, Uryu T, Nakashima H, Yamamoto N, Mimura T, et al. Synthesis and In-Vitro Inhibitory Effect of L-Glycosyl-Branched Curdlan Sulfates on Aids Virus-Infection. Macromolecules 1994;27(22):6272-6276.

[82] Yoshida T, Yasuda Y, Mimura T, Kaneko Y, Nakashima H, Yamamoto N, et al. Synthesis of Curdlan Sulfates having Inhibitory Effects In-Vitro Against Aids Viruses Hiv-1 and Hiv-2. Carbohydrate Research 1995;276(2):425-436.

[83] Gao Y, Katsuraya K, Kaneko Y, Mimura T, Nakashima H, Uryu T. Synthesis, enzymatic hydrolysis, and anti-HIV activity of AZT-spacer-curdlan sulfates. Macromolecules 1999;32(25):8319-8324.

[84] Ungurenasu C, Pinteala M. Syntheses and characterization of water-soluble C-60-curdlan sulfates for biological applications. Journal of Polymer Science Part A-Polymer Chemistry 2007;45(14):3124-3128.

[85] Browder W, Williams D, Lucore P, Pretus H, Jones E, Mcnamee R. Effect of Enhanced Macrophage Function on Early Wound-Healing. Surgery 1988;104(2):224-230.

[86] Delatte SJ, Evans J, Hebra A, Adamson W, Othersen HB, Tagge EP. Effectiveness of beta-glucan collagen for treatment of partial-thickness burns in children. Journal of Pediatric Surgery 2001;36(1):113-118.

Chapter 7

Complexes of Polysaccharides and Glycyrrhizic Acid with Drug Molecules – Mechanochemical Synthesis and Pharmacological Activity

A. V. Dushkin, T. G. Tolstikova, M. V. Khvostov and G. A. Tolstikov

Additional information is available at the end of the chapter

1. Introduction

1.1. Using poly- and oligo-saccharides for drug delivery; Possibility for mechanochemical synthesis of supramolecular systems

Providing efficient ways of delivering active drug molecules to their destinations in target organisms, the so-called drug delivery, is among major challenges in today's pharmacy. An important relevant issue is to enhance the efficacy and safety of pharmaceutical compounds by correcting their solubility [1-3]. Polysaccharides (e.g., derivatives of cellulose, chitosan, and alginic and hyaluronic acids) make part of compositions with controlled or retarded drug release [4-6], while oligosaccharides (alpha-, beta-, and gamma-cyclodextrins and their derivatives) are broadly used to increase solubility and dissolution rates as they can form guest-host supramolecular complexes with poorly soluble drugs [7, 8]. Until recently little was known whether complexes of this kind may result from the activity of natural plant-derived or synthetic water-soluble polysaccharides though these are common elements in dietary supplements or drugs. Polysaccharides have aroused no interest in this respect, possibly because the technology for producing supramolecular complexes requires liquid phases (solutions or melts): The complexes form by molecular interaction in the liquid, and the solid phase is extracted then on drying (solutions) or cooling (melts). However, being easily soluble in water, polysaccharides are almost insoluble in other solvents and, moreover, decompose on heating rather than melt. The target drugs, instead, often dissolve rapidly in non-aqueous solvents but are poorly soluble or insoluble in water. Therefore, the liquid-phase synthesis of polysaccharide-drug complexes has been impeded by the lack of co-solubility.

This difficulty may be surmountable with solid-state chemistry approaches, specifically, with mechanochemical transformations in mixtures of solids [9-11]. Unlike the liquid-phase synthesis, mechanochemical treatment is a simpler single-stage process going without solvents or melts and respective additional procedures. The flow chart in Fig. 1 shows a simplified sequence of transformations the powder mixtures experience during dry milling in various mills.

There may be three types of main products relevant to our study, depending on the properties of starting materials:

1. "molecular dispersions", or solid solutions of drugs in excess filling (dispersion medium);
2. supramolecular complexes or products of chemical reactions between the components;
3. composite materials: aggregates of powdered particles.

In fact, they all are solid dispersions that form supramolecular structures (complexes or micelles) that enclose drug molecules and provide their solubility.

Generally, solid-phase processes have a number of advantages in laboratory and technological uses as they yield, in a shorter time, materials which the classical liquid-phase technology can never provide and allow avoiding problems associated with melts or solvents and side reactions. The high potentiality of mechanical activation was proven in our previous studies [12-14], e.g., on quick-dissolving pharmaceutical compositions [15-18] and synthesis of polyfluorinated aromatic compounds [19, 20].

In this synopsis we present techniques for synthesizing supramolecular complexes of poorly soluble drugs with water-soluble polysaccharides or with glycyrrhizic acid (a plant-derived glycoside), describe physicochemical properties of their solid forms and solutions, and report the results of pharmacological testing.

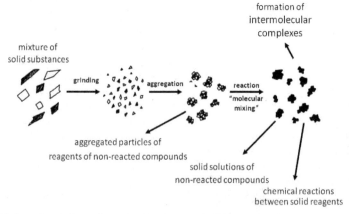

Figure 1. Mechanochemical transformations in mixtures of solids organic substances.

2. Physicochemical description of supramolecular systems including polysaccharides and glycyrrhizic acid with drug molecules

2.1. Synthesis of supramolecular systems of water-soluble polysaccharides and their stability in aqueous solutions

In the course of this study, different conditions of mechanochemical synthesis and various complexing agents have been tried and compared in terms of efficacy. The complexing agents were: arabinogalactan (AG), a water-soluble larch polysaccharide derived from *Larix sibirica* Ledeb. and *Larix gmelinii* (Rupr.), fibregum (FG), a glycoprotein of acacia gum, fruit pectin (PC), hydroxyethyl starch (HES200/0,5), dextrans (D) 10. 40. 70, and ß-cyclodextrin (CD), the latter chosen as standard for being widely used in pharmaceutics. The mixtures of powdered components (polysaccharides/complexing agents and drugs) were dispersed in ball mills at greater or lower intensities in laboratory planetary- and rotary-type mills, respectively. Milder rotary milling was predominantly applied because the molecular-level mixing in a planetary mill, common to laboratory studies of mechanochemical modification of drugs [9], may partly destroy the material and pose scaling problems. The materials processed by nondestructive rotary milling [10, 11, 21], instead, interact to produce solid dispersed systems of components (composite aggregates of superdispersed particles), and the process is easily scaled onto industrial flow mills.

The obtained compositions were checked for drug contents to avoid unwanted chemical reactions. Formation of supramolecular complexes was identified from changes in solubility of drugs in the water solution of the compositions [22].

Dissolution and complexing of poorly water-soluble drugs can be illustrated by such simplified equations:

$$\text{Drug}_{solid} \longleftrightarrow \text{Drug}_{solution} \tag{1}$$

$$\text{Drug}_{solution} + \text{CA}_{solution} \longleftrightarrow (\text{Drug} \cdot \text{CA})_{solution} \tag{2}$$

Equilibrium, according to (2), is given by

$$K_{DCA} = [(\text{Drug} \cdot \text{CA})_{solution}] \, / \, [\text{Drug}_{solution}] \times [\text{CA}_{solution}], \tag{3}$$

where Drug_{solid} is the drug in a crystalline solid phase, in equilibrium with the solution; $\text{Drug}_{solution}$ is the drug existing in the free form in the solution; $\text{CA}_{solution}$ is the free complexing agent in the solution; $(\text{Drug} \cdot \text{CA})_{solution}$ is the complexing agent-drug complex in the solution; K_{DCA} is the constant of supramolecular complexing.

The value $\text{Drug}_{solution}$ corresponds to the thermodynamic equilibrium solubility in the absence of complexing agents. In the case of complexing, the total concentration of the dissolved drug C_{drug} equals the sum its free and bound forms.

$$C_{Drug} = [\text{Drug}_{solution}] + [(\text{Drug} \; \text{CA})_{solution}] \tag{4}$$

Thus, the solubility increase of a drug in the solution (X) in the presence of a complexing agent is

$$X = C_{Drug} / [Drug_{solution}] = 1 + K_{DCA} \cdot [CA_{solution}] \qquad (5)$$

In our view, X is a good proxy of binding strength in the supramolecular complexes drugs may form with various water-soluble polymers.

All poorly soluble drugs we studied have shown a notable solubility increase when became incorporated into compositions with complexing agents. Table 1 below shows solubility data reported in [22-32] as far as published for the first time in this review.

The binding strength in the complexes grows in the series "dextran 70 < dextrans 40 and 10 ~ < HES < ß-cyclodextrin, fibregum < pectin < arabinogalactan". Complexing of pectin with mezapam and clozapine most probably occurs by acid-base reactions, which accounts for quite a high binding strength. However, other complexing agents lack acid-base groups and the interaction mechanism is most likely "hydrophobic", as in the case of cyclodextrin complexes. Thus, the mechanochemical treatment strengthens considerably the drug binding in compositions. The solubility of drugs increases, depending on the way of mixing, in the series "mixing without milling < high-rate milling < low-rate milling.

The obtained compositions were analyzed by X-ray powder diffraction and thermal methods. All non-processed mixtures showed X-ray and thermal features typical of crystalline drugs, which disappeared or decreased markedly after milling. Therefore, drugs in the ground mixtures partly or fully loose their crystallinity, possibly, as their solid phase becomes disordered and their molecules are dispersed into the excess solid phase of complexing agents, with formation of solid solutions or supramolecular complexes. In the latter case, the solubility changes evidence that the analyzed compositions form more strongly bound complexes when form in the solid phase than in the solution.

2.2. Molecular dynamics and structure of arabonigalactan complexes

AG-drug systems were investigated by ^{1}H NMR spectroscopy [22] for the molecular dynamics of complexes and the mobility of arabinogalactan (AG) molecule fragments. NMR relaxometry is applicable to molecular complexes as the spin-lattice and spin-spin relaxation times (T_1 and T_2, respectively) are highly sensitive to interactions and diffusion mobility of molecules. As a molecule becomes bound in a complex, its diffusion mobility slows down, and the proton relaxation times decrease notably. In the case of rapid complex-solution molecular exchange, the NMR signal decays according to the mono-exponential law. Otherwise, if the exchange is slower than the relaxation time, the kinetics is biexponential:

$$A(t) = P_1 \cdot \exp(-t / T_{21}) + P_2 \cdot \exp(-t / T_{22}) \qquad (6)$$

API	Complexing agent/drug mass ratios mass	Solubility pure drug [Drugsolution} g/l / Solubility by complexation CDrug g/l	Solubility increase, X^1	Reference
Diazepam	Arabinogalactan (1/10)[2]	0.048/0.058	1.2	[22,24]
	Arabinogalactan (1/10)[3]	0.048/0.115	2.4	[22,24]
	Arabinogalactan (1/10)[4]	0.048/2.31	48.2	[23]
	Pectin (1/10)[3]	0.048/0.67	14.2	[23]
	Hydroxyethylstarch (1/10)[4]	0.048/0.075	1.53	[23]
	Beta-cyclodextrin (1/10)[3]	0.048/0.086	1.8	[23]
	Glycyrrhizic acid (1/10)[3]	0.048/0.16	3.4	[29]
	Dextran 10[4]	0.048/0.09	1.9	[23]
	Dextran 40[4]	0.048/0.092	1.9	[23]
	Dextran 70[4]	0.048/0.057	1.2	[23]
Indomethacin	Arabinogalactan (1/10)[2]	0.04/0.044	1.1	[22,24]
	Arabinogalactan (1/10)[3]	0.04/0.396	9.9	[22,24]
	Arabinogalactan (1/10)[4]	0.04/1.59	39.7	[23]
	Hydroxyethylstarch (1/10)[4]	0.04/0.54	13.5	[23]
	Beta-cyclodextrin (1/10)[3]	0.04/0.096	2.4	[23]
Mezapam	Arabinogalactan (1/10)[2]	0.02/0.98	4.9	[22,24]
	Arabinogalactan (1/10)[3]	0.02/0.382	19.1	[22,24]
	Arabinogalactan (1/10)[4]	0.02/2.81	140.6	[23]
	Pectin (1/10)[3]	0.02/1.54	77.1	[23]
	Hydroxyethylstarch(1/10)[4]	0.02/0.04	2.0	[23]
Clozapine	Arabinogalactan (1/10)[2]	0.04/0.176	4.4	[22,24]
	Arabinogalactan (1/10)[3]	0.04/0.82	20.5	[22,24]
	Arabinogalactan (1/10)[4]	0.04/4.32	107.9	[23]
	Pectin (1/10)[3]	0.04/1.63	40.8	[23]

	Hydroxyethylstarch(1/10)[4]	0.04/0.222	5.5	[23]
	Beta-cyclodextrin (1/10)[3]	0.04/0.60	15.1	[23]
	Glycyrrhizic acid (1/10)[3]	0.04/0.088	2.2	[29]
Nifedipine	Arabinogalactan (1/10)[3]	0.18/1.24	6.9	[26]
	Arabinogalactan (1/20)[3]	0.18/2.46	13.7	[26]
	Glycyrrhizic acid (1/10)[3]	0.18/0.92	5.1	[30]
Dihydro-quercitin	Arabinogalactan (1/10)[4]	0.65/3.75	5.9	[23,27]
	Hydroxyethylstarch(1/10)[4]	0.65/1.97	3.0	[23]
	Fibregum(1/10)[4]	0.65/5.72	8.8	[23,27]
Quercitin	Arabinogalactan (1/10)[3]	0.019/0.21	11.6	[28]
	Arabinogalactan (1/20)[3]	0.019/1.28	71.0	[28]
Ibuprofen	Arabinogalactan (1/10)[2]	0.03/0.036	1.2	[29]
	Arabinogalactan (1/10)[4]	0.03/0.85	28.4	[29]
	Hydroxyethylstarch (1/10)[4]	0.03/0.08	2.6	[29]
	Glycyrrhizic acid (1/10)[3]	0.03/0.441	14.7	[29]
Beta-Carotene	Arabinogalactan (1/40)[3]	< 0.001/2.65	> 2000	[25], This article
Warfarin	Arabinogalactan (1/40)[3]	0.021/0.111	5.3	[31]
Contaxantine	Arabinogalactan (1/40)[3]	< 0.001/2.64	> 2000	[25], This article
Albendazol	Arabinogalactan (1/10)[4]	0.003/0.174	58.0	[32]
	Hydroxyethylstarch(1/10)[4]	0.003/0.094	31.3	[32]
Carbenazim	Arabinogalactan (1/10)[4]	0.009/0.146	16.2	[32]
	Hydroxyethylstarch(1/10)[4]	0.009/0.020	2.1	[32]
Simvastatin	Glycyrrhizic acid (1/10)[3]	0.0012/0.314	260	This article
	Arabinogalactan (1/10)[3]	0.0012/0.044	36,7	This article

1 – To determine the solubility of the drug, machined mixture of complexing agent/drug, in amounts of 0.4 grams, as well as the linkage of individual substances which are equivalent to their content in the above mixture was dissolved in 5 ml of water while stirring with a magnetic stirrer at +25 ° C till reaching constant concentration. The concentration of drug in the solution was analyzed by HPLC.
2 – mixing without mechanical treatment;
3 – treatment in a planetary mill, acceleration 40 g;
4 – treatment in a rotary ball mill, acceleration 1 g;

Table 1. Increase in water solubility of some drugs as a result of complexing.

The fast component P_1 and the slow component P_2 correspond, respectively, to the shares of molecules in the complex and in the solution. Typical T_2 values are 0.5-1 s for molecules in the solution and 0.03-0.09 s for those bound in the complex. Shorter T_{21} times mean lower mobility of drug molecules in the latter case.

Similar considerations apply to the mobility within polymers when parts of a macromolecule differ in mobility, possibly, controlled by their spin and conformations.

2.2.1. T_2 measurements

Arabinogalactan shows biexponential relaxation patterns. The calculated parameters for arabinogalactan and AG-drug complexes are listed in Table 2.

Sample	P_1 %	T_{21} msec	P_2 %	T_{22} msec
Arabinogalactan, native[2]	80	17	22	250
Arabinogalactan, treated in planetary mill[2]	65	25	35	250
Clozapine/Arabinogalactan 1/20 w/w No treatment[3]	88	90	12	1000
Clozapine/Arabinogalactan 1/20 w/w Mixture treated in planetary mill[3]	90	40	10	1000
Mezapam/Arabinogalactan 1/20 w/w No treatment[3]	55	50	45	250
Mezapam/Arabinogalactan 1/20 w/w Mixture treated in planetary mill[3]	90	30	10	250
Diazepam/Arabinogalactan 1/20 w/w No treatment[3]	mono	150	-	-
Diazepam/Arabinogalactan 1/20 w/w Mixture treated in planetary mill[3]	20	60	80	800
Indomethacin/Arabinogalactan 1/20 w/w No treatment[3]	58	50	42	900
Indomethacin/Arabinogalactan 1/20 w/w Mixture treated in planetary mill[3]	67	40	33	900

1 - T2 measurements were performed for the aromatic protons of drug molecules, to an accuracy of ± 10%
2 – solvent D_2O;
3 - solvent 70% D_2O + 30% CD_3OD;

Table 2. Spin-spin relaxation times of protons for arabinogalactan and drug molecules in solutions[1]

The short relaxation times may correspond to the interior protons and the long times may represent the exterior protons of the polymer compound. Mechanical activation in a planetary mill increases the molecular mobility of the interior fragments but decreases their percentage. A relatively narrow ~ 6 kHz band in the ¹H NMR spectra of AG powder, which stands out against a broad line associated with dipole-dipole interaction non-averagable in solids, represents a mobile phase with its integral intensity up to ~ 15% of the number of hydrogen nuclei in the sample. The mobile phase may correspond to fragments of AG macromolecules, possibly, side chains, as one may reasonably hypothesize given that water content in AG never exceeds 2 wt.%. This very fact appears to facilitate AG-Drug molecular complexing on mechanical activation of solids.

AG-Drug systems most often exhibit distinct biexponential kinetics as evidence that the drug molecules are either free or bound in complexes with AG. The bound molecules are more abundant and less mobile in milled samples, while the free ones keep almost invariable NMR relaxation times. The characteristic ^{1}H NMR bands of clozapine and mezapam move to low field on complexing, possibly because the molecules become protonated at the account of minor remnant uronic acid present in AG, the shift being likewise greater in the milled samples. However, no complexing-related shifting appears in the cases of indomethacin and diazepam. The life time of drug molecules in complexes with AG must to be $\sim \geq 100$ ms, judging from the conditions of slow exchange.

The system AG-diazepam offers an illustrative example. Solutions of these mixtures not subjected to mechanical treatment show mono-exponential relaxation behavior, but with shorter times than in free diazepam, likely as a result of rapid solution-complex molecular exchange. The milled mixtures, on the contrary, have biexponential kinetics corresponding to slower exchange of molecules and stronger binding.

Thus, dynamic NMR spectroscopy of all Drug-AG solutions indicates formation of supramolecular drug-polysaccharide complexes, like the data on solubility increase. Most likely, the complexing sites are at side chain spaces in the branching macromolecules. Unlike cyclodextrins, ensembles of polysaccharide molecules (including arabainogalactan) are micro-heterogeneous in mass and structure. As a result, molecular modeling of the complex is very difficult. The binding mechanism appears to lie mainly with hydrophobic interactions [33, 34] which are typical of guest-host cyclodextrin complexes. A certain support to this hypothesis comes from stronger binding of highly lipophilic drugs which are almost insoluble in water. In this case, the branched structure of AG macromolecules [35, 36] is especially favorable for complexing. However, Coulomb interactions may contribute as well in the presence of acid-base groups in polysaccharides and drugs [37].

2.3. Transformations of polysaccharides in solid state and in solutions

Macromolecules characteristically have broadly varying molecular weights, from $\sim 10^3$ to $\sim 10^7$ Da. Macromolecules in polymers involved in technological production of various materials may experience mechanical action and partial destruction (breakdown of chains) whereby their molecular weight becomes ever more heterogeneous and diminishes on average [38]. The destructive change may be especially prominent in "dry" technological processes, such as pulverization, pelleting, or mixing, e.g., in mechanochemical solid-state complexing of drugs with water-soluble polymers (polysaccharides). Partial destruction of polymers may change their toxicological properties which have to be cautiously monitored when making new drugs and food products.

Molecular weight patterns were studied [39] in polysaccharides (dextrans 10, 40, and 70, HES 200/0.5, and larch AG and acacia FG gum) by gel permeation chromatography (GPC) [40] of samples treated in rotary and planetary mills; the obtained materials were tested for their toxicity.

See Fig. 2 for example chromatograms of AG and Table 3 for calculated molecular weights of the analyzed polysaccharides before and after mechanical treatment.

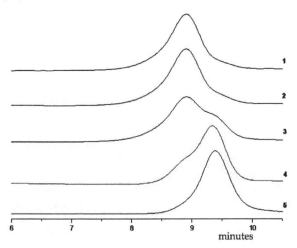

Figure 2. GPC chromatograms of 0.02 wt.% arabinogalactan water solution. 1 = native; 2 – 5 = subjected to mechanical treatment: ball mill, 2 hours (2), ball mill, 24 hours (3), planetary mill, 10 min (4), extremely intense treatment, mixed ball loading (5); Eluent: $H_2O/0.1$ N $LiNO_3$

Polysaccharides ground in a high-rate planetary mill diminish markedly in molecular weight and change slightly their polydispersity index Mw/Mn. The mechanical destruction is stronger in polymers with larger molecular weights, which agrees with published evidence [39]. Note that highly branching macromolecules (HES and AG) break down into roughly equal fragments. The Mw/Mn ratios in polysaccharides do not grow much, possibly, because destruction mostly affects their high-molecular fractions. Destruction is apparently controlled by the structure of polysaccharide molecules and physicochemical chain breakdown mechanisms. According to a model for linear synthetic polymers [41], the chains that occur in the middle of macromolecules are especially prone to failure. Destruction of HES and AG is qualitatively similar to that model, though dextrans and fibregum may deform by a different mechanism.

The results for larch arabinogalactan are worth of special consideration. High-rate treatment in a planetary mill, especially with mixed ball loading reduces strongly the AG molecular weight. According to chromatograms (Fig. 2), its Mw 17.3 kDa macromolecules split quantitatively into two almost equal parts of Mw = 8.3 kDa, while their Mw/Mn ratio decreases to 1.08 [23]. Therefore, the native AG molecules may consist of two relatively weakly bonded fragments of equal molecular weights and easily break down on milling

[23,39]. Note that AG macromolecules with MM (Molecular Mass) ~9 kDa are likewise the main product of chemical destruction of Canadian larch AG [42].

Furthermore, the analyzed polysaccharides experience almost no mechanical failure on low-rate grinding in a rotary mill (Table 3). Thus, ball rotary milling appears to be most often preferable, as molecular mass changes in technologically produced polymers are commonly unwanted in view of their further use in dietary supplements and drugs, otherwise additional tests and standardization may be required.

Sample	Treatment	Mn, kDa	Mw, kDa	Mw/ Mn	Weight shares of macromolecules, kDa	
					10%	90%
Fibregum	Native	146.6	256.7	1.8	<75.9	<528.2
	Planetary mill, 20g, 20min	31.4	55.2	1.8	<16.3	<113.5
	Ball mill, 1g, 4hours	120.3	231.6	1.9	<60.3	<478.4
Arabinogalactan	Native	13.9	17.3	1.2	<9.0	<27.9
	Planetary mill, 20g, 20min	9.3	11.2	1.2	<6.1	<18.4
	Ball mill, 1g, 4hours	13.1	16.3	1.2	<8.1	<26.2
Hydroxyethyl starch 200/0,5	Native	47.9	116.9	2.4	<20.9	<265.3
	Planetary mill, 20g, 20min	26.6	55.2	2.1	<12.7	<118.9
	Ball mill, 1g, 4hours	45.6	105.5	2.3	<20.0	<237.6
Dextran 70	Native	30.9	76.4	2.5	<14.0	<174.7
	Planetary mill, 20g, 20min	22.7	54.8	2.4	<10.4	<123.5
	Ball mill, 1g, 4hours	29.6	73.5	2.5	<13.4	<169.2
Dextran 40	Native	24.6	38.0	1.5	<13.3	<72.3
	Planetary mill, 20g, 20min	19.5	31.9	1.6	<10.4	<61.3
	Ball mill, 1g, 4hours	24.3	37.4	1.5	<13.0	<71.2
Dextran 10	Native	8.3	13.4	1.6	<4.2	<26.4
	Planetary mill, 20g, 20min	8.0	12.1	1.5	<41.9	<22.7
	Ball mill, 1g, 4hours	8.3	13.4	1.6	<4.2	<26.3

Table 3. Molecular mass distribution of polysaccharides

Toxicological tests of the milled polysaccharides show that a single intragastric injection administered at doses from 500 to 6000 mg/kg body weight caused no death in experimental animals. Their appearance, behavior, and state were within the background over the whole dose range; no statistically significant changes in body temperature relative to the control was observed, and body weight growth was uniform in all groups. Injections of the tested polysaccharides neither induced any considerable effect on the central nervous system of the mice. Patomorphological postmortem examination of mice in 14 days after polysaccharide administration revealed no pathology in thoracic and abdominal cavities. The median lethal dose LD_{50} for all polysaccharides was over 6000 mg/kg body weight on single intragastric injection.

2.4. Supramolecular structures of glycyrrhizic acid (GA) and poorly soluble drugs water solutions: Synthesis and properties

Biosynthetic and natural plant-derived carbohydrate-bearing metabolic agents have been increasingly used for obtaining complexes (clathrates) with drugs in drug delivery research. The mechanism of GA-pharmacon interaction in solutions may consist in involving drug molecules into self-associates (micelles) that exist in a wide range of concentrations in GA solutions. Until recently however, the existence of micelles in GA solutions had no direct proof but was either inferred from measured concentration dependences of solution viscosity [43] or was studied by dynamic NMR spectroscopy in water-methanol solutions [44]; the latter (30% concentration) were at the same time used as solvent for technical reasons. Thus, the molecular mechanism of drug complexing remained unclear, whether it was incorporation into micelles or supramolecular complexes with GA in water solutions, without organic solvents which change notably the GA-pharmacon reactions.

GA water solutions, with and without the presence of poorly soluble drugs, were investigated in [29] using gel permeation chromatography, which allows detecting self-associates/micelles and estimating their sizes and concentrations. On the other hand, solid GA-drug dispersions were obtained with the mechanochemical approach developed earlier [10, 11]. The binding strengths in GA-pharmacon supramolecular complexes or GA micelles in water solutions were compared using the criterion of solubility increase in poorly soluble drugs [22] and studied in terms of pharmacological activity.

Chromatograms of GA water solutions (Fig. 3) show peaks of high-molecular (~ 46-67 kDa) forms over all studied concentration ranges, while the GA molecular weight is 836.96 Da (Table 4).

The peak areas, being proportional to the solution concentrations and calculated relative to the known amounts of standard dextrans, show that almost all GA is stored in the solution. This, in our view, is evidence for the existence of GA self-associates (micelles). The critical concentration of micelle formation (CMC) was estimated earlier [43] from viscosity change in GA solutions to be 0.004 wt. % (0.05 mM). In our case estimating exact CMC is difficult for the limited sensitivity of refractometric detector and for dilution in the chromatographic column. However, it may be inferred from the time to the chromatographic peak (~0,5 min) and the elution rate. The solution we studied underwent about 10-fold dilution, and the derived CMC is $\sim\leq$ 0.0001 wt.%, (0.001mM), or far less than in water-methanol solution (0.04-0.08 wt.% or 0.5-1.0 mM) [44]. In diluted 0.01-0.001 wt. % solutions, there is only one type of micelles (~ 66 kDa) with a very low Mw/Mn ratio of 1.08-1.06. As the GA solutions reach concentrations of 0.5 wt.%, micelles decrease in weight to form ~ 46 kDa bodies and increase in Mw/Mn ratios. Therefore, almost all glycyrrhizic acid in water solutions from 0.0001 to 0.5 wt.% exists in the self-associated form of micelles, out of which the ones with MM= ~ 66 kDa consisting of about 80 GA molecules are most stable.

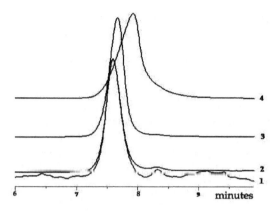

Figure 3. GPS chromatograms of glycyrrhizic acid water solution. 1 - concentration - 0.001, 2 - 0.01, 3 - 0.1, 4 - 0.5 wt.%.

COMPOSITION	Molecular masses	Solution concentration, wt. %			
		0.001	0.01	0.1	0.5
GLYCYRRIZIC ACID	Mw/Mn kDa	65.93/61.0	66.2/62.5	60.7/57.3	45.3/36.5
GLYCYRRIZIC ACID/ /IBUPROFEN 10/1 w/w Treated in planetary mill 3 min	Mw/Mn kDa	69.0/67.2	69.4/65.3	65.2/61.7	48.6/39.0

Table 4. Molecular-mass characteristics of micelles in solutions of glycyrrhizic acid and its composition with ibuprofen

2.4.1. GA solid dispersions with poorly water-soluble drugs

Solid GA dispersions with ibuprofen, phenylbutazone, clozapine, and diazepam were obtained by milling with GA (10:1 mass, or 2.5/1 – 4/1 molar ratios). Their thermal analysis data indicate that the crystalline phase of the drugs becomes disordered, until complete loss of crystallinity. In our view, the molecules of drugs may disperse into the excess solid GA with formation of solid solutions. Other investigated systems behave in a similar way.

As the dispersions dissolve, the drugs become more soluble in water (Table 1), this being evidence of the efficiency of GA as a solubilizing agent and the mechanochemical treatment as a tool for synthesis of water-soluble solid dispersions. GA has a nearly intermediate solubilizing effect higher than HES but lower than AG.

2.4.2. GPC of dissolved GA-drug solid dispersions

The GPC data for GA-ibuprofen dispersions in water are shown in Table 4. Similar results of MM increase in micelles were obtained for the GA-diazepam, GA- phenylbutazone, and GA-clozapine systems. The peak areas, proportional to the concentrations of the analyzed solutions, indicate that they bear almost the entire mass of GA-drug samples. Thus, the dissolved drugs in the GA-drug complexes are likewise self-associated in micelles which are stable in a broad range of concentrations, as well as in the solutions of native GA. Therefore, we suggest that poorly soluble drugs increase their solubility by incorporating into GA micelles/self-associates. GA molecules contain a hydrophilic (two glucuronide residues) and a hydrophobic (triterpene) components (Fig. 4).

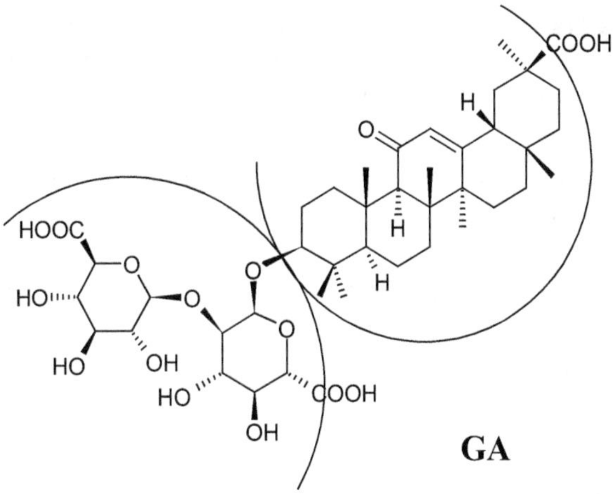

Figure 4. The structure of glycyrrhizic acid

In micelles, the latter are most likely oriented inward and the former toward the outer surface of the self-associate, while the drug molecules may occur either in the interior hydrophobic part or complex with the exterior hydrophilic part of the micelles. Unfortunately, the experimental evidence is insufficient to judge about these subtle GA-drug interaction mechanisms. Generally, the MM of GA-pharmacon micelles are 5-7% higher than those in GA water solutions, all over the studied range of concentrations. Another possibility is that drug molecules may substitute for some GA molecules while the micelles generally grow in size. As the solution concentration increases, the size difference of micelles grows correspondingly. Note that the region of high GA concentrations is proximal to the conditions in which the solubility of drugs was measured, this being additional evidence for the suggested mechanism of water solubility increase. An explanation for the decreasing MM differences on dilution may be that drug molecules can escape from GA micelles during GPC process to diluted solutions and appear in chromatograms as individual substances eluted at different rates [45]. Anyway, further investigation is needed to gain more insights into the GA-drug interaction.

3. Pharmacological activity of supramolecular complexes

The pharamacological activity of the complexes was investigated *in vivo* on females and males of outbred white mice and on Wistar rats, obtained from the SPF vivarium of the Institute of Cytology and Genetics, Novosibirsk. All animal procedures and experiments followed the 1986 Convention on Humane Care and Use of Laboratory Animals. The activity was determined using standard pharmacological tests [46]. All studied complexes, with some exceptions, are AG-Drug or GA-Drug compositions of 10:1 weight ratio, which was found out to be of greatest efficacy [22].

3.1. Arabinogalactan-drug complexes

As a special study has shown, AG complexing with nonsteroidal anti-inflammatory drugs and non-narcotic analgesics of various action mechanisms can reduce the required dose for 5-100 times and, hence, avoid the side effects typical of the drugs.

For instance, complexing with indomethacin allows 10 to 20 times dose reduction relative to the standard and induces twice fewer cases of lesion to gastric mucous membrane, with the same high anti-inflammatory activity [22, 24].

Administration of phenylbutazone in a complex with AG, at ten times lower pharmacon content, stimulated analgesic activity, both in the model of chemical effect and on thermal action, which may expand its applicability scope (Table 5).

Compounds	Hot plate, s	Acetic acid writhing model, amount
Control	28.73±2.34	4.76±0.26
Phenylbutazone:AG 1:10. 120 mg/kg per os	42.20±5.20*	2.00±0.09*
Phenylbutazone, 12 mg/kg per os	11.40±0.80	4.25±0.50
Ibuprofen:AG 1:10. 200 mg/kg per os	23.50±2.43	0.75±0.01*
Ibuprofen, 200 mg/kg per os	19.5±2.45	1.50±0.03*
Metamizole sodium: AG 1:10. 50 mg/kg per os	32.50±3.70	1.75±0.08
Metamizole sodium, 50 mg/kg per os	30.40±3.20	0.63±0.10
Metamizole sodium, 5 mg/kg per os	17.70±2.30	4.25±0.75
*p <0.05 relative to control		

Table 5. Analgesic activity of AG complexes with non-narcotic analgesics

AG complexes with 10 times smaller doses of ibuprofen and metamizole sodium showed a high analgesic activity in visceral pain models (the former) and in two models of test pain.

Studies in this line were continued, with the above approach and proceeding from the obtained results, on AG complexes with drugs that involve the central nervous system. Specifically, AG complexing with diazepam allowed reducing the dose for ten times and enhanced the anxiolytic effect. AG complexes with mezapam acted as standard anxiolytics at 20 times lower doses of the phamacon. The AG-clozapine complexing provides a two-fold dose decrease at a higher sedative action.

Another objective was to study drugs involving the blood coagulation and cardiovascular systems. The AG-warfarin (WF) complex was tested on intragastric injection in females of Wistar rats, with prothrombin time (PT, in seconds) as the principal criterion of the action. PT is the classical laboratory test of the exterior blood coagulation pathway used to evaluate the system of hemostasis in general and the efficacy of warfarin therapy in particular. The complex was administered once, in a dose of 20 mg/kg body weight, which is equivalent to 2 mg/kg warfarin. In 24 hours after the injection, PT increased considerably (to 30 s against the 11.63 s for the intact control). With free warfarin, this time was 42 s, or 28.5% longer than with the AG-warfarin complex, but it equalized (21 s) for both agents in 48 hours after a single injection (Fig. 5,6).

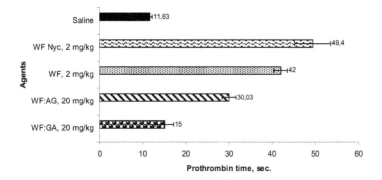

Figure 5. Prothrombin time, 24 hours after single dosing of WF:AG and WF:GA.

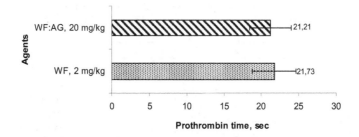

Figure 6. Prothrombin time, 48 hours after single dosing of WF:AG and WF.

The difference in pharmacokinetics between the AG:WF complex and free warfarin was further explored after a single administration of 20 and 2 mg/kg body weight, respectively. Blood was sampled in 1, 8, 10, 12, 24, 48, and 72 hours after injection on decapitation. Fig. 7 shows average plasma warfarin contents, and pharmacokinetic parameters are listed in Table 6. The concentrations of the compounds increase in a similar way but free warfarin reaches C_{max} seven hours sooner (T_{max}) than in complexes with AG, and the concentrations become equal in 24 hours after single dosing. Excretion, on the contrary, is slower in pure warfarin than in the complex, which is consistent with 27 % higher clearance (CL) in the complex than in the free drug. Thus, warfarin increases more smoothly when bound with AG than in the free form and thus poses lower bleeding risks associated with its abrupt rise during dosage adjustment. Furthermore, shorter mean retention times (MRT) for the AG:WF complex may secure the following injection and accelerate warfarin excretion in the case of drug withdrawal.

Thus, complexing of warfarin with AG increases its safety and reduces unwanted bleeding risks in the case of anti-coagulant treatment.

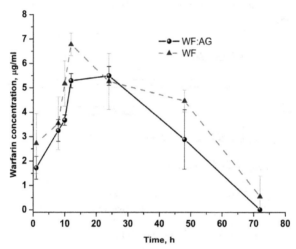

Figure 7. Mean plasma concentration–time profile of WF:AG and blank WF after single oral administration at a dose of 20 mg/kg (dose of WF is equal to 2 mg/kg) and 2 mg/kg, respectively.

Compounds	WF, 2 mg/kg	WF:AG, 20 mg/kg
CL, ml/h	1.52±0.03	1.93±0.18*
MRT, h	31.39±1.82	21.81±2.38*
Terminal half life, $T_{1/2}$, h	5.11±0.24	6.38±2.55
T_{max}, h	11.00±1.41	18.00±8.49
C_{max}, μg/ml	6.47±0.91	5.64±0.19
AUC, μg h/ml	263.01±0.02	208.34±20.03*
*$p<0.05$ against WF		

Table 6. Pharmacokinetic parameters of WF and WF:AG

Another drug in which we studied the pharmacological effect of complexing with AG was nifedipine (NF). NF is a dihydropyridine blocker of slow calcium channels that dilates coronary and peripheral vessels and reduces the oxygen demand in myocardium.

Nifedipine exerts a minor negative inotropic effect and a very weak antiarrhythmic action. Intravenous injection of 3.5 mg/kg AG:NF complex (0.35 mg/kg NF) caused 26 % drop of blood pressure, measured via a carotid cannula, while 0.35 mg/kg NF can provide only a 9% decrease.

Being aware that nifedipine has a pleiotropic antiarrhythmic effect, besides the basic hypotensive action, the NF:AG complex was investigated in this respect on intravenous injection in a model of arrhythmia induced with 250 mg/kg calcium chloride. The complex administered in a dose of 0.175 mg/kg body weight (0.0175 mg/kg NF) arrested lethal heart rate disorder in 100% and 65% of cases, respectively, when applied prior to and after exposure to the arrhythmogen. Pure NF at 0.0175 mg/kg had no antiarrhythmic effect in the model of calcium chloride arrhythmia.

Thus, the NF:AG complex has demonstrated a stronger hypotensive and antiarrhythmic action than pure NF on intravenous administration, while its effective hypotensive dose is ten times smaller. Arabinogalactan itself does not induce any statistically significant decrease in blood pressure and cardiac rates. Furthermore, it is important that the new method adds another water-soluble form of NF as there is the only soluble nifedipine (adalate) available in the market.

A similar hypotensive effect was obtained with nisoldipine, another dihydropyridine.

Complexing of hypoglycemic drugs (metformin, rosiglitazone, insulin) with AG increased their solubility but allowed no dose reduction, though it improved notably the state of animals exposed to a toxic dose of alloxan, an agent simulating trial hypergclycemia.

The reported pharmacological data on the AG complexes with these pharmacons agree with the results on complexing with terpenoids (glycirrhizic acid, stevioside, and rebaudioside). In both cases, complexing increases the basic activity of drugs, allows dose reduction and forms new properties. We suggest to call this effect complexing or clathration of pharmacons with plant-derived carbohydrate-bearing metabolic agents.

3.2. Glycyrrhizic acid (GA)-drug complexes

3.2.1. Nonsteroidal anti-inflammatory drugs (NSAID)

GA:NSAID complexes with acetylsalicylic acid (ASA), diclofenac (OF), phenylbutazone (BD), and indomethacin (IM) were synthesized in solutions [24] and in solid state [23]. Complexing was confirmed by spectrometry. In IR spectra, the bands of hydroxyl and carbonyl groups of the glycoside were shifted to short wave numbers.

All mentioned complexes show anti-inflammatory action in smaller doses than the primary drug, and have 3-11 times larger therapeutic index (LD_{50} /ED_{50}) [47,48] (Table 7).

GA complexes with aspirin and diclofenac (GA:ASA, GA:OF) exerted a prominent anti-inflammatory action in six models of acute inflammation induced by carrageenan, formalin, histamine, serotonin, Difko's agar, and trypsin, as well as in the cases of chronic inflammation (cotton and pocket granulomas) in intact and adrenalectomized animals [48].

Compounds	Dose range of complex*	Dose range of free NSAID
GA : ASA (1:1)	4500/82=54.8	1900/98=19.4
GA: OF (1:1)	1750/12.5=140	310/8=33.7
GA : BD (1:1)	3150/62=50.8	880/56=15.7
GA : AN (1:1)	8000/68=117.6	570/55=10.3

Table 7. Anti-inflammatory effects of GA complexes and free NSAID.*LD_{50}/ED_{50}; ED_{50} – effective dose

The anti-inflammatory action of the GA:IM complex is stronger than of the free drug at equal dosing (10 mg/kg body weight). GA:NSAID complexing also potentiates other (analgesic, antipyretic) biological activities [29]. The GA:OF complex exerted a more prominent anesthetic action than diclofenac in electric and thermal stimulation (57.5±2.0 and 43.2±2.6) and exceeded the amidopyrine effect in the case of thermoalgesic stimulation (23.4±1.1 and 18.5±1.4). The anesthetic activity of the GA: metamisole sodium (GA:AN) complex is 11.4 times higher relative to metamisole sodium (AN) alone [23] while and that of the GA:ASA complex exceeds the effect of aspirin in animals exposed to thermoalgesic stimuli. The GA complexes with ASA and OF are 3 and 2.3 times more potent pain relievers than the respective NSAID in acetyl choline writhing model. The GA:ASA complex demonstrates a 4 times higher therapeutic index than aspirin in the acetic acid writhing model. The GA:ASA and GA:OF complexes show high antipyretic activity, twice larger than in the pure pharmacons [47,48].

Thus, water-soluble GA:ASA and GA:OF complexes evoke prominent anti-inflammatory and antipyretic effects, their spectrum of pharmacological activities and therapeutic ratio being larger than in the respective NSAID. The complexes also induce a marked membrane-stabilizing effect and reduce accumulation of primary and secondary products of lipid peroxidation in animals with chronic inflammation.

Complexing of glycyrrhizic acid with ibuprofen increases the analgesic action of the latter at twice lower doses.

Furthermore, GA complexes irritate less strongly the gastric mucosal membrane than their NSAID counterparts. For instance, the GA:ASA complex promotes reparation of ulcers though the ulcerogenic activity of GA:OF is minor. Both complexes diminish E1 and E2 blood prostaglandins in animals with chronic inflammation. The complexes can be recommended for clinical trials as anti-inflammatory agents, including for patients that suffer from ulcer of stomach and duodenum. The acute toxic effect of GA:NSAID complexes is 2 to 14 times as low as in the respective pharmacons (Table 8) [47, 48].

GA:NSAID complexing obviously exerts a synergetic effect of higher biological activity along with lower toxicity and weaker ulcerogenic action on the gastrointestinal tract, and has a higher water solubility. It is evident on comparing the pharmacological activities of the ASA:GA and OF:GA supramolecular complexes obtained by dissolution and solid-state mehcanochemical synthesis that the latter opens a promising perspective of a technologically preferable and saving way of producing highly active NSAI drugs [23].

Complexes	LD_{50}, mg/kg	Pharmacon LD_{50}, mg/kg
GA : ASA (1:1)	4500	1900
GA: OF (1:1)	1750	310
GA : BD (1:1)	3150	880
GA : AN (1:1)	8000	570

Table 8. Acute toxicity of GA:NSAID complexes in mice (per os)

The GA:Rofecoxib complex at doses 50 and 100 mg/kg body weight shows no active anti-inflammatory and analgesic effects but is more highly soluble.

3.2.2. Prostaglandins

Prostaglandins have been of broad use in human and veterinary medicine for their ability, in small doses, of stimulating womb muscles.

Veterinary uses of prostaglandins have been especially important: prostaglandin-based drugs are employed in swine breeding for farrow synchronization and are highly potent in preparing cattle and horse females for artificial insemination; these drugs allowed solving many problems with puerperal complications in cows and horses.

Kloprostenol, one main veterinary drug, is made by multistage synthesis, like other prostaglandins, which imposes its high price. It is urgent to reduce the effective dose and at the same time to increase the stability of labile prostaglandins in finished drugs. Both solutions have been found through complexing E and F prostaglandins (PGE1, PGE2, PGF2α), sulprostone (SP) and kloprostenol with GA. The complexes were synthesized and tested for uterotonic activity (Table 9).

In experiments on rats and guinea pigs, the GA:PGE1 (1:1) and GA:PGF2α (1:1) complexes changed the amplitude of uterine actions twice more strongly than the same concentration of PGE1 sodium (10-8 g/ml). SP and PGE2 as complexes with GA (1:1) induced three times greater amplitudes and increased uteric tonicity [49].

An efficient veterinary drug of klatraprostin has been developed on the basis of GA complexing with kloprostenol, a well known synthetic luteolitic prostaglandin, at doses five times as low as in the world practice [48, 49]. The drug is cheaper than its imported analogs while its action is more physiological than in the best foreign counterparts.

Klatiram, which contains an aminoacid (tyrosine) besides GA and kloprostenol [48,49], has a still greater potency. It is more effective than estrofan though bears 100 times less prostaglandin (kloprostenol).

Compound	Change of uterine contraction amplitude, %	P	Change in uterus tonus, %	P
GA: PGE1	53.4±5.0	<0.002	49.4±1.2	<0.002
PGE1	24.3±1.5	<0.05	30.7±2.2	<0.05
GA : SP	150.0±11.0	<0.001	135.0±10.0	<.001
SP	50.0±5.0	<0.001	115.0±9.5	<0.001
GA : PGE2	63.5±6.0	<0.001	40.7±4.0	<0.002
PGE2	20.0±2.8	<0.05	33.5±2.4	<0.05
GA: PGF2α	55.6±5.0	<0.001	61.0±5.6	<0.001
PGF2α	27.8±1.5	<0.05	39.4±5.3	<0.02

Table 9. Uterotonic activity of prostaglandins and their GA complexes (1:1) in rats *ex vivo*, phosphate buffer C = 10-8 g/ml. PGE1 , PGE2, SP, PGF2α.

3.2.3. Cardiovascular drugs and anticoagulant warfarin

Pharmacological activity was also investigated in GA complexes with antiarrhythmic Lappaconitine hydrobromide (LA) and antihypertensive nifedipine (NF) drugs.

Lappaconitine hydrobromide belongs to the group of clinically used antiarrhythmic drugs and is administered to patients with various rhythm disorders, especially, ventricular arrhythmia, paroxysmal ciliary arrhythmia, and monofocal atrial tachycardia, but it has a drawback of high toxicity.

In special experiments on antiarrhythmic action of LA:GA complexes, the one patented as alaglizin [50] showed the highest efficacy. Alaglizin, being ten times less toxic than LA, induced antiarrhythmic effects in models of calcium chloride and aconitine arrhythmia and had the highest antiarrhythmic therapeutic index (LD_{50}/ED_{50}) among all available drugs of this kind. When administered at 0.125 mg/kg and 0.250 mg/kg body weight, alaglizin causes no influence on electrocardiogram parameters, according to an extended study in models of calcium chloride and adrenal arrhythmia. Intravenous injection of 0.125 mg/kg alaglizin prior to exposure to a lethal dose of calcium chloride blocked arrhythmia in 80% of rats; 0.250 mg/kg of alaglizin applied after arrhythmogenic $CaCl_2$ stopped the already developed arrhythmia in 50% of animals. A single injection of 0.125 mg/kg and 0.250 mg/kg alaglizin in a model of adrenal arrhythmia prevented full development of arrhythmia; 50% and 100 % of animals recovered normal ECG parameters on receiving 0.125 mg/kg and 0.250 mg/kg alaglizin, respectively. Alaglizin in the model had an ED_{50} of 0.125 mg/kg body weight against 0.290 mg/kg in LA and thus contained 14 times lower lappaconitine.

The NF:GA complex exerted hypotensive action on intravenous injection of its water solution in rats at a dose with ten times lower NF [51, 52]. The hypotensive effect of GA complexing with nisoldipine (another dihydropyridine) was similar to that of complexes with AG.

Complexing can have an important consequence of amplifying the pleiotropic effect of pharmacons (see above), such as NF. The wanted antiarrhythmic effect of NF can be only achieved with a dose that causes almost critical blood pressure drop, but the NF:GA complex induces the same effect with 29 times lower NF than the hypotensive dose.

Thus, the NF:GA complex is a promising parenteral drug with universal activity against hypertensive crises attendant with arrhythmia.

A presumable action mechanism of the NF:GA complex was studied *in vitro* on neurons isolated from peripharyngeal ganglia of *Lymnacea stagnalis* molluscs. The provoked responses are arrested completely with an NF concentration as high as 3.0 mM but with 30 times as low concentration (0.1 mM) of NF:GA. The responses to NF are blocked faster and recover sooner after neuron washing than the responses induced by the complex, which indicates stronger binding of the latter with receptors and its more prolonged action. The higher receptor affinity of the complex is corroborated by comparing the NF and NF:GA effects on calcium channels [51] (Fig. 8)

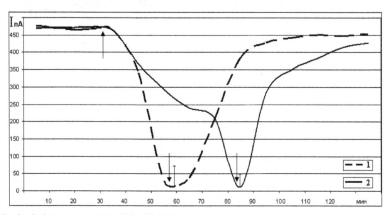

1. amplitude of calcium currents induced by Nifedipine.
2. amplitude of calcium currents induced by Nifedipine clathrate with glycyrrhizic acid.
Upward arrows show start point of the blocker action. Downward arrows show start point of blockers washout.
X axis set as time in minutes.
Y axis set as amplitude of incoming current in pA.

Figure 8. Averaged changes in calcium current amplitude induced by action and washout of blockers within a group of neurons (12 cells in each group).

The propranolol-GA complex studied in terms of hypotensive activity ensured 14% blood pressure drop in normotensive animals at a dose of 0.2 mg/kg body weight (minimum dose 0.0025 mg/kg gave a 11% decrease). Pure propranolol decreases blood pressure for 11% in a dose of 0.2 mg/kg; a similar pressure drop may be achieved by the minimum dose 0.0025 mg/kg, but it is statistically unconfident. Therefore, complexing enhances and stabilizes the hypotensive action of propranolol and allows 12.5-fold reduction of its dose.

Furthermore, complexing amplifies the antiarrhythmic action (pleiotropic effect) of propranolol. Namely, 100% of animals exposed to 0.3 mg/kg arrhythmogenic 0.1 % adrenalin survived on administration of 0.0025 mg/kg GA:Propranolol, while only 40 %

survived among those who received 0.0002 mg/kg propranolol (this being the dose contained in the complex). Control animals responded to adrenalin by fatal disrhythmia grading into ventricular fibrillation, and 100% of them died [53].

The complex of GA with anticoagulant warfarin was investigated in a dose of 20 mg/kg body weight, which corresponds to 2 mg/kg WF. Prothrombin time was first measured six hours after a single intragastric injection. The interval was chosen proceeding from pharmacokinetics of warfarin: its plasma metabolites reach the maximum in 6-12 hours in rats [54]. In our experiments, PT has shown a confident change only in the positive control group with 2 mg/kg WF, but it is not clinically significant as it fails to ensure the required increase in coagulation time. A significant PT increase was observed on single intragastric administration of WF, while the WF:GA complex induced a smooth PT rise as late as in 30 hours (two injections) after the beginning of the experiment and the values corresponding to the positive control (WF) were reached only in 54 hours (three injections). Thus, WF:GA complexing made WF more soluble in water but slowed down its wanted anticoagulant action, possibly because GA molecules screened its active centers (Fig. 9).

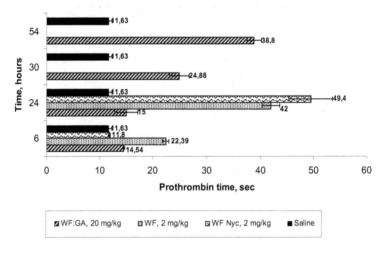

Figure 9. Prothrombin time of WF:GA

3.2.4. Psychotropic drugs

The complexation effect was discovered in a pharmacological study of GA complexes with antidepressant fluoxetine and with anxiolytic gamma-amino-beta-phenylbutirate hydrochloride. Fluoxetine (FL), N-methyl-3-phenyl-3-[4-(trifluoromethyl)phenoxy]propan-1-amine acts as a depression antagonist by inhibiting serotonin neuronal uptake in CNS. Antidepressants are known to have many drawbacks, such as large doses, a narrow activity spectrum, high toxicity and prolonged elimination that involves liver cells and exerts deleterious effects on the renal function.

The study of fluoxetine complexes with GA (hereafter fluaglizin) aimed primarily at checking the possibility to alleviate the side effects of the pharmacon. In preliminary testing, the 1:10 FL:GA complex showed the highest activity. The complex, with 0.072 pharmacon parts in its one weight part, was patented under the name of fluaglizin (FG) [55,56]. Note that its LD_{50} exceeds 5000 mg/kg body weight against LD_{50} =248 mg/kg in fluoxetine.

Fluaglizin exerts a more prominent antidepressant effect than fluoxetine on single-dose administration in the Porsolt test and is a stronger serotonin uptake inhibitor. For instance, it suppressed the action of chloral hydrate more effectively than fluoxetine in a test with 5-hydroxytryptophan. Like fluoxetine, fluaglizin lacks anxiolytic activity.

The antidepressive effect of fluaglizin was identical to that of fluoxetine in a model of social depression in mice, namely, the animals became twice more communicative toward familiar and unfamiliar partners (had twice greater frequency and time of contacts). The dose of fluoxetine in the FL:GA complex is 1.08 mg/kg body weight against the standard 15 mg/kg. Fluaglizin, like fluoxetine, prevents blood glucose drop and attenuates peroxidation normalizing the antioxidant status of depressed individuals [55, 56].

In order to understand the mechanism of FL:GA action and compare it with fluoxetine, we studied its effect on contents of catecholamines and their precursors in different brain parts on single and therapeutic 25 mg/kg administration. The 17 times lower dose of fluoxetine in FL:GA complexes induced a weaker effect on serotonin uptake and triggered dophamine exchange in brain [57]. The nootropic activity of fluoxetine, first mentioned in [48], shows up also in the FL:GA complex.

Fluoxetine (30 μM) is known to suppress epileptiform activity which is evident in 50% lower-amplitude oscillations of electric potential (reflecting the activity of nerve cells) in response to pulse stimulation of hippocampus on the background of picrotoxin action.

Like fluoxetine, fluaglizin inhibits bikukulin-induced epileptiform activity in hippocampus sections in rats and acts as epilepsy antagonist [48].

Phenybut (FB), 4-amino-3-phenyl-butyric acid, is a nootropic and tranquilizing drug which relieves stress and anxiety and improves sleep. It is used in clinical practice against asthenia, neurotic anxiety, and sleep disorders, prior to surgery, and for preventing naupathia. It, however, has drawbacks of inducing somnolence and allergic reactions.

The FB:GA complex is twice less toxic than FB and stimulates cognitive activity in the same way as the pharmacon and GABA but, unlike the two latter, it provides a 20% increase in memorizing abilities in animals and attenuates sedative effects [48].

3.2.5. Anticancer drugs

GA complexes (1:1) with 5-fluorouracil, tegafur, and daunorubicin hydrochloride were synthesized in solution [58]. Complexing decreases the toxicity of the drugs and increases their solubility in water. The GA complex with fluorouracil exerts antineoplastic action with respect to Pliss lymphosarcoma, B-16 melanoma, and Geren carcinoma. The effective indices are, respectively, 3.05; 2.11, and 1.76. The tumor growth inhibition is 67.2%; 53.4 %; 87.1 %.

3.3. Antimicrobial drugs

GA complexes with antibiotics (chloramphenicol) and sulfanilamides (sulfamethoxypyridazine, saladimetoxine, sulfamonomethoxine, sulfadimidine, and sulfaguanidine), as well as with isoniazid and nitrofural, 1:1, were likewise synthesized in solution and compared in terms of their microbial activity. The highest survival rate (90%) was observed on the 10th day after exposure to infection in animals with staphylococcosis that received 50 mg/kg GA-chloramphenicol, while only 30% survived when pure chloramphenicol was given. The percentage of survived animals with *Pseudomonas aeruginosa*, *Proteus vulgaris*, and *E. coli* infections reached 80% on GA: chloramphenicol administration but it was only 20-50% on treatment with free chloramphenicol. The complex was also shown to stimulate humoral and cell immune responses [59].

3.4. Simvastatin (hypocholesterolemic drug)

Inhibitors of 3-hydroxy-3-methylglutaryl coenzyme A (3HMG-CoA) reductase, the so-called statins, are known to successfully reduce low-density lipoproteins, and are used for this in atherosclerosis therapy. However, most statins cause side effects and have to be replaced by safer drugs, with a more prolonged action. NMR spectroscopy of the behavior of simvastatin (SMS) in solutions with GA indicates formation of stable complexes. The synthesized SMS:GA complex is stable in water solutions at GA above 0.2 mM [60]. The complex patented as simvaglysine (SMG) [61] demonstrates uncompetitive inhibition of 3HMG-CoA reductase. SMG, in doses with three times lower statin, turns out to be more potent and safer than SMS.

Thus, simvaglysine acts as an uncompetitive inhibitor/proinhibitor of the 3HMG-CoA reductase reaction by inducing inhibition of cholesterol synthesis in liver microsomal fraction of rats *in vitro*, being no less potent than simvastatin. Within the inhibition constants from 100 to 300 nM, SMS inhibits 37.7- 42.0% mevalonate formation, while SMG provides a 31-33% total blood cholesterol decrease after 14 days of administration in rats with hypercholesterolemia, at doses with 3 times lower SMS, which is as effective as the therapeutic dose of simvastatin. The greater safety of SMG is confirmed by a lower blood CPK (creatine phosphokinase) increase after 14 days of treatment than in the case of SMS: 2.3 lower than with 2-5 times larger SMS doses [60, 61].

4. Conclusion

Thus, water-soluble molecular complexes of various polysaccharides and glycyrrhizic acid with drugs that normally dissolve poorly in water have been synthesized and tested for binding strength and pharmacological activity in comparison with the constituent drugs.

The suggested mechanochemical synthesis of solid dispersed systems "drug-complexing agent" ensures high strength of the complexes, on condition of low-energy nondestructive treatment of polysaccharide macromolecules.

Arabinogalactan, a water soluble polysaccharide of *Larix sibirica* Ledeb. and *Larix gmelinii* (Rupr.), when used as a complexing agent, provides the highest solubility among other studied poly- and oligosaccharides. Another advantage of arabinogalactan is its exceptional, almost infinite, raw material source in the Northern Hemisphere (in Russia and Canada), as well as the availability of extraction and purification technologies [62].

The molecules of poorly water soluble drugs can be carried by micelles that form in water solutions of glycyrrhizic acid. Complexing with polysaccharides, in the same way as with GA, increases the solubility of pharmacons and allows reducing significantly (up to ten times) the effective dose avoiding, at the same time, some unwanted side effects.

Thus, a new perspective is opening of obtaining highly effective and safe drugs and new ways of drug delivery.

Author details

A. V. Dushkin
Institute of Solid State Chemistry and Mechanochemistry, Siberian Branch,
Russian Academy of Sciences, Russia

T. G. Tolstikova, M. V. Khvostov and G. A. Tolstikov
N. N. Vorozhtsov Novosibirsk Institute of Organic Chemistry, Siberian Branch,
Russian Academy of Sciences, Russia

Acknowledgement

This work is supported in part by the program of partnership fundamental research of the SB RAS, Grant 19.

5. References

[1] Anil J Shinde (2007) Solubilization of Poorly Soluble Drugs: A Review. Pharmainfo.net. Available: http://www.pharmainfo.net/reviews/solubilization-poorly-soluble-drugs-review. Submitted 2007 Nov 13.
[2] Payghan S, Bhat M, Savla A, Toppo E, Purohit S   (2008) Potential Of Solubility In   Drug   Discovery   And   Development.   Pharmainfo.net.   Available:

http://www.pharmainfo.net/ reviews/potential-solubility-drug-discovery-and-development. Submitted 2008 Oct 21.

[3] Yellela S.R. Krishnaiah (2010) Pharmaceutical Technologies for Enhancing Oral Bioavailability of Poorly Soluble Drugs. Journal of Bioequivalence & Bioavailability. 2(2): 028-036.

[4] Robert H. Marchessault, François Ravenelle, Xiao Xia Zhu editors (2006) Polysaccharides for Drug Delivery and Pharmaceutical Applications. V.934. ACS Symposium Series. 365p.

[5] Gordon A. Morris, M. Samil Kok, Stephen E. Harding and Gary G. Adam (2010) Polysaccharide drug delivery systems based on pectin and chitosan. Biotechnology and Genetic Engineering Reviews. 27: 257-284.

[6] Sabyasachi Maiti, Somdipta Ranjit, Biswanath Sa (2010) Polysaccharide-Based Graft Copolymers in Controlled Drug Delivery. International Journal of PharmTech Research. 2(2):1350-1358.

[7] Mark E.Davis, Marcus E. Brewster (2004) Cyclodextrin-based pharmaceutics: past, present and future. Nature reviews I Drug discovery. 3: 1023-1035.

[8] Erem Bilensoy, editor (2011) Cyclodextrins in Pharmaceutics, Cosmetics, and Biomedicine: Current and Future Industrial Applications. WILEY. 440p.

[9] A.M.Dubinskaya (1999) Transformations of organic compounds under the action of mechanicalal stress. Russian Chemical Reviews. 68(8): 637- 652.

[10] Dushkin A.V. (2004) Potential of Mechanochemical Technology in Organic Synthesis and Synthesis of New Materials. Chemistry for Sustainable Development. 12(3): 251 – 273.

[11] Dushkin A.V. (2010) Mechanochemical synthesis of organic compounds and rapidly soluble materials : in M.Sopicka-Lizer, editor, High-energy ball milling. Mechanochemical processing of nanopowders. Woodhead Publishing Limited. Oxford, pp. 249-273.

[12] A.V. Dushkin, E.V.Nagovitsina, V.V.Boldyrev, A.G.Druganov (1991) Mechanochemical reactions of solid organic compounds. Siberian Chemical J. 5: 75 -81.

[13] A.V.Dushkin,_I.B.Troitskaya,V.V.Boldyrev and I.A.Grigorjev (2005) Solid-phase mechanochemical incorporation of a spin label into cellulose. Russian Chemical Bulletin, International Edition, 54(5): 1155—1159.

[14] E.S.Meteleva, A.V.Dushkin, V.V.Boldyrev (2007) Mechanochemical obtaining of chitosan's derivatives. Chemistry for Sustainable Development. 15: 127-133 (rus).

[15] A. V. Dushkin, L. M. Karnatovskaya and others. (2001) Synthesis and evaluation of ulcerogenic activity of instant disperse solid systems based on acetylsalicylic acid and biologically active licorice components. Pharmaceutical Chemistry Journal. 35(11): p.605-607.

[16] A,V.Dusjkin, N.V.Timofeeva (2001) Method of producing quick soluble tabletted form of acetyl-salicylic acid, Russia Patent 2170582.

[17] A.V.Dushkin, S.A.Guskov, V.N.Bugreev (2005), Method of preparing powder-like water-soluble sparking composition, Russia patent 2288594.

[18] Gus'kov S, Dushkin A and Boldyrev V (2007) Physical-chemical bases of mechanochemical obtaining of rapidly soluble disperse systems. Chemistry for Sustainable Development. (15): 35-43 (rus).

[19] Dushkin A, Karnatovskaia L and others (2000), Solid-state mechanochemical synthesis of fluoroaromatic compounds, Doklady Chemistry, 371, 632-635.

[20] Dushkin A, Karnatovskaia L and others. (2001b), Solid-phase mechanochemical synthesis of fluoroaromatic compounds, Synth. Commun., 31, 1041-1045.

[21] A.V.Dushkin, Z.Yu.Rykova, V.V.Boldyrev, T.P.Shaktshneider (1994) Aggregation processes in the reactivity of mechanicalally activated organic solids. Int. J. Mechanochem. Mechanochem. Alloying. 1: 1 – 10.

[22] A.V.Dushkin, E.S.Meteleva, T.G.Tolstikova and others (2008) Mechanochemical preparation and pharmacological activities of water-soluble intermolecular complexes of arabinogalactan with medicinal agents. Rus. Chem. Bull. 6: 1299-1307.

[23] A.V.Dushkin, E.S.Meteleva, T.G.Tolstikova, M.V.Khvostov, G.A.Tolstikov (2010) Mechanochemical Preparation and Properties of Water-Soluble Intermolecular Complexes of Polysaccharides and β-Cyclodextrin with Pharmaceutical Substances. Chemistry for Sustainable Development. 18(6):719-728.

[24] A.V.Dushkin, T.G.Tolstokova and others (2008) Water-soluble pharmaceutical composition and method for its preparation. Russia patent 2337710.

[25] N.E.Polyakov, T.V.Leshina, E.S.Meteleva, A.V.Dushkin and others (2009) Water Soluble Complexes of Carotenoids with Arabinogalactan. J. Phys. Chem. B. 113 (1): 275-282.

[26] T.G.Tolstikova, G.A.Tolstikov and others (2008) Drugs for treatment of hypertension and heart rhythm disorders. Russia patent 2391980.

[27] A.V.Dushkin, E.S.Meteleva and others (2011) The composition of high pharmacological activity on the basis of dihydroquercetin and plant polysaccharides (options). Russia patent 2421215.

[28] L.N.Pribytkova, S.A.Gus'kov, A.V.Dushkin, S.I.Pisareva (2011) Mechanochemical Obtaining of Water-soluble Composition on the Base of Quercetin. Chemistry of Natural Compounds. 3: 333-336.

[29] A. V. Dushkin, E. S. Meteleva, T. G. Tolstikova and others (2010) Complexation of Pharmacons with Glycyrrhizic Acid as a Route to the Development of the Preparations with Enhanced Efficiency. Chemistry for Sustainable Development. 18(4): 517-525.

[30] T.G.Tolstikova, M.V.Khvostov and others (2010) Improvement of pharmacological values of the nifedipine by means of mechanochemical complexion with glycyrrhizic acid. Biomedical chemistry. 56(2): 187-194 (rus).

[31] Tolstikova T. G.,   Khvostov M. V. And others (2011) Alteration of Warfarin's Pharmacologic Properties in Clathrates with Glycyrrhizic Acid and Arabinogalactan. Letters in Drug Design & Discovery. 8(3): 201-204.

[32] S.S.Khalikov, A.V.Dushkin, M.S. Khalikov and others (2011) Mechanochemical modification of properties of anthelmintic preparations. Chemistry for Sustainable Development. 19(6): 699-703 (rus).

[33] T. Loftsson, M.E. Brewster (1996) Pharmaceutical Applications of Cyclodextrins. 1. Drug Solubilization and Stabilization. J.Pharm.Sci. 85(10): 1017-1025.

[34] Ramnik Singh, Nitin Bharti, Jyotsana Madan, S. N. Hiremath. (2010) Characterization of Cyclodextrin Inclusion Complexes – A Review. Journal of Pharmaceutical Science and Technology. 2 (3): 171-183.

[35] Clarcke A.E., Anderson R.L., Stone B.A. (1979) Form and function of arabinogalactans and arabinogalactan-proteins. Phytochemistry. 18: 521–540.

[36] Marta Izydorczyk, Steve W. Cui, and Qi Wang (2005) Polysaccharide Gums: Structures, Functional Properties, and Applications. Food Carbohydrates Chemistry, Physical properties and Applications. Available: http://uqu.edu.sa/files2/tiny_mce/plugins/filemanager/files/4300270/1/2/1574_C006.pdf.

[37] Philippe H. Hünenberger, Ulf Börjesson, Roberto D. Lins (2001) Electrostatic Interactions in Biomolecular Systems. Chimia. 55: 861–866.

[38] C Oprea, F Dan (2006) Macromolecular Mechanochemistry Polymer Mechanochemistry. Cambridge International Science Publishi. (1)1: 495p.

[39] A.V.Dushkin, E.S.Meteleva, T.G.Tolstikova, M.V.Khvostov (2012) GPC and toxicology investigation of mechanochemical transformation of water-soluble polysaccharides. Pharmaceutical Chemistry Journal. 2012(7) In press.

[40] A.M.Striegel, J.J.Kirkland, W.W.Yau, D.D.Bly (2009) Modern Size Exclusion Liquid Chromatography, Practice of Gel Permeation and Gel Filtration Chromatography, 2nd ed.; Wiley: NY. 494p.

[41] N.Grassie, G.Scott (1985) Polymer degradation and stabilization. Cambridge University Press.(1985: 201-207.

[42] J.H.Prescott, P.Enriquez, Chu Jung, E.Menz and others (1995) Larch arabinogalactan for hepatic drug delivery: isolation and characterization of a 9 kDa arabinogalactan fragment. Carbohydrate Research. 278: 113-128.

[43] T.V.Romanko, Yu.V.Murinov (2001) Some features of the flow of dilute solutions of glycyrrhizic acid. Rus.J.Phys.Chem A. 75(9): 1459-1462.

[44] N.E.Polyakov, V.K.Khan, M.B.Taraban, T.V.Leshina (2008) Calcium Receptor Blocker Nifedipine with Glycyrrhizic Acid. J. Phys. Chem. B. 112: 4435-4440.

[45] H.Determan. (1969) Gel Chromatography. Springer Verlag, New York. 252p.

[46] R.U.Khabriev (2005) Guidance for experimental (preclinical) study of new pharmacological compounds, Medicine, Moscow, 832p. (rus).

[47] T.G.Tolstikova, M.V.Khvostov, A.O.Bryzgalov (2009) The Complexes of Drugs with Carbohydrate-Containing Plant Metabolites as Pharmacologically Promising Agents. Mini-Reviews in Medicinal Chemistry. 9: 1317-1328.

[48] G.A.Tolstikov, L.A.Baltina, V.P.Grankina, R.M.Kondratenko, T.G. Tolstikova (2007) Licorice: biodiversity, chemistry and application in medicine, Academic Publishing House "GEO", Novosibirsk, 311p. (rus).

[49] G.A.Tolstikov, Yu.I.Murinov and others (1991) Complexes of β-glycyrrhizic acid with prostaglandins – new class of uterotonic agents. Pharm. Chem. J. 3: 42-44 (rus).

[50] A.O.Bryzgalov, T.G.Tolstikova, I.V.Sorokina and others (2005) Antiarrhythmic activity of alaglizin. J. Expert Clin. Pharm. 68(4): 24-28 (rus).

[51] T.G.Tolstikova, A.O.Bryzgalov, I.V.Sorokina, M.V.Hvostov and others (2007) Increase in Pharmacological Activity of Drugs in their Clathrates with Plant Glycosides. Letters in Drug Design & Discovery. 4: 168-170.

[52] T.G.Tolstikova, M.V.Khvostov, A.O.Bryzgalov, A.V.Dushkin, E.S.Meteleva (2009) Complex of Nifedipine with Glycyrrhizic Acid as a Novel Water-Soluble Antihypertensive and Antiarrhythmic Agent. Letters in Drug Design & Discovery. 6: 155-158.

[53] T.G.Tolstikova, G.I.Lifshits, M.V.Khvostov, A.O.Bryzgalov, I.V.Sorokina. (2007) Complexing in the development of low-dose beta-adrenoceptor antagonists (experimental study). Herald of the NSU. Series: biology, clinical medicine. 2(5): 70-73 (rus).

[54] X.Zhu,. W.G.Shin (2005) Gender Differences in Pharmacokinetics of Oral Warfarin in Rats. Biopharm. Drug Dispos. 26: 147-150.

[55] D.F.Avgustinovich, I.L.Kovalenko, I.V.Sorokina, T.G.Tolstikova, A.G.Tolstikov (2004) Ethological study of antidepressant effect of fluaglisine in conditions of chronic social stress in mice. Bul. Exp. Biol. and Med. 137(1):99-103.

[56] T.G.Tolstikova, I.V.Sorokina, I.L.Kovalenko and others (2004) Influence of clathrate formation to the pharmacon activity in complexes with glycyrrhizic acid. Doklady Akademii Nauk. 394(2):707-709.

[57] G.T.Shishkina, N.N.Dygalo, A.M.Yudina and others (2005) Effects of Fluoxetine and Its Complexes with Glycyrrizhinic Acid on Behavior and Brain Monoamine Levels in Rats. I. P. Pavlov Journal of Higher Nervous Activity. 2: 207-212 (rus).

[58] L.A.Baltina, Yu.I.Murinov, A.F.Ismagilova, G.A.Tolstikov and others (2003) Production and antineoplastic action of complex compounds of β-glycerrhizic acid with some antineoplastic drugs. Pharm. Chem. J. 37: 3-4 (rus).

[59] R.M.Kondratenko, L.A.Baltina, G.A.Tolstikov and others (2003) Complex compounds of glycerrhizic acid with antimicrobial drugs. Pharm. Chem. J. 37: 32-35 (rus).

[60] V.A.Vavilin, N.F.Salakhutdinov, Yu.I.Ragino and others (2008) Hypocholesteremic properties of complex compound of simvastatin with glycerrhizic acid (simvaglizin) in experimental models. Biomed. Chem. 54: 301-313.

[61] Yu.I.Ragino, V.A.Vavilin, N.F.Salakhutdinov and others (2008) Examination of the hypocholesteremic effect and safety of simvaglizin on the model of hypercholesterolemia in rabbits. Bull. Exp. Biol. Med. 145: 285-287.

[62] E. N. Medvedeva, V. A. Babkin, L. A. Ostroukhova (2003) Larch arabinogalactan - features and prospects, Review Chemistry of Vegetable Raw Materials [Khimiya rastitel'nogo syr´y], 1: 27-37 (rus).

Chapter 8

Polysaccharides from Red Algae: Genesis of a Renaissance

María Josefina Carlucci, Cecilia Gabriela Mateu,
María Carolina Artuso and Luis Alberto Scolaro

Additional information is available at the end of the chapter

1. Introduction

The Red Queen Effect is an evolutionary hypothesis [1]. This evolutionary concept is named for the Red Queen´s comment to Alice in Through the Looking Glass that "it takes all the running you can do, to stay in the same place". It posits that multicellular organisms with long life cycles must constantly change, adaptation process driven by the changing conditions of the environment, in order to survive the onslaught of potentially lethal pathogens which have much shorter life cycles and can thus evolve orders of magnitude faster.

Viruses play a relevant role in these new evolutionary mechanisms by transferring of genes to and from the hosts they parasite [2]. Over the past three decades, it has become apparent that viruses are ubiquitous, abundant and ecologically important in the environment [3]. As phylogenetic analysis shows, nearly all organisms of all kingdoms have become infected by viruses since the beginning of life. The great impact that viruses can have on the genetic systems is well illustrated by the evolution of mitochondria. In reference [4] have shown, the existence of a strong selection pressure has pushed for the replacement of cellular enzymes by viral ones in mitochondria and chloroplasts. In both organelles, this replacement has been associated with profound modifications in the mechanism of DNA replication and chromosome structure [5]. The fact that viruses are probably very ancient allows better understanding their extraordinary diversity, explaining why most viral proteins inferred from genome sequencing have no cellular homologues [6]. Besides, the existence in the biosphere of an unlimited reservoir of viral proteins has provided opportunities at different steps of the evolutionary process, to introduce new functions into organisms.

At the present, it remains controversial the inclusion of viruses in the "tree of life". Several authors assume viruses are non-living organisms and believe their properties are driven solely by thermodynamically spontaneous reactions while others give priority to the fact that phylogenetic tree is based on the genomic content of its components, not the physical manifestations of these genomes. Moreover, the fact that viral genomes carried inside virions encode gene products that allow for adaptation and response to changing intracellular and extracellular conditions favors the inclusion of these agents in the tree of life [7,8].

The oligosaccharides chains (glycans) attached to cell surface and extracellular proteins and lipids are known to mediate many important biological roles [9,10]. However, for many glycans, there are still no evident functions that are of obvious benefit to the organism that synthesizes them. In 1949, Haldane postulated "Now every species of mammal and bird so far investigated has shown quite surprising biochemical diversity by serological tests. The antigens concerned seem to be proteins to which polysaccharides are attached. We do not know their functions in the organism, though some of them seem to be part of the structure of the cell membrane. I wish to suggest that they may play a part in disease resistance, a particular race of bacteria or virus being adapted to individuals of a certain range of biochemical constitutions, while those of other constitutions are relatively resistant" [11]. In [12], suggested that glycan diversification in complex multicellular organisms is driven by evolutionary selection pressures of both endogenous and exogenous origin. They also argued that exogenous selection pressures mediated by viral and microbial pathogens and parasites that recognize glycans have played a more prominent role, favoring intra-and interspecies diversity.

2. Red algae and carrageenans

Red algae (Division: Rhodophyta) are one of the oldest and largest groups of eukaryotic algae with more than 10000 species described (Figure 1). They are distributed worldwide but grow best in waters of near 15°C. They have the characteristic of all eukaryotes including the nuclei which in some algae are smaller than their plastids. However, their cells lack of flagella so they need the water movement to carry masculine cells to the oocyte. They also have disorganized chloroplasts lacking of external endoplasmic reticulum and containing unstacked thylakoids. Their red colour is due to the presence of the phycoerythrin pigment which reflects red light and absorbs blue/green ones. Since blue light penetrates water to a greater depth than light of longer wavelengths, red algae are able to photosynthesize and live in water of 260 m in deep which receive 0.1% of surface irradiance; this means one thousand times less light than the surface. Those rhodophytes that have small amounts of this pigment might seem green or bluish from the chlorophyll and other pigments present in them.

Over the last 2.45 billion years, algae have been diversifying [13] in order to survive in competitive ecological niches. This adaptation led to evolution of a large and diverse array of biochemical constituents.

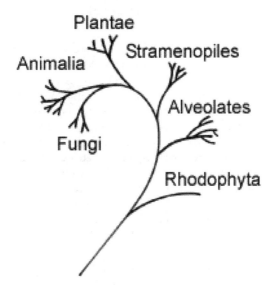

Figure 1. Rhodophyta branched off very early in the tree of life

Red algae contain large quantities of polysaccharides in the cellular wall, thereof, most are sulfated galactans. These galactans are generally constituted by alternatively repeated units of bonds 1,3-α-galactopyranose and 1,4-β-D-galactopyranose can defer in the level and pattern of sulfation, in the substitution by methoxy and/or pyruvate groups and other sugars such as mannose and xylose. They also defer in the 3,6-anhydrogalactose content and the 1,3 -α-galctopyranose residues configuration.

Among these galactans, the carrageenans (CGNs) may be mentioned, which have similar structures to the pattern observed in the galactosaminoglycans. They comprise a wide group of structures that may be divided in two families: the κ-family, defined by the presence of a sulfated C4 group in the unit β-D, and formed by CGNs-κ/ι and the carrageenans-μ/ν, and the λ-family, characterized by a sulfate-C2 group and constituted by all the varieties of λ structures. The λ- and ι-carrageenan types are more strongly sulfated than the most of the heparan sulfate (HS) derived from tissues [14]. In general, this type of carrageenans exhibits a viral inhibitory potential a little greater than the κ-carrageenans.

Polysaccharides are composed of building-blocks that although being not numerous, their almost infinite combination led to an array of polysaccharides with an important structural complexity [15]. The diversity of polysaccharides can be further increased by acetylation, methylation and, more commonly in the case of many marine algal polysaccharides, sulfation [16]. Moreover, many algal polysaccharides are metabolically active, either as a storage molecule which undergo structural changes during their life cycles or as a structural component [17].

The sulfated polysaccharides are highly abundant and accessible compounds that may be isolated from various natural sources. Micro- and macroalgae are under investigation for numerous commercial food, agri- and horticultural, pharmaceutical, cosmetic and bioenergy applications [18-21]. Polysaccharides are also known for their wide and variable physicochemical properties which make them suitable for different applications in the fields of medicine and pharmacology. They have proved to be useful tools due to their immune-modulator and antitumoral activity, their interference in the clotting system and in the inflammatory processes, in dermatology, in dietary programs and moreover by affecting the viral replication. Among the natural sources where they can be found are the cell walls from algae. Depending on the type of algae, those with similar structures to the glycosaminoglycans (GAGs) and wide antiviral activity can be isolated [22] (Table 1). Antiviral activity has been documented for retrovirus: human immunodeficiency virus type 1 and 2 (HIV-1 and HIV-2), herpesvirus: herpes simplex virus (HSV) type 1 and 2, human cytomegalovirus (HCMV); pseudorabies virus; flavivirus: dengue virus type 2; smallpox virus: variola virus; hepadnavirus: hepatitis B virus (HBV); orthomyxovirus: influenza A virus (inf A); paramyxovirus: respiratory syncytial virus (RSV) and parainfluenza virus; rhabdovirus: vesicular stomatitis virus (VSV); arenavirus: Junin virus, Tacaribe virus and togavirus: Sindbis virus, Semliki Forest virus and against some naked viruses, such as encephalomyocarditis virus, Hepatitis A virus [23] and papilloma virus (HPV) [24], of both DNA and RNA viral types (Table 1). For most of these viruses the initial bond of the virus to the cells would be mainly mediated by the interaction of virus with a GAG of the cellular surface known as HS [14]. In general, sulfated polysaccharides have a chemical structure very similar to the HS. Thus, they might block viral infection by competing against virion attachment to the cell surface.

GAGs are linear polysaccharides constituted of successive repetition of a disaccharide unit which may be sulfated. GAGs can be divided in two groups; Glucosaminoglycans, like HS and galactosaminoglycans like chondroitin sulfate. The initial incorporation of saccharide units of N-acetylglucosamine or N-acetylgalactosamine, respectively, gives their names. An important difference between these groups is that Glucosaminoglycans are attached by 1,4 union while galactosaminoglycans are attached by 1,3 and 1,4. GAGs are found mainly in the cell surface and in much of the intracellular matrix of the mesodermic tissue as is shown in Table 2 (connective, cartilage, muscle and bone). Frequently, they are linked to a core protein and one or more covalently attached glycosaminoglycan chains, known it as Proteoglycans.

GAGs are negatively charged molecules that can have a physiological significance like hyaluronic acid, dermatan sulfate, chondroitin sulfate, heparin, HS and keratan sulfate.

Algae	Compound	Virus	Reference
Red algae			
Schizymenia pacifica	λ- carrageenan	HIV-1, AMV	[25]
Schizymenia dubyi	Sulfated galactans with uronic acid	HIV-1, HSV-1, HSV-2, VSV	[26]
Nothogenia fastigiata	(Xylo)mannans	HIV-1,HIV-2, SIV, HSV-1, HSV-2, HCMV, Inf A, RSV, Junin, Tacaribe	[27,28]
Aghardiella tenera	Sulfated Agarans	HIV-1, HIV-2, HSV-1, HSV-2, HCMV, VSV, Inf A, RSV, togavirus, parainfluenza virus, smallpox	[29]
Digenea simples	sp non-characterized	HIV	[30]
Nothogenia fastigiata	Xylogalactans	HSV-1, HSV-2	[31]
Pterocladiella capillacea	Sulfated Agarans and hybrid DL- galactans	HSV-1, HSV-2, HCMV	[32]
Gigartina skottsbergii	λ-,κ/ι- and μ/ν- carrageenans	HSV-1, HSV-2	[33,34]
Cryptopleura ramosa	Sulfated agarans	HSV-1, HSV-2	[35]
Stenogramme interrupta	Carrageenans	HSV-1, HSV-2	[36]
Asparagopsis armata	Sulfated agarans	HIV	[37]
Bostrychia montagnei	Sulfated agarans	HSV-1, HSV-2	[38]
Gymnogongrus torulosus	Hybrid DL- galactans	HSV-2, virus dengue 2	[39]
Gracilaria corticata	Sulfated agarans	HSV-1, HSV-2	[40]
Brown algae			
Pelvetia fastigiata	Fucans	HBV	[41]
Fucus vesiculosus	Fucans	HIV-1	[42]
Sargassum horneri	Fucans	HSV-1, HCMV HIV-1	[43]
Leathessia difformis	Fucans	HSV-1, HSV-2	[44]
Adenocystis utricularis	Fucans	HSV-1, HSV-2	[45]
Microalga			
Cochlodinium polykrikoides	Extract	HIV-1, RSV, Inf A, Inf B	[46]
Porphyridium sp	Extract	HSV-1, HSV-2	[47]
Green algae			
Monostroma latissinum	Sulfated Rhamnans	HSV-1, HCMV, HIV-1	[48]

Table 1. Antiviral activity of sulfated polysaccharides extracted from marine algae

GAG	Location	Comments
Hialuronates	Synovial fluid, vitreous humor, extracellular matrix with loss of connective tissue (vasculogenesis).	Long polymers (containing no sulfates), shock-absorbing.
Chondroitin sulfate	cartilages, bone and cardiac valves.	More abundant GAGs.
Heparan sulfate	Basal Membrane and components of the cellular surface.	
Heparin	Components of the intracellular granules of the mastocytes, coating of the lung arteries, liver and skin.	More sulfated than the Heparan sulfate.
Dermatan sulfate	Skin, cardiac valves and blood vessels.	Long polymers (no sulfates), shock-absorbing.
keratan sulfate	Cornea, bone and cartilage.	More abundant GAGs.

Table 2. Normal distribution of GAGs in the body.

3. Relationship among glycosaminoglycans, carrageenans and viruses

The discovery that viruses are highly abundant in natural waters initiated renewed research on the impact of viral infection and lysis on aquatic microorganisms [49]. It is believed that viruses influence the composition of marine communities and are a major force behind biogeochemical cycles. Each infection has the potential to introduce new genetic information into an organism or progeny virus, thereby driving the evolution of both, host and virus [50].

The eukaryotic algae represent the oldest known eukaryote for which there exist clear geological data [51] and all classes of algae have their specific DNA virus. The HSV is an ancient DNA virus which is widespread in nature and has coevolved with its hosts. Many viruses interact with their host polysaccharides present on the cell wall; HSV uses HS.

The basic structural motifs and modifications of HS glycosaminoglycans seem to have been conserved for several hundred million years of evolution [52,53].

One suggested explanation is that endogenous heparan sulfate-binding proteins may have developed different binding specificities with evolutionary time. At first glance, this may seem an exception to the suggestion made here, in view that extensive glycan diversification has accompanied species evolution. However, HS can generate numerous intrinsic structural variations, and there are currently inadequate data about the extent of species-specific differences in the specificities of the binding proteins and/or the expression of structural motifs in different cell lineages.

HS is highly sulfated and it is thought to be the most biologically active GAG. The sulfated monosaccharide sequences within HS determine the protein binding specificity and regulate fundamental biological functions including growth control, signal transduction, cell adhesion, homeostasis, morphogenesis, lipid metabolism and pathophysiology [14]. Numerous viruses including herpesviruses utilize cell surface HS as receptor to infect target cells.

It has been reported that in the course of an inflammation, an infection or tissue damage, the proteoglycan HS is cleaved causing fragments of soluble HS [54]. On the other hand, in healthy tissues, no significant fractions of soluble HS are found, though they can be found in the fluids of damaged tissues –at concentrations within the required ranges to stimulate dendritic cells [55] and in the infected individuals urine [56].

HSV attaches to cells by an interaction between the envelope glycoprotein C and cell surface HS. The virus-cell complex is formed by ionic interactions between the anionic (mainly sulfate) groups in the polysaccharide and basic amino acids of glycoproteins, and non-ionic ones depending on hydrophobic amino acids interspersed between the basic ones in the glycoprotein-binding zone [57]. This interaction is a decisive step in virus multiplication and may be differentiated but not dissociated from an evolutive point of view.

CGNs resemble to some extent the naturally occurring GAGs owing to their backbone composition of sulfated disaccharides are believed to be of potential therapeutic importance because they can mimic with GAGs present in cell membranes.

Natural CGNs, extracted from red seaweeds, are well known as potent and selective inhibitors of HSV-1 and HSV-2. CGNs chemical structures are similar to that of HS that serves as a primary receptor for adsorption of HSV onto cells. Mode of antiviral action is mediated by the interference with HSV attachment to cells, blocking the interaction virus-HS, a mandatory step during the multiplication cycle to achieve a productive infection that involves viral glycoproteins. In this work were used the κ and ι CGNs, their structures are present in the compound named 1C3 CGN, which is an "hybrid" κ/ι-and partially cyclized μ/ν CGN (Figure 2).

Since most pathogens and the toxins they produce bind to specific sugar sequences to initiate infection and disease, it is reasonable to assume that at least some glycan variation must have arisen from this selection pressure [12]. On this basis, pressure of selection *in vitro* with an antiviral drug like HS in the case of HSV may be employed to shorten the time necessary for attenuation. Moreover, in this last case we may speculate that if herpesviruses which are extensively spread in the environment are exposed to sulfated polysaccharide (its natural receptor), in the form of CGN, the appearance of virus variants would readily occur as a consequence of an intense virus-host interaction. Our results indicate that attenuation is a common trait of HSV obtained under selective pressure of CGN.

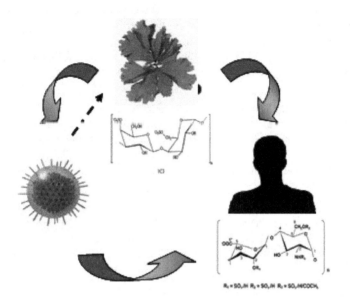

Figure 2. 1C3 CGN isolated from the red algae *Gigartina skottbergii* has a chemical structure similar to cellular HS. Attenuated HSV can be isolated from viral populations grown in the presence of increasing concentrations of CGN. This procedure may reflect an accelerated evolution process for HSV where biological modification of the viral particle can be demonstrated.

4. Viruses: Friend or enemy

It is tempting to postulate that the driving forces of evolutionary novelty are not randomly derived from chance mutations of the genetic text, but from a precise genome editing by omnipresent viruses [58]. For decades, non-coding regions of the genome have been ignored or declared as "junk"-DNA. Recently, scientists have realized that these regions incorporate decisive higher-order regulatory functions. New research has shown that these non-coding repetitive sequences originated primarily from retroviral RNA [59,60].

For a long time, viruses were interpreted as causing acute infections of susceptible organisms, using the host cellular machinery to reproduce, and achieving their lytic nature only in order to infect other cells. Although this narrative remains valid, it merely represents a case of viruses that were unable to reach persistent or chronic status of infection [61]. Most viruses, however, are stable, persistent agents that are able to establish a complex relationship with the host cell and in many cases this interaction lasts for the entire lifespan of the cell even when a competent immune system is present. The immune system is of crucial importance in defense against infection. It has to cope with a large number of different pathogens that relentlessly develop new ways to avoid recognition or elimination. Yet most infections are cleared. Immune-system genes must evolve to keep pace with increasingly sophisticated evasion by pathogens. To remain effective, defense demands creativity and competence because many pathogens have sophisticated and rapidly evolving evasion mechanisms [62]. This defense response is triggered by the presence of pathogens within the host, however there exists an immune response that comprises a set of autoreactive "natural antibodies" that do not rely on exogenous antigen stimulation to be synthesized by autoantibody-secreting B lymphocytes [63,64]. On the other hand, these autoreactive natural antibodies are reactive against components of the host's immune system (i.e. cytokines) exacerbating ongoing infectious diseases or predisposing host to infection [65,66].

Individual resistance to pathogens depends on the combination of receptors on cells from the immune system although non-immune genes also influence resistance [67]. Signs of natural selection in a human population are especially illustrative, when a mutation in a certain gene is dangerous in normal conditions but confers resistance to infections widespread in the region. Among the better known are the mutations in hemoglobin and glucose-6-phosphate-dehydrogenase affecting red blood cells and conferring resistance to malaria [68]. Another example is the deletion at the 5′end of the CCR5 chemokine receptor conferring resistance to HIV infection. This molecule serves as the principal co-receptor, with CD4, for HIV-type 1. The allele with the deletion was intensely selected in Europe probably because it also provided resistance to plague and smallpox [69]. More subtle, but nonetheless important, relationship between cell and virus is that associated to changes in cell physiology due to viral infection that regulates cell death, transformation, secretory pathways, cell stress response, etc [70-72]. This panorama may account for a "symbiotic evolution" of cell and virus, although viruses have much shorter generation times than cells. Studies of genomic polymorphism of HSV-1 suggest that the evolution of this virus would be very slow and host-dependent [73].

5. Effect of carrageenans on the virus

In our laboratory, viral variants of HSV-1 (strain F) and HSV-2 (strain MS) were obtained by successive passages in Vero cells under selective pressure with the 1C3 CGN (is an "hybrid" κ/ι-and partially cyclized μ/ν-CGN) and ι or κ-CGNs respectively in order to test the ability of the CGN to generate resistant variants during the selection process and to study the CGN-virus modulation. Different clones were plaque purified and pretested to exclude reversion to wild type (Figure 3). 1C314-1 and 1C317-2 are viral variants derived from HSV-1 strain F (1C314-1 means, "1C3" is the type of CGN that was used for the selection of the variant, "14" number of pasagge that we chose for cloning, and "1" is the number of selected clon). κ22-12, κ22-13, ι22-9 and ι22-12 are variants derived from HSV-2 strain MS. In this case, κ and ι CGNs were used for the selection of these variants, respectively.

After the viral selection and cloning, all the variants showed a syncytial phenotype on Vero cells, this phenotype was also observed on mouse lung and genitals primary cell cultures [74,75].

Polysaccharide concentration

Figure 3. Variants of HSV were obtained by successive passages in Vero cells under selective pressure with different types of CGNs with increasing concentration of them.

In order to characterize the obtained viral variants, the susceptibility to CGNs was assessed. All the viral variants showed low or middle levels of resistance to CGNs, heparin, Aciclovir (ACV), and Foscarnet (PFA) (Figure 4).

The results are shown as relative resistance (RR): IC50 viral variant/IC50 parental strain.(IC50 : inhibitory concentration 50%, is the concentration in µg/ml required to reduce plaque number by 50%).

Although some variants showed 15 or 17 fold of RR, those values are not significant compared with viral variants selected with ACV (ACVp6-F and ACVp6-MS), that showed higher values, > 50 RR, after only 6 passages with ACV. (unpublished data).

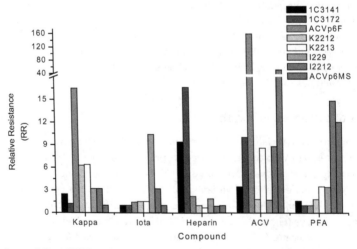

Figure 4. Susceptibility of HSV viral variants to several compounds with known antiherpetic activity. IC50values were determinate on Vero cells by a reduction plaque assay; the RR values were calculated as (IC50: viral variants/ IC50 parental strain). Kappa: κ CGN; iota: ι CGN; ACV: Aciclovir; PFA: Foscarnet.

Mucosal surface represents the primary site for replication and spread of several viruses including HSV type 1 or 2. In order to resemble natural routes of infections, virulence of the viral variants was assessed in a mouse model by two routes of infection, intravaginal or intranasal using either BALB/c or C57BL/6 mice.

The HSV-1 variants were avirulent for BALB/c mice infected by intravaginal route, although the parental strain F was highly lethal (Table 3). The attenuation of the variants correlated with low levels of pro-inflammatory cytokines (IL-6 and TNF-α) in vaginal lavages with respect to the parental strain, despite that viral titers were similar between the viral variants and the F strain. Nevertheless, the variants were highly lethal for BALB/c mice inoculated by the intranasal route, with a generalized organ spreading of virus.

On the other hand all the clones of HSV-2 were less virulent for mice intravaginally and intranasally inoculated, particularly for C57BL/6 mice, whereas MS strain produced 100% of mortality. In contrast to HSV-1 variants, the attenuation correlated with high levels of pro-inflammatory cytokines (TNF-α and IL-6) detected in vaginal lavages (Table 3).

Despite the differences in the virulence of variants, the levels of infectivity in the site of inoculation did not differ from those observed for the parental strains. These results suggested that the observed attenuation was not due by a lower viral replication and it could be explained by a differentiated immunological response.

Viral Characterization	HSV-1 (F)		HSV-2 (MS)	
	Parental strain	Viral variants	Parental strain	Viral variants
Cytopathic effect (*in vitro*)	Cell rounding	Syncytial	Cell rounding	Syncytial
Intravaginal virulence	High	Avirulent	High	Low
Intranasal virulence	High	High	High	Low
Levels of IL-6 and TNF-α	High	Low	Low	High

Table 3. Characterization of HSV variants obtained by selective pressure with CGN. Viral variants obtained under pressure of selection with carrageenans were characterized regarding their phenotype, *in vivo* virulence for BALB/c mice and immune response. The parental strains were also assessed for final comparison.

Moreover, by the same methodology, variants of HSV-2 obtained by selective pressure with nonsulfated compounds (from the condensation of mandelic acid) have been obtained. Viral variants were not syncytial and showed resistance to the same drug, heparin and the carrageenan 1C$_3$ in the order of 2.6 to 6.7 times with respect to the control virus. However, in vivo, these variants showed no difference of pathogenicity and mortality with the parental strain.

6. The evolutive beliefs

About seven hundred and fifty million years ago, the first multicellular organisms appeared on earth. Since then, cells from multicellular organisms have found a way to become smarter. Multicellular life forms were initially isolated communities or "colonies" of unicellular organisms. However, the evolutionary advantages of living in a community soon led to communities composed of millions and billions of individual cells socially interactive. The evolutionary trend towards increasing complexity in the community is but a reflection of the biological imperative of survival. The better an organism perceives the surrounding environment, the more chances to survive. Unfortunately, we "forget" that cooperation is conveniently necessary for the evolution when Charles Darwin brought out a radically different theory about the origin of life. One hundred and fifty years ago, Darwin concluded that living organisms are involved in a constant "struggle for survival". In the final chapter of The Origin of Species by Means of Natural Selection, or The maintenance of favored races in the struggle for existence, Darwin speaks of an unstable "struggle for existence" and that evolution is conditioned by "the war of nature, from famine and death". This concept is related to the Darwinian notion that evolution occurs at random [76]. While Darwin is the most famous evolutionist, the first scientist to consider evolution as a scientific fact was the distinguished French biologist Jean Baptiste de Lamarck [77,78]. Even Ernst Mayr, the principal agent of the "neo-Darwinism," a modernization of Darwin's theory that incorporates molecular genetics of the twentieth century, admits that Lamarck was the pioneer [79]. Lamarck presented his theory not only fifty years before Darwin, but also offered a far less violent theory of evolutionary mechanisms. Lamarck's theory suggests that evolution is based on a cooperative and "instructive" interaction between organisms and the environment that allows living things to survive and evolve in a dynamic world. His idea was that organisms acquire and pass on the necessary adaptations to survive in a changing environment. One of the reasons why scientists are rethinking the theories of Lamarck is that evolutionists continue to remind the invaluable role of cooperation in maintaining life in the biosphere. In the book Darwin's Blind Spot written in 2002 by British physicist Frank Ryan has been recorded a number of these relationships [80]. Today, the understanding of cooperation in nature is much deeper than that obtained by simple observation. Biologists are increasingly aware that the animals have coevolved with different sets of microorganisms necessary for a healthy and "normal" development [81]. Recent advances in genetics have revealed an additional mechanism of cooperation between species. It has been discovered that genes are shared not only among individual members of a species (sexual reproduction) but also between members of different species. The distribution of information through gene transfer accelerates the process of evolution, in view that organisms can learn lessons "learned" by others [82-85]. Because of this distribution of genes, organisms can no longer be considered as isolated entities, there are no walls between species [86]. The distribution of genes is not a "by chance" mechanism. Is the method that nature uses to increase the survival of the biosphere. Genes are nothing more than a physical memory of the experiences learned by the organisms, the exchange of these genes between species spreads these "memories" and, consequently, influences the survival of all organisms that constitute a living community. Timothy Lenton has shown that evolution is more dependent on the interaction between species than on the interaction among individuals of the same species [87]. The evolution thus becomes a question of survival of fittest groups, not of individuals better adapted.

7. Genesis of a renaissance – Cooperative biocommunication

Darwinian theory emphasizes competition and selection of the individual as a main guiding force in evolution (not cooperation or group membership). Evidence supports the fact that viruses and other genetic parasites are key elements in the evolution of all living organisms. In "Life: The Communicative Structure", Dr. Witzany suggests that the genesis of new species, genera, and realms of organisms would not occur in any neo-Darwinistic sense via "chance mutations" and selection, but via a kind of innovation code (evolution code, creation code, text generating code), which is capable of DNA/RNA text editing [88]. It turns out that the genetic code that encodes proteins -practically the sole subject of current bioengineering- is only a kind of structuring vocabulary, and not a complete structure in itself, and is subjected to a high-order regulatory code that lies hidden in the nonprotein-coding regions of the DNA, which have been identified as RNA agents many years later [89]. There is increasing evidence that all cellular life is colonized by exogenous and/or endogenous viruses in a nonlytic but persistent lifestyle. A persistent lifestyle in cellular life-forms most often seems to derive from an equilibrium status reached by at least two competing genetic and the immune function of the host that keeps them in balance. If we imagine that humans and one of the simplest animals, *Caenorhabditis elegans*, share a nearly equal number of genes (ca. 20,000) it become obvious that the elements that create the enormous diversity are not the protein coding genes but their higher order regulatory network that is processed by the mobile genetic elements, such as transposons and retroposons and noncoding RNAs [90]. If we consider the important role of the highly structured and ordered regulatory network of noncoding RNAs as not being randomly derived, one of the most favorable models with explanatory power is the virus-first thesis [91].

For many decades it was common practice to speak about the "genetic code" with its inherent language-like features. The concept postulated by Manfred Eigen that nucleic acid sequences are comparable to and function like a real language, coherent with a (molecular) syntax, linguistic and a vocabulary, was commonly used in genetics, cell biology, and molecular biology.

In contrast with the evolutionary paradigm of random assemblies of nucleic acids that constitute the genetic text we do not know any real-life languages or codes which emerged as a randomly derived mixture of characters. Every language is based on signs, whether they are signals or symbols. In humans and other animals they are transported auditively, visually, or tactilely. In nonhuman living beings they are transported by small molecules in crystallized, fluid, and gaseous form. Additionally these signs can be combined coherently with combinatorial rules (syntax). Signs are not generated and used by themselves, but in real-life languages by living beings. These sign-generating and sign-using agents live *in vivo* in continued changing interactions and environmental circumstances (interaction HSV-cells-carrageenans). This is the context (pragmatics) in which a living being is interwoven. This context determines the meaning (semantics) of the signs in messages that are used to communicate and to coordinate single as well as group behavior (changes in HSV-glycoproteins in contact with cell surface modificate innate response). Therefore, we may understand that the same sentence, or the same syntactic sequence order, of any language or code can have different, and in extreme cases, opposite meanings and therefore transport different messages. The important consequence of this fact is that it is not possible to extract the meaning of an information content solely out of the syntactic structure, but someone has to identify the context within which the living being uses this syntactic structure. The primary agents are not the sequences of signs, nor the rules which determine sequences, but the living agents. Without living agents there are no signs, no semiotic rules, no signalling, and no communication, no living agents could coordinate growth and development. If we assume the genetic code to function language-like, knowing that no language which has been observed functions by itself, then we have to postulate living agents that are competent to use signs coherent with syntactic, pragmatic, and semantic rules. Adapted to the genetic code, this means that there must be living agents competent in generation and integration of meaningful nucleotide sequences, and meaningful nucleotide sequences are not a randomly derived mixture of nucleotides. In accord with Dr. Witzany, this view could change the construction of research projects, that is, shifting the focus from mutational (random) changes of nucleotide sequences to investigating nucleotide sequences from the perspective of viral-derived sequences that now play important roles in the regulation of cellular functions, for example, HSV asymptomatic infection where the virus have reached an equilibrium status balance by the immune response of the infected host to achieve a latent lifestyle. Their status within one of many addiction modules (genetic and genomic innovations) together with the host immune system, each of them a unique culture-dependent habitat can be changed by nonbeneficial circumstance for the cell (e.g., stress) and they may become lytic again, resulting in a variety of diseases [59].

8. Conclusion

This work invites us to think about a possible alternative to attenuate virus for basic science study, therapeutical or prophylatic applications, employing natural compounds with chemical structures already "seen" by the pathogen and present in the host as essential cellular components widely distributed in nature. Besides, selective reexpression of viral ligands (using different type of polysaccharides) in conjunction with pathogenesis experiments will allow the testing of predictions about putative protective roles played by some glycans in certain tissues. Studying the comparative glycobiology of closely and distantly related species should also help, by ascertaining the rates of glycan diversification during evolution.This strategy could be considered as a natural evolutionary process where the virus contributes with valuable "updated" information gathered from previous ancestral infections and making it available for "new" actual hosts, generating a reciprocal benefit between host and virus.

Author details

Carlucci María Josefina*, Mateu Cecilia Gabriela,
Artuso María Carolina and Scolaro Luis Alberto
*Departamento de Química Biológica- IQUIBYCEN (CONICET-UBA),
Laboratorio de Virología, Facultad de Ciencias Exactas y Naturales, Ciudad Universitaria,
Universidad de Buenos Aires, Argentina*

Acknowledgement

Research in the authors' laboratory was supported by Agencia Nacional de Promoción Científica y Tecnológica y Consejo Nacional de Investigaciones Científicas y Tecnológicas (CONICET).

9. References

[1] Van Valen (1974) Two modes of evolution. Nature, 252, 298-300.

[2] Villarreal L.P, Witzany G (2010) Viruses are essential agents within the roots and stem of the tree of life. J.Theor.Biol. 262, 698-710.

[3] Breitbart M, Rohwer F (2005) Here a virus, there a virus, everywhere the same virus? Trends Microbiol. 13, 278-84.

[4] Gray M.W, Lang B.F (1998) Transcription in chloroplasts and mitochondria: a tale of two polymerases. Trends Microbiol. 6, 1-3.

[5] [5] Forterre P (2006) The origin of viruses and their possible roles in major evolutionary transitions.Virus Res. 117, 5-16.

[6] Daubin V, Ochman H (2004) Start-up entities in the origin of new genes. Curr. Opin. Genet. Dev. 14 ,616-619.

[7] Hegde N.R, Maddur M.S, Kaveri SV, Bayry, J (2009) Reasons to include viruses in the tree of life. Nat. Rev. Microbiol. 7, 615.

[8] Ludmir E.B, Enquist L.W (2009) Viral genomes are part of the phylogenetic tree of life. Nat. Rev. Microbiol. 7, 615.

[9] Gahmberg C.G, Tolvanen M (1996) Why mammalian cell surface proteins are glycoproteins. Trends Biochem. Sci. 21,308-311.

[10] Varki A, Marth J (1995) Oligosaccharides in vertebrate development. Semin. Dev. Biol. 6, 127-138.

[11] Haldane J.B (1949) Disease and evolution. In Symposium sui fattori ecologi e genetici della specilazione negli animali, Supplemento a la Ricerca Scientifica Anno 19th, 68-75.

[12] Gagneux P, Varki A (1999) Evolutionary considerations in relating oligosaccharide diversity to biological function. Glycobiology 9, 747-755.

[13] Rasmussen B, Fletcher I.R, Brocks J.J, Kilburn M.R (2008) Reassessing the first appearence of eukaryotes and cyanobateria. Nature; 255:1101–4.

---

* Corresponding Author

[14] Esko J. D, Selleck S. B (2002) Order out of chaos: assembly of ligand binding sites in heparan sulfate. Annu Rev Biochem 71, 435-471.

[15] Stengel D. B, Connan S, Popper Z. A (2011) Algal chemodiversity and bioactivity: sources of natural variability and implications for commercial application. Biotechnol Adv 29, 483-501.

[16] Wijesekara I, Pangestuti R, Kim S-K (2011) Biological activities and potential health benefits of sulphated polysaccharides derived from marine algae. Carbohydr Polymers 84:14–21.

[17] Craigie J.S (1990) Cell walls. In: Cole KM, Sheath RM, editors. Biology of the red algae. Cambridge: Cambridge University Press; pp. 221–57.

[18] Demirbas A, Demirbas M.F(2011) Importance of algae oil as a source of biodiesel. Energy Convers Manage 52:163–70.

[19] Milledge J.J (2010) Commercial application of microalgae other than as biofuels: a brief review. Rev. Environ. Sci. Biotechnol. 10:31–41.

[20] Plaza M, Herrero M, Cifuentes A, Ibanez E (2009) Innovative natural functional ingredients from microalgae. J. Agric. Food Chem. 57:7159–70.

[21] Smit A.J (2004) Medicinal and pharmaceutical uses of seaweed natural products: a review. J. Appl. Phycol. 16:245–62.

[22] Pujol C.A, Carlucci M.J, Matulewicz M.C, Damonte E.B (2007) Natural sulfated polysaccharides for the prevention and control of viral infections. Spring-Verlag Top Heterocycl. Chem. 11: 259-281.

[23] Girond S, Crance J. M, Van Cuyck-Gandre H, Renaudet J, Deloince R (1991) Antiviral activity of carrageenan on hepatitis A virus replication in cell culture. Res Virol 142, 261-270.

[24] Buck C. B, Thompson C. D, Roberts J. N, Muller M, Lowy D. R, Schiller J. T (2006) Carrageenan is a potent inhibitor of papillomavirus infection. PLoS Pathog 2, e69.

[25] Nakashima H, Kido Y, Kobayashi N, Motoki Y, Neushul M, Yamamoto N (1987) Purification and characterization of an avian myeloblastosis and human immunodeficiency virus reverse transcriptase inhibitor, sulfated polysaccharides extracted from sea algae. Antimicrob. Agents Chemother. 31, 1524-1528.

[26] Bourgougnon N, Roussakis C, Kornprobst JM, Lahaye M (1994) Effects in vitro of sulfated polysaccharide from Schizymenia dubyi (Rhodophyta, Gigartinales) on a non-small-cell bronchopulmonary carcinoma line (NSCLC-N6). Cancer Lett. 85(1):87-92.

[27] Kolender A. A, Pujol C. A, Damonte E. B, Matulewicz M. C, Cerezo A. S (1997) The system of sulfated alpha-(1-->3)-linked D-mannans from the red seaweed Nothogenia fastigiata: structures, antiherpetic and anticoagulant properties. Carbohydr. Res. 304, 53-60.

[28] Damonte E, Neyts J, Pujol CA, Snoeck R, Andrei G, Ikeda S, Witvrouw M, Reymen D, Haines H, Matulewicz MC, et al (1994) Antiviral activity of a sulphated polysaccharide from the red seaweed Nothogenia fastigiata..Biochem Pharmacol. 47(12):2187-92.

[29] De Clercq E (2000) Current lead natural products for the chemotherapy of human immunodeficiency virus (HIV) infection. Med Res Rev. 20(5):323-49.

[30] Sekine H, Ohonuki N, Sadamasu K, Monma K, Kudoh Y, Nakamura H, Okada Y, Okuyama T (1995) The inhibitory effect of the crude extract from a seaweed of Dygenea simplex C. Agardh on the in vitro cytopathic activity of HIV-1 and it's antigen production. Chem Pharm Bull (Tokyo) 43, 1580-1584.

[31] Damonte E. B, Matulewicz M. C, Cerezo A. S, Coto C. E (1996) Herpes simplex virus-inhibitory sulfated xylogalactans from the red seaweed Nothogenia fastigiata. Chemotherapy 42, 57-64.

[32] Pujol CA, Errea MI, Matulewicz MC, Damonte EB (1996) Antiherpetic activity of S1, an algal derived sulphated galactan. Phytother Res. 10(5):410-413.

[33] Carlucci M. J, Pujol C. A, Ciancia M, Noseda M. D, Matulewicz M. C, Damonte E. B, Cerezo, A. S (1997) Antiherpetic and anticoagulant properties of carrageenans from the red seaweed Gigartina skottsbergii and their cyclized derivatives: correlation between structure and biological activity. Int J Biol Macromol 20, 97-105.

[34] Carlucci M. J, Ciancia M, Matulewicz M. C, Cerezo A. S, Damonte E. B (1999) Antiherpetic activity and mode of action of natural carrageenans of diverse structural types. Antiviral Res 43, 93-102.

[35] Carlucci MJ, Scolaro LA, Matulewicz MC, Damonte EB (1997) Antiviral activity of natural sulphated galactans on herpes virus multiplication in cell culture. Planta Med. 63 :429-432.

[36] Caceres P. J, Carlucci M. J, Damonte E. B, Matsuhiro B, Zuniga E. A (2000) Carrageenans from chilean samples of Stenogramme interrupta (Phyllophoraceae): structural analysis and biological activity. Phytochemistry 53, 81-86.

[37] Haslin C, Lahaye M, Pellegrini M, Chermann J. C (2001) In vitro anti-HIV activity of sulfated cell-wall polysaccharides from gametic, carposporic and tetrasporic stages of the Mediterranean red alga Asparagopsis armata. Planta Med 67, 301-305.

[38] Duarte M. E, Noseda D. G, Noseda M. D, Tulio S, Pujol C. A, Damonte E. B (2001) Inhibitory effect of sulfated galactans from the marine alga Bostrychia montagnei on herpes simplex virus replication in vitro. Phytomedicine 8, 53-58.

[39] Pujol C. A, Estevez J. M, Carlucci M. J, Ciancia M, Cerezo A. S, Damonte E. B (2002). Novel DL-galactan hybrids from the red seaweed Gymnogongrus torulosus are potent inhibitors of herpes simplex virus and dengue virus. Antivir Chem. Chemother. 13, 83-89.

[40] Mazumder S, Ghosal P. K, Pujol C. A, Carlucci M. J, Damonte E. B, Ray B (2002) Isolation, chemical investigation and antiviral activity of polysaccharides from Gracilaria corticata (Gracilariaceae, Rhodophyta). Int. J. Biol. Macromol. 31, 87-95.

[41] Venkateswaran P. S, Millman I, Blumberg B. S (1989) Interaction of fucoidan from Pelvetia fastigiata with surface antigens of hepatitis B and woodchuck hepatitis viruses. Planta Med. 55, 265-270.

[42] Beress A, Wassermann O, Tahhan S, Bruhn T, Beress L, Kraiselburd E. N Gonzalez L. V, de Motta G. E, Chavez P. I (1993) A new procedure for the isolation of anti-HIV compounds (polysaccharides and polyphenols) from the marine alga Fucus vesiculosus. J. Nat. Prod. 56, 478-488.

[43] Hoshino T, Hayashi T, Hayashi K, Hamada J, Lee J. B, Sankawa U (1998) An antivirally active sulfated polysaccharide from Sargassum horneri (TURNER) C. AGARDH. Biol. Pharm. Bull. 21, 730-734.

[44] Feldman S. C, Reynaldi S, Stortz C. A, Cerezo A. S, Damonte E. B (1999) Antiviral properties of fucoidan fractions from Leathesia difformis. Phytomedicine 6, 335-340.

[45] Ponce N. M, Pujol C. A, Damonte E. B, Flores M. L, Stortz, C. A (2003) Fucoidans from the brown seaweed Adenocystis utricularis: extraction methods, antiviral activity and structural studies. Carbohydr. Res. 338, 153-165.

[46] Hasui M, Matsuda M, Okutani K, Shigeta S (1995) In vitro antiviral activities of sulfated polysaccharides from a marine microalga (Cochlodinium polykrikoides) against human immunodeficiency virus and other enveloped viruses. Int J Biol Macromol. 17, 293-297.

[47] Huheihel M, Ishanu V, Tal J, Arad S. M (2002) Activity of Porphyridium sp. polysaccharide against herpes simplex viruses in vitro and in vivo. J Biochem Biophys. Methods 50, 189-200.

[48] Lee J. B, Hayashi K, Hayashi T, Sankawa U, Maeda M (1999) Antiviral activities against HSV-1, HCMV, and HIV-1 of rhamnan sulfate from Monostroma latissimum. Planta Med 65, 439-441.

[49] Bergh O, Borsheim K. Y, Bratbak G, Heldal M (1989) High abundance of virus found in aquatic environments. Nature 340, 467-468.

[50] Carlucci M. J, Damonte E B, Scolaro L A (2011) Virus driven evolution: a probable explanation for "Similia Similibus Curantur" philosophy. Infect Genet Evol 11, 798-802.

[51] Knoll A. H (1992) The early evolution of eukaryotes: a geological perspective. Science 256, 622-627.

[52] Cassaro C. M, Dietrich C. P (1977) Distribution of sulfated mucopolysaccharides in invertebrates. J Biol Chem 252, 2254-2261.

[53] Dietrich C. P, Nader H. B, Straus A. H (1983) Structural differences of heparan sulfates according to the tissue and species of origin. Biochem Biophys Res Commun 111, 865-871.

[54] Ihrcke N. S, Parker W, Reissner K. J, Platt J. L (1998) Regulation of platelet heparanase during inflammation: role of pH and proteinases. J Cell Physiol 175, 255-267.

[55] Kainulainen V, Wang H, Schick C, Bernfield M (1998) Syndecans, heparan sulfate proteoglycans, maintain the proteolytic balance of acute wound fluids. J Biol Chem 273, 11563-11569.

[56] Oragui E. E, Nadel S, Kyd P, Levin M (2000) Increased excretion of urinary glycosaminoglycans in meningococcal septicemia and their relationship to proteinuria. Crit Care Med 28, 3002-3008.

[57] Damonte E. B, Matulewicz M. C, Cerezo A. S (2004) Sulfated seaweed polysaccharides as antiviral agents. Curr Med Chem 11, 2399-2419.

[58] Witzany G (2006) Natural Genome-Editing competences of viruses. Acta Biotheor. 54, 235-253 .

[59] Villarreal L.P (2009) Origin of Group Identity. Springer. New York,USA

[60] Ryan F.P (2006) Genomic creativity and natural selection: A modern synthesis. Biol. J. Linn. Soc. 88, 655-672

[61] Villarreal L.P (2005) The dilemma of the transition in Evolution: the Eukaryotes, in: Viruses and the Evolution of life. Washington. American Society for Microbiol. Press, pp.101-142 .

[62] Trowsdale J, Parham P (2004) Defense strategies and immunity-related genes. Eur J Immunol 34,7-17.

[63] Kohler H, Bayry J, Nicoletti A, Kaveri S.V (2003) Natural Autoantibodies as tools to predict the outcome of immune response? Scand. J. Immunol.58, 285-9

[64] Ochsenbein A. F, Fehr T, Lutz C, Suter M, Brombacher F, Hengartner H, Zinkernagel R. M (1999) Control of early viral and bacterial distribution and disease by natural antibodies. Science 286, 2156-2159.

[65] Maddur M.S, Vani J, Desmazes-Lacroix S, Kaveri S, Bayry J (2010) Autoimmunity as a predisposition for infectious diseases. Plos Pathog. 6,11 e1001077.

[66] van de Vosse E, van Dissel J, Ottenhoff T (2009) Genetic deficiencies of innate immune signalling in human infectious disease. Lancet Infect Dis 9, 688-698.

[67] Danilova N (2006) The evolution of immune mechanisms. J. Exp. Zool. (Mol. Dev. Evol). 306:,496-520.

[68] Bamshad M, Wooding S.P (2003) Signatures of natural selection in the human genome. Nat. Rev. Genet. 4,99-111 .

[69] Galvani A.P, Slatkin M (2003) Evaluating plague and smallpox as historical selective pressures for the CCR5-Delta 32 HIV-resistance allele. Proc Natl Acad Sci. USA 100, 15276-79.

[70] Bureau J-F, Le Goff S, Thomas D, Parlow A.F, de la Torre J.C, Homann D, Brahic M, Oldstone M.B.A (2001) Disruption of differentiated functions during viral infection in vivo. V. Mapping of a locus involved in susceptibility of mice to growth hormone deficiency due to persistent lymphocytic choriomeningitis virus infection. Virology 281, 61-66.

[71] Shadan F.F, Villarreal L.P (1993) Coevolution of persistently infecting small DNA viruses and their hosts linked to host-interactive regulatory domains. Proc. Natl. Acad. Sci. USA. 1,(90) 4117-21.

[72] Villarreal L.P (2009) The source of self: genetic parasites and the origin of adaptative immunity. Ann. NY Acad. Sci. 1178, 194-232.

[73] Sakaoka H, Kurita K, Iida Y, Takada S, Umene K, Kim Y.T, Ren C.S, Nahmias A.J (1994) Quantitative analysis of genomic polymorphism of herpes simplex virus type 1 strains from six countries: studies of molecular evolution and molecular epidemiology of the virus. J. Gen. Virol. 75, 513-527.

[74] Carlucci M.J, Scolaro L.A, Damonte E.B (2002) Herpes simplex virus type 1 variants arising after selection with antiviral carrageenan: lack of correlation between drug-susceptibility and syn phenotype. J. Med. Virol. 68, 82-91.

[75] Mateu C, Perez Recalde M, Artuso M, et al (2011) Emergence of HSV-1 syncytial variants with altered virulence for mice after selection with a natural carrageenan. J Sex. Transm. Dis. 38(6), 548-554

[76] Lipton B.H (2008) The Biology of belief. Hay House Publishers India

[77] Lamarck J.B (1809) Philosophie zoologique, on exposition des considerations relatives á l'histoire naturelle des animaux. Libraire, Paris.

[78] Lamarck J.B (1963) Zoological philosophy. Hafner Publishing Co, New York. USA.

[79] Mayr E (1976) Evolution and the diversity of life: selected essays. The Belknap Press of Harvard University Press, Cambridge, Mass.

[80] Ryan F (2002) Darwin's blind spot: Evolution beyond natural selection, Houghton Mifflin, NewYork, USA.

[81] Ruby E, Henderson B, et al (2003) We get by with a little help from our (little) friends. Science 303:1305-1307.

[82] Nitz N, Gomes C, et al (2004). Heritable integration of kDNA minicircle sequences from Tripanosoma cruzi into the Avian genome: Insights into human Chagas disease. Cell 118:175-186.

[83] Pennisi E (2004) Researchers trade insights about gene swapping. Science 305:334-335.

[84] Boucher Y, Douady C.J, et al (2003) Lateral gene transfer and the origins of prokaryotic groups. Annual Review of Genetics 37:283-328.

[85] Gogarten J.P (2003) Gene Transfer: Gene swapping craze reaches eukaroytes. Current Biology, 13:R53-R54.

[86] Pennisi E (2001) Sequences reveal borrowed genes. Science 294:1634-1635.

[87] Lenton T (1998) Gaia and natural selection. Nature 394:439-447.

[88] Witzany G (2000) Life: the communicative structure: a new philosophy of biology. Norderstedt: Libri Books on Demand.

[89] Witzany G (2009) Noncoding RNAs: persistent viral agents as modular tools for cellular needs. Ann. NY Acad. Sci. 1178:244-267.

[90] Claverie JM (2005) Fewer genes, more noncoding RNA. Science 309:1529-1530.

[91] Villarreal L.P (2005) Viruses and the Evolution of life. ASM Press.Washington, USA.

Chapter 9

Bioactive Polysaccharides of American Ginseng *Panax quinquefolius* L. in Modulation of Immune Function: Phytochemical and Pharmacological Characterization

Edmund M. K. Lui, Chike G. Azike, José A. Guerrero-Analco, Ahmad A. Romeh, Hua Pei, Sherif J. Kaldas, John T. Arnason and Paul A. Charpentier

Additional information is available at the end of the chapter

1. Introduction

Ginseng has a long history of use as a traditional medicine; and it is one of the top selling medicinal herbs in the world. It is a multi-action herb with a wide range of pharmacological effects on the central nervous system, cardiovascular system and endocrine secretion, and the reproductive and immune systems [1]. Ginseng is a deciduous, perennial plant of the Araliaceae family. There are two major species of ginseng: *Panax ginseng and Panax quinquefolius*; and the roots are primarily used for medicinal benefits. Ginseng's wide range of pharmacological activities is believed to be due to the presence of a host of bioactive compounds. The primary ones are the ginsenosides, which are steroidal saponins conjugated to different sugar moieties and polysaccharides (PS) which account for 10-20% by weight of ginseng.

Polysaccharide components of ginseng have received much attention recently because of the emergence of different biological activities, such as immunomodulatory, antibacterial, anti-mutagenic, radioprotective, anti-oxidative, anti-ulcer, antidepressant, anti-septicaemic and anti-inflammatory activities [2]. Specifically, the polysaccharide fraction of ginseng has been shown to have immunomodulatory effects in both preclinical and clinical studies [3-6], although they are poorly characterized.

Several polysaccharides have been identified in *P. ginseng* and *P. notoginseng* but these compounds, including arabinogalactan, pectins, and acidic polysaccharides, have been rarely studied in the North American species. They are made up of a complex chain of monosaccharides rich in L-arabinose, D-galactose, L-rhamnose, D-galacturonic acid, D-glucuronic acid and D-galactosyl residues [7]. The actual structural characteristics and the heterogeneity of PS components are poorly understood due to a lack of methodologies for separation as well as quantitative and qualitative analysis.

Most studies have focused on Asian ginseng PS and mostly *in vitro* experimental models. In this chapter, we will focus on *Panax quinquefolius* (American ginseng). The use of gel permeation chromatography (GPC) with multiple detectors has provided enhanced resolution for phytochemical analysis. And we have used both *in vivo* and *in vitro* models to evaluate its immunomodulatory activity.

2. Materials

Ginseng. Four-year-old American ginseng roots collected in 2007 from five different farms in Ontario, Canada were provided by the Ontario Ginseng Growers Association. Ginseng extracts from each farm were prepared individually and combined to produce composite extracts which were used for phytochemical and pharmacological studies [6]. The AQ extract has no detectable endotoxin contamination as determined by Limulus test.

2.1. Chemicals and biologicals

Sephadex G75 was purchased from GE Healthcare Bio-Sciences AB (Sweden). The Diethylaminoethyl (DEAE)-Cellulose and monosaccharide standards were purchased from Sigma (Oakville, Ontario). All other chemicals were of analytical grade and used as received. Cell culture medium and reagents were purchased from Gibco laboratories (USA). BD OptEIA ELISA kits tumour necrosis factor-α and interleukin-6 (BD Biosciences, USA). LPS from *Escherichia coli* and Griess reagent were purchased from Sigma-Aldrich (USA).

Animals. Adult male Sprague-Dawley rats (200–250g; Charles River, St. Constant, QC, Canada) were used. The Animal Ethics Review Committee of the Western University approved the study (Protocol No: 2009-070).

3. Methods

Preparation of the aqueous (AQ) and polysaccharide (PS) ginseng extracts. Dried ginseng root samples were shipped to Naturex (USA) for extraction. Samples were ground between ¼ and ½ inch and used to produce the AQ extract [6]. Briefly, 4kg ground ginseng roots were soaked three times during five hours in 16L of water solution at 40°C. After extraction, the solution was filtered at room temperature. The excess solvent was then removed by a rotary evaporator under vacuum at 45°C. The three pools were combined and concentrated again until the total solids on a dry basis were around 60%. These concentrates were lyophilized with a freeze dryer (Labconco, USA) at -50°C under reduced pressure to produce the AQ ginseng extract in powder form. Yield of the powder extracts from the concentrates was about 66%. The yields of the final extract (mean ± standard deviation of % extractive) from the initial ground root were 41.74±4.92.

A solution of AQ extract in distilled water (10g/10mL) was prepared, and the crude PS was precipitated by the addition of four volumes of 95% ethanol. The PS fraction was collected by centrifugation at 350×g (Beckman Model TJ-6, USA) for 10 minutes and lyophilized to produce the crude PS extract.

To prepare the water soluble polysaccharide extract (WSPE), ginseng roots (500 g) were extracted with 7.0 L of MilliQ (EMD Millipore) water at 100 ºC for 4 h and filtered through sheets of glass fiber. The solid material was extracted twice under identical conditions. The filtrates were combined, then centrifuged to remove water insoluble materials and supernatants were concentrated (1.0 L) and precipitated by the addition of 95% ethanol (4 to 1 volumes). After centrifugation, the precipitate was washed and dried by solvent exchange, first using 95% ethanol, and then absolute ethanol. Crude water soluble polysaccharide extract (WSPE) was obtained with a 20.0 % yield (relative to dry weight of plant material) (Figure 1).

3.1. Chromatography of ginseng extracts

High performance liquid chromatography (HPLC) analysis for ginsenoside determination

HPLC analysis on the composition of ginsenosides in the AQ extracts (100mg/ml methanol) was performed with a Waters 1525 HPLC System with a binary pump and UV detector [6]. A reversed-phase Inspire C18 column (100mm×4.6 mm, i.d. 5μm) purchased from Dikma Technologies (USA) was used for all chromatographic separations. Gradient elution consisted of [A] water and [B] acetonitrile at a flow of 1.3mL/min as follows: 0min, 80-20%; 0-60min, 58-42%; 60-70min, 10-90%; 70-80min, 80-20%. Absorbance of the eluates was monitored at 203nm.

Sephadex G-75 chromatography

AQ ginseng extract (500mg) was dissolved in 5mL distilled water and then fractionated by loading to a calibrated Sephadex G-75 column (47×2.5cm) equilibrated and eluted with distilled water mobile phase at 4°C with a flow rate of 1mL/min [6]. Absorbance of the eluates was monitored at 230nm. Fractions were collected and lyophilized for the study of bioactivity distribution.

Preparation of the de-proteinated water soluble polysaccharide extracts (DWSPE)

WSPE (50 g) were re-dissolved in 1.5 L of MilliQ water and partitioned five times with Sevag reagent (1:4 n-butanol:chloroform, v/v, 500 mL each) to remove proteins [8]. Polysaccharides were precipitated again by ethanol and dried by solvent exchange. This procedure yielded 46.0 g of the de-proteinated water soluble polysaccharide fraction (DWSPE). The procedure for the preparation of DWSPE from *P. quinquefolius* is shown in Figure 1.

Total fractionation of the DWSPE by ion exchange chromatography on DEAE-Cellulose

DWSPE fraction (20 g) was dissolved in MilliQ water (200 mL) and loaded on a DEAE-Cellulose column (10.0 X 20 cm, Cl-) pre-equilibrated with MilliQ water. The column was eluted first with 4.0 L of MilliQ water at a flow rate of 10 mL/min (4 bar column pressure) to obtain the unbound or neutral fraction (DWSPE-N) and then with 4.0 L of 0.5 M NaCl to obtain the bound or acidic fraction (DWSPE-A). The fractions were concentrated, dialyzed (cut off pore size of 2 KDa) against MilliQ water and freeze dried to give 15.0 g (13.8 %) of the DWSPE-N and 0.9 g (0.83 %) of the acidic fraction DWSPE-A. DWSPE-A (0.6 g) was dissolved in 50 mL MilliQ water and loaded on a DEAE-Cellulose column (10 X 20 cm, Cl-). The column was eluted by a stepwise gradient with 2.0 L of NaCl aqueous solutions (0.0, 0.1, 0.2, 0.3 and 0.5 M each) at a flow rate of 10 mL/min (4 bar column pressure). A total of 120 eluate fractions were collected (50 mL each), dialyzed and lyophilized (Figure 2).

High performance gel permeation chromatography-evaporative light scattering detection (HPGPC-ELSD) analysis

HPGPC was carried out at 40 ºC using a TSK-gel G-3000PWXL column (7.8 X 300 mm, TOSOH, Japan) connected to a HPLC system coupled with Diode Array and Evaporative Light Scattering Detectors (DAD-ELSD). The column was calibrated with standard dextrans (5 to 410 KDa range, Figure 7-B). Ten microliters of 20 mg/mL solutions of DWSPE, DWSPE-N and DWSPE-A were separately injected and eluted with HPLC grade water at a flow rate of 0.8 mL/min and monitored using ELSD with a temperature setting at 80 °C.

High performance gel permeation chromatography-multi-detector analysis

AQ and PS ginseng extracts were analysed at 40°C with TSK-gel PWXL G-4000PWXL column (7.8 X 300 mm, TOSOH, USA) connected to a Viscotek (Varian Instruments, USA) gel permeation chromatography system with Omnisec software (version 4.5, Viscotek, USA) for data acquisition. Solutions of AQ and PS extract (1mg/mL) were filtered with 0.2μm nylon filter and used for analysis. Each sample (100μl) was injected and eluted with 0.3M sodium chloride (NaCl) mobile phase at a flow rate of 1mL/min and monitored using a multiple detectors system for light scattering, refractive index and viscosity. Pullulan polysaccharide reference standard was analyzed as a positive control.

3.2. Analysis of the monosaccharides composition in WSPE by HPLC-ELSD

Carbohydrate analysis represents a challenge in analytical chemistry since neutral or acidic saccharides (mono, oligo and poly) have little UV activity. In our study, evaporative light scattering detection (ELSD) was used. The ELSD does not require the solutes of interest to have any optical properties; and the only requirement is that the eluent be more volatile than the solutes. WSPE (20 mg) was dissolved in 10 mL of 2N HCl solution and was boiled for 2 h. The hydrolyzed product was neutralized (pH 6-7) and centrifuged before analysis.

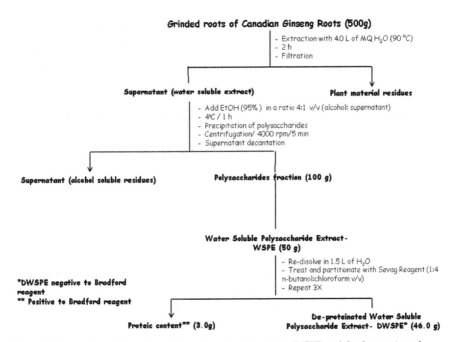

Figure 1. Preparation of the water soluble polysaccharides extract (WSPE) and the de-proteinated extract from the roots of 4-year-old Ontario-grown Panax quinquefolius.

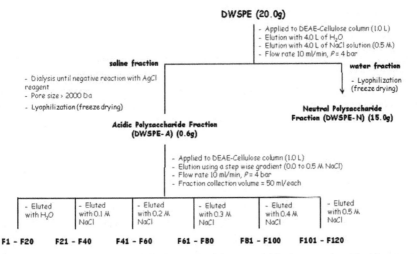

Figure 2. Step wise fractionation of the de-proteinated water soluble ginseng polysaccharide extract (DWSPE).

HPLC analysis was conducted using an 1100 series HPLC-DAD-ELSD system (Agilent Technologies Inc., Santa Clara, CA, USA). To enhance resolution of monosaccharides, analysis was performed using two separate columns (Figure 2). Glucose, galactose, arabinose, mannose and xylose were eluted using a Rezex RPM Monosaccharide PB+2 (8%) (Phenomenex, Torrance, California) column with a mobile phase of 100 % water (Chromasolv Plus, HPLC grade) isocratically at 80 °C and a flow rate of 0.6 mL/min. The ELSD temperature was set to 80 °C. Galacturonic acid and rhamnose content were examined using a Luna 5 μ NH$_2$ 100Å column (Phenomenex, Torrance, California) and eluted with a mobile phase of acetonitrile and water (80:20) at 40 °C and a flow rate of 3 mL/min. The ELSD temperature was set to 44 °C.

3.3. Pharmacological evaluation

3.3.1. In vivo study

Adult male rats (250-300 gm) were treated with 125 mg/kg of AQ extract or crude PS fraction dissolved in saline by gastric garvage (10 ml/Kg bwt) once daily for 3 or 6 consecutive days, and examined 24 hr after the last dose. Animals were anesthetized with i.m. injection (80 and 5 mg/kg b.wt. ketamine and xylazine, respectively) and the trachea was cannulated for lung bronchoalveolar lavage (BAL) with Dulbecco's phosphate-buffered saline (PBS) to collect alveolar macrophages.

Blood was collected into heparinized tubes from rats by intracardiac puncture, samples were immediately centrifuged at 3000 rpm for 10 minutes and the plasma was separated, aliquoted and stored at -20 °C until use.

3.3.2. In vitro study

Cell culture

Rat alveolar macrophages were collected by BAL using cannulated 10-ml syringe with three 10ml washes of PBS. Fluid recovered from BAL was centrifuged at 1000 rpm for 5 minutes. Cells were cultured in RPMI-1640 medium supplemented with 10% Fetal Bovine Serum (FBS), 25mM HEPES, 2mM Glutamine, 100IU/ml penicillin and 100μg/ml streptomycin in 96-well tissue culture plates, at a density of 2×10^5 cells per well at 37°C maintained in a humidified incubator with 5% CO$_2$.

Cell treatment

Immuno-stimulatory effect

Experiments to evaluate dose-related stimulation of inflammatory mediators profile *in vitro* were carried out by treating and incubating rat alveolar macrophages with 0, 50, 100 and 200μg/ml of ginseng extracts for 24 hours and washed before challenging with LPS (1μg/mL) was used as positive control. The 24 hours-production of NO, TNF-α and IL-6 in culture medium was determined.

LPS-induced immuno-suppression

To examine the direct inhibitory effect of ginseng extracts on LPS-stimulated immune function, we pre-treated the macrophages with 0, 50, 100 or 200μg/ml of ginseng extracts for 24 hours and washed before challenging with LPS (1 μg/ml). The 24-hour cytokine production induced by LPS was determined by measuring NO, TNF-α and IL-6 levels in the culture medium.

3.3. Quantification of NO, TNF-α and IL-6

TNF-α and IL-6 concentrations in supernatants from cultured cells and plasma were analyzed with ELISA [6]. Samples were evaluated with rat cytokine-specific BD OptEIA ELISA kits (BD Biosciences, USA) according to the manufacturer's protocol. NO production was analyzed as accumulation of nitrite in the culture medium. Nitrite in culture supernatants was determined with Griess reagent (Sigma-Aldrich, USA) as previously described [6].

3.4. Statistical analysis

In vivo and *in vitro* experiments were performed at least three separate times. All statistical analyses were performed with GraphPad prism 4.0a Software (GraphPad Software Inc., USA). Data were presented as the mean ± standard deviation (SD) of triplicates from three independent experiments. Data sets with multiple comparisons were evaluated by one-way analysis of variance (ANOVA) with Dunnett's *post-hoc* test. $P<0.05$ was considered to be statistically significant.

4. Results

4.1. Phytochemistry

Ginsenoside Compositon. AQ extract contained a total ginsenoside content of 13.87% dry weight of extract and showed characteristics of *Panax quinquefolius* with Rb1 and Re as the predominant ginsenosides, with no detectable Rf and minimal levels of Rg1 (Fig.3).

Crude PS extract. The yield of crude PS fraction by four volumes of 95% ethanol precipitation was 10 % dry weight of root materials. A representative G-75 chromatographic profile of the crude PS extract is shown in Figure 4 [6]. The major PS peak (with a elution volume of 100 ML) had an estimated average molecular weight of 73kDa, while there were two minor, less well- resolved peaks (Fig 4).

Monosaccharide composition of the WSPE determined by HPLC-ELSD. Representative chromatograms showing individual monosaccharides in water soluble polysaccharide extract (WSPE) are shown in Figure 5 and 6. Glucose was found to be the major neutral monosaccharide present in WSPE with amounts ranging from 77 to 86 % (w/w). Galactose and arabinose were present in similar amounts with galactose being present at levels of 6.8

to 7.5 % (w/w) and arabinose being present at levels between 4.5 to 5.9 % (w/w). Galacturonic acid was also identified on WSPE at levels ranging from 8.7 to 9.5 % (w/w). Mannose and xylose were also monitored but were not detected in the sample.

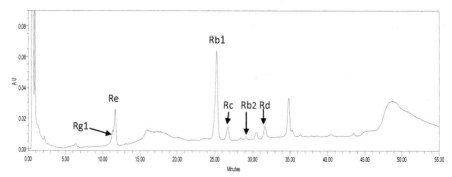

Figure 3. HPLC ginsenoside profile of Ontario-grown American ginseng aqueous extract. A reversed-phase Inspire C18 column was used with gradient elution consisted of [A] water and [B] acetonitrile: 0min, 80-20%; 0-60min, 58-42%; 60-70min, 10-90%; 70-80min, 80-20%. Absorbance of the eluates was monitored at 203nm.

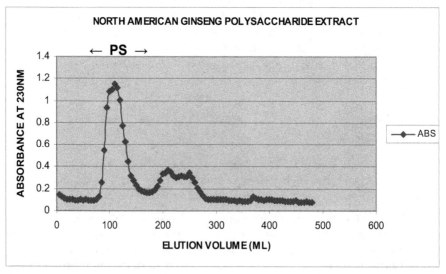

Figure 4. Sephadex G-75 (47×2.5cm) chromatographic fractionation of PS extracts of ginseng. A calibrated column was loaded with 500mg of the extract, and then eluted with distilled water at a flow rate of 1mL/min. The y-axis is the absorbance at 230nm while the x-axis represents the elution volume (mL).

Total fractionation of the DWSPE by ion exchange chromatography on DEAE-Cellulose. To further analyse the WSPE, the material was de-proteinated using the Sevag method, giving a

DWSPE with a yield of 92% and the protein content yield of 1.8% (relative to the dry weight of WSPE). Fractionation of WSPE by a combination of anion-exchange on DEAE-cellulose and gel permeation chromatographies with a procedure shown in Figure 1 revealed the elution of a neutral fraction with water (N-DWSPE, 75 % relative to the dry weight of DWSPE) and an acidic fraction (A-DWSPE, 4.5%) with a 0.5 M NaCl solution (Figure 7). Both fractions presented a wide and complex molecular weight distribution ranging from 5 to 410 KDa. The acidic fraction A-DWSPE containing most likely uronic acids was subjected to a second fractionation by DEAE-cellulose chromatography.

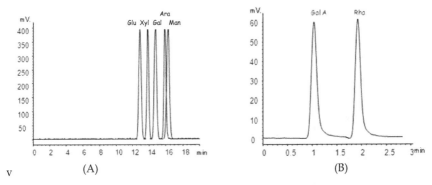

Figure 5. Representative chromatograms of the novel analytical method by HPLC-ELSD for the analysis of seven monosaccharides including glucose (Glu), galactose (Gal), arabinose (Ara), Xylose (Xyl), Mannose (Man) (A), galacturonic acid (Gal A) and rhamnose (Rha) (B).

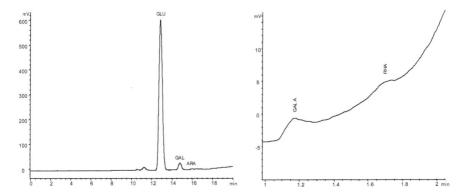

Figure 6. Representative HPLC-ELSD Chromatograms of monosaccharide detected in WSPE, including glucose (Glu), galactose (Gal), arabinose (Ara) (A), galacturonic acid (Gal A) and rhamnose (Rha) (B).

Secondary Fractionation with the A-DWSPE on DEAE-Cellulose column chromatography. The elution of 0.6 g of A-DWSPE on a DEAE-cellulose chromatography was carried out using a stepwise gradient of NaCl. With these elution steps, A-DWSPE was separated into six fractions (Figure 2): A-DWSPE-1 (F1-F20, 96 mg, 16 %), A-DWSPE-2 (F21-F40, 54mg, 9%), A-DWSPE-3 (F41-F60, 162 mg, 27%), A-DWSPE-4 (F61-F80, 120mg, 20%), A-DWSPE-5 (81-F100, 78mg, 13%) and A-DWSPE-6 (F101-F120, 60mg, 10%) corresponding to the elution with 0.0, 0.1, 0.2, 0.3, 0.4 and 0.5 M NaCl solutions, respectively.

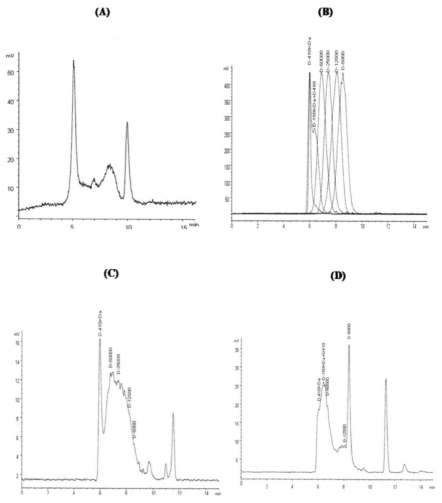

Figure 7. Representative chromatrograms by HPGPC-ELSD analysis; (A) DWSPE sample, (B) Dextran standards mixture, (C) DWSPE-N and (D) DWSPE-A. All samples were eluted under the same conditions.

In order to further analyze the unique NA ginseng polysaccharide samples, GPC with multiple detectors was utilized. Figure 8 shows the results of multiple pullulan polysaccharide standards ranging from 1800 Da to 1,050,000 Da. The utilized column gave good resolution of the individual standards over this range. Figure 8 shows the results of the multi-detector system containing 4 different detectors, i.e. refractive index (RI), right angle light scattering, low angle light scattering and viscometer. In Figure 9, the crude polysacchride extract is observed to have three major peaks with M_w values of 1092 kDa, 135 kDa and 12 kDa (Table 1). The major peak at 1092 kDa accounts for 66% of the weight fraction. Figure 9 shows the deproteinated polysaccharide fraction, in which the M_w values are largely unchanged (Table 1). However, the protein content fraction (PTF) contains only a small amount of carbohydrates with the majority of the high molecular weight fraction removed. The acid fraction contains 3 peaks, of generally lower M_w values, while the neutral fraction obtained from anionic exchange of DWSPE contains only one broad peak (Figure 9).

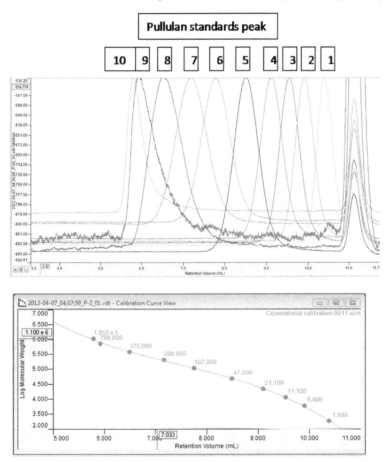

Figure 8. a) Resolution of individual pullulan standards: peak 1, 1800; peak 2, 5900; peak 3, 11,100; peak 4, 21,100; peak 5, 47,000; peak 6, 107,000; peak 8, 375,000; peak 9, 708,000; peak 10, 1,050,000 and b) calibration curve of pullulan standards through G-4000PWXL column.

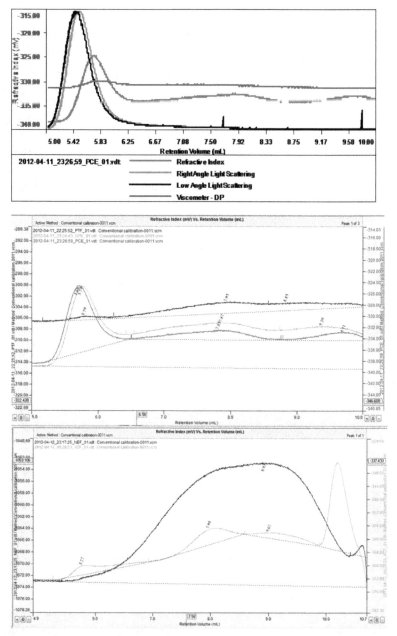

Figure 9. a) GPC traces showing traces from the individual detectors of the PCE (Crude extract of polysaccharides), b) RI plot for the PCE (red), DWSPE (green) and PTF (black) samples. The high

molecular weight peak at retention volume 5.79mL has very low intensity in the PTF sample, which shows that this sample has a very low content of polysaccharide., c) RI plot of the neutral extract (black) and the acidic extract (green) from the ion exchange column, the neutral extract shows much bigger RI area, therefore much higher sample recovery.

PCE	1	2	3
Mn (Dalton)	995,589	106,001	9,502
Mw(Dalton)	1,092,000	135,019	12,037
Mz(Dalton)	1,188,000	172,473	15,433
RI area (mVmL)	5.84	2.03	1.03
Weight fraction (% of RI area)	65.6%	22.8%	11.6%

Table 1. Molecular weight data of the PCE Extract.

5. Immunomodulatory activity

In vivo effect. Treatment with AQ extract for 3-6 days produced marked stimulation of alveolar macrophages as determined by increased production of NO and TNF- and IL-6 following culturing for 24 hrs, reaching activities that were 50-100% of the positive (LPS) control (Figure 10). This immunostimulatory effect was also reflected in the elevation of plasma TNF- and IL-6 levels of treated animals (Figure 11). However, the responsiveness of macrophages collected from ginseng treated animals to LPS stimulation *ex vivo* showed >50% to 100% reduction in NO, TNF-α and IL-6 production as compared to those non-ginseng treated controls, especially with 6 days of ginseng treatment (Figure 10). These data showed that orally administered AQ extract had both immuno-stimulatory and anti-inflammatory effect. Data presented in Figure 12 showed that this immunomodulatory activity could be extended to the PS extract on the basis of its effect on macrophage NO production and the LPS responsiveness.

(i) NO	(ii) TNF-α	(iii) IL-6

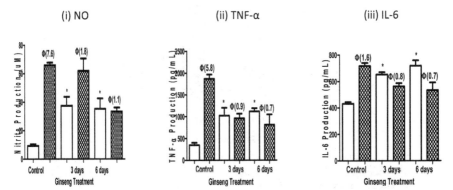

Figure 10. Figure 10. Orally administered ginseng AQ extract: elevated cytokine production and reduced LPS-stimulatd cytokine production in cultured alveolar macrophages. Alveolar macrophages of rats treated orally with 0 and 125mg/kg ginseng AQ extract for 3 and 6 Days were cultured for 24 hours to measure production of NO and cytokines (by ELISA) ▭. To determine responsiveness to

LPS stimulation, ginseng treated macrophages were exposed to 1ug/ml LPS in culture to determine changes in 24 hr NO and cytokine production ▆▆▆. Three independent experiments were performed and the data were shown as mean ± SD. Datasets were evaluated by ANOVA. * Values P<0.05 compared to the untreated control were statistically significant. Φ values in bracket denote fold increase in LPS-stimulated cytokine production over control.

(i) TNF-α   **(ii) IL-6**

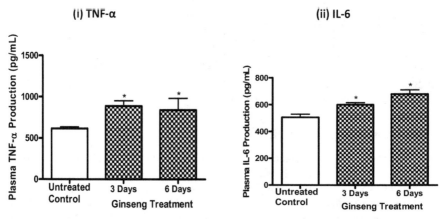

Figure 11. Ginseng AQ extract treatment elevated plasma (i) TNF-α and (ii) IL-6 levels. Rats were treated orally with 125mg/kg ginseng AQ extract for 3 and 6 days▆▆▆. Plasma cytokine concentrations were determined by ELISA. Three independent experiments were performed and the data were shown as mean ± SD. Datasets were evaluated by ANOVA. * Values P<0.05 compared to the untreated control were statistically significant.

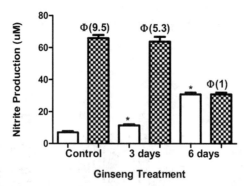

Figure 12. Orally administered ginseng PS extract (125 mg/kg) up-regulated NO production ▭ and reduced responsiveness to LPS (LPS 1 µg/ml) stimulation ▆▆▆ in cultured alveolar macrophages. Cells from untreated controls were treated with LPS 1 µg/ml as positive control for macrophage responsiveness. NO was determined by Griess reaction assay. Three independent experiments were performed and the data were shown as mean ± SD. Datasets were evaluated by ANOVA. * Values P<0.05 compared to the untreated control were statistically significant. Φ values in bracket denote fold increase in LPS-stimulated NO production over control.

(a) NO   **(b) TNF-α**

(i) Aqueous (AQ) extract

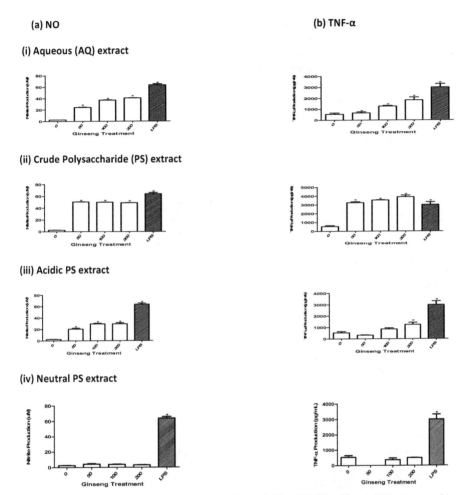

(ii) Crude Polysaccharide (PS) extract

(iii) Acidic PS extract

(iv) Neutral PS extract

Figure 13. Immuno-stimulatory effects *in vitro* of AQ, crude PS, acidic PS and neutral PS extracts on 24 hours macrophage production of (a) NO and (b) TNF-α. Alveolar macrophages isolated from control rats were treated with 0, 50, 100 and 200µg/ml of ginseng extracts for 24 hours, and the culture supernatants were analysed for NO and TNF-α by Griess reaction assay and ELISA, respectively. Cells treated with LPS (1 µg/ml) were used as positive controls. Three independent experiments were performed and the data were shown as mean ± SD. Datasets were evaluated by ANOVA. * Values P < 0.05 compared to the untreated (vehicle) control were statistically significant.

In vitro effect

Both AQ, crude PS and acidic PS showed stimulation of NO and TNF-α production by alveolar macrophages *in vitro* (Figure 13). The magnitude of the response to PS was greater than those induced by acidic PS or AQ extract. Neutral PS was devoid of activity. The lack of concentration-dependent effect of PS was probably due to its high potency and inducing its maximum effect at the concentration studied. Data presented in Figure 14 showed how pretreatment with various extracts for 24 hrs altered the subsequent response to LPS challenge. Since prior LPS treatment was known to cause desensitization of subsequent response to LPS, this was used as a positive control to evaluate the immunosuppressive effect of ginseng extracts. It was apparent that PS was the most effective in reducing the NO and TNF-α response to LPS, while the AQ and acidic PS extracts were similar, and neutral PS was inactive.

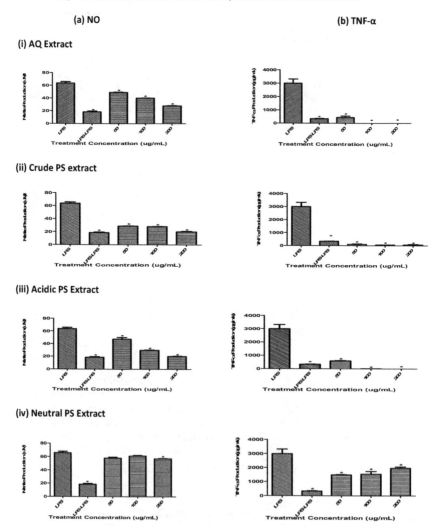

Figure 14. *In vitro* effects of AQ, crude PS, acidic PS and neutral PS extracts on LPS-stimulated 24 hours macrophage production of (a) NO and (b) TNF-α. Rat Alveolar macrophages were pre-treated with

ginseng extracts (0, 50, 200 μg/ml) for 24 hours and were washed before challenged with LPS 1 μg/ml. 24 hr-pretreatment with LPS prior to LPS stimulation was used as positive control to demonstrate desensitization of macrophage responsiveness (LPS-LPS). NO and TNF-α were determined by Griess reaction assay and ELISA, respectively. Three independent experiments were performed and the data were shown as mean ± SD. Datasets were evaluated by ANOVA. * Values P < 0.05 compared to the LPS positive control were statistically significant.

6. Discussion

Medicinal plants have been in use for human health for thousands of years, yet polysaccharides have only been recognized recently as a major contributor to the bioactivity of these traditional medicines. Polysaccharides from plant sources with immunomodulatory, anti-tumor, anti-viral, anti-bacterial, anti-inflammatory, anti-oxidant, and anti-diabetogenic activities have been reported [2, 9-15]. And a few polysaccharides, including lentinan, *Astragalus* polysaccharide, polyporus polysaccharide and *Achyranthes bidentata* polysaccharide have been licensed for clinical application in China [16]. In Canada, a polysaccharide-enriched American ginseng extract (Cold Fx®) has been licensed in 2007 as a natural health product to 'help reduce the frequency, severity and duration of cold and flu symptoms by boosting the immune system' with an estimated annual sale of over $48M [4, 5]. Our findings on the paradoxical effects of AQ and PS extract on macrophage function *in vivo* may have significant implication in the use of American ginseng polysaccharides in several clinical applications.

The structure and biological activities of the polysaccharides from the roots, leaves and fruits of *Panax ginseng* have recently been reviewed by Sun [2]. There is limited information on American ginseng. Our study *in vitro* has demonstrated the up-regulation of inflammatory mediators production by AQ and crude PS extracts in rat alveolar macrophages (Figure 13) which validated what we have previously reported [3 and 6]. In addition we have demonstrated specificity of PS in that acidic but not neutral species of the PS was bioactive. Following sub-acute oral administration, both PS and AQ extracts were also immuno-stimulatory based on elevation of plasma cytokine levels and increase in the function of alveolar macrophages *ex vivo*. The immuno-stimulatory dosage used in the present *in vivo* study was comparable to those proven to be effective for cardiovascular health and for protection against diabetic retinopathy, neuropathy and cardiomyopathy reported by other investigators using identical ginseng extracts and in the same animal species [17-19]. However, the magnitude of the immunostimulatory effect *in vivo* was smaller than the *in vitro* response. This may be related to the lower bioavailability of the orally-administered ginseng extract. The pharmacokinetics of oral ginseng PS is not known, but the recovery of major ginsenosides in plasma after oral administration was quite low [20]. Result of our *in vitro* and *in vivo* studies was supportive of what was reported for CVT-E002 (a patented, poly-furanosyl-pyranosyl polysaccharide-rich extract of the root of North American ginseng): stimulation of normal mouse spleen cells and immunoglobulin G production as well as activation of peritoneal exudate macrophages leading to enhanced cytokine stimulation in treated mice.

In addition to the well-recognized immuno-stimulatory activity of ginseng, an anti-inflammatory effect was shown in the present study as reflected in the reduced responsiveness of alveolar macrophages collected from ginseng-treated animals to LPS challenges *ex vivo* (Figure 10). This apparent anti-inflammatory effect of ginseng PS is different from what we have previously reported for a specific component(s) of the alcoholic extract of American ginseng as well as the anti-inflammatory effects that have been ascribed to some ginsenosides and their metabolites [6, 21-22]. This potential anti-inflammatory mechanism is being validated by evaluating changes in LPS–induced inflammatory response following polysaccharide pretreatment *in vivo* in our on-going research. The intent is to determine whether ginseng PS causes desensitization of immune cells as reported for LPS [23]. This action of ginseng, when proven, may be particularly relevant to bacterial infection and related toxicemia. An anti-inflammatory effect of ginseng polysaccharide has been reported by Zhao et al. using a model of auto-immune disease, as evidenced by the reduction in the expression of TNF-α and IFN-γ in lymphocytes in the enteric mucosal immune system of rats with collagen induced arthritis [24]. In view of the diverse immunomodulatory effects of ginseng polysaccharide, the identification of specific polysaccharides with unique property and biological action will be of great interest.

It appears that AQ and PS ginseng extracts have a paradoxical effect on macrophage function: stimulation under normal condition, but reduction when the biological system is under a pro-inflammatory state. In the context of sepsis, AQ and PS immuno-stimulatory effect will be beneficial as a first line of defense during the initial infection stage of bacterial infection by rendering macrophages to be cytotoxic [25] whereas the immune-suppressive activity may be effective in antagonizing the cytokine storm at the later stages of infection by suppressing LPS activation of macrophages.

Carbohydrates analysis represents a major challenge in analytical chemistry since neutral or acidic saccharides (mono, oligo and poly) have little UV activity. The refractive index (RI) detector, which is commonly used in HPLC analysis, has issues with baseline stability and sensitivity. The ELSD used in the present study has the advantage of its independence of any optical properties in the solutes of interest. Our HPLC-based analysis allowed the measurement of 7 mono-saccharides within a 30 minute total run time. The monosaccharide composition of *P. quinquefolius* reported in this study was similar to that described previously [3] with the addition of galacturonic acid. The monosaccharide composition provides insight into the types of polysaccharides which may be found in *P. quinquefolius*. Polysaccharides structures in *P. quinquefolius* have not previously been thoroughly studied, though several polysaccharides have been isolated from *P. ginseng* and *P. notoginseng*. Glucose and galacturonic acid were the most prominent monosaccharides detected. Previously, polysaccharide fractions from *P. ginseng* with high levels of glucose have been determined to contain starch-like glucans and arabinogalactans and fractions with high levels of galacturonic acid have been shown to contain pectins with several linked galacturonic acid domains [26]. It is possible that polysaccharides similar to these may be present in Ontario-grown American ginseng though further work will have to be done to characterize their structures.

Polysaccharides are very complex with a wide range of MW, varying monosaccharide composition and conformation (degree of branching or linearity), which contribute to their diverse structure and biological activities, these can also hamper the study of their structure–function relationships [16]. The multi-detector GPC instrument provided additional information on the polysaccharide structure. This instrument uses Triple detection with a concentration detector (refractive index detector), viscometer and light scattering detector, with each detector providing different although complementary information [20]. Pullulan standards of up to 1,000 kDa showed that the measured ginseng polysaccharide molecular weights for the crude, water soluble and deproteinated extracts are within the range of standards. The light scattering detector, which is based on fluctuations in interference between macromolecules (e.g. polysaccharides) scattering a coherent monochromatic laser beam, is considered an absolute detector for Mw values, confirming the reported values of Table 1. The viscometer is sensitive to branching effects, although the trace in Figure 9 shows no significant branching effect for the ginseng polysaccharides. The RI detector was rather sensitive to the carbohydrate fractions, showing 3 major polysaccharide components for each of the three measured extracts. The high molecular weight peak has a very low intensity for the PTF sample, indicating that the deproteination step removed the high molecular weight fraction. The neutral extract gave similar molecular weights as the acidic polysaccharide extract, although with a higher sample recovery.

7. Conclusions and future directions

Our study has revealed a paradoxical immunomodulatory effect of AQ and PS extracts isolated from Ontario-grown *Panax quinquefolius*. This activity may be relevant to clinical application involving bacterial infection and toxemia. Challenges in development of botanical polysaccharides for clinical application include lack of methodology for structure identification and characterization, isolation and purification, and product quality control, has been recognized [11]. Our research will focus on the application of new methodologies to characterize the polysaccharide structures from ginseng and to elucidate the structure-biological activity relationship.

Author details

Edmund M. K. Lui*, Chike G. Azike and Hua Pei
Ontario Ginseng Innovation and Research Consortium
Department of Physiology & Pharmacology, Schulich School of Medicine and Dentistry, Canada

José A. Guerrero-Analco and John T. Arnason
Ontario Ginseng Innovation and Research Consortium
Department of Biology, University of Ottawa, Ottawa Ontario, Canada

---

* Corresponding Author

Ahmad A. Romeh
Department of Chemical and Biochemical Engineering, Faculty of Engineering,
University of Western Ontario, London Ontario, Canada

Sherif J. Kaldas
Department of Biology, University of Ottawa, Ottawa Ontario, Canada

Paul A. Charpentier
Ontario Ginseng Innovation and Research Consortium,
Department of Chemical and Biochemical Engineering, Faculty of Engineering,
University of Western Ontario, London Ontario, Canada

Acknowledgement

This research was supported by Ontario Ginseng Research & Innovation Consortium (OGRIC) funded by the Ministry of Research & Innovation, Ontario Research Funded Research Excellence program for the project 'New Technologies for Ginseng Agriculture and Product Development'(RE02-049 awarded to EMK Lui). We acknowledge the contribution of PolyAnalytik London Ontario, Canada in providing instrument for the gel permeation chromatography analysis of ginseng extracts.

8. References

[1] Attele AS, Wu JA, Yuan CS (1999) Ginseng Pharmacology; Multiple Constituents and Multiple Actions. Biochem. pharmacol. 58:1685–1693.

[2] Sun Y (2011) Structure and Biological Activities of the Polysaccharides from the Leaves, Roots and Fruits of *Panax ginseng* C.A. Meyer: An Overview. Carbohyd. polym. 85: 490-499.

[3] Assinewe VA, Arnason JT, Aubry A, Mullin J, Lemaire I (2002): Extractable Polysaccharides of *Panax quinquefolius* L. (North American ginseng) Root Stimulate TNF-α Production by Alveolar Macrophages. Phytomedicine 9:398–404.

[4] Biondo PD, Goruk S, Ruth MR, O'Connell E, Field CJ (2008) Effect of CVT-E002™ (COLD-fX®) Versus a Ginsenoside Extract on Systemic and Gut-Associated Immune Function. Int. immunopharmacol. 8:1134–1142.

[5] Predy GN, Goel V, Lovlin RE, Basu TK (2006) Immune Modulating Effects of Daily Supplementation of COLD-FX (a Proprietary Extract of North American Ginseng) in Healthy Adults. J Clin Biochem Nutr 39: 162-167.

[6] Azike CG, Charpentier PA, Hou J, Pei H, Lui EMK. (2011). The Yin and Yang actions of North American ginseng root in modulating the immune function of macrophages Chin Med. 6, 1-12.

[7] Wang M, Guilbert LJ, Li J, Wu Y, Pang P, Basu TK, Shan JJ (2004) A Proprietary Extract from North American Ginseng (*Panax quinquefolium*) Enhances IL-2 and IFN-gamma Productions in Murine Spleen Cells Induced by Con-A. Int. immunopharmacol. 4:311–315.

[8] Sevag M G, Lackman D B, Smolens J (1938). The isolation of the components of streptococcal nucleoproteins in serologically active form. Journal of Biological Chemistry, 124, 425–436.

[9] Igor A. Schepetkin, Mark T. Quinn (2006). Botanical polysaccharides: Macrophage Immunomodulation & therapeutic potential. International Immunopharmacology 6: 317–333.

[10] Meiqi Wang, Larry J. Guilbert, Lei Ling, Jie Li, Yingqi Wu, Sharon Xu, Peter Pang and Jacqueline J. Shan (2001). Immunomodulating activity of CVT-E002, a proprietary extract from North American ginseng (*Panax quinquefolium*). Journal of Pharmacy and Pharmacology 53: 1515–1523.

[11] Jie-Young Song, Seon-Kyu Han, Eun-Hwa Son, Suhk-Neung Pyo, Yeon-Sook Yun, Seh-Yoon, Induction of secretory and tumoricidal activities in peritoneal macrophages by ginsan, Yi International Immunopharmacology 2 (2002) 857–865.

[12] Yoo DG, Kim MC, Park MK, Park KM, Quan FS, Song JM, Wee JJ, Wang BZ, Cho YK, Compans RW, Kang SM (2012). Protective effect of ginseng polysaccharides on influenza viral infection. PLoS One. 2012;7(3):e33678.

[13] Ahn JY, Song JY, Yun YS, Jeong G, Choi IS (2006). Protection of *Staphylococcus aureus*-infected septic mice by suppression of early acute inflammation and enhanced antimicrobial activity by ginsan. FEMS Immunol Med Microbiol; 46:187-97.

[14] Dianhui Luo, Baishan Fang, Structural identification of ginseng polysaccharides and testing of their antioxidant activities (2008). Carbohydrate Polymers 72 376–381.

[15] Xie JT, Wu JA, Mehendale S, Aung HH, Yuan CS (2004). Anti-hyperglycemic effect of the polysaccharides fraction from American ginseng berry extract in ob/ob mice. Phytomedicine. 11:182-7.

[16] He X, Niu X, Li J, Xu S , Lu A (2012). Immunomodulatory Activities of Five Clinically Used Chinese Herbal Polysaccharides. J. exp. integr. med. 2:15-27.

[17] Wu Y, Lu X, Xiang FL, Lui EMK, Feng Q (2011) Ginseng protects the heart from ischemia and reperfusion injury via up regulation of endothelial nitric oxide synthase. Pharmacological Research. 64:195-202.

[18] Subhrojit S, Chen S, Feng B, Wu Y, Lui E, Chakrabarti S (2012) Preventive effects of North American ginseng (*Panax quinquefolius*) on diabetic retinopathy and cardiomyopathy. Phytotheraphy Res. (In Press)

[19] Subhrojit S, Chen S, Feng B, Wu Y, Lui E, Chakrabarti S (2012) Preventive effects of North American ginseng (*Panax quinquefolius*) on diabetic neuropathy. Phytomedicine. 19: 494– 505.

[20] Zhou D, Tong L, Wan M, Wang G, Ye Z, Wang Z, Lin R (2011). An LC-MS method for simultaneous determination of nine ginsenosides in rat plasma and its application in pharmacokinetic study. Biomed Chromatogr, 25::720-6.

[21] Lee DC, Yang CL, Chik SC, Li JC, Rong JH, Chan GC, Lau AS (2009) Bioactivity-Guided Identification and Cell Signalling Technology to Delineate the Immunomodulatory Effects of *Panax ginseng* on Human Promonocytic U937 cells. J. transl. med. 34:1-10.

[22] Oh GS, Pae HO, Choi BM, Seo EA, Kim DH, Shin MK, Kim JD, Kim JB, Chung HT (2004). 20(S)-Protopanaxatriol, one of ginsenoside metabolites, inhibits inducible nitric

oxide synthase and cyclooxygenase-2 expressions through inactivation of nuclear factor-kappaB in RAW 264.7 macrophages stimulated with lipopolysaccharide. Cancer Lett, 205:23-29.

[23] Leon-Ponte M, Kirchhof MG, Sun T, Stephens T, Singh B, Sandhu S, Madrenas J. (2005). Polycationic lipids inhibit the pro-inflammatory response to LPS. Immunol Lett. 96:73-83.

[24] Zhao H, Zhang W, Xiao C, Lu C, Xu S, He X, Li X, Chen S, Yang D, Chan ASC, Lu A (2011) Effect of Ginseng Polysaccharide on TNF-α and INF-γ Produced by Enteric Mucosal Lymphocytes in Collagen Induced Arthritic Rats. J. med. plants res. 5:1536-1542.

[25] Lim DS, Bae KG, Jung IS, Kim CH, Yun YS, Song JY (2002) Anti-Septicaemic Effect of Polysaccharide from *Panax ginseng* by Macrophage Activation. J. infection. 45: 32-38.

[26] Zhang X, Yu L, Bi H, Li X, Ni W, Han H, Li N, Wang B, Zhou Y, Tai G (2009) Total Fractionation and characterization of the Water-Soluble Polysaccharides Isolated from *Panax ginseng* C.A. Meyer. Carbohyd. polym. 77: 544-552.

Chapter 10

The Chitosan as Dietary Fiber: An *in vitro* Comparative Study of Interactions with Drug and Nutritional Substances

Máira Regina Rodrigues, Alexandre de Souza e Silva and Fábio Vieira Lacerda

Additional information is available at the end of the chapter

1. Introduction

Obesity is considered a disease that has grown significantly in the last two decades worldwide. The concern with this increase is justified because obesity develops and remains due to different factors, causing several other diseases and sometimes, can lead to death. Several strategies have been searched to control its progress usually through therapies and, in critical cases, surgeries. Despite all efforts, statistical studies show that obesity is still growing consequently generates the necessity of further studies on the subject (Dacome, 2005).

Epidemiological researches have studied the impact of overweight and obesity on the risk of chronic disease, as coronary heart disease, type 2 diabetes mellitus, hypertension, stroke, dyslipidemia, insulin-resistance, glucose intolerance, metabolic syndrome, and cancers of the breast, endometrium, prostate and colon (Aslander-van Vliet et al., 2007). Health consequences and compromised quality of life associated with obesity provide incentives to abate the obesity epidemic. However, despite recognition of these effects, the epidemic of obesity and overweight is not reversed (Johnson et al., 2007).

In general, treatments for obesity are based on regular exercise, nutritional reeducation, pharmacological treatment, behavioral therapy and use of dietary fibers that promote the reduction of fat absorption (Aslander-van Vliet et al., 2007). Much research has been conducted on the dietary supplements that promote the reduction of body weight and fat mass (Saper et al., 2004). These ingredients reportedly act as a fiber to increase satiety and also to decrease the absorption of fat by binding to it (Kumirska et al., 2010).

A natural substance that helps in these anti-obesity treatments that has been highly recommended to control obesity is chitosan (Hennen, 2005). Chemically speaking, chitosan (Figure 1) is a linear polysaccharide of $\beta(1\rightarrow4)$-linked-2-amino-2-deoxy-D-glucopyranose obtained by deacetylation of chitin, the main component of the exoskeleton of insects and crustaceans (Kumirska et al., 2010). It has many important properties, such as non-toxicity, biocompatibility, biodegradability, antimicrobial activity, chemical reactivity (Cummings et al., 2010), industrial applications (Hennen, 2005), as well as carrier for body fat (Ni Mhurchu et al., 2004; Ni Mhurchu et al., 2005; Jull et al., 2008; Lois & Kumar, 2008), cholesterol and triglyceride (Razdan & Petterson, 1994; Liu et al., 2008; Zhang et al., 2008). Many mechanisms (Tapola et al., 2008; Prajapati, 2009) to explain the carriers and absorptive properties of microenvironment produced by chitosan in solution have been proposed.

However, the use of chitosan is still controversial, and studies in favor and against the use of chitosan have been constantly reported. Many studies have confirmed the hypocholesterolemic activity of chitosan (Sugano et al., 1978; Liao et al., 2007; Yao et al., 2008; Liu et al., 2008; Zhang et al., 2008). The same way, works have reported that the triglyceride and cholesterol absorption have been inhibited and the cholesterol concentration of mice fed with a high fat diet plus chitosan has been decreased (Razdan & Petterson, 1994; Liu et al., 2008; Zhang et al., 2008). Other studies reported that chitosan is efficacious in facilitating the reducing body fat and weight loss in obese individuals (Schiller et al., 2001; Kaats et al., 2006).

On the other hand, studies have shown that oral administration of chitosan has weak action on the reduction of triglyceride and plasma cholesterol in rabbits (Hirano & Akiyama, 1995). Other works have reported that the effect of chitosan on body weight is minimal and unlikely to be of clinical significance (Ni Mhurchu et al., 2004; Ni Mhurchu et al., 2005; Lois & Kumar, 2008, Jull et al., 2008), as well as that the fat trapped was clinically insignificant in studies with overweight adults treated with chitosan capsules before each meal (Pittler et al., 1999; Pittler & Ernst, 2004; Gades & Stern, 2005).

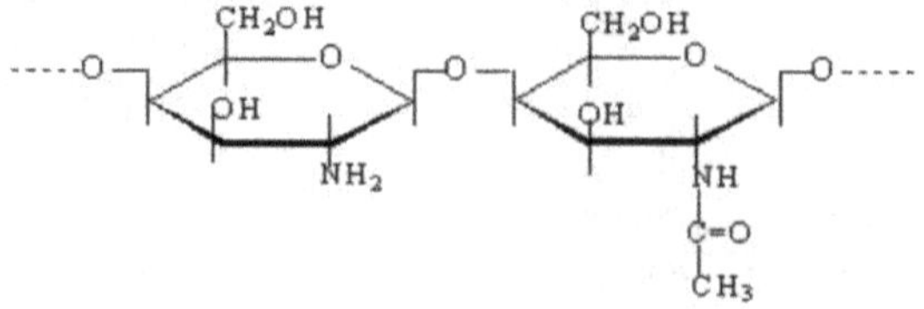

Figure 1. Chemical structure of chitosan.

Is well known that chitosan produces microenvironments with carriers and absorptive properties in acidic aqueous solution. These begin to form above a certain concentration, critical aggregate concentration, CAC (Rodrigues, 2005). The mechanism of solubilization of molecules is well known (Rodrigues, 2005; Rodrigues et al., 2008) however, the process by which chitosan acts as a carrier of fat is not yet fully understood and two mechanisms have been suggested (Prajapati, 2009; Tapola et al., 2008). One of these mechanisms describe the effect of chitosan fiber network, were chitosan also binds neutral lipids like cholesterol and triglycerides through hydrophobic bonds (Tapola et al., 2008; Prajapati, 2009). In other mechanism, the positive charges (NH_3^+ group generated by stomach acids) on chitosan attract and binds to fatty and bile acids (both negatively charged). This complex is indigestible by the body and excreted in the feces (Tapola et al., 2008; Prajapati, 2009).

Regardless of the solubilization mechanism, nutrients can also be solubilized in chitosan microenvironments, as reported in some studies. Works demonstrated that chitosan causes significant decrease in protein digestibility (Deuchi et al., 1994) and its effect on nutrient digestibility (Ho et al. 2001). Nevertheless, studies on the interaction of chitosan with nutrients are still rare and inconclusive (Gades & Stern, 2005; Hennen, 2005; Kaats et al., 2006; Barbosa et al., 2007; Tapola et al., 2008).

In this context, we present a comparative study of interactions of the chitosan with molecules of two vitamins and one drug. To each molecule, the study was conducted in acid aqueous solution, condition similar to the stomach environment, where occurs formation of chitosan gel responsible for solubilizing molecules.

Drug fluoxetine was chosen for this study. The need for anti-depressive drugs with few side effects, as anticholinergic activity and cardiovascular accidents, boosted the development of new anti-depressant compounds (Böer et al., 2010), as fluoxetine, which inhibits the uptake of serotonin by the neurons in the brain, enhances serotonin neurotransmission and had the longest half-life that other selective serotonine reuptake inhibitors (SSRIs) (Rizo et al., 2011). The precise mechanism of action is not clear but it has less cardiovascular, sedative and anticholinergic effects than the tricyclic antidepressant drugs (Shah et al., 2008). The main indications for the prescription of fluoxetine are for obsessive-compulsive disorder, depression therapy, bulimia nervosa, alimentary disorders and obesity (Suarez et al., 2009). Besides drug, the nutritional reeducation and intake of dietary fibers as chitosan has been recommended in treatments for obesity (Aslander-van Vliet et al., 2007). Based on the possible concurrent use of fluoxetine and chitosan, it is important to evaluate the interactions between both substances.

Vitamins chosen for this study were the B2 and B12. Vitamin B2 or riboflavin is a vitamin B complex that participates in numerous metabolic reactions and physiological functions (United States Pharmacopeia, 2007). Vitamin B12 or cyanocobalamin is an essential component in human diet, plays a key role in cell nucleus, enzymatic processes in the mitochondria, and cytoplasm; it is necessary for the synthesis of red blood cells, for the maintenance of the nervous system, and for the growth and development in children (Wang et al., 2007). Both vitamins are not produced by the body and are consumed only in small quantities (Sommer, 2008); deficiency can cause many diseases (Sun et al., 2007).

The interactions between chitosan-vitamin and chitosan-drug have been verified by monitoring the photophysical properties of these components. For this, fluorescence and UV-Vis absorption measurements were initially evaluated in acid aqueous solution and after in weakly acidic solution of chitosan given information about the interactions between this chemical component in conditions that approaches the stomach chemical environment.

2. Experimental

2.1. Chemicals

Chitosan and fluoxetine hydrochloride were purchased from Aldrich Chemical Co. (St. Louis, MO, USA), and vitamin B2 (riboflavin, 96%) and vitamin B12 (cyanocobalamin,, USP Grade) from Veter Co. (Duque de Caxlas, RJ, Brazil).and Merck Co. (Darmstadt, Hessen, Federal Republic of *German), respectively.* Other chemicals were ultraviolet/high-performance liquid chromatography grade and used without further purification; ultrapure water was supplied by a Milli-Q system.

2.2. Spectroscopic measurements

Previous studies have shown that the best conditions to solubilize chitosan are: chitosan 1% (w/v) dissolved in aqueous solution of glacial acetic acid 1% (v/v) under stirring (Signini & Campana Filho, 1999; Rodrigues, 2005; Rodrigues et al., 2008), so measurements in presence of chitosan were conducted in these conditions.

Chitosan has no fluorescence emission or absorption in the experimental conditions. The absorption spectra of chemicals (fluoxetine, B2 and B12) were measured using quartz cuvettes with 1 cm of optical pathway. The fluorescence measurements were performed at λexcit= 275 nm and λemis= 305 nm for vitamin B12 (Li and Chen, 2000); at λexcit= 440 nm and λemis= 305 nm for vitamin B2 (United States Pharmacopeia, 2007; Association of Official Analytical Chemists, 2005); and at λexcit= 230 nm and λemis= 290 nm for fluoxetine (United States Pharmacopeia, 2007; Association of Official Analytical Chemists, 2005).

Absorbance measurements were taken at maximum of the absorption spectra and performed at room temperature.

Initially, variations of both fluorescence and absorption spectra of the chemicals (fluoxetine, B12 and B2) were taken as a function of their concentration in acid aqueous solution and after in different concentrations of chitosan in aqueous acid solution, at the same range of chemicals concentration. The variation in the spectra of the chemicals (fluoxetine, vitamins B2 and B12) was also studied by keeping fixed the concentration of vitamin and varying the concentration of chitosan in acid aqueous solution).

3. Results and discussion

Absorption spectra and chemical structures of chemicals are shown in Figure 2. Spectral data in Figure 2.a shows that vitamin B12 absorb significantly within 425–600 nm range, as described in the literature (Zheng & Lu, 1997; British Pharmacopoeia, 1998) and the present work chose the absorption peak at ~550 nm to assess the spectral behavior. Similar graphs have been obtained for vitamin B2 and fluoxetine. Figure 2.b shows that the absorption maximum for the vitamin B2 occurs at ~440 nm, data consistent with the literature ((United States Pharmacopeia, 2007). Figure 2.c. shows the absorption spectrum of fluoxetine consistent with the literature (Fregonezi-Nery et al., 2008) that exhibits two absorption maxima at 270 and 275 nm. The last one maximum was chosen to monitor the spectral behavior of fluoxetine.

The chitosan-chemicals (fluoxetine, vitamins B2 and B12) interaction have been studied in aqueous acid solution by the monitoring the fluorescence and UV-visible spectra of chemicals, each monitored separately.

In three cases, the increase in concentration of chemical causes an increase in both absorption and fluorescence intensities due to the increase of species that absorb and emit light and this increases is linear profile always indicating that self-aggregation processes are not occurring in this concentration range (data not shown).

Subsequently, chemicals (fluoxetine, vitamins B2 and B12) were studied in the absence and the presence of chitosan, at concentrations 0.050 g.L^{-1}, 0.60 g.L^{-1} and 1.0 g.L^{-1} of polysaccharide, keeping fixed the chemicals concentration (8.5x10^{-5} mol.L^{-1}). With the increase of chitosan concentration both fluorescence and absorption intensities of chemicals are increased.

Figures 3, 4 and 5 show the behavior of fluorescence intensities to fluoxetine, vitamin B12 and B2, respectively. In all graphics, fluorescence intensities significantly increase when chitosan concentration goes from zero to 1.0 g.L^{-1}. This is a common behavior of fluorescent molecules when they migrate from the solution for environment of different polarity (Kalyanasundaram, 1987) and is due to the influence of microenvironment formed by chitosan on the photophysics of the chemical that is changed due to spatial hindrance that it suffers and due to loss of part of rotational freedom of substituent groups, (Kalyanasundaram & Thomas, 1977; Valeur, 2001). Then, with increase concentration of chitosan, the microenvironment becomes more rigid and the lifetime of the chemicals (fluoxetine, vitamins B2 and B12) in the excited states are living longer (Kalyanasundaram, 1987). However, fluorescence intensities of vitamin B12 and fluoxetine show similar increase rate while vitamin B2 is markedly lower. The increase of fluorescence intensities with the polysaccharide concentration has been observed also to vitamin in pharmaceutical formulations containing dextran (Alda et al., 1996).

Absorption intensities of chemicals (fluoxetine, vitamins B2 and B12) increase with the chitosan concentration similarly to the of fluorescence intensities, Figure 6. The reason for this behavior is the increased stiffness of environment generated by chitosan chains. However, in this case, intensities show the following increasing order: vitamin B2, fluoxetine and vitamin B12.

In chemical structure of all chemicals (fluoxetine, vitamins B2 and B12) there are rings with double bonds and polar groups that can interact strongly with the similar groups of chitosan. There are also OH groups in the molecular structure of chitosan favor hydrogen-bonding type interactions with polar groups of chemicals. These interactions can influence the absorption and the emission processes of radiation of molecules reducing the rotational degrees of freedom of the molecule (Ramamurthy, 1991).

In general way, results demonstrated that three chemicals (fluoxetine, vitamins B2 and B12) are transferred to microenvironment generated by weakly acidic solution of chitosan but in different proportions, due to the structural feature and solubility of each. Table 1 describes the relative increase of the fluorescence and absorption intensities of the chemicals when chitosan concentration ranges from zero to 1.0 g.L^{-1}.

Both vitamins belong to the class of hydro soluble vitamins (Sun et al., 2007), while the fluoxetine drug is slightly soluble in water (Darwish, 2005). The low solubility promotes some molecules of fluoxetine to migrate from aqueous environment to the more rigid environment generated by chitosan (the higher the concentration) causing a proportionately greater increase of absorbance and fluorescence intensities. However, the fluorescence intensities of vitamin B12 also increase in the same proportion and the absorbance intensities in proportion even higher, despite the hydro soluble nature of this vitamin.

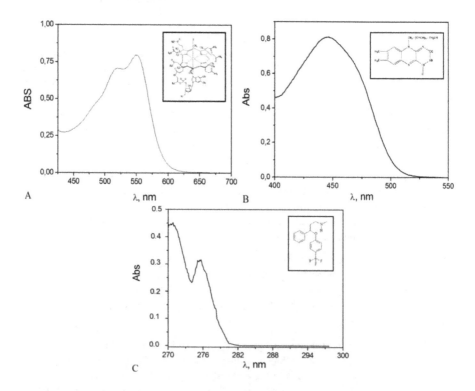

Figure 2. Absorption spectra and chemical structures of vitamin B12 (A), vitamin B2 (B) and fluoxetine (C).

These results demonstrate that the transfer process of chemicals from aqueous environment to the chitosan aggregates is influenced by solubility of the molecules in water and/or by the molecular structure. Particularly, the structure molecular of vitamin B12 seems to have an interesting effect in this case. Vitamin B12 belong to the cobalamins, a class of octahedral Co(III) complexes which contain a planar framework called a corrin, with the metal center coordinated in the equatorial position by the four corrin nitrogens. Her energies of excited states are sensitive to the nature of the ligands of center coordinated and are influenced by water content of the surrounding environment (Solheim et al., 2011). These characteristics may be the reason for the more significant increase of the spectral properties of the vitamin B12, with increasing concentration of chitosan (lower water content), compared with the vitamin B2.

In fact, some molecules of fluoxetine or vitamins B2 or B12, are transferred to microenvironment generated by weakly acidic solution of chitosan. Among the three, the vitamin B2 is transferred in a smaller proportion. However, for all of them is expected that its loss in the diet, caused by administration of chitosan, is not so significant.

From our observations, possible risks to the patient should be considered when prolonged treatment with chitosan is prescribed and perhaps extra care should be taken when chitosan and fluoxetine are prescribed together in slimming diets. In the case of vitamins, essential for many physiological functions, there must be some precautions to minimize the impacts generated by this therapeutic, as the replacement of nutrients in the diet of patient.

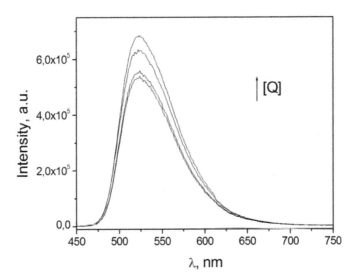

Figure 3. Fluorescence spectra of fluoxetine in acid aqueous solution of chitosan.
Chitosan concentrations: 0.00; 0.050; 0.60 and 1.0 (from the base to the top).

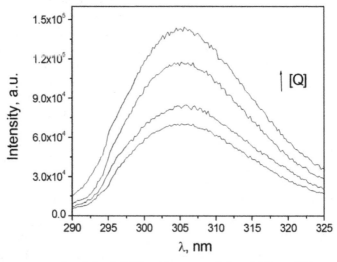

Figure 4. Fluorescence spectra of vitamin B12 in acid aqueous solution of chitosan. Chitosan concentrations: 0.00; 0.050; 0.60 and 1.0 (from the base to the top).

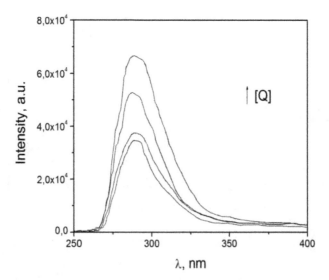

Figure 5. Fluorescence spectra of vitamin B2 in acid aqueous solution of chitosan. Chitosan concentrations: 0.00; 0.050; 0.60 and 1.0 (from the base to the top).

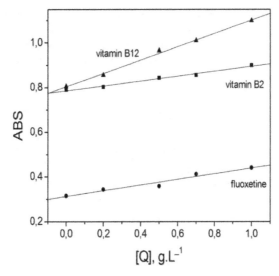

Figure 6. Behavior of absorption intensities of fluoxetine, vitamin B12 and B2.

Chemicals	I	ABS
Vitamin B12	107 %	80 %
Vitamin B2	30 %	14 %
fluoxetine	109 %	40 %

Table 1. Relative increase on the fluorescence (I) and absorption (ABS) intensities of the fluoxetine, vitamins B2 and B12 when chitosan concentration ranges from zero to 1.0 g.L^{-1}.

This paper seeks to warn to possible problems connected with the excessive loss of vitamins and other nutrients by the body during prolonged treatment with chitosan, as well as due the concomitant use of chitosan and fluoxetine.

4. Conclusions

Innumerous studies have described the formation of aggregates in naturals (Kim et al., 2000; Pelletier et al., 2000; Zhbankov et al., 2003), as chitosan ((Rodrigues, 2005; Hennen, 2005; Rodrigues et al., 2008) and synthetic (Kalyanasundaram, 1987; Neumann & Rodrigues, 1994; Neumann et al., 1995; Gomes et al., 2006; Gomes et al., 2007; Sur, 2010) polymers solutions and these molecular structures occur due to intra- and intermolecular interactions.

Chitosan is a polysaccharide precursor of materials suitable to release and/or dissolve drugs into the human body (Hennen, 2005), among other uses. However, the study of spectral properties of chemicals, fluoxetine, vitamins B2 and B12, demonstrated that the microenvironment generated by weakly acidic solution of chitosan also is able to sequester some B2, B12 and fluoxetine molecules, despite the hydro soluble nature of vitamins.

The results described in the current work demonstrated the wide range of possibilities in the studies of interaction between chitosan, used in diets as anti-obesity supplement, and molecules that are present in the human body, as well as with other drugs. Moreover, our work demonstrates the need for more studies on the subject as a means of providing information on the use of chitosan in diets as anti-obesity supplement.

Author details

Máira Regina Rodrigues*
Universidade Federal Fluminense, Polo Universitário de Rio das Ostras, Rio das Ostras, RJ, Brazil

Alexandre de Souza e Silva and Fábio Vieira Lacerda
Centro Universitário de Itajubá, Instituto de Ciências Biológicas, BairroVarginha, Itajubá, MG, Brazil

Acknowledgement

The authors thank CNPq and FAPESP for the financial support.

5. References

Alda, M.J.L.; Parceiro, P.; Goicoechea, A.G. (1996). Rapid zero-order and second-derivative UV spectrophotometric determination of cyanocobalamin in pharmaceutical formulations containing dextran and preservatives. *Journal of Pharmaceutical and Biomedical Analysis*, 14:363-365.

Aslander-van Vliet, E.; Smart, C.; Waldron, S. (2007). Nutritional management in childhood and adolescent diabetes. *Pediatric Diabetes*, 8:323-339.

Association of Official Analytical Chemists. (2005). Official Methods of Analysis of the Association Analytical Chemists. 18th ed. Gaithersburg, MD: AOAC.

Barbosa, L.S.; Pazdziora, A.Z.; Kojima, A.S.G.; Martins, M.S.F.; Ferreira, C.L.P.; Macêdo, G.S.; Latorraca, M.Q.; Gomes-da-Silva, M.H.G. (2007). Effects of chitosan over some nutritional and biochemical parameters in rats. *Journal of Brazilian Society of Food and Nutrition*, 32:1-13.

Böer, U.; Noll, C.; Cierny, I.; Krause, D.; Hiemke, C.; Knepel, W. (2010). A common mechanism of action of the selective serotonin reuptake inhibitors citalopram and fluoxetine: Reversal of chronic psychosocial stress-induced increase in CRE/CREB-directed gene transcription in transgenic reporter gene mice. *European Journal of Pharmacology*, 633:33-38.

British Pharmacopoeia. (1998). Her Majesty's *Stationery* Office, London, p.413.

Cummings, J.H.; Roberfroid, M.B.; Andersson, H.; Barth, C.; Ferro-Luzzi, A.; Ghoos, Y.; Gibney, M.; Hermansen, K.; James, W.P.T.; Korver, O.; Lairon, D.; Pascal, G.; Voragen,

* Corresponding Author

A.G.S. (2010). A new look at dietary carbohydrate: chemistry, physiology and health. *European Journal of Clinical Nutrition*, 64:334-339.

Dacome, L. (2005). Useless and pernicious matter: corpulence in eighteenth-century britain. In: *Cultures of the abdomen: diet, digestion and fat in the modern world*. Eds: A. Carden-Coyne and C. Forth. Palgrave, New York.

Darwish, I.A. (2005). Development and validation of spectrophotometric methods for determination of fluoxetine, sertraline, and paroxetine in pharmaceutical dosage forms. *Journal of AOAC International*, 88:38-45.

Deuchi, K.; Kanauchi, O.; Imasato, Y.; Kobayashi, E. (1994). Decreasing effect of chitosan on the apparent fat digestibility by rats fed on a high-fat diet. *Bioscience Biotechnology & Biochemistry*, 58:1613-1616.

Fregonezi-Nery, M.M.; Baracat, M.M.; Casagrande, R.; Machado, H.T.; Miglioranza, B.; Gianotto, E.A.S. (2008). Validação de métodos para determinação de fluoxetina em cápsulas. *Química Nova*, 31:1665-1669.

Gades, M.D.; Stern, J.S. (2005). Chitosan supplementation and fat absorption in men and women. *Journal of the American Dietetic Association*, 105:72-77.

Gomes, A.J.; Faustino, A.S.; Machado, A.E.H.; Zaniquelli, M.E.; Rigoletto, T.P.; Lunardi, C.N. (2006). Characterization of PLGA microparticle as a drug carrier for the compound 3-ethoxycarbonyl-2H-benzofuro[3,2-d]-1-benzopyran-2-one. Ultra structural study of cellular up take and intracellular distribution. *Drug Delivery*, 13:447–454.

Gomes, A.J.; Assunção, R.M.N.; Rodrigues Filho, G.; Espreafico, E.M.; Machado, A.E.H. (2007). Preparation and characterization of poly(D,L-lactic-co-glycolicacid) nanoparticles containing 3-(benzoxazol-2-yl)-7-(N,N-diethylamino)cromen-2-one. *Journal of Applied Polymer Science*, 105:964–972.

Hennen, W.J. (2005). *Chitosan*. Woodland Publishing Inc., Pleasant Grove, Utah.

Hirano, S.; Akiyama, Y. (1995). Absence of a hypocholesterolaemic action of chitosan in high-serum-cholesterol rabbits. *Journal of the Science of Food and Agriculture*, 69:91-94.

Ho, S.C.; Tai, E.S.; Eng, P.H.; Tan, C.E.; Fok, A.C. (2001). In the absence of dietary surveillance, chitosan does not reduce plasma lipids or obesity in hypercholesterolaemic obese Asian subjects. *The Singapore Medical Journal*, 42:6–10.

Johnson, R.J.; Segal, M.S.; Sautin, Y.; Nakagawa, T.; Feig, D.I.; Kang, D.H.; Gersch, M.S.; Benner, S.; Sánchez-Lozada, L.G. (2007). Potential role of sugar (fructose) in the epidemic of hypertension, obesity and the metabolic syndrome, diabetes, kidney disease, and cardiovascular disease. *The American Journal of Clinical Nutrition*, 86:899-906.

Jull, A.B.; Ni Mhurchu, C.; Bennett, D.A.; Dunshea-Mooij, C.A.; Rodgers, A. (2008). Chitosan for overweight or obesity. *Cochrane Database System Review*, 16:CD003892.

Kaats, G.R.; Michalek, J.E.; Preuss, H.G. (2006). Evaluating eficacy of a chitosan product using a double-blinded, placebo-controlled protocol. *Journal of the American College of Nutrition*, 25:389-394.

Kalyanasundaram. K.; Thomas, J.K. (1977). Environment effects on vibronic band intensities in pyrene monomer fluorescence and their application in studies of micellar systems. *Journal of the American Chemical Society*, 99:2039–2044.

Kalyanasundaram, K. (1987). *Photochemistry in microheterogeneous systems*. Academic, New York.

Kim, I.S.; Jeong, Y.L.; Kim, S.H. (2000). Core-shell type polymeric nanoparticles composed of poly(l-lacticacid) and poly(N-isopropylacrylamide). *The International Journal of Pharmaceutics*, 205:109-115.

Kumirska, J.; Czerwicka, M.; Kaczyński, Z.; Bychowoka, A., Brzozowski, K.; Thöming, J.; Stepnowski, P. (2010). Application of spectroscopic methods for structural analysis of chitin and chitosan. *Marine Drugs*, 8:1567–1636.

Li, H.B.; Chen, F. (2000). Determination of vitamin B12 in pharmaceutical preparations by a highly sensitive fluorimetric method. *Fresenius' Journal of Analytical Chemistry*, 368:836–838.

Liao, F.H.; Shieh, M.J.; Chang, N.C.; Chien, Y.W. (2007). Chitosan supplementation lowers serum lipids and maintains normal calcium, magnesium, and iron status in hyperlipidemic patients. *Nutrition Research*, 27: 146-151.

Liu, J.; Zhang, J.; Xia, W. (2008). Hypocholesterolaemic effects of different chitosan samples in vitro and in vivo. *Food Chemistry*, 107:419-425.

Lois, K.; Kumar, S. (2008). Pharmacotherapy of obesity. *Therapy*, 5:223-235.

Neumann, M.G.; Rodrigues, M.R. (1994). Photochemical determination of aggregation in polyelectrolytes and the binding constants of quencher molecule. *Journal of Photochemistry and Photobiology A: Chemistry*, 83:161-164.

Neumann, M.G.; Tiberti, L.; Rodrigues, M.R. (1995). Binding of metallic cations to poly(styrenesulphonate)polymers and copolymers. *Journal of the Brazilian Chemical Society*, 6:179-195.

Ni Mhurchu, C.; Poppitt, S.D.; McGill, A.T.; Leahy, F.E.; Bennett, D.A.; Lin, R.B.; Ormrod, D.; Ward, L.; Strik, C.; Rodgers, A. (2004). The effect of the dietary supplement, chitosan, on body weight: a randomised controlled trial in 250 overweight and obese adults. *International Journal of Obesity*, 28:1149–1156.

Ni Mhurchu, C.; Dunshea-Mooij, C.; Bennett, D.; Rodgers, A. (2005). Effect of chitosan on weight loss in overweight and obese individuals: a systematic review of randomized controlled trials. *Obesity Reviews*, 6:35–42.

Pelletier, S.; Hubert, P.; Lapicque, F.; Payan, E.; Dellacherie, E. (2000). Amphiphilic derivatives of sodium alginate and hyaluronate: synthesis and physic-chemical properties of aqueous dilute solutions. *Carbohydrate Polymers*, 43:343-349.

Pittler, M.H.; Abbot, N.C.; Harkness, E.F.; Ernst, E. (1999). Randomized, double-blind trial of chitosan for body weight reduction. *European Journal of Clinical Nutrition*, 53:379-381.

Pittler, M.H.; Ernst, E. (2004). Dietary supplements for body-weight reduction: a systematic review. *American Journal of Clinical Nutrition*, 79:529-536.

Prajapati, B.G. (2009). Chitosan a marine medical polymer and its lipid lowering capacity. *The internet Journal of Health*, 9:1-7.

Ramamurthy, V. (1991). Use of photophysical techniques in the study of organized assemblies. In: *Photochemistry in organized and constrained media*. Eds: C. Bohne, RW Redmond, JC Scaiano. VCH, New York.

Razdan, A.; Petterson, D. (1994). Effect of chitin and chitosan on nutrient digestibility and plasma lipid concentration in broiler chickens. *British Journal of Nutrition*, 72:277-288.

Rizo, C.; Deshpande, A.; Ing, A.; Seeman, N. (2011). A rapid, web-based method for obtaining patient views on effects and side-effects of antidepressants. *Journal of Affective Disorders*, 130:290-293.

Rodrigues, M.R. (2005). Synthesis and investigations of chitosan derivatives formed by reaction with acyl chlorides. *Journal of Carbohydrate Chemistry*, 24:41-54.

Rodrigues, M.R.; Lima, A.; Codognoto, C.; Villaverde, A.B.; Pacheco, M.T.T.; de Oliveira, H.P.M. (2008). Detection of polymolecular associations in hydrophobized chitosan derivatives using fluorescent probes. *Journal of Fluorescence*, 8:973–977.

Saper, R.B.; Eisenberg, D.M.; Phillips, R.S. (2004). Common dietary supplements for weight loss. *American Family Physician*, 70:1731-1738.

Schiller, R.N.; Barrager, E.; Schauss, A.G.; Nichols, E.J. (2001). A randomized, double-blind, placebo-controlled study examining the effects of a rapidly soluble chitosan dietary supplement on weight loss and body composition in overweight and mildly obese individuals. *Journal of the American Nutraceutical Association*, 4:42-49.

Shah, J.; Jan, R.; Rehman, F. (2008). Flow injection spectrophotometric determination of fluoxetine in bulk and in pharmaceutical preparations. *Journal of the Chilean Chemical Society*, 3:53-59.

Signini, R.; Campana Filho, S.P. (1999). On the preparation and characterization of chitosan hydrochloride. *Polymer Bulletin*, 42:159–163.

Solheim, H.; Kornobis, K.; Ruud, K.; Kozlowski, P.M. (2011). Electronically Excited States of Vitamin B12 and Methylcobalamin: Theoretical Analysis of Absorption, CD, and MCD Data. *The Journal of Physical Chemistry B*, 115:737–748.

Sommer, A. (2008). History of nutrition - Vitamin A deficiency and clinical disease: An historical overview. *Journal of Nutrition*, 138:1835-1839.

Suarez, W.T.; Sartori, E.R.; Batista, É.F.; Fatibello-Filho, O. (2009). Flow-injection turbidimetric determination of fluoxetine hydrochloride in pharmaceutical formulations. *Química Nova*, 32:2396-2400.

Sugano, M.; Fujikawa, T.; Hiratsuji, Y.; Hasegawa, Y. (1978). Hypocholesterolemic effects of chitosan in colesterol-fed rats. *Nutrition Reports International*, 18:531-537.

Sun, J.; Zhu, X.; Wu, M. (2007). Hydroxypropyl-β-Cyclodextrin Enhanced Determination for the Vitamin B12 by Fluorescence Quenching Method. *Journal of Fluorescence*, 17:265–270.

Sur, U.K. (2010). Stimuli-responsive bio-inspired synthetic polymer nanocomposites. *Current Science*, 98:1562-1563.

Tapola, N.S.; Lyyra, M.L.; Kolehmainen, R.M.; Sarkkinen, E.S.; Schauss, A.G. (2008). Safety aspects and cholesterol-lowering efficacy of chitosan tablets. *Journal of* the American College of Nutrition, 27:22–30.

United States Pharmacopeia (USP). (2007). United States Pharmacopeial Convention. 30 th ed. Rockville.

Valeur, B. (2001). *Molecular fluorescence: principles and applications*. Wiley-VCH, Verlag.

Wang, X.; Wei, L.; Kotra, L.P. (2007). Cyanocobalamin (vitamin B12) conjugates with enhanced solubility. *Bioorganic & Medicinal Chemistry*, 15:1780–1787.

Yao, H.T.; Huang, S.Y.; Chiang, M.T. (2008). A comparative study on hypoglycemic and hypocholesterolemic effects of high and low molecular weight chitosan in streptozotocin-induced diabetic rats. *Food and Chemical Toxicology,* 46:1525-1534.

Zhang, J.; Zhang, J.; Liu, J.; Li, L.; Xia, W. (2008). Dietary chitosan improves hypercholesterolemia in rats fed high-fat diets. *Nutrition Research,* 28:383-390.

Zhbankov, R.G.; Firsov, S.P.; Grinshpan, D.D.; Baran, J.M.K.; Marchewkac, Ratajczak, H. (2003). Vibrational spectra and noncovalent interactions of carbohydrates molecules. *Journal of Molecular Structure,* 645:9-16.

Zheng, D.; Lu, T. (1997). Electrochemical reactions of cyanocobalamin in acidic media. *Journal of Electroanalytical Chemistry,* 429:61-65.

Chapter 11

The Future of Synthetic Carbohydrate Vaccines: Immunological Studies on *Streptococcus pneumoniae* Type 14

Dodi Safari, Ger Rijkers and Harm Snippe

Additional information is available at the end of the chapter

1. Introduction

Studies on synthetic carbohydrates to be used as potential vaccine candidates for polysaccharide encapsulated bacteria were started in the mid-1970s. They were the logical follow-up to studies being performed at that time on the immunogenicity of antigens composed of carrier proteins and synthetic hapten groups. Hapten-carrier complexes were first introduced in immunology by Karl Landsteiner in the early 1900s [1]. He discovered that (i) small organic molecules with a simple structure, such as phenyl arsonates and nitrophenyls, do not provoke antibodies by themselves, but (ii) if those molecules are attached covalently, by simple chemical reactions, to a protein carrier, then antibodies against those small organic molecules are evoked. Since their introduction, these hapten-carrier complexes have become excellent tools to elucidate the role of different antigen-reactive cells in the immune response [2]. The key players in this immunological process are thymus-derived T cells and bone marrow-derived B cells. The former group of lymphoid cells is responsible for various phenomena of cell-mediated immunity, e.g. delayed hypersensitivity, allograft-, and graft-versus-host reactions, and reacts with specific determinants on the carrier protein (T cell epitopes). The latter group of lymphoid cells (B cells) give rise to the precursors of antibody-secreting cells, and reacts with both the carrier protein and the synthetic haptenic determinants. This results in antibody formation to both the carrier and the hapten.

The reason to apply the above concepts and techniques to carbohydrate antigens was to address an immunological problem: polysaccharide molecules are classified as so-called thymus-independent (TI) antigens, because they do not require T cells to induce an immune response of B cells. As a result, the antibodies formed are mainly of the IgM class and have a low avidity. Moreover, no immunological memory is generated and the antigens are poorly immunogenic in infants. Latter characteristic has major implications for development of vaccines against polysaccharide encapsulated bacteria. It was hypothesized that by linking small carbohydrates (oligosaccharides) to a carrier protein, the immunogenic behavior would change to that of a thymus-dependent (TD) antigen. Therefore, the studies of both Goebel [3, 4] and Campbell and Pappenheimer [5], who first isolated the antigenic determinant of *Streptococcus pneumoniae* type 3, were combined and extended. The hapten-inhibition studies by Mage and Kabat [6] demonstrated that the antibody-combining site of type 3 pneumococcal polysaccharide consists of two to three cellobiuronic acid units. In the dextran-anti-dextran system extensively studied by Kabat and colleagues [7] the upper size limit of the antibody-combining site appeared to be a hexa- or heptasaccharide and the lower limit was estimated to be somewhat larger than a monosaccharide. Snippe and colleagues [8] proved in 1983 that small synthetic oligosaccharides (tetra- and hexasaccharides) of *S. pneumoniae* type 3 could be transformed into TD antigens by conjugating them to a protein carrier. This opened the way to explore the synthesis and immunogenicity of numerous oligosaccharide-carrier protein conjugates of different pneumococcal serotypes. Those studies culminated in 2004 in the large-scale synthesis and introduction of a synthetic oligosaccharide vaccine for *Haemophilus influenzae* type b for use in humans in Cuba [9]. The recent exploration of gold nanoclusters coated with synthetic oligosaccharides and peptides as a vaccine are a promising platform towards the development of fully synthetic carbohydrate-based vaccines [10].

2. Streptococcus pneumoniae

Streptococcus pneumoniae (*S. pneumoniae* or pneumococcus) is a leading cause of bacterial pneumonia, meningitis, and sepsis in children worldwide. It is estimated that 1.6 million people die from these infections each year, of whom one million are children [11, 12]. *S. pneumoniae* are lancet-shaped, gram-positive, and alpha-hemolytic bacteria that colonize the mucosal surfaces of the upper respiratory tract [13]. Three major surface layers can be distinguished from the inside to the outside: the plasma membrane, the cell wall, and the capsule (Fig. 1) [14]. The cell wall consists of a triple-layered peptidoglycan backbone that anchors the capsular polysaccharide, the cell wall polysaccharide, and also various proteins such as pneumococcal surface protein A (pspA) and hyluronate lyase (Hyl) (Fig. 1). The capsule is the thickest layer, completely concealing the inner structures of exponentially growing *S. pneumoniae* bacteria.

3. Capsular polysaccharide

Capsular polysaccharides are well known as the major virulence factors of *S. pneumoniae*. Today more than 92 serotypes have been identified based on the different chemical structures of these polysaccharides [16, 17]. This diversity determines the ability of the serotypes to survive in the bloodstream and very likely the ability to cause invasive disease, especially in the respiratory tract [14, 16]. Recently, new *S. pneumoniae* serotypes have been identified, e.g. serotype 6C [17], 6D [18, 19], and 11E [20]. Capsular polysaccharides are large polymers (0.5-2×10^6 Da), composed of multiple repeating units of up to eight sugar residues [14]. The capsular polysaccharides are generally synthesized by the Wzx/Wzy-dependent pathway, except for type 3 and 37 which are synthesized by the synthase pathway [21, 22] (Fig. 2). In the synthase pathway capsule is produced through processive transferase activity [23, 24].

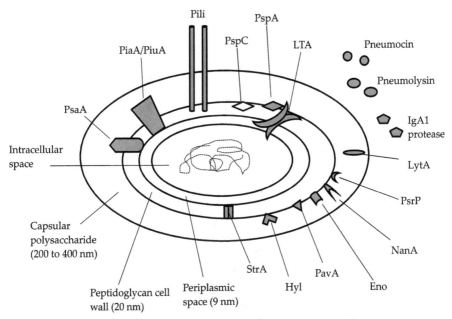

Figure 1. Schematic structure of *S. pneumoniae*. StrA=sortase A. Hyl=hyluronate lyase. PavA=pneumococcal adhesion and virulence. Eno=enolase. NanA=neuraminidase. PsrP=pneumococcal serine-rich repeat protein. LytA=autolysin. LTA=lipoteichoic acid. PspA=pneumococcal surface protein A. PspC=pneumococcal surface protein C. PiaA/PiuA=pneumococcal iron acquisition and uptake. PsaA=pneumococcal surface antigen A. (Adopted from van der Poll, T. and Opal, S.M. [15] and de Velasco, E.A. et al [14])

Many studies have demonstrated that antibodies directed against the capsular polysaccharide are essential for protection against pneumococcal disease [25-27]. However, the native capsular polysaccharides are well-known thymus-independent type-2 (TI-2) antigens that lack T-helper epitopes and therefore mainly induce IgM antibodies, and to a lesser degree IgG [28]. The TI-2 characteristics of polysaccharides can be altered by conjugation of polysaccharide to a protein carrier (glycoconjugate) resulting in a switch to an anti-polysaccharide antibody response with characteristics of a T-cell-dependent response. This is reflected by the generation of memory B and T cells and the induction of high titers of anti-polysccharide IgG antibodies after booster immunization [29].

It should be noted that not all polysaccharides behave as TI-2 antigens. Zwitterionic polysaccharides such as *S. pneumoniea* type 1 polysaccharide: [→3)-α-AATGal-(1→4)-α-D-GalpA-(1→3)-α-D-GalpA-(1→]n with a right-handed helix with repeated zwitterionically charged grooves elicit potent T cell responses *in vivo* and *in vitro* [30, 31].

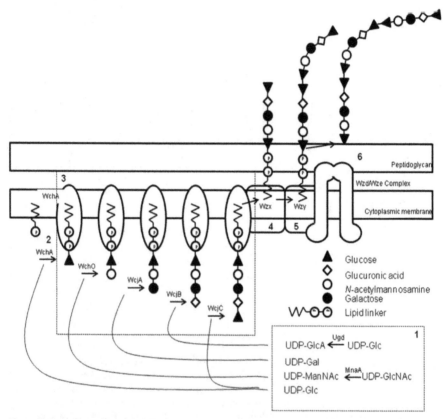

Figure 2. Representation of the Wzx/Wzy-dependent pathway for biosynthesis of CPS 9A (Adopted from Bentley. S.D. et al [21]). Representation of the Wzx/Wzy-Dependent Pathway Pictured is a hypothetical model for capsule biosynthesis in *S. pneumoniae* based on a mixture of experimental evidence and speculation.

1. Non-housekeeping nucleotide sugar biosynthesis.
2. The initial transferase (WchA in this case) links the initial sugar as a sugar phosphate (Glc-P) to a membrane-associated lipid carrier (widely assumed to be undecaprenyl phosphate).
3. Glycosyl transferases sequentially link further sugars to generate repeat unit.
4. Wzx flippase transports the repeat unit across the cytoplasmic membrane.
5. Wzy polymerase links individual repeat units to form lipid-linked CPS.

6.   Wzd/Wze complex translocates mature CPS to the cell surface and may be responsible
     for the attachment to peptidoglycan.

4. Development of pneumococcal vaccines

Although the first pneumococcal vaccines, including the application of the principle of
conjugate vaccination, were already initiated in the beginning of the previous century, most
of these developments stopped when antibiotics were introduced. Existing vaccines were
even withdrawn from the market. By now, in many parts of the world, the antibiotic
resistance of S. *pneumoniae* bacteria has increased: America [32, 33], Africa [34], Europe [35,
36], Asia [37-39], and Australia [40]. This makes treatment of pneumococcal infections more
difficult and stresses the importance of the development of effective vaccines as a strategy to
reduce morbidity and mortality caused by S. *pneumoniae* infection worldwide.

4.1. Pneumococcal polysaccharide-based vaccines.

Currently two vaccine types against S. *pneumoniae* are commercially available: a
pneumococcal polysaccharide vaccine (PPV) and a pneumococcal conjugate vaccine (PCV)
[41]. The first multivalent pneumococcal polysaccharide vaccine (PPV) contains 23 purified
capsular polysaccharides (25 µg of each capsule type; Pneumovax®, PPV23: 1, 2, 3, 4, 5, 6B, 7,
8, 9N, 9V, 10A, 11A, 12F, 14, 15B, 15F, 18C, 19A, 19F, 20, 22F, 23F, 33F) which is licensed for
use in adults and children older than 2 years of age [42]. This vaccine was shown to be
moderately effective in young adults [43] but not in young children [44] and elderly [45] and
also not in immunocompromised patients, e.g HIV infected people [46, 47].

In early 2000, a polysaccharide-protein conjugate vaccine targeting seven pneumococcal
serotypes was licensed in the United States for use in young children (Prevnar®, PCV7: 4, 6B,
9V, 14, 18C, 19F, 23F). The polysaccharides are conjugated to the non-toxic cross reactive
material from diphtheria toxin, CRM197 and each dose contains 2µg of each capsule type,
except for 6B, for which 4 µg is included in every vaccine dose[48]. The PCV7 vaccine
produces a significant effect regarding prevention of invasive pneumococcal disease in
children younger than 24 months (based on a meta-analysis of published data from trials on
pneumococcal vaccine) [49]. Large scale introduction of PCV7 has resulted in an overall
decline in infectious pneumococcal disease (IPD). However, IPD caused by the non-vaccine
serotypes serotypes 1, 19A, 3, 6A, and 7F has increased (replacement disease), highlighting
the need for inclusion of these serotypes in future improved vaccine formulations [50].
Apart from the CRM197 based PCV7, several new candidate pneumococcal conjugate
vaccines have been developed to cover more serotypes with different protein carriers and
most of them are in clinical trials, such as PCV10 vaccine (1, 4, 5, 6B, 7F, 9V, 14, 18C, 19F,
23F) [51, 52] and PCV13 vaccine (1, 3, 4, 5, 6A, 6B, 7F, 9V, 14, 18C, 19A, 19F, 23F) [53].

4.2. Pneumococcal protein-based vaccines

An alternative vaccine strategy focuses on the use of pneumococcal surface-associated
proteins which are to be assumed to elicit protection in all age groups against all, or nearly

all, pneumococcal serotypes (Fig. 1). Protection induced by the proteins should be serotype-independent and possibly cheaper and thus within reach of developing countries [54]. Currently, several surface pneumococcal proteins are investigated as a candidate vaccine against *S. pneumoniae* infection with single or combination of recombinant proteins, such as pspA family fusion protein [55]; pneumolysis and pspA1/pspA2 combined [56]. Recently new candidate protein antigens were discussed at the 8th International Symposium on Pneumococci and Pneumococcal Diseases at Iguaçu Falls, Brazil (2012), phtD (pneumococcal histidin triad protein D) and PcpA (pneumococcal choline binding protein A) [57].

4.3. Pneumococcal synthetic oligosaccharide-based vaccines

The current polysaccharide conjugate vaccines are based on natural polysaccharides, purified form bacterial cultures. Synthetic oligosaccharide–protein conjugates (neoglycoconjugate), involving functional mimics of the natural polysaccharide antigens have emerged as an attractive option [58]. The advantages of neoglycoconjugates are well-defined chemical structures (chain length, epitope conformation, and carbohydrate/protein ratio) as well as a lack of the impurities present in polysaccharides obtained from bacterial cultures [59, 60].

The chemical synthesis of oligosaccharide fragments however is complex. According to the sequence in the natural polysaccharide, monosaccharide residues have to be linked in such a way that they form an oligosaccharide with the required stereospecificity (epitope). Various methodologies and strategies for synthesis of carbohydrates have successfully been used for production of experimental neoglycoconjugates, as reviewed by Kamerling [16]. In 2001, the first automated synthesis of oligosaccharides was reported by Plante, O.J. et al [61].

Neoglycoconjugates have been prepared for saccharides of different microorganisms. In 2004, Verez Bencomo et al., reported the large-scale synthesis and the introduction of a synthetic oligosaccharide vaccine for *Haemophilus influenzae* type b for use in humans in Cuba [9]. The immunogenicity of the synthetic oligosaccharide fragment of the O-specific polysaccharide (O-PS) of *Vibrio cholera* O1, serotype Ogawa, conjugated to bovine serum albumin has been investigated in a mouse model [62, 63]. A multimeric bivalent synthetic hexasaccharide fragment of the O-specific polysaccharide of *Vibrio cholera* O1, serotype Ogawa, in combination with Inaba:1 or a synthetic disaccharide tetrapeptide peptidoglycan fragment as adjuvant were prepared and conjugated to recombinant tetanus toxin H(C) fragment as protein carrier [64]. The immunogenicity of synthetic oligosaccharides mimicking the O-antigen of the *Shigella flexneri* 2a lipopolysaccharide (LPS) was also investigated in mice [65, 66]. Immunization of mice with synthetic hexasaccharide of glycosylphosphatidylinositol malarial toxin conjugated to a protein carrier was reported to protect the mice from an otherwise lethal dose of malaria parasites [67]. A fully synthetic carbohydrate-based antitumor candidate vaccine for the common T-synthase was recently reported [68].

Meanwhile we and other groups have been working on improving the immunogenicity of neoglycoconjugates against different *S. pneumoniae* serotypes in animal models: Di-, tri-, and tetrasaccharides related to polysaccharide type 17F conjugated to keyhole limpet hemocyanin (KLH) protein[69, 70] and tri- and tetrasaccharides related to type 23 conjugated to KLH protein [71]; Di-, tri-, and tetrasaccharides related to type 6B conjugated to KLH protein [72]; Di-, tri-, and tetrasaccharide related to type 3 conjugated to the cross-reactive material of diphteria toxin (CRM197) protein [60] and most recently overlapping oligosaccharide varying from tri- to dodecasaccharides related to polysaccharide type 14 conjugated to CRM197 protein [73, 74].

5. Immunogenicity of synthetic oligosaccharide based vaccines

This review focuses on the *S. pneumoniae* type 14 capsular polysaccharide (Pn14PS) which consists of biosynthetic repeating units of the tetrasaccharide {6)-[β-D-Galp-(1→4)-]β-D-GlcpNAc-(1→3)-β-D-Galp-(1→4)-β-D-Glcp-(1→}n [75] (Fig. 3).

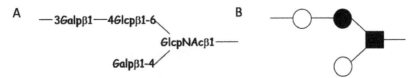

Figure 3. A branched tetrasaccharide repeating unit of *S. pneumoniae* type 14 capsular polysaccharide (A) and its nomenclature symbol (B): filled circle = glucose (Glc); open circle = galactose (Gal), and filled square = N-acetylglucosamine (GlcNAc)

5.1. Identification of the minimal structure of oligosaccharide capable in evoking anti-Pn14PS antibodies.

It was reported that a synthetic branched tetrasaccharide, corresponding to a single structural repeating unit of Pn14PS conjugated to the cross-reactive material of diptheria toxin (CRM197), was found to induce anti-polysaccharide type 14 antibodies by Mawas, F. et al [74]. We continued to investigate further how small the minimal structure in Pn14PS can be and still produce specific antibodies against native polysaccharide type 14 [73]. 16 overlapping oligosaccharide fragments of Pn14PS were synthesized as described previously [76-79] and were conjugated to the protein carrier CRM197. The mice immunization studies were performed to investigate the immunogenicity of the neoglycoconjugates. We found that the fragments with a linear and/or incomplete branched structure did not elicit specific antibodies against native Pn14PS (Fig. 4: JJ118, JJ42, JJ141, DM65, JJ153, JJ9, JJ6 and DM35) [73]. High titer of anti-Pn14PS IgG antibodies was observed when the complete branched structure fragments, conjugated to the protein carrier were used in the mouse model (Fig. 4: JJ1, DM66, DM36, ML1, ML2, and CRM197-Pn14PS as a positive control), excepted for JJ5 and JJ10 which elicited low titer of anti-Pn14PS antibodies.

We also tested the phagocytic capacity of mice sera by human polymorph nuclear cells and a mouse macrophage cell line. We found that the sera containing antibodies against Pn14PS were also capable of promoting the phagocytosis of *S. pneumoniae* type 14. Conjugates that did not evoke specific antibodies against polysaccharide type 14 also did not display phagocytic capacity [73].

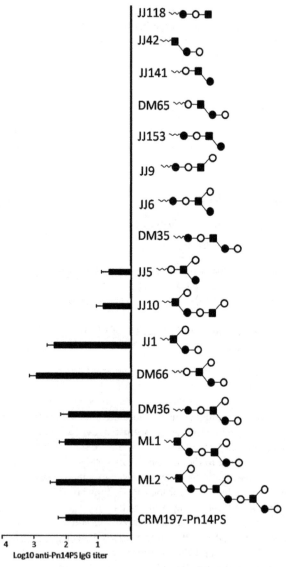

Figure 4. **Level of anti-Pn14PS antibodies and schematic structure of overlapping synthetic oligosaccharide fragments of Pn14PS** (Adopted from Safari et al 2008 [73]). The oligosaccharides were conjugated to CRM197 protein and the immunogenicity of those conjugates were studies in a mouse model. Mice were immunized with polysaccharide type 14 conjugated to CRM197 (CRM197-Pn14PS) as a positive control. Enzyme-linked immunosorbent assay was employed to measure specific anti-Pn14PS IgG antibodies after the booster immunization. Antibody titers were expressed as the log10 of the dilution Filled circle = glucose (Glc); open circle = galactose (Gal), and filled square = N-acetylglucosamine (GlcNAc).

In conclusion, the present study has shown that the branched trisaccharide Glc-(Gal-)GlcNAc is the core structure inducing Pn14PS-specific antibodies and that the neighboring galactose at the non-reducing end significantly contributes to the induction of phagocytosis-promoting antibodies [73]. Our study provides evidence that the branched tetrasaccharide Gal-Glc-(Gal-)GlcNAc is a prime candidate for a synthetic oligosaccharide conjugate vaccine against infections caused by *S. pneumoniae* type 14 [73].

5.2. Relationship between polysaccharide of Pn14PS and GBSIII

We also determined the minimal epitope in group B streptococcus type III polysaccharide (GBSIIIPS), using both a panel of anti-Pn14PS mouse sera and sera of humans vaccinated with either Pn14PS or GBSIIIPS as reported by Safari et al [80]. Native Pn14PS is structurally related to and has cross-reactivity with GBSIIIPS [81]. The branched structures of Pn14PS and GBSIIIPS differ only in the absence (in Pn14PS) or presence (in GBSIIIPS) of the ($\alpha2\rightarrow3$)-linked sialic acid N-acetylneuraminic acid (Neu5Ac) in their side chains: {$\rightarrow$4)-β-D-Glcp-(1$\rightarrow$6)-[$\pm\alpha$-Neu5Ac-(2$\rightarrow$3)-β-D-Galp-(1$\rightarrow$4)-]β-D-GlcpNAc-(1$\rightarrow$3)-β-D-Galp-(1$\rightarrow$}n [82]. We reported that type-specific Pn14PS antibodies which recognize the branched structure of Pn14PS have a low affinity for the native GBSIIIPS and do not promote opsonophagocytosis of GBSIII, however desialylation of GBSIIIPS, however, resulted in dramatically higher affinity of anti-Pn14PS antibodies in mice when GBSIIIP was treated by nurimindase (desialylation) [80]. These results revealed that GBSIII bacteria are protected from binding of antibodies against Pn14PS by a residue of ($\alpha2\rightarrow3$)-linked sialic acid, as described previously [83, 84].

5.3. Booster immunization either with either neoglycoconjugate or native polysaccharide

We investigated further the immune response to a neoglycoconjugate of Pn14PS (GC) on the outcome of sustained immunity to *S. pneumoniae* type 14 in a mouse model after the booster injection with either (GC) or native Pn14PS (PS) [85]. We found, as we expected, that the amount of specific IgG antibodies against Pn14PS increased substantially when a GC booster was given to mice previously primed with the same GC [85]. The induced antibodies were capable to opsonise *S. pneumoniae* type 14. Boosting with PS following a primary conjugate vaccine injection did not result in IgG antibody formation to Pn14PS (Table 1).

In order to explain these phenomena we investigated how a booster immunization with a GC or PS affects the cell-mediated immune response by measuring the production profile of a panel of cytokines [85]. We observed a high level of IL-5 in serum after a booster injection with GC (GC-GC or GC-GC-GC). Boosting with PS did not result in the induction of IL-5 nor of any of the other tested cytokines (Table 1; GC-PS and GC-PS-PS). We conclude that induction of the cytokine IL-5 in serum is an early sign of a successful booster immunization and is a prerequisite for the production of specific anti-polysaccharide IgG antibodies [85]. In-vitro spleen cell cultures were also used to investigate the effect of a booster injection on activation of memory T cells. IL-5 which well known Th2 cytokines, were evoked by the GC in spleen cell cultures of mice previously primed and boosted with the same GC [85]. In conclusion, the inability of polysaccharide to boost primed mice might be due to the incapability to induce the cytokines.

Immunization[1]	IgG titer (Log$_{10}$)[2]	Level of Cytokine IL-5 (pg/ml)	
		In serum[3]	After stimulation[4]
GC-GC	2.18±0.22	1022.3±275.2	571.2±20.0
GC-PS	0.34±0.47	0.3±0.5	66.1±0.4
GC-GC-GC	3.02±0.17	2700.4±112.3	1172.8±7.1
GC-PS-PS	0.0	0.0	664.9±221.
Saline	0.0	6.9±1.1	0.0

[1]Five mice per group were immunized with a CRM-neoglycoconjugate (GC), a synthetic branched tetrasaccharide of Pn14PS that is conjugated to a CRM$_{197}$ protein. Booster doses containing either a GC (GC-GC and GC-GC-GC) or a native polysaccharide of Pn14PS (PS) (GC-PS, GCGC-PS, and GC-PS-PS) were injected at Weeks 5 and 10.
[2]ELISA was employed to measure specific anti-Pn14PS IgG antibodies, and expressed as the log10 of the sera dilution
[3]Cytokine levels in sera from mice receiving booster injection. Sera were collected on Day 1 after the primary immunization
[4]Splenocytes were isolated 7 days after the first booster injection. Spleen cells were cultured in vitro and stimulated with CRM-neoglycoconjugate and supernatants were collected 72 h after culture initiation.

Table 1. Effect of booster immunization either with with either the same neoglycoconjugate or a native polysaccharide (Adopted from Safari, D. el at [85] with permission)

5.4. Improvement of anti-Pn14PS antibodies level by coadjuvant administration

The immunogenicity of neoglycoconjugate was increased with adjuvant coadministration [73, 86]. We set out to investigate in a mouse model the effect of adjuvant coadministration i.e. Quil-A, MPL, DDA, CpG and Alum on both the antibody- and cell-mediated immune response against a neoglycoconjugate as reported by Safari et al [87]. In the absence of adjuvant, immunization with neoglycoconjugate leads after a booster merely to IgG1 antibodies against PnP14PS. Coadministration of adjuvant had multiple effects: a diversified anti-Pn14PS IgG antibody response (also other IgG subclasses than IgG1 were evoked), an enhanced avidity and increased opsonic activity of these antibodies [87]. We found that next to Quil-A also DDA as a single dose or in combination with CpG had similar effects on the diversification of eliciting a broader variety of anti-Pn14PS IgG antibody subclasses. Meanwhile, CpG or alum on their own showed in majority IgG1 antibodies after booster immunization in a same pattern as in non adjuvant groups [87]. Compared to other adjuvants, codelivered Quil-A strongly improved the antibody avidity and enhanced the phagocytosis of S. pneumoniae type 14 [87].

6. Future researches

In this review, synthetic oligosaccharide-protein conjugates are proven to be effective vaccines in mice model. A logical next step would be a feasibility and immunogenicity study in human volunteers. Before that, a study should be started with synthetic oligosaccharide-protein conjugates for at least the pneumococcal serotypes 1, 4, 5, 9V and 18C and should even have been completed, because the minimal epitopes for these polysaccharides are still unknown.

To improve the immunogenicity of oligosaccharide-protein conjugates co-delivery of adjuvants are required. As an alternative to the addition of adjuvants, studies should be initiated to direct oligosaccharide-protein conjugates to dendritic cells by incorporation of specific ligands. Targeting to and activation of dendritic cells by TLR5 is a possibility to be explored.

Author details

Dodi Safari*
Eijkman Institute for Molecular Biology, Jakarta, Indonesia

Ger Rijkers
Department of Sciences, Roosevelt Academy, Middelburg and Department of Medical Microbiology and Immunology, St. Antonius Hospital, Nieuwegein, The Netherlands

Harm Snippe
Department of Medical Microbiology, University Medical Center Utrecht, Utrecht, The Netherlands

Acknowledgement

This review is a part of the PhD thesis of Dodi Safari with the title: Development of new synthetic oligosaccharide vaccines: the immunogenicity of oligosaccharide- CRM197 neoconjugates and oligosaccharide/peptide hybrid gold nanoparticles based on the capsular polysaccharide structure of *Streptococcus pneumoniae* type. 14. Utrecht University. 2010. This work was supported by a grant from the European Commission under Contract No: MRTN-CT-2004-005645.

7. References

[1] Landsteiner K, Jacobs J. Studies on the sensitization of animals with simple chemical compounds. J.Exp.Med. 1935;61:643-56.

[2] Mitchison NA; Rajewsky K; Taylor RB. Sterzl J and Riha I, editors.Prague symposium on developmental aspects of antibody formation and structure. Praha: Academia Publishing House; 1970.

[3] Goebel WF. Studies on antibacterial immunity induced by artificial antigens. I. Immunity to experimental pneumococcal infection with an antigen containing cellobiuronic acid. J.Exp.Med. 1939;69:353-64.

[4] Goebel WF. Studies on antibacterial immunity induced by artificial antigens. II. Immunity to experimental pneumococcal infection with antigens containing saccharides of synthetic origin. J.Exp.Med. 1940;72:33-48.

[5] Campbell JH, Pappenheimer AM. Quantitative studies of the specificity of anti-pneumococcal polysaccharide antibodies, type III and VIII. I. Isolation of

* Corresponding Author

oligosaccharides from acid and from enzymatic hydrolysates. Immunochemistry 1966;3:195-212.

[6] Mage RG, Kabat EA. Immunochemical studies on dextrans: III. The specificities of rabbit antidextrans. Further findings on antidextrans with 1,2- and 1,6-specificities. J.Immunol. 1963;(91):633-40.

[7] Kabat EA; Mayer MM. Experimental immunochemistry II , USA: Springfield; 1940.

[8] Snippe H, van Dam JEG, van Houte AJ, Willers JMN, Kamerling JP, Vliegenthart JFG. Preparation of a semisynthetic vaccine to *Streptococcus pneumoniae* type 3. Infect.Immun. 1983;42(2):842-4.

[9] Verez-Bencomo V, Fernandez-Santana V, Hardy E, Toledo ME, Rodriguez A, Baly A, Harrera L, Izquierdo M, Villar A, Valdes Y, Cosme K, Deler ML, Montane M, Gracia E, Ramos A, Aguilar A, Medina E, Torano G, Sosa I, Hernandez I, Martinez R, Muzachio A, Carmenates A, Costa L, Cardoso F, Campa C, Diaz V, Roy R. A synthetic conjugate polysaccharide vaccine against *Haemophilus influenzae* type b. Science 2004;305:522-5.

[10] Safari D, Maradi M, Chiodo F, Dekker HA, Shan Y, Adamo R, Oscarson S, Rijkers GT, Lahmann M, Kamerling JP, Penades S, Snippe H. Gold nanoparticles as carriers for a synthetic *Streptococcus pneumoniae* type 14 conjugate vaccine. Nanomedicine 2012;doi:10.2217/NNM.11.151

[11] Bryce J, Boschi-Pinto C, Shibuya K, Black RE. WHO estimates of the causes of death in children. Lancet 2005;365(9465):1147-52.

[12] World Health Organization. Pneumococcal conjugate vaccine for childhood immunization--WHO position paper. Wkly.Epidemiol.Rec. 2007;82(12):93-104.

[13] Kadioglu A, Weiser JN, Paton JC, Andrew PW. The role of *Streptococcus pneumoniae* virulence factors in host respiratory colonization and disease. Nat.Rev.Microbiol 2008;6(4):288-301.

[14] de Velasco EA, Verheul AF, Verhoef J, Snippe H. *Streptococcus pneumoniae*: virulence factors, pathogenesis, and vaccines. Microbiol.Rev. 1995;59(4):591-603.

[15] van der PT, Opal SM. Pathogenesis, treatment, and prevention of pneumococcal pneumonia. Lancet 2009;374(9700):1543-56.

[16] Kamerling JP. Tomasz A, editors. *Streptococcus pneumoniae*, molecular biology & mechanisms of disease. New York: Mary Ann Liebert; 1999;Pneumococcal polysaccharides: A chemical view. p. 81-114.

[17] Park IH, Pritchard DG, Cartee R, Brandao A, Brandileone MC, Nahm MH. Discovery of a new capsular serotype (6C) within serogroup 6 of *Streptococcus pneumoniae*. J Clin.Microbiol 2007;45(4):1225-33.

[18] Bratcher PE, Kim KH, Kang JH, Hong JY, Nahm MH. Identification of natural pneumococcal isolates expressing serotype 6D by genetic, biochemical and serological characterization. Microbiology 2010;156:555-60.

[19] Oftadeh S, Satzke C, Gilbert GL. Identification of the newly described *Streptococcus pneumoniae* serotype 6D using Quellung reaction and PCR. J Clin.Microbiol 2010;

[20] Calix JJ, Nahm MH. A new pneumococcal serotype, 11E, has a variably inactivated wcjE gene. J Infect.Dis. 2010;202(1):29-38.

[21] Bentley SD, Aanensen DM, Mavroidi A, Saunders D, Rabbinowitsch E, Collins M, Donohoe K, Harris D, Murphy L, Quail MA, Samuel G, Skovsted IC, Kaltoft MS, Barrell B, Reeves PR, Parkhill J, Spratt BG. Genetic analysis of the capsular biosynthetic locus from all 90 pneumococcal serotypes. PLoS.Genet. 2006;2(3):e31

[22] Mavroidi A, Aanensen DM, Godoy D, Skovsted IC, Kaltoft MS, Reeves PR, Bentley SD, Spratt BG. Genetic relatedness of the *Streptococcus pneumoniae* capsular biosynthetic loci. J.Bacteriol. 2007;189(21):7841-55.

[23] Llull D, Garcia E, Lopez R. Tts, a processive beta-glucosyltransferase of *Streptococcus pneumoniae*, directs the synthesis of the branched type 37 capsular polysaccharide in Pneumococcus and other gram-positive species. J Biol.Chem 2001;276(24):21053-61.

[24] Arrecubieta C, Lopez R, Garcia E. Type 3-specific synthase of *Streptococcus pneumoniae* (Cap3B) directs type 3 polysaccharide biosynthesis in Escherichia coli and in pneumococcal strains of different serotypes. J Exp.Med 1996;184(2):449-55.

[25] Breukels MA, Rijkers GT, Voorhorst-Ogink MM, Zegers BJ, Sanders LA. Pneumococcal conjugate vaccine primes for polysaccharide-inducible IgG2 antibody response in children with recurrent otitis media acuta. J.Infect.Dis. 1999;179(5):1152-6.

[26] Rennels MB, Edwards KM, Keyserling HL, Reisinger KS, Hogerman DA, Madore DV, Chang I, Paradiso PR, Malinoski FJ, Kimura A. Safety and immunogenicity of heptavalent pneumococcal vaccine conjugated to CRM197 in United States infants. Pediatrics 1998;101(4 Pt 1):604-11.

[27] Prymula R, Kriz P, Kaliskova E, Pascal T, Poolman J, Schuerman L. Effect of vaccination with pneumococcal capsular polysaccharides conjugated to Haemophilus influenzae-derived protein D on nasopharyngeal carriage of *Streptococcus pneumoniae* and *H. influenzae* in children under 2 years of age. Vaccine 2009;28(1):71-8.

[28] French N. Use of pneumococcal polysaccharide vaccines: no simple answers. J.Infect. 2003;46(2):78-86.

[29] Ada G, Isaacs D. Carbohydrate-protein conjugate vaccines. Clin.Microbiol.Infect. 2003;9(2):79-85.

[30] Kalka-Moll WM, Tzianabos AO, Bryant PW, Niemeyer M, Ploegh HL, Kasper DL. Zwitterionic polysaccharides stimulate T cells by MHC class II-dependent interactions. J Immunol. 2002;169(11):6149-53.

[31] Choi YH, Roehrl MH, Kasper DL, Wang JY. A unique structural pattern shared by T-cell-activating and abscess-regulating zwitterionic polysaccharides. Biochemistry 2002;41(51):15144-51.

[32] Karnezis TT, Smith A, Whittier S, Haddad J, Saiman L. Antimicrobial resistance among isolates causing invasive pneumococcal disease before and after licensure of heptavalent conjugate pneumococcal vaccine. PLoS One. 2009;4(6):e5965

[33] Valenzuela MT, de Quadros C. Antibiotic resistance in Latin America: a cause for alarm. Vaccine 2009;27 Suppl 3:C25-C28

[34] Vlieghe E, Phoba MF, Tamfun JJ, Jacobs J. Antibiotic resistance among bacterial pathogens in Central Africa: a review of the published literature between 1955 and 2008. Int.J Antimicrob.Agents 2009;34(4):295-303.

[35] Siira L, Rantala M, Jalava J, Hakanen AJ, Huovinen P, Kaijalainen T, Lyytikainen O, Virolainen A. Temporal trends of antimicrobial resistance and clonality of invasive *Streptococcus pneumoniae* isolates in Finland, 2002 to 2006. Antimicrob.Agents Chemother. 2009;53(5):2066-73.

[36] Vila-Corcoles A, Bejarano-Romero F, Salsench E, Ochoa-Gondar O, de DC, Gomez-Bertomeu F, Raga-Luria X, Cliville-Guasch X, Arija V. Drug-resistance in *Streptococcus pneumoniae* isolates among Spanish middle aged and older adults with community-acquired pneumonia. BMC.Infect.Dis. 2009;9:36

[37] Imai S, Ito Y, Ishida T, Hirai T, Ito I, Maekawa K, Takakura S, Iinuma Y, Ichiyama S, Mishima M. High prevalence of multidrug-resistant Pneumococcal molecular epidemiology network clones among *Streptococcus pneumoniae* isolates from adult patients with community-acquired pneumonia in Japan. Clin.Microbiol.Infect. 2009;

[38] Shibl AM. Distribution of serotypes and antibiotic resistance of invasive pneumococcal disease isolates among children aged 5 years and under in Saudi Arabia (2000-2004). Clin.Microbiol.Infect. 2008;14(9):876-9.

[39] Yang F, Xu XG, Yang MJ, Zhang YY, Klugman KP, McGee L. Antimicrobial susceptibility and molecular epidemiology of *Streptococcus pneumoniae* isolated from Shanghai, China. Int.J.Antimicrob.Agents 2008;32(5):386-91.

[40] Gottlieb T, Collignon PJ, Robson JM, Pearson JC, Bell JM. Prevalence of antimicrobial resistances in *Streptococcus pneumoniae* in Australia, 2005: report from the Australian Group on Antimicrobial Resistance. Commun.Dis.Intell. 2008;32(2):242-9.

[41] Barocchi MA, Censini S, Rappuoli R. Vaccines in the era of genomics: the pneumococcal challenge. Vaccine 2007;25(16):2963-73.

[42] O'Brien KL, Hochman M, Goldblatt D. Combined schedules of pneumococcal conjugate and polysaccharide vaccines: is hyporesponsiveness an issue? Lancet Infect.Dis. 2007;7(9):597-606.

[43] Singleton RJ, Hennessy TW, Bulkow LR, Hammitt LL, Zulz T, Hurlburt DA, Butler JC, Rudolph K, Parkinson A. Invasive pneumococcal disease caused by nonvaccine serotypes among alaska native children with high levels of 7-valent pneumococcal conjugate vaccine coverage. JAMA 2007;297(16):1784-92.

[44] Douglas RM, Paton JC, Duncan SJ, Hansman DJ. Antibody response to pneumococcal vaccination in children younger than five years of age. J Infect.Dis. 1983;148(1):131-7.

[45] Ortqvist A, Hedlund J, Burman LA, Elbel E, Hofer M, Leinonen M, Lindblad I, Sundelof B, Kalin M. Randomised trial of 23-valent pneumococcal capsular polysaccharide vaccine in prevention of pneumonia in middle-aged and elderly people. Swedish Pneumococcal Vaccination Study Group. Lancet 1998;351(9100):399-403.

[46] Teshale EH, Hanson D, Flannery B, Phares C, Wolfe M, Schuchat A, Sullivan P. Effectiveness of 23-valent polysaccharide pneumococcal vaccine on pneumonia in HIV-infected adults in the United States, 1998--2003. Vaccine 2008;26(46):5830-4.

[47] Veras MA, Enanoria WT, Castilho EA, Reingold AL. Effectiveness of the polysaccharide pneumococcal vaccine among HIV-infected persons in Brazil: a case control study. BMC.Infect.Dis. 2007;7:119

[48] Whitney CG, Farley MM, Hadler J, Harrison LH, Bennett NM, Lynfield R, Reingold A, Cieslak PR, Pilishvili T, Jackson D, Facklam RR, Jorgensen JH, Schuchat A. Decline in invasive pneumococcal disease after the introduction of protein-polysaccharide conjugate vaccine. N.Engl.J.Med. 2003;348(18):1737-46.

[49] Pavia M, Bianco A, Nobile CG, Marinelli P, Angelillo IF. Efficacy of pneumococcal vaccination in children younger than 24 months: a meta-analysis. Pediatrics 2009;123(6):e1103-e1110

[50] Isaacman DJ, McIntosh ED, Reinert RR. Burden of invasive pneumococcal disease and serotype distribution among *Streptococcus pneumoniae* isolates in young children in Europe: impact of the 7-valent pneumococcal conjugate vaccine and considerations for future conjugate vaccines. Int.J.Infect.Dis. 2009;14(3):e197-209.

[51] Knuf M, Szenborn L, Moro M, Petit C, Bermal N, Bernard L, Dieussaert I, Schuerman L. Immunogenicity of routinely used childhood vaccines when coadministered with the 10-valent pneumococcal non-typeable Haemophilus influenzae protein D conjugate vaccine (PHiD-CV). Pediatr.Infect.Dis.J. 2009;28(4 Suppl):S97-S108

[52] Bermal N, Szenborn L, Chrobot A, Alberto E, Lommel P, Gatchalian S, Dieussaert I, Schuerman L. The 10-valent pneumococcal non-typeable Haemophilus influenzae protein D conjugate vaccine (PHiD-CV) coadministered with DTPw-HBV/Hib and poliovirus vaccines: assessment of immunogenicity. Pediatr.Infect.Dis.J. 2009;28(4 Suppl):S89-S96

[53] Meng C, Lin H, Huang J, Wang H, Cai Q, Fang L, Guo Y. Development of 5-valent conjugate pneumococcal protein A - Capsular polysaccharide pneumococcal vaccine against invasive pneumococcal disease. Microb.Pathog. 2009;47(3):151-6.

[54] Bogaert D, de Groot R, Hermans PWM. *Streptococcus pneumoniae* colonisation: the key to pneumococcal disease . Lancet.Infect.Dis. 2004;4:144-54.

[55] Xin W, Li Y, Mo H, Roland KL, Curtiss R, III. PspA family fusion proteins delivered by attenuated Salmonella enterica serovar Typhimurium extend and enhance protection against *Streptococcus pneumoniae*. Infect.Immun. 2009;77(10):4518-28.

[56] Francis JP, Richmond PC, Pomat WS, Michael A, Keno H, Phuanukoonnon S, Nelson JB, Whinnen M, Heinrich T, Smith WA, Prescott SL, Holt PG, Siba PM, Lehmann D, van den Biggelaar AH. Maternal antibodies to pneumolysin, but not pneumococcal surface protein A, delay early pneumococcal carriage in high-risk Papua New Guinean infants. Clin.Vaccine Immunol. 2009;16(11):1633-8.

[57] Ljutic B, Ochs M, Messham B, Ming M, Dookie A, Harper K, Ausar SF. Formulation, stability and immunogenicity of a trivalent pneumococcal protein vaccine formulated with aluminum salt adjuvants. Vaccine 2012;Epub ahead of print

[58] Pozsgay V. Recent developments in synthetic oligosaccharide-based bacterial vaccines. Curr.Top.Med Chem 2008;8(2):126-40.

[59] Jansen WT, Snippe H. Short-chain oligosaccharide protein conjugates as experimental pneumococcal vaccines. Indian.J.Med.Res. 2004;119:7-12.

[60] Benaissa-Trouw B, Lefeber DJ, Kamerling JP, Vliegenthart JF, Kraaijeveld K, Snippe H. Synthetic polysaccharide type 3-related di-, tri-, and tetrasaccharide- CRM$_{197}$ conjugates

induce protection against *Streptococcus pneumoniae* type 3 in mice. Infect.Immun. 2001;69(7):4698-701.

[61] Plante OJ, Palmacci ER, Seeberger PH. Automated solid-phase synthesis of oligosaccharides. Science 2001;291(5508):1523-7.

[62] Saksena R, Ma X, Wade TK, Kovác P, Wade WF. Length of the linker and the interval between immunizations influences the efficacy of *Vibrio cholerae* O1,Ogawa hexasaccharide neoglycoconjugates. FEMS Immunol.Med.Microbiol. 2006;47:116-28.

[63] Saksena R, Ma X, Wade TK, Kovac P, Wade WF. Effect of saccharide length on the immunogenicity of neoglycoconjugates from synthetic fragments of the O-SP of *Vibrio cholerae* O1, serotype Ogawa. Carbohydr.Res. 2005;340:2256-69.

[64] Bongat AF, Saksena R, Adamo R, Fujimoto Y, Shiokawa Z, Peterson DC, Fukase K, Vann WF, Kovac P. Multimeric bivalent immunogens from recombinant tetanus toxin H(C) fragment, synthetic hexasaccharides, and a glycopeptide adjuvant. Glycoconj.J. 2009;27(1):69-77.

[65] Phalipon A, Tanguy M, Grandjean C, Guerreiro C, Belot F, Cohen D, Sansonetti PJ, Mulard LA. A synthetic carbohydrate-protein conjugate vaccine candidate against Shigella flexneri 2a infection. J.Immunol. 2009;182(4):2241-7.

[66] Said HF, Phalipon A, Tanguy M, Guerreiro C, Belot F, Frisch B, Mulard LA, Schuber F. Rational design and immunogenicity of liposome-based diepitope constructs: application to synthetic oligosaccharides mimicking the Shigella flexneri 2a O-antigen. Vaccine 2009;27(39):5419-26.

[67] Schofield L, Hewitt MC, Evans K, Siomos MA, Seeberger PH. Synthetic GPI as a candidate anti-toxic vaccine in a model of malaria. Nature 2002;418(6899):785-9.

[68] Jeon I, Lee D, Krauss IJ, Danishefsky SJ. A new model for the presentation of tumor-associated antigens and the quest for an anticancer vaccine: a solution to the synthesis challenge via ring-closing metathesis. J.Am.Chem.Soc. 2009;131(40):14337-44.

[69] de Velasco EA, Verheul AF, Veeneman GH, Gomes LJ, van Boom JH, Verhoef J, Snippe H. Protein-conjugated synthetic di- and trisaccharides of pneumococcal type 17F exhibit a different immunogenicity and antigenicity than tetrasaccharide. Vaccine 1993;11(14):1429-36.

[70] Jansen WT, Verheul AFM, Veeneman GH, van Boom JH, Snippe H. Revised interpretation of the immunological results obtained with pneumococcal polysaccharide 17F derived synthetic di-, tri- and tetrasaccharide conjugates in mice and rabbits. Vaccine 2002;20:19-21.

[71] de Velasco EA, Verheul AF, van Steijn AM, Dekker HA, Feldman RG, Fernandez IM, Kamerling JP, Vliegenthart JF, Verhoef J, Snippe H. Epitope specificity of rabbit immunoglobulin G (IgG) elicited by pneumococcal type 23F synthetic oligosaccharide- and native polysaccharide-protein conjugate vaccines: comparison with human anti-polysaccharide 23F IgG. Infect.Immun. 1994;62(3):799-808.

[72] Jansen WT, Hogenboom S, Thijssen MJL, Kamerling JP, Vliegenthart JFG, Verhoef J, Snippe H, Verheul AFM. Synthetic 6B di-, tri-, and tetrasaccharide-protein conjugates contain pneumococcal type 6A and 6B common and 6B-specific epitopes that elicit protective antibodies in mice. Infect.Immun. 2001;69(2):787-93.

[73] Safari D, Dekker HA, Joosten JA, Michalik D, de Souza AC, Adamo R, Lahmann M, Sundgren A, Oscarson S, Kamerling JP, Snippe H. Identification of the smallest structure capable of evoking opsonophagocytic antibodies against Streptococcus pneumoniae type 14. Infect.Immun. 2008;76(10):4615-23.

[74] Mawas F, Niggemann J, Jones C, Corbel MJ, Kamerling JP, Vliegenthart JFG. Immunogenicity in a mouse model of a conjugate vaccine made with a synthetic single repeating unit of type 14 pneumococcal polysaccharide coupled to CRM197. Infect.Immun. 2002;70(9):5107-14.

[75] Lindberg B, Lonngren J, Powel DA. Structural studies on the specific type 14 pneumococcal polysaccharide. Carbohydr.Res. 1977;58:177-86.

[76] Joosten JA, Lazet BJ, Kamerling JP, Vliegenthart JF. Chemo-enzymatic synthesis of tetra-, penta-, and hexasaccharide fragments of the capsular polysaccharide of Streptococcus pneumoniae type 14. Carbohydr.Res 2003;338(23):2629-51.

[77] Joosten JA, Kamerling JP, Vliegenthart JF. Chemo-enzymatic synthesis of a tetra- and octasaccharide fragment of the capsular polysaccharide of Streptococcus pneumoniae type 14. Carbohydr.Res 2003;338(23):2611-27.

[78] Michalik D, Vliegenthart JFG, Kamerling JP. Chemoenzymic synthesis of oligosaccharide fragments of the capsular polysaccharide of Streptococcus pneumoniae type 14. J.Chem.Soc., Perkin Trans.1 2002;1973-81.

[79] Sundgren A, Lahmann M, Oscarson S. Block synthesis of Streptococcus pneumoniae type 14 capsular polysaccharide structures. J.Carbohydr.Chem. 2005;24(4):379-91.

[80] Safari D, Dekker HA, Rijkers G, van der Ende A, Kamerling JP, Snippe H. The immune response to group B streptococcus type III capsular polysaccharide is directed to the -Glc-GlcNAc-Gal- backbone epitope. Glycoconj.J. 2011;28:557-67.

[81] Guttormsen HK, Baker CJ, Nahm MH, Paoletti LC, Zughaier SM, Edwards MS, Kasper DL. Type III group B streptococcal polysaccharide induces antibodies that cross-react with Streptococcus pneumoniae type 14. Infect.Immun. 2002;70(4):1724-38.

[82] Kadirvelraj R, Gonzalez-Quteirino J, Foley BL, Beckham ML, Jennings HJ, Foote S, Ford MG, Woods RJ. Understanding the bacterial polysaccharide antigenicity of Streptococcus agalactiae versus Streptococcus pneumoniae. Proc.Natl.Acad.Sci.USA 2007;103(21):8149-54.

[83] Jennings HJ, Lugowski C, Kasper DL. Conformational aspects critical to the immunospecificity of the type III group B streptococcal polysaccharide. Biochemistry 1981;20(16):4511-8.

[84] Miernyk KM, Butler JC, Bulkow LR, Singleton RJ, Hennessy TW, Dentinger CM, Peters HV, Knutsen B, Hickel J, Parkinson AJ. Immunogenicity and reactogenicity of pneumococcal polysaccharide and conjugate vaccines in alaska native adults 55-70 years of age. Clin.Infect.Dis. 2009;49(2):241-8.

[85] Safari D, Dekker HA, de Jong B, Rijkers G, Kamerling JP, Snippe H. Antibody- and cell-mediated immune responses to a synthetic oligosaccharide conjugate vaccine after booster immunization. Vaccine 2011;29(38):6498-504.

[86] Lefeber DJ, Benaissa-Trouw B, Vliegenthart JFG, Kamerling JP, Jansen WTM, Kraaijeveld K, Snippe H. Th1-directing adjuvants increase the immunogenicity of

Permissions

The contributors of this book come from diverse backgrounds, making this book a truly international effort. This book will bring forth new frontiers with its revolutionizing research information and detailed analysis of the nascent developments around the world.

We would like to thank Professor Desiree Nedra Karunaratne, for lending her expertise to make the book truly unique. She has played a crucial role in the development of this book. Without her invaluable contribution this book wouldn't have been possible. She has made vital efforts to compile up to date information on the varied aspects of this subject to make this book a valuable addition to the collection of many professionals and students.

This book was conceptualized with the vision of imparting up-to-date information and advanced data in this field. To ensure the same, a matchless editorial board was set up. Every individual on the board went through rigorous rounds of assessment to prove their worth. After which they invested a large part of their time researching and compiling the most relevant data for our readers. Conferences and sessions were held from time to time between the editorial board and the contributing authors to present the data in the most comprehensible form. The editorial team has worked tirelessly to provide valuable and valid information to help people across the globe.

Every chapter published in this book has been scrutinized by our experts. Their significance has been extensively debated. The topics covered herein carry significant findings which will fuel the growth of the discipline. They may even be implemented as practical applications or may be referred to as a beginning point for another development. Chapters in this book were first published by InTech; hereby published with permission under the Creative Commons Attribution License or equivalent.

The editorial board has been involved in producing this book since its inception. They have spent rigorous hours researching and exploring the diverse topics which have resulted in the successful publishing of this book. They have passed on their knowledge of decades through this book. To expedite this challenging task, the publisher supported the team at every step. A small team of assistant editors was also appointed to further simplify the editing procedure and attain best results for the readers.

Our editorial team has been hand-picked from every corner of the world. Their multi-ethnicity adds dynamic inputs to the discussions which result in innovative outcomes. These outcomes are then further discussed with the researchers and contributors who give their valuable feedback and opinion regarding the same. The feedback is then collaborated with the researches and they are edited in a comprehensive manner to aid the understanding of the subject.

Apart from the editorial board, the designing team has also invested a significant amount of their time in understanding the subject and creating the most relevant covers. They scrutinized every image to scout for the most suitable representation of the subject and create an appropriate cover for the book.

The publishing team has been involved in this book since its early stages. They were actively engaged in every process, be it collecting the data, connecting with the contributors or procuring relevant information. The team has been an ardent support to the editorial, designing and production team. Their endless efforts to recruit the best for this project, has resulted in the accomplishment of this book. They are a veteran in the field of academics and their pool of knowledge is as vast as their experience in printing. Their expertise and guidance has proved useful at every step. Their uncompromising quality standards have made this book an exceptional effort. Their encouragement from time to time has been an inspiration for everyone.

The publisher and the editorial board hope that this book will prove to be a valuable piece of knowledge for researchers, students, practitioners and scholars across the globe.

List of Contributors

Rosa M. Raybaudi-Massilia and Jonathan Mosqueda-Melgar
Institute of Food Science and Technology, Central University of Venezuela, Lomas de Bello Monte, Calle Suapure, Caracas, Venezuela

Alireza Alishahi
The University of Agriculture Sciences and Natural Resources of Gorgan, Gorgan, Iran

Amit K. Ghosh and Prasun Bandyopadhyay
Unilever R & D Bangalore, Whitefield, Bangalore, India

Marina Dello Staffolo
CIDCA (Centro de Investigación y Desarrollo en Criotecnología de Alimentos), CON-ICET-CCT La Plata, Fac. Cs. Exactas, Universidad Nacional de La Plata, La Plata, Argentina

Alicia E. Bevilacqua
CIDCA (Centro de Investigación y Desarrollo en Criotecnología de Alimentos), CON-ICET-CCT La Plata, Fac. Cs. Exactas, Universidad Nacional de La Plata, La Plata, Argentina
Departamento de Ingeniería Química, Facultad de Ingeniería, UNLP

María Susana Rodríguez and Liliana Albertengo
Instituto de Química del Sur (INQUISUR), CO NICET-UNS Bahía Blanca, Departamento de Química, Universidad Nacional del Sur, Argentina

María Victoria Busi and Diego F. Gomez-Casati
Centro de Estudios Fotosintéticos y Bioquímicos (CEFOBI-CONICET), Suipacha 570, Rosario, Argentina
IIB-INTECH, Universidad Nacional de General San Martín (UNSAM), San Martín, Buenos Aires, Argentina

Mariana Martín
Centro de Estudios Fotosintéticos y Bioquímicos (CEFOBI-CONICET), Universidad Nacional de Rosario, Suipacha 570, Rosario, Argentina

Mohit S. Verma and Frank X. Gu
Chemical Engineering, University of Waterloo, Waterloo, ON, Canada
Waterloo Institute for Nanotechnology, University of Waterloo, Waterloo, ON, Canada

A. V. Dushkin
Institute of Solid State Chemistry and Mechanochemistry, Siberian Branch, Russian Academy of Sciences, Russia

T. G. Tolstikova, M. V. Khvostov and G. A. Tolstikov
N. N. Vorozhtsov Novosibirsk Institute of Organic Chemistry, Siberian Branch, Russian Academy of Sciences, Russia

Carlucci María Josefina, Mateu Cecilia Gabriela, Artuso María Carolina and Scolaro Luis Alberto
Departamento de Química Biológica- IQUIBYCEN (CONICET-UBA), Laboratorio de Virología, Facultad de Ciencias Exactas y Naturales, Ciudad Universitaria, Universidad de Buenos Aires, Argentina

Edmund M. K. Lui, Chike G. Azike and Hua Pei
Ontario Ginseng Innovation and Research Consortium Department of Physiology & Pharmacology, Schulich School of Medicine and Dentistry, Canada

José A. Guerrero-Analco and John T. Arnason
Ontario Ginseng Innovation and Research Consortium Department of Biology, University of Ottawa, Ottawa Ontario, Canada

Ahmad A. Romeh
Department of Chemical and Biochemical Engineering, Faculty of Engineering, University of Western Ontario, London Ontario, Canada

Sherif J. Kaldas
Department of Biology, University of Ottawa, Ottawa Ontario, Canada

Paul A. Charpentier
Ontario Ginseng Innovation and Research Consortium, Department of Chemical and Biochemical Engineering, Faculty of Engineering, University of Western Ontario, London Ontario, Canada

Máira Regina Rodrigues
Universidade Federal Fluminense, Polo Universitário de Rio das Ostras, Rio das Ostras, RJ, Brazil
Alexandre de Souza e Silva and Fábio Vieira Lacerda Centro Universitário de Itajubá, Instituto de Ciências Biológicas, BairroVarginha, Itajubá, MG, Brazil

Dodi Safari
Eijkman Institute for Molecular Biology, Jakarta, Indonesia

Ger Rijkers
Department of Sciences, Roosevelt Academy, Middelburg and Department of Medical Microbiology and Immunology, St. Antonius Hospital, Nieuwegein, The Netherlands

Harm Snippe
Department of Medical Microbiology, University Medical Center Utrecht, Utrecht, The Netherlands

T0257457

Encyclopedia of Insecticides:

Volume IV

Encyclopedia of Insecticides: Pest Engineering
Volume IV

Edited by **Nancy Cahoy**

New York

Published by Callisto Reference,
106 Park Avenue, Suite 200,
New York, NY 10016, USA
www.callistoreference.com

Encyclopedia of Insecticides: Pest Engineering
Volume IV
Edited by Nancy Cahoy

International Standard Book Number: 978-1-63239-265-7 (Hardback)

Contents

Preface

The purpose of the book is to provide a glimpse into the dynamics and to present opinions and studies of some of the scientists engaged in the development of new ideas in the field from very different standpoints. This book will prove useful to students and researchers owing to its high content quality.

This book provides substantial information regarding insecticides and their applications in the field of pest engineering. The first section on insecticides mode of action focuses on toxicity of organic and inorganic insecticides, organophosphorus insecticides, toxicity of fenitrothion and permethrin, and dichlorodiphenyltrichloroethane (DDT). The second section is dedicated to vector control using insecticides, biological control of mosquito larvae by Bacillus thuringiensis, metabolism of pyrethroids by mosquito cytochrome P40 susceptibility status of Aedes aegypti, etc. The book will attract readers to make rational decisions regarding the usage of pesticides.

At the end, I would like to appreciate all the efforts made by the authors in completing their chapters professionally. I express my deepest gratitude to all of them for contributing to this book by sharing their valuable works. A special thanks to my family and friends for their constant support in this journey.

Editor

Part 1

Insecticides Mode of Action

Insecticide

A. C. Achudume
Institute of Ecology and Environmental Studies
Obafemi Awolowo UniversitY, Ile-Ife,
Nigeria

1. Introduction

Insecticides are organic or inorganic chemicals substances or mixture of substances intended for preventing, destroying, repelling or mitigating the effect of any insect including crawling and flying insects which may occur naturally or is synthesized (pyrethroids) e.g. organic perfumed and hydrocarbon oil and pyrethrins. There are various forms of insecticides. Most of the synthesized insecticides are by their nature are hazardous on health under the condition in which it is used. Insecticides therefore, range from the extremely hazardous to those unlikely to produce any acute hazard. Most are repellants and or insect growth regulators used in agriculture, public health, horticulture, food storage or other chemical substances used for similar purpose.

It is evident that insecticides have been used to boost food production to a considerable extent and to control vectors of disease. However, these advantages that are of great economic benefits sometimes come with disadvantages when subjected to critical environmental and human health considerations. Many insecticides are newly synthesized whose health and environmental implications are unknown.

Insecticides have been used in various forms from hydrocarbon oils (tar oils), arsenical compounds, organochlorine, organ phosphorous compounds carbonates, dinitrophenols, organic thiocynates, sulfur, sodium fluoride, pyethroids ,rotenone to nicotine, in solid or liquid preparation. Interestingly, most of these have been withdrawn due to the deleterious effect of the substances. Analysis of these formulations, their by- products and residues had in the past aided objective re–evaluation and re-assessment of these substances on a benefit-risk analysis basis and their subsequent withdrawal from use when found to be dangerous to human health and the environment. The quality and sophistication of these analyses have grown and very minute quantities of these insecticides or their residues can be analysed these days with a high degree of specify, precision and accuracy.

1.1 Inorganic and organ metal insecticides

The sequential organomentals and organometalloids insecticides are described in connection with the corresponding inorganic compounds. The highly toxic and recalcitrant compounds e.g. trichloro-bis-chlorophenyl ethane (DDT) and bis-chlorophenyl aqcetic acid (DDA) are formed unintentionally. The organic combination usually changes the absorption and distribution of a toxic metal and thus changes the emphasis of its effects, while the basic mode of action remains the same. The toxic effects of insecticides depend on the elements that characterize it as inorganic or organmetal insecticides and on the specific properties of one form of the element or one of its components or merely on an inordinately high dosage. Some highly toxic elements such as iron, selenium, arsenic and fluorine are essential to normal development. The organometals and organometalloids are here described in connection with the corresponding inorganic compounds. Organic combination usually changes the absorption and distribution of a toxic metal and as a result changes the emphasis of its effects, but the basic mode of action remains the same.

The distinction between synthetic compounds and those of natural origin somewhat artificial. In practice, related compounds are assigned to one category or the other, depending on whether the particular compound of the group that was first known and used was of synthetic or of natural origin. For example, pyrethrum and later the naturally occurring pyrethrins were well known for years before the first synthetic parathyroid was made; as such, pyrethroid are thought of as variants of natural compounds, even though they have not been found in nature and are unlikely to occur.

1.2 Pyrethrum and related compounds

The insecticidal properties of pyrethrum flowers (chrysanthemum cinerarae- folum) have been recognized as insect powder since the middle of last century (McLaughlin 1973). In addition to their insect-killing activity, an attractive feature of the natural pyrethrins (pyrethrum) as insecticides was their lack of persistence in the environment and their rapid action whereby flying insects very quickly become incapacited and unable to fly. Prior the development of DDT, pyrethrum was a major insecticides for both domestic and agricultural use, despite its poor light stability. Development of synthetic pyrethroid with increased light stability and insecticidal activity allows it to be used as foliar insecticide while the natural pyrethrins are now used mainly as domestic insecticides.(Elliot, 1979).

1.3 Mode of Action

Pyrethrum and the synthetic pyrethroids are sodium channel toxins which, because of their remarkable potency and selection have found application in general pharmacology as well as toxicology (Lazdunski et al., 1985). Pyrethroids have a very high affinity for membrane sodium channels with dissociation constants of the order of 4×10^{-8} M (Sodeland,1985), and produce subtle changes in their function. By contrast, inexcitable cells are little affected by pyrethroids. The pyrethroids are thus referred to as open channel blockers.

1.4 Metabolism

The relative resistance of mammals to the pyrethriods is almost wholly attributable to their ability to hydrolyze the pyrethroids rapidly to their inactive acid and alcohol components, since direct injection into the mammalian CNS leads to susceptibility similar to that seen in insects (Lawrence and Casida, 1982). Metabolic disposal of the pyrethroids is very rapid (Gray et al., 1980), which means that toxicity is high by intravenous route, moderate by slower oral absorption, and often immeasurably moderate by slower oral absorption.

1.5 Poisoning syndromes

The pyrethroids are essentially functional toxins, producing their harmful effects largely secondarily, as a consequence of neuronal hyperexitability (Parker et al.1985). Despite this dependence on a relatively well-defined mode of action, the pyrethroids are capable of generating a bewildering variety of effects in mammals and insects, which although showing some analogies with those produced by other sodium channel toxins (Gray, 1985; Lazdunski et al., 1985) and with DDT (Narahashi, 1986), have many unique characteristics (Ray, 1982b). Thus, toxicity of pyrethroids is divided into two groups Table 1. Type 1 pyrethroids produce the simplest poisoning syndrome and produce sodium tail currents with relatively short time constants (Wright et al., 1988). Poisoning closely resembles that produced by DDT involving a progressive development of fine whole-body tremor, exaggerated startle response, uncoordinated twitching of the dorsal muscles, hyperexcitability, and death (Ray, 1982b). The tremor is associated with a large increase in metabolic rate and leads to hyperthermia which, with metabolic exhaustion, is the usual cause of death. Respiration and blood pressure are well sustained, but plasma noradrenalin, lactate, and adrenaline are greatly increased (Cremer and Servile 1982). Type 1 effects are generated largely by action on the central nervous system, as shown by the good correlation between brain levels of cismethrin and tremor (White et al., 1976). In addition to these central effects, there is evidence for repetitive firing in sensory nerves (Staatz-Benson and Hosko, 1986).

Type I	Intermediate	Type II
Allethrin	Cyhenothrin	Cyfluthrin
Barthrin	Fenproponate	Cyhalothrin
Bioalethrin	Flucythrinate	Cypermethrin
Cismethrin		Deltamethrin
Fenfluthrin		Fenvalerate
Trans-fluorocyphenothrin		Cis-fluorocyphenothrin
Kadethrin		
Permethrin		
Phenothrin		
Pyrethrin I		
Pyrethrin II		
Resmethrin		
Tetramethrin		

Table 1. I Acute toxicity of pyrethroids (Wright et al., 1988; Forshow and Ray, 1990).

The type 11 pyrethroid produces a more complex poisoning syndrome and act on a wider range of tissues. They give sodium tail currents with relative longterm constants (Wright, et al., 1988). At lower doses more suble repetitive behavior is seen (Brodie and Aldridge, 1982). As with type I pyrethroids, the primary action is on the central nervous system, since symptoms correlate well with brain concentrations (Rickard and Brodie, 1985). As might be expected, both classes of parathyroid produce large increases in brain glucose utilization (Cremer et al. 1983). A final factor distinguishing type 11 pyrethroids is their ability to depress resting chloride conductance, thereby amplifying any sodium or calcium effects (Forshaw and Ray, 1990).

Intermediate signs representing a combination of type I and type 11 are produced by some pyrethroids. These appear to represent a true combination of the type I and 11 classes (Wright et al., 1983) and thus represent a transitional group. Evidence in support of this is given by measurement of the time constants of the sodium after potential produced by the pyrethroids. Since pyrethroids appear to be essentially functional toxins, they produce few if any specific neuropathological effects.

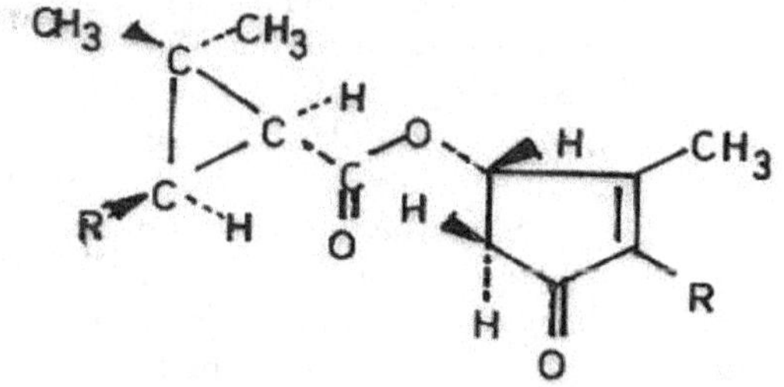

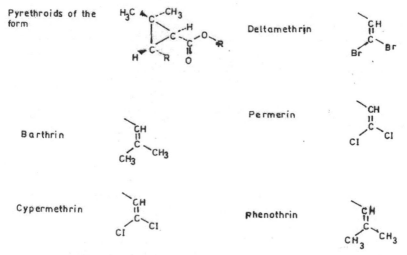

Fig. 1. Pyrethrins of the form.

Fig. 2. Pyrethroids of the form.

1.6 Identity, properties and uses

The six known insecticidal active compounds in pyrethrum are esters of two acids and three alcohols. Insect powder made from "Dalmatian insect flower" (*Chrysanthemum cinerariaefolium*) is called pyrethrum powder or simply pyrethrum. The powder itself was formerly used as an insecticide, but now it is usually extracted. The six active ingredients are:

- Pyrethrin I – pyrethrolone ester of chrysanthemic acid
- Pyrethrin II- pyrethrolone ester of pyrethric acid
- Cinerin I- cinerolone ester of chrysanthemic acid
- Cinerin II- cinerolone ester of pyrethric acid
- Jasmolin I- jasmolone ester of chrysanthemic acid
- Jasmolin II- jasmolone ester of pyrethric acid

The six known insecticidally active compounds in pryrethrum are esters of two acids and three alcohols. Specifically, pyrethrins 1 is the pyrethrolone ester of chrysanthermic acid, - pyrethrin II is the pyrethrolone ester of pyrethria acid, cinerin 1 is the cinerolone ester of chrysanthemic acid, cinerin II is the cinerolone ester of pyrethric acid, jasmolin I is the jasmolone ester of chrysanthemic acid and jasmolin II is the jasmolone ester of pyrethric acid. There is much evidence indicating that the biological activity of these molecules depends on their configuration (Elliot, 1969, 1971).

The six active ingredients are known collectively as pyrethrins; those based on chrysanthemic acid are called pyrethrin I, and those based on pyrethric acid are called pyrethrins II. Pyrethrins, generally combined with a synergist, are used in sprays and aerosols against a wide range of flying and crawling insects. Usually about 0.5% active pyrethrum principles are formulated. They are equally effective for control of head lice and flea in dogs and cats.

1.7 Raid as insecticide

The insecticide 'Raid' belongs to a group of chemically stable pyrethrin, has widespread use in control of insects. Chemical stability, insecticide and organic phosphorus hydrocarbon have been shown to accumulate rapidly in tissues causing death and have profound effect on growth (Nebeker et al., 1994). Insecticide raid shows no observable effects on mortality and growth at lower test concentrations in rats. At higher concentration of 430 and 961 $\mu g/g$, survival decreased as concentration increases. In addition, mean total body weight of animals fed insecticides raid with concentrations of 430 and 961 $\mu g/g$ were significantly decreased ($P<0.05$) than the controls. Conclusively, the higher the concentration of the insecticide Raid, the more hazardous it has on cell death (Achudume et al., 2008)(Table II).

Bioaccumulation factor of insecticide Raid was observed in lipids, up to three times that of the feed at the first concentration and gradually decreases as the concentrations increase (Table III), whereas accumulation factor in the muscle (0.7), brain (0.5), and liver (0.3) was about the indicated number times that of the feed. At higher concentration of 961 $\mu g/g$, bioaccumulation factor decreased in the lipid to 1.2 and 0.6 in the muscle, 0.03 in the brain, and 0.08 in the liver. Using the mean of insecticide in feed, the tissues accumulate the insecticide in the following ascending order: brain < liver < muscle < lipid. Similarly, Table III indicates the estimated detectable levels of toxicity in rat tissues exposed to the insecticide Raid. The brain shows mild decrease in toxicity of the enzymes glucose-6-phosphatase and lactic acid dehydrogenase, whereas significant decreases were noticeable in the muscle and liver (Achudume et al. 2008).

Long-term exposure of insecticide had been reported to result in systemic toxicity such that may impair the function of the nervous system and increase the risk of acute leukemia in children (Menegaux et al., 2006). Also, pesticides including organ phosphorus insecticides used against crawling and flying insects in homes have the potential of being carcinogens (Peter and Cherion, 2000). The adverse effect of insecticide Raid was demonstrated in a study by increase in alkaline phosphates activity in both plasma and liver which is a known measure of hepatic toxicity, and confirms "Raid" as a hepatotoxicant. The significant increase in alkaline phosphates activity (Table IV) may be due to hepatocellular necrosis which causes increase in permeability of cell membrane resulting in the release of this enzyme into the blood stream. The insecticide Raid significantly decreased reduced glutathione levels especially in the liver and this has implications for the ability of the animal to withstand oxidative stress. Studies have shown that GSH deficiency in cells is associated with markedly decreased survival (Kohlmeier et al., 1997), thus, chemically stable lipid-soluble, organophosphorus insecticides are hazarddous to health through mechanisms including depletion of GSH (Menegaux et al., 2006).

Means SD concentrations of insecticide "Raid" in feed (Mg/g)	Mortality	Means: SD body weight (g)
0.00	Nil	135=5.4
25.0=2.4	Nil	135=21.7
54.0=9.5	Nil	132=2.9
108.2=12.5	Nil	129=3.2
216.2=14.6	Nil	128=19.8
430.0=20.2	1	118=20.5
961.2=70.5	2	116=5.3

Table 1. II mortality and growth of wistar rats exposed to different concentrations of "Raid".

Raid Concentration in Wistar Rats (µg/g)[a] and Bioaccumulation Factor (BAF)				
Mean±SD Insecticide "Raid" in Feed (µg/g) Lipid	Muscle	Brain	Liver	
00.0	-	-	-	-
25.0±2.4	72.5± (2.9)	17.5(0.7)	12.5(0.5)	7.5(0.3)
54.0±9.2	86.4(1.6)	21.7(0.4)	16.4(0.3)	9.4(0.2)
108.2±12.5	172.8(1.6)	30.4(0.3)	19.5(0.2)	10.8(0.10)
216.2±14.6	280.8(1.3)	45.8(0.2)	22.9(0.1)	19.8(0.09)
430.0±20.6	324.0(0.8)	86.4(0.2)	25.8(0.06)	37.3(0.09)
961.2±70.5	1153.2(1.2)	576.6(0.6)	28.8(0.03)	76.9(0.08)

Table 1. III Tissue total raid concentrations and bioaccumulation factors (BAF) in wistar rats.

Raid concentrations Tissue In feed (µg/g)		Alk pase activity µgml^{-min-L}	GSH level mg/ml	Glucose level mg/g liver
430±20.2	Control	0.08±0.04	0.18±0.02	0.96±0.04
	Plasma	0.06±0.09	0.15±0.6	0.90±0.04
	Control	0.08±0.04	0.18±0.02	0.94±0.01
	Liver	0.06±0.02*	0.15±0.01	1.05±0.12
961.2±70.5	Control	0.09±0.05	0.19±0.05	0.96±0.52
	Plasma	0.06±0.01	0.11±0.05	1.09±0.52
	Control	0.08±0.08	0.19±0.02	0.96±0.06
	Liver	0.05±0.08*	0.09±0.03*	1.66±0.04

Data values are mean±SD
*Statistically significant $p < 0.05$

Table 1. IV Effect of Raid concentrations in feed on hepatic enzyme activity, reduced glutathione and glucose levels.

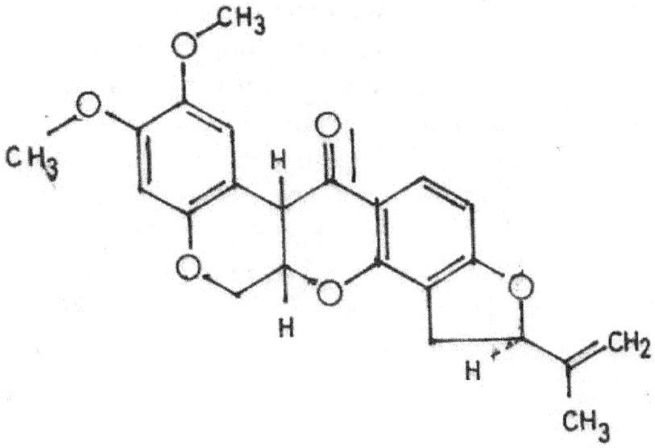

Fig. 3. Structure of rotenone.

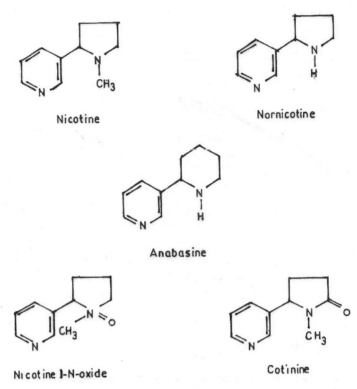

Fig. 4. Nicotine, narnicotine and anabasine with two important metabolites of nicotine.

Some other studies confirm that glutathione deficiency is associated with impaired survival in HIV disease (Herzenbery et al., 1997). Glutathione may be consumed by conjugation reaction, which mainly involve metabolism of zenobiotic agent. However, the principle mechanisms of hepatocyte glutathione turnover are known to be by cellular efflux (Sies et al., 1978). Glutathione reducatase is a known defense against oxidative stress, which in turn needs glutathione as co-factors. Catalase is an antioxidant enzyme which destroys H_2O_2 that can form a highly reactive radical in the presence of iron as catalyst (Gutter ridge, 1995).

Achudume et al., 2008 showed that bioaccumulation factor of insecticides raid was observed in lipid. Lipid peroxidation is a chemical mechanism capable of disrupting the structure and function of the biological membranes that occurs as a result of free radical attack on lipids. Some study confirms that insecticide raid increased lipid peroxidation, oxidative stress and hepatotoxicity due to reduced antioxidant system.

In addition, SOD is family of metalloid enzyme which is considered to be stress protein which decreases in response to oxidative stress (McCord, 1990). It is evident that decrease of SOD in the tissue is a confirmation of its protection from damage caused by insecticide Raid.

2. Classes of insecticides

- The classification of insecticides is done in several different ways (Hayes, 1982),(Heam 1973, Lehman, 1954, Martin and 1977).

- Systemic insecticides are incorporated by treated plants. Insect ingest the insecticide while feeding on the plants.
- Contact insecticides are toxic to insects by direct contact. Efficacy is often related to the quality of pesticide application in aerosols which often improve performance.
- Natural insecticides, such as pincotine, pyrethrum and neem extracts are from plants as defences against insects.
- Inorganic insecticides are manufactured with metals e.g. Heavy metals
- Organic insecticides are synthetic chemicals which comprise the largest numbers of pesticides available.

Insecticides are pesticides used to control insects many of these insecticides are very toxic to insects and many others are relatively harmless to other organism except fish.

Insecticides decompose readily so the residues do not accumulate on crops or in the soil. Insecticides include ovicides and larvicides used against the eggs and larvae of insects respectively.

The use of insecticides is believed to be one of the major factors behind the increase in agricultural productivity (McLaughlin,1973, van Emden and Pealall, 1996). Nearly all insecticides have the potential to significantly alter ecosystems; many are toxic to humans; and others are concentrated in the food chain (WHO 1962, 1972). Selected inorganic metals are discussed in the next section followed by individual insecticides organ metals.

2.1 Barium

Barium is an alkaline earth metal in the same group as magnesium, calcium, strontium and radium. It valence is two. All are water-and acid soluble compounds. They are poisonous. Barium carbonate is a rat poison. It is used in ceramics, paints, enamels, rubber and certain plastics.

Absorption, Distribution, Metabolism Excretion (ADME): Barium carbonate is highly insoluble in water. It is partially solubilized by acid in the stomach. The danger of the insecticide is through ingestion. Various barium compounds can cause pneumoconiosis. It is absorbed from gastrointestinal tracts of rat rapidly and completely. It is stored in bone and in other tissues (Hayes 1982, Castagnou et al 1957, Dencker et al., 1976, 83). Excretion takes place rapidly in urine and feces in 24hr (Bauer et al. 1956).

Mode of action: Barium stimulates striated cardiac and smooth muscle, regardless of the innervation.

2.2 Chromium

Chromium is a metal somewhat like iron and separated in the periodic table by manganese. Only hexavalent chromium compounds (chromates) are important as pesticides. They are also the most toxic. Chromate is absorbed by the lung (Baetjer et al., 1959), gastrointestinal tract and skin. It is widely distributed in the liver, kidney, bone and spleen (Mackenzie et al 1958). Acute poisoning may produce death rapidly through shock or renal tubular damage and uremia (Steffee and Baetjer 1965).

2.3 Mercury

Mercury is toxic no matter what its chemical combination. It is widely distributed in the environment, and traces of it occur in food, water and tissues even in the absence of occupational exposure. Inhaled mercury vapour diffuses across the alveolar regions of the lung into the blood stream. Mercury vapour is a monatomic gas which is highly diffusible and lipid soluble (Berlin et al 1969a , Hush,1985). Once in the bloodstream mercury vapour enters the blood cells where it is oxidized to divalent inorganic mercury under the influence of catalase (Halbach and Clarkson 1978). Mercury is widely distributed with the highest concentrations in the kidney.

2.4 Thallium

Thallium stands between mercury and lead in the periodic table, and compounds of these metals show marked similarities. All of them may produce immediate local irritation followed by delayed effects in various organs, notable the nervous system. Thallium sulphate has been more widely used as pesticide than any other compound of thallium. It has produced many cases of poisoning and serves as good example of the toxicity of thallium generally (Lund, 1956b).

Thallium is easily absorbed by the skin as well as by the respiratory and the gastrointestinal tracts. Thallium accumulates in hair follicles and much less in those in the resting phase. Excretion is slow and is entirely by urine in humans but in rats via faeces (Barclay et al., 1953, Lund, 1956a)

2.5 Lead arsenate

Lead arsenate includes acid lead arsenate, dibasic lead arsenate, dilead orthoarsenate, diplumbic hydrogen arsenate, lead hydrogen arsenate and standard lead arsenate. Lead arsenate is used as an insecticide. it is used to control moths, leaf rollers and other chewing insects and in soil for the treatment of Japanese and Asian beetles in lawn. Absorption is generally via gastrointestinal. Dermal absorption is extremely small. Lead and arsenate are distributed separately in the body. lead is stored in highest concentration in the bone with much lower concentrations in soft tissues. Arsenic is stored in the liver and in some instances in the kidney at higher concentrations than those for lead (Fairhall and Miller, 1941. Lead is transferred to the fetus of animals humans (Heriuchi et al., 1959).

2.6 Antimony potassium tartrate

This compound serves as a poison in baits to control insects, especially thrips, and as an emetic in bait to control rodents. Ingestion of the compound usually leads to repeated vomiting. Excretion is mainly urinary (Fairhall and Hyslop, 1947).

2.7 Sodium selenate

Sodium selenate is an insecticide used in horticulture for control of mites, aphids and mealybugs. Various compounds of selenium are freely absorbed from the respiratory and gastrointestinal tracts. Dermal absorption is less important. Selenium is stored more in the liver, kidney, spleen, pancreas, heart and lung than in other organs (Underwood, 1977). Selenium is excreted chiefly in the urine but about 3-10% is metabolized and excreted by the lungs and through faecal excretion.

2.8 Sodium fluoride

Sodium fluoride is toxic to all forms of life. It has been used as an insecticide, rodenticide and herbicide and as fungicide for preservation of timber. Its toxicity to plants generally has restricted its use as an insecticide to bait formulations (who, 1970). Sodium fluoride concentrates more in the plasma and liver and is excreted in urine.

3. Miscellaneous elements

3.1 Boric acid
Boric acid and borax have been used as an insecticide, both mainly for the control of cockroaches. Boric acid also is known as boracic acid and as orthoboric acid. Absorption from the gastrointestinal tract is rapid and virtually complete. Its peak concentration is in brain and less in other tissues. Boric acid is excreted unchanged in the urine (wong et al., 1964).

3.2 Insecticides derived from living organisms and other sources:
Different groups of insecticide; derived from living organisms are entirely unrelated chemically and pharmacologically. They range from relatively simple alkaloids such as nicotine, with a molecular weight of only 162.2, through proteinaceous poisons to virulent living organism. They range in toxicity from harmless and fragile pheromones, which are used as a chemical warfare agent.

The distinction between synthetic compounds and those derived from living organisms is somewhat artificial. In practice, related compounds are assigned to one category or the other, depending on whether the particular compound of the group that was first known and used was of synthetic or of natural origin. For example, pyrethrum and later the naturally occurring pyrethriums were well known for years before the first synthetic pyrethroid was made; as a result, pyrethroids are thought of as various of natural compounds, even though they have not been found in nature and are unlikely to occur. By contrast, synthetic sodium fluoroacetate acquired a reputation as a rodenticids and was explored as a synthetic insecticide before it was realized that the potassium salt is the active principal of a poisonous plant. Thus pyrethroids are discussed extensively.

Perhaps the only unifying feature of the diverse array of poisons derived from living organism is the popular view that "natural" substances are harmless. On this matter of safety, an expert committee of the world health organisation pointed out that "all the most poisonous materials so far know are, in fact, of natural origin" (WHO,1967).

3.3 Pyrethrum and related compounds
The insecticidal properties of pyrethrum flowers (genus chrysanthemum) have been recognized since the middle of 1st century, when commercial sale of "insect powder" from Dalmatian pyrethrum flower heads began (McLaughlin, 1973). In addition to their insect-killing activity, their lack of persistence in the environment and rapid "knock down" activity whereby flying insects become uncoordinated and unable to fly makes it very useful. Pyrethrum used to be a major insecticide for both domestic and agricultural use despite its poor light stability. Its usefulness was extended by introduction of piperonyl butoxide and other compounds as synergists, which greatly reduced the unit cost of crop treatment. Development of synthetic pyrethroids with increased stability and insecticidal activity (Elliot 1977) reduced the use of pyrethrum. However, natural pyrethrins are now used mainly as domestic insecticides, while the synthetic pyrethroids represented 20-25% of the world foliar insecticide market in 1983 (Herve's 85) and the proportion is increasing

steadily. Thousands of new synthetic pyrethroids have been synthesized, some showing complete divergence from the original pyrethrins (casida et al., 1973). Table 2.1 and Table 2.2.

4. Mode of action

Pyrethrum and the synthetic pyrethroid are sodium channel toxins which, because of their remarkable potency and selectivity, have found application in general toxicology (Lazdunski et al 1985). Their actions on the nerve membrane sodium channel are well understood. Pyrethroids have a very high affinity for membrane sodium channel, they have little effect on inactive sodium channels or close channels and produce subtle changes in their functions. After modification by prethroids, sodium channels continue in many of their normal functions, retaining their selectivity for sodium ions and link with membrane potential (Narahashi, 1986). The pyrethroids are thus known as open channel blockers. Detailed studies can be found in Narahishi 1986, Jacques et al., 1980, and Gray 1985.

4.1 Metabolism
The relative resistance of mammals to the pyrethroids is almost wholly attributable to their ability to hydrolyze the pyrethroids rapidly to their inactive acid and alcohol components, since direct injection into the mammalian CNS leads to susceptibility similar to that observed in insects (Lawrence and Casida, 1982). Some additional resistance of homoeothermic organisms can be attributed to the negative temperature coefficient of action of the pyrethroids (Van den Bercken et al., 1973) which are thus less toxic at mammalian body temperature but the major effect is metabolic.

The metabolic pathways for the breakdown of the pyrethroids vary little between mammalian species but vary somewhat with structure. This literature has been ably summarized by Leahy (1985), and further references to the metabolism of specific pyrethroids are given in the sections on individual compounds. Generally pyrethrum and allethrin are broken down mainly by oxidation, whereas for the other pyrethroids ester hydrolysis predominates. These reactions can take place in both liver and plasma and are followed by hydroxylation and conjugation to glucuronides or sulphates, which are then excreted in the urine (Gray 1985).

4.2 Individual insecticides
Other known insecticides pyrethroids under organophosphates are listed below only selective ones are discussed.

Allethrin	Permethrin
Bifenthrin	
Cyhalothrin, Lambda-cyhalothrin	
Cypermethrin	Phenothrin
Cyfluthrin	Prallethrin
Deltamethrin	Resmethrin
Ftofenprox	Tetramethrin
Fenvalerate	Transfluthrin

Table 4. I Other known insecticides.

4.3 Cypermethrin
Cypermethrin (R, S)-∞- cyano-3-pheno-xybenzyl-2, 2-dimethyl. There are eight isomeric forms. It was introduced commercially in 1977 as an emulsifiable concentrate to be used against a wide range of insect pest (Elliot, 1977).

4.4 Deltamethrin
Deltamethrin S-∞- cyano-3-phenoxybenzyl-(IR)-cis-3-(2, 2-dibromovinyl)2,2-dimethcyclo-propane carboxylate. It is a single isomer. It is used against a wide range of insect pests. It produces a typical type II motor symptom in mammals (Barnes and verschoyle, 1974. Metabolism of deltamethrin involves rapid ester cleavages and hydroxylation (Shono eta al; 1977).

4.5 Fenproponate
Fenproponate(∞-cyano-3-phenoxybenzyl-2,2,3,3-tetra-methylcyclopropanecarboxylate). There are eight isomerism forms. Fenpropathrin is another common name, was first developed by sumitomo and commercialized in 1980 as an emulsifiable concentrate to be used against a wide range of insect pests. Fenproponate produces intermediate or mixed motor symptoms in mammals (Wright et al., 1988).

4.6 Fenvalerate
Fenvalerate (R,S)- ∞- cyano-3-phenoxy-benzyl (IR,IS)-2-(4-chlorophenyl)-3-methyl-1-butyrate. There are four isomeric forms. It should be noted that fenvalerate is not based on a cyclopropane ring structure. It was introduced commercially to be used against a wide range of insect pests fenvalrate produces typical type II motor symptoms in mammals (Verschoyle and Aldrige, 1980).

4.7 Phenothrin
Phenothrin (3-phenoxybenzy-(IR,IS)-cis,trans-3(2-methylprop-1-enyl)-2,2 dimethyicyclo-propane carboxylate). There are four isomeric forms. It is used as a domestic insecticide in a partially resolved mixture rich in the IR isomer (Sumithrin) and for grain protection. Phenothrin produces typical type 1 moto symptoms in mammals (Lawrence and Casida, 1982)

4.8 Rotenone and related materials
Rotenone-bearing plants have longed being used as a fish poison by many ancient different indigenous people, nut their use as an insecticide is probably more than a century old. Plants known to produce rotenone and other rotenoids belong to at least 68 species of the family Leguminosae, the same as that for peas and beans. The genera most exploited so far are Derris, native to southeast Asia, and lonchocarpus to south America (shepard 1951).
Rotenone and other active principles often occur chiefly in the roots of rotenone bearing plants but may be in the leaves (as in *Tephrosia vogeli*), seeds (as in *Milletia pachycarpa*), or bark (as in *Mundulea serica*).
Regardless of the genus or the particular part of the plant involved, the active constituents of rotenone-bearing plants may be extracted with ether or acetone as resin.
Rotenone is (2R,6a 5,12a 5)-1,2,6,6a,12,12a-hexahydro-2-isopropenyl-8,9-dimethoxy-chromeno (3,4-b) furo(2,3-h) chromen-6-one. Its structure is depicted in fig. 3. Although

rotenone generally is considered to be the active ingredient in all resins isolated, the other constituents show considerable insecticidal activity (Metcalf, 1955).

Rotenone is readily oxidized and racemized in the presence of light and the process is accelerated in alkaline solution (Cheng et al., 1972). It is active as a nonsytemic pesticide against a wide variety of insects, arachnids and molluscs. Its rapid photodecomposition means that it is active only for about 1 week on plants or 2-6 days in water and this limits its commercial use though still finds use as a domestic garden insecticide.

Rotenone is a highly potent mitochondrial poison, blocking NADH oxidation, this property dominates its actions in animals (Heikkila et al., 1985).

Rotenone is metabolized rather effectively by the liver in isolated rat liver mitochondria, the aerobic oxidation of pyruvate is almost completely inhibited by rotenone (Haley 1978).

4.9 Nicotine and related compounds

Three closely related compounds (nicotine, nornicotine and anabasine fig4) were commonly used as insecticides, although only the most potent, nicotine, is now used to any extent. Nicotine is usually obtained from the dried leaves of nicotiana tabacum, but it also occurs in *N. rustica* and *Duboisia*, another genus of the solanaceae, and in three other taxonomically diverse general, namely Asclepia (Asclepidaceae), Equisetaceae (Equisetaceae), and Lycopodium (Lycopodiaceae); Nicotine (S-3-(1-methyl pyrrolidin-2-yl) pyridine) is used as nicotine sulphate as a stomach poison for leaf eating insects (Haigh and Haigh 1980). Nicotine is rapidly absorbed from all mucosal surfaces, including those of the mouth, gastro-intestinal tract, and lung. Since nicotine readily forms salts in acid solution, its penetration through biological membranes is strongly pH dependent (Schievelbein, 1982).

The metabolism of nicotine is highly complex and reviewed by Gorrod and Jenner (1975) and schievelbein (1982). Metabolism mainly by cytochrome P.450 linked microsomal oxidative pathways in the liver. Cotinine (Fig4) is major metabolite, which then undergoes further oxidation. Nicotine stimulates the action of acetylcholine at nicotinic receptors in the central nervous system, autonomics ganglia and some pheripheral nerves. It central actions result in tremor and convulsions, stimulation and then depression of ventilation and induction of vomiting by a direct action on the medulla. Ventilation is stimulated by peripheral actions on the aortic and carotid chemoreceptors, and adrenal catecholamine. Secretion is increased at low doses. Heart rate and blood pressure are largely dominated by sympathetic effects and show increases compounded by adrenal catecholamines. The gastrointestinal tract is dominated by parasympathetic effects and shows hypersecretion followed by block as well as increased tone and peristalsis. Death is usually a result of block of neuromuscular transmission in the respiratory muscles or a consequence of seizures. In addition to its action on cholinergic transmission, nicotine can act at noncholinergic sites and also activate receptors on sensory nerve endings and vagal C fibers (Martin, 1986).

The carcinogenic potential of tobacco is well established, but there is debate about the role of nicotine, which, although probably not carcinogenic itself can be converted to carcinogens such as N'-nitrosonornicotine and 4-(methyl-nitrosamino)-1-(3-pyriyl)-1-butanone. The metabolites cotinine and nicotine 1-N-oxide are not carcinogenic although they do produce hyperplasia of the bladder epithelium (Hoffmann et al., 1985).

5. Living organisms as pesticides

The use of biological control agents has many potential advantages over chemical control, not least the possibility of high selectivity for the predators and other beneficial species. Several microorganisms or microbial products have been identified as potential insecticides (Miller et al, 1983). Most successful attempts have been directed against insects, as biological control of vertebrates has met with little success due to cross-infection problems. The world Health Organisation has investigated viruses, bacteria fungi and nematodes as potential insect control agents since all play a part in limiting the growth of natural insect populations.

5.1 Viral insecticides
Viral insecticides are still in the experimental stage but many are under investigation, as reviewed by Miller et al., 1983. Bacterial insecticides represent the largest and widest used group and reviewed by Burges (1982) and Lysenko(1985). All of those used are spore-formers, since the spores can be readily stored in dried form and applied by conventional means as wettable powders or dusts. Many form a crystalline toxin within the spore which enhances their pathogenicity to insects. The most widely used is *Bacillus thuringiensis*. A closely similar bacterium, *Bacillus papilliae* has been used against Japanese beetle. It has the advantage that once spores are introduced into the environment the bacterial population is sustained by reinfection of the insect hosts, but the disadvantage that spore production requires expensive in-vivo production using insect pupae and is now of declining importance. It is highly specific, does not infect vertebrates, and despite production of a crystal toxin is nontoxic to mammals by repeated oral administration (Burges, 1982).

5.2 Fungal insecticides
Fungal insecticides are commercially produced for a variety of specific applications. Their importance in controlling natural insect populations has been recognized since 1834, Aschersonia has been used to control Floridian white fly on citrus since the early 1900s. Fungi have the advantages of forming a stable population in the insect environment and are capable of infection through the insect cuticle, not by ingestion as bacteria. A disadvantage is their susceptibility to widely used fungicides. Examples include *Beauveria basiana* is marketed as Boverin and used against Colorado beetle and corn borer in Russia and China. *Metarhizium anisopliae* was used against a range of insects as metaquino. *Hirsutella thompsoni*, is used to control citrus rust in the united states as myear and *Vecticillium lecani* is used as vertalec or mycotal for aphid control in united kingdom. Some fungi such as *Beauveria bassiana* produce toxins which may be involved in their pathogenicity. *Culicinomyces clavosporus* and *lagenidium giganteum* are mosquito pathogens (Miller et al. 1983).

5.3 Nematate insecticides
Nematate insecticides have been isolated from mosquito larvae at low natural population densities. They are reared in vivo, which is expensive, and there some resistant mosquito population Nametodes are tolerant of many insecticides and insect growth regulators and can be used in combined malaria control programs and are rapidly broken down by human gastric juice (Gajana et. al; 1978).

6. Conclusion

Given the enthusiasm of the proponents of biological insect control and the limited role that these agents play in current pest control may be perhaps surprising. There are however, a number of difficulties in sustaining a usefully large population of the control agent on crops, or in the case of mosquitoes at the water surface, and in agriculture difficulties associated with the very high host specificity of some agents. More fundamental problems are the potential risk from replicating agents which can increase in the environment and the possibility of transfer of toxin encoding genes from invertebrate to vertebrate bacteria or viruses. It is clear, however, that current experience with biological control agents is very encouraging and that they can be expected to play an important part in integrated pest control programs in the future (Laird, 1985).

While animals as well as humans may be adversely affected mainly by ingestion of the active ingredients, the effect of propellant chemical cannot be ignored. Inflammatory activation might be an important mechanism underlying toxicity effects in the tissue (Mense et al., 2006). The role of propellant in the toxicity of insecticide Raid may not be cleared. A comprehensive assessment of the risk associated with environmental use of insecticide Raid was determined in various tissues as it affects the basal biochemical molecules of cells (Achudme et al., 2008).

7. References

Achudume, A.C., Nwoha, P.C., Ibe, J.N. (2008) Toxicity and Bioaccumulation of insecticide "Raid" in Wistar Rats Inter Environ Toxicity 24(4); 357-361.

Barcley, R.K;peacock, W.C. and Karnofsky,G.A. (1953). Distribution and excretion of radioactive thallium in the chick embryo, rat and man.J. Phaemacol.Exp.Ther 107, 178-187.

Barnes, J.M. and Verschoyle, R.D. 1974 Toxicity of new Pyrethroid insecticide. Nature (London) 248-711.

Bauer, G.C.; Carlsson, A. and Lindquist, B. (1956). A comparative ^{45}Ca in rats. Biochem. J. 63, 535-542.

Berlin, M.H; Nordbery, G.F. and Serenius, FR. (1969a). on the site and mechanism of mercuric nitrate Hg. 203.Arch.Environ. Health 18, 42-50.

Brodie, M.E. and Aldridge, W.N 1982 Elevated cerebellar cyclic GMP levels during the deltamethrin induced motor syndrome. Neurobehav. Toxicol. Teratol 4, 109-113.

Burges, H.D. (1982). Control of insects by bacteria. Parasitology 84, (symp), 79-117.

Casida, J.E. (1973). Biochemistry of the pyrethrins. In "pyrethrum: The Natural insecticide" (j.E. casida, ed). Academic press, New York and London.

Castagnous, R; Paolett; C. And Larcebeau, S. 1957. Absorption and distribution of barium administered intravenously or orally to rats C.R. Hebd, seanes ser Acad Sci D 244, 2994-2996. (in French).

Cheng, P.Y., Buster, D., Hommock, B.D., Roe, R.M, and Alford, A.R. 1987. *Bacillus thuringiensis* van. Israelenisio-enctotosian Evidence of Neurotoxic action. Pestic Biochem Physiol. 27, 42-49.

Cremer, J.E and Seville, M.P. 1982-Comparative effects of two Pyrethroids, deltamethrin and cismethrin on plasma catecholamines and on blood glucose and lactate. Toxicol. Appl Pharmacol. 66, 124-133

Cremer, J.E., Cunningham, V.J. and Seville, M.P. 1983. Relationship between extraction and metabolism of glucose, blood flow, and tissue blood volume in regions of rat brain. J. Cereb. Blood Flow Metab. 3, 291-3002.

Dencker, L; Danielsson, B; khayal, A, and Lindren, A. (1983). Deposition of metals in the embryo and fetus. In "Reproductive and Developmental Toxicity of metals" (T.W. Clarkson, G.F Nordberg, and P.R. sage eds), pp 607-631. Plenum, New York.

Dencker, L; Nillson, A; Ronnback, C. And walinder, G. (1976). Nptake and retention of ^{133}Ba and ^{140}Ba-^{140}La in mouse tissue. Acta Radiol: Ther; Phys; Biol (N.S) 15, 273-28.

Elliot, M. 1977. "Synthetic Pyrethroids", ACS Symp. Ser. No. 42 Am. Chem. Soc., Washington, D.C.

Elliot, M., 1971. The relationship between the structure and the activity. Chem. Ind. (London) 24 776-791.

Fairhall, L.T. and Hyslop, F. (1947). The toxicology of antimony. Public Health Rep; suppl. 195.

Fairhall, L.T. and miller, J.W. (1941). A study of the relative toxicity of the molecular components of lead arsenate public Rep 56, 1610-1625.

Forshaw, P.J and Ray, D. E. 1990. A novel action of deltamethrin on membrane resistance in Mammalian skeletal muscle and non-myelinated nerve fibres. Neuropharmcol. 29, 75-81.

Gajana, A; kazimi, S.J; Bheemarao U.S., Suguna, S.G; and chandrahas, R.K 1978 studies on a nematode parasite (Romano-mermis sp; mermithidae) of mosquito larvae isolated in Pondicherry Indian J. Med Res. 68, 242-247.

Gorrod, J.W, and Jenner, P. 1975. The metabolism of tobacco alkaloids. Essays Toxical 6,35-78 schievelbein, H 1982.

Gray, A.J. 1985 Pyrethroid structure-toxicity relationships in mammals. Neurotoxicology 3, 25-35

Gray, A. J., Connors, T.A., Hoellinger, H. and Nguyen-Hoang-Nam. 1980. The relationship between the Pharmacolcinetics of intravenous cismethrin and bioresmethrin and their mammalian toxicity. Pestic. Biochem. Physiol. 13, 281-293.

Haigh, J.C. and Haigh, J.M. 1980, immobilizing drug emergencies in humans Vet Hum Toxical 22, 1-5.

Halbach, S. And Clarkson, T.W.(1978). Enzymic oxidation of mercury vapour by erythrocytes. Biochim Biophys Acta 523, 522-531.

Haley, T.J. 1978. A review of the literal of rotenone 1, 2,12,12a tetrahydro-8-9- dimethoxy-(2-(-1 methyl ethenyl)-1-benzo-pyrano (3,5-B)fluoro (2,3-H) (1)-benzo-pyran-6(6h), one J. Enviror pathol. Toxicol 1, 315-337.

Hayes, W.J., Jr (1982). "Pesticides Studies in Man". Williams & Wilkins, Baltimore, Maryland.

Hearn, C.E.D. (1973). A review of Agricultural Pesticide Incidents in Man in England and Wales, 1952-71, Br. J. Ind. Mect 30, 253 -258

Heikkila, E, Nicklas, W.J,vyas,I, and Duvoisin, R.C. 1985. Dopeminergic toxicity of rotenone and the 1 methyl-4-phenylpridium ion after stereotoxic administration to rats:implication for the mechanism of 1-methyl-4-phenyl-1,3,3,6-tetrahydropyridine toxicity. Neurosci. LeH. 62, 389-394.

Herve, J.J. (1985). Agricultural, public health health and animal health usage. In "The pyrethroid insecticides" (J.P. Leahey, ed) Taylor & Francis, London and Philadelphia.

Hoffmann, D; Laboie, E.J. and Hecht, S. 5,1985 Nicotine A precursor for carcinogens. Cancer letter, 26,67-75.

Horiuchi, K, Horiguchi S and suekane, M (1959). Studies on the industrial lead poisoning .I. Absorption, transportation, deposition and excretion of lead 6. The lead contents in organ tissues of the normal Japanese Osaka city med J. 5, 41-70.

Hursh, J.B.(1985). Partition coefficients of mercury (^{203}Hg) vapour between air and biological fluids.J.Appl Toxicol 5,327-332.

Jacques, Y., romey, G., Cavey, M.T., Kartalovski, B. And Lazdunski, M. 1980. Interaction of Pyrethroids with the Na$^+$ channel in mammalian neuronal cells in culture. Biochim. Biophys. Acta 600, 882-897.

Lawrence, L.J. and Caside, J.E. 1982 Pyrethroid toxicology: Mouse intracerebral structure-toxicity relationships. Pestic Biochem Physiol 18, 9-14.

Lazdunski, M., Barhanin, J., Borsotto, M., Frelin, C., Hugues, M.n Lombet, A., Pauron, D., Renaud, J., Schmid, A. and Vigne, P. 1985 Markers of membrane ionic channels In "Vascular Neuroeffector Mechanisms (J.A Bevan et al. Eds) Elsevier, Amsterdam.

Lehman, A.J, 1954. A toxiocological Evaluation of household insecticides Q.Bull-Assoc. Food Drug Off. 18, 3-13

Lund, A.(1956a). Distribution of thallium in the organism and its dimination. Acta pharmacol. Toxicol 12,251-259.

Lund, A. (1956b). The effect of various substances on the excretion and the toxicity of thallium in the rat. Acta Pharmacol. Toxicol 12, 251-259

Martin, B.R. (1986). Nicotine receptors in the central nervous system. In "The receptors (P.M. comm., ed), vol.3 pp. 393-415. Pergamon Oxford.

Martin, H. And Worthing, C.R; eds (1977). The Pesticide Manual," 5th ed. Br. Crop Rot. Counc; Molvern, Worcestershire, England.

McLaughlin, G.A. 1973. History of Pyrethrum. In "Pyrethrum: The Natural Insecticide" (J.E. Casida, ed), pp 3-15. Academic Press, New York and London.

Metcalf, R.L. (1955). "organic Insecticides" Wiley (intersciences), New York.

Miller, L.K., Lingg, A.J. and Bulla, L.A. 1983. Bacterial, viral and fungal insecticides science 219, 715-721.

Molaughlin, G.A. (1973). Histgory of Pyrethrum. In "Pyrethrum: The National Insecticide" (J.E Casida, ed), pp 3-15. Academic Press, New York and London.

Narahashi, R, 1986. Mechanisms of action of Pyrethroids on sodium and calcium channel gating. In "Neuropharmacology of Pesticide Action" (M.G. Ford, G.G. Lunt, R.C

Reay, and P.N.B. Usherwood, eds), pp 36-40. Ellis Horword, Chichester, U.K.

Nebeker, A.V., Dunn, K.D, Griffis, W.H., Schuytema, G.S. 1994. Effects of dieldrin in food and growth and bioaccumulation in Mallard ducklings Arch Environ Contam. Toxicol 26; 29-32.

Parker, C.M., Albert, J.R., Van Gelder, G.A; Patterson, D.R. and Taylor, J.L. 1985. Neuropharmacologic and neuropathologic effect of Fenvalerate in mice and rats. Fundam.

Rickard, J. and Brodie, M.E. 1985 Correlation of blood and brain levels of the neurotoxic Pyrethroid deltamethrin with the onset of symptoms in rats. Pestic, Biochem. Physiol 23, 143-156.

Schieveibein, H. 1982, Nicotine, resorption and fate pharmacol. Ther 18, 233-248.

Shepard, H.H. 1951. "The Chemistry and action of Insecticides". 1st ed. McGraw-Hil, New York.

Shono, T., Ohsawa, K. And Casida, J.E. 1979. Metabolism of trans- and cis-cypermetrin and decamethrin by microsomal enzymes J. Agric Food Chem. 27, 316-325.

Staatz-Benson, C.G., and Hosko, M.J. 1986. Intataction of Pyrethroids with mammalian spinal neurons. Pestic. Biochem physiol 75, 19-30

Steffeee, C.H and Baetjer, A.M. (1965). Histopathologic effects of chromate chemical Arch. Environ.Health 11,66-75.

Underwood, E.J. (1977)."Trace elements in Human and Animal Nutrition". 4th ed. Academic press, New York.

Verschoyle, R.D. and Aldridge, W.N. 1980 Structure-activity relationship of some Pyrethroids in rats. Pestic. Biochem. Physiol. 2, 308-311.

Van den Bercken, J., Akkermann, L.M.A and van der Zalm, J.J, 1973. DDT. Like action of Allethrin in the sensory nervous system of Xenopus laevis. Eur. J. Pharmacol. 21, 95-106.

Van Emden HF, Pealall DB (1996) Beyond Silent Spring, Chapman & Hall, London, pp 322.

White, I.N.H., Verschoyle, R.D., Moradian, M.H., and Barnes, J.M. 1976. The relationship between brain levels of cismethrin and bioresmethrin in female rats and neurotoxic effects. Pestic. Biochem. Physiol. 6, 491-500.

WHO 1962 Accidental Food Poisoning with Agrosan. Communication to World Health, Organisation from S.A Raz Ali. WHO Inf. Circ. Toxic. Pestic. Man No 9, P. 23

WHO 1972, "IARC Monographs on the Evaluation of Carcinogenic Risk of Chemicals to Man," Vol. 1, Int. Agency Res. Cancer, Lyon, France.

Wong, L.C; Heimbach, M.D; Truscolt, D.R. and Duncan, B.D. (1964). Boric acid poisoning. Report of 11 cases. Can.med. Assoc J. 90, 1023.

World Health Organisation (WHO) (1970). "Fluuorides and Human Health". World Health organ. Geneva

World Health Organization (WHO) (19670) "safe use of pesticides in public Health", Who Tech. Rep ser. No 356 world Health Organ, Geneva.

Wright, C.D.P., Forshaw, P.J. and Ray, D.E. 1988 Classification of the actions of two Pyrethroid insecticides in the rat, using the trigeminal reflex and skeletal muscle as test systems. Pestic. Biochem, Physiol. 30, 79-86.

2

The Toxicity of Fenitrothion and Permethrin

Dong Wang, Hisao Naito and Tamie Nakajima
Nagoya University Graduate School of Medicine
Japan

1. Introduction

Fenitrothion: An organophosphorus insecticide, fenitrothion (*O,O*-dimethyl *O*-4-nitro-*m*-tolyl phosphorothioate; CAS No. 122-14-5), which is a yellow-brown liquid with an unpleasant odor at room temperature, was introduced in 1959 by both Sumitomo Chemical Company and Bayer Leverkusen, and later by American Cyanamid Company (Hayes, 1982; Hayes and Laws, 1990; Worthing and Walker, 1987).

Organophosphorus insecticides began with the massive development of agriculture and agribusiness after World War II. At that time, parathion, one of the famous organophosphorus insecticides, was used in large quantities for preventing rice-stem borer worldwide. However, because of the high acute toxicity, parathion was thought to be an extremely hazardous substance. In man, an oral dose of 3-5 mg/kg is usually fatal. The following case report additionally came across the high toxicity and persistence in humans (Clifford and Nies, 1989). A 25-year-old worker in a pesticide-formulating plant was contaminated after accidentally spilling a 76% parathion solution on his groin and scrotal areas. Although he showered and changed clothes immediately, the resulting nausea and diarrhea made him consult a doctor two days later. The worker placed the parathion-saturated uniform in a plastic bag to be burned. But the contaminated clothing was laundered, and then was used in succession by a second worker, who wore it until he had complaints similar to the first worker. The coveralls were again laundered and used by still a third intoxicated worker. Totally, three workers suffered from toxic reaction to parathion. This case shows the toxic nature of parathion and its persistence on clothing even after successive laundering. Moreover, Etzel et al. (1987) reported that 49 persons in Sierra Leone were acutely poisoned by parathion in May and June 1986, 14 of whom later died. The case-control study of the employed 21 cases and 22 household controls was undertaken to explore which factors were associated with the development of the symptoms such as excess salivation, excess tearing, increased urination, diarrhea, convulsions, and loss of consciousness. Each case and control were questioned about foods and beverages that had been consumed during the 4 hours before becoming ill (for cases) or on the day of a case's illness (for controls). The odds ratio of cases (12.7; 95% confidence interval (CI), 2.4-83.8) for taking bread was significantly increased, suggesting that cases were more likely than controls to have eaten bread within the 4 hours before becoming ill. In addition, when stratified by age, the odds ratio was far higher in children under 18 years (odds ratio, 21.7; 95% CI, 2.4-264.6) than adults (odds ratio, 2.3; 95% CI, 0.02-195.9). This may be due to the higher consumption of parathion based on body weight or higher susceptibility to the insecticide in the former than the latter. Parathion was detected from residue floor on the truck that had brought the wheat flour from the milling factory to the general store where the baker purchased it, suggesting that the flour had been contaminated during transport. The authors estimated that 10-15 ml of parathion may have spilled onto a 22.5 kg bag of flour in the truck. Besides these, many cases of parathion intoxication have been reported to date (Aardema et al., 2008; Eyer et al., 2003; Hoffmann and Papendorf, 2006; Laynez et al., 1997).

In light of this background, fenitrothion was developed in place of parathion for its highly selective toxicity to insects over humans and animals. While the structure of fenitrothion is similar to that of parathion, its residual effects and acute toxicity are lower than parathion (Miyamoto, 1969). An oral LD50 of parathion is approximately 6 mg/kg for rats, against 330 mg/kg for fenitrothion, which is more rapidly broken down and does not persist in areas where they are used. Fenitrothion is effective against a wide range of pests on rice, cereals, fruits, vegetables, stored grains, cotton and forests, and also in public health programs for control of flies, mosquitoes and cockroaches. Fenitrothion is produced at the rate of 15,000 to 20,000 tons per year worldwide, and is available in emulsifiable concentrates, ultra-low-volume concentrates, powders, granules, dustable powders, oil-based sprays and in combination with other pesticides.

Permethrin: Permethrin (3-phenoxybenzyl (1RS,3RS;1RS,3SR)-3-(2,2-dichlorovinyl)-2,2-dimethylcyclopropanecarboxylate; CAS No. 52645-53-1) was first synthesized in 1973 and marketed in 1977 as a photostable pyrethroid (Elliott et al., 1973). Approximately 600 tons per year is at present used worldwide not only in agriculture but also in forestry, household settings, and public health programs.

Pyrethroids represented a major advancement as a high insecticide potential, but showed relatively low potential for mammals. Their development was especially timely with the identification of problems with DDT use. Pyrethroids consist first of identifying components of pyrethrum, which were extracted from East African chrysanthemum flowers, long known to have insecticidal properties. Pyrethrum rapidly knocks down flying insects, but has low mammalian toxicity and negligible persistence, which are good for the environment but yield poor efficacy when applied in the field. In the 1960s, 1st-generation pyrethroids, including bioallethrin, tetramethrin, resmethrin and bioresmethrin, were developed. They are more active than natural pyrethrum, but are unstable in sunlight. Then, permethrin, cypermethrin and deltamethrin were discovered as a 2nd generation of more persistent compounds. They are substantially more resistant to degradation by light and air, thus making them suitable for use in agriculture.

Permethrin is highly effective for protection of stored grains, cotton and other crops, and the control of body lice and household noxious insects. Technical products, which are a brown or yellowish-brown liquid, are a mixture of *cis* and *trans* isomers in the ratio of 40:60 or 25:75, and are available in emulsifiable concentrates, ultra-low-volume concentrates, wettable powders, and dustable powders.

2. Absorption, metabolism and excretion in laboratory animals and humans

In mammals, fenitrothion and permethrin are absorbed via gastrointestinal or respiratory tract and skin, and are rapidly metabolized and excreted.

Fenitrothion: After uptake into the body, fenitrothion is metabolized by hepatic cytochrome P450 (CYP) to form fenitrooxon, which is thought to have a higher potential acute neurotoxicity than the parent compound. Fenitrooxon is further metabolized to dimethylphosphate and 3-methyl-4-nitrophenol (MNP) by paraoxonase 1 (PON1). MNP and methylphosphate are also produced by glutathione-S-aryltransferase (GST) and PON1. In another pathway, fenitrothion is directly metabolized to MNP and dimethylthiophosphate by PON1 or MNP and methylthiophosphate by GST and PON1 (Figure 1). Interestingly, its major metabolic route differs between mammals and birds as mentioned later. Most of the metabolites are excreted in urine within 24 hours in humans (Nosal and Hladka, 1968), and within 2-4 days in the rat, guinea-pig, mouse, and dog (Miyamoto et al., 1963; Miyamoto, 1964). Species and sexes differences are observed in the composition of the metabolites. MNP, which is also contained in diesel exhaust emissions, has potential adverse effects on the reproductive systems in mammals and birds. Fenitrothion at doses of 0.18 and 0.36 mg/kg per day was administered to 12 human volunteers for 4 days (Meaklim et al., 2003). Pharmacokinetic parameters could only be determined at the high dosage, because the blood levels of fenitrothion at the low dosage were below the detectable level. Fenitrothion concentrations showed a wide range of interindividual variability, with peak blood levels achieved 1-4 hours after dosing, and the half-life ranged from 0.8 to 4.5 hours. Serum concentrations of fenitrothion were measured in 15 patients after acute fenitrothion intoxication, who admitted to the hospital 0.5-12 hours after the ingestion of 5-50 g fenitrothion (Koyama et al., 2006). The serum fenitrothion concentrations ranged from undetectable (< 0.01 µg/ml) to 9.73 µg/ml. Serum fenitrothion concentrations were less than 7 µg/ml in the patients with mild intoxication, while in the severe cases, the levels were more than 7 µg/ml. The elimination half-lives in the mild cases were 9.9 ± 7.7 hours (mean $\pm$ SD), and the serum fenitrothion concentrations declined below the detectable level in 48 hours. The elimination half-lives relating to two severe cases were 5.3 and 6.7 hours in the alpha phase (under direct hemoperfusion), and 35 and 52 hours in the beta phase, respectively. The serum fenitrothion concentrations fell below the detectable level in 300 hours.

Permethrin: Regarding its metabolism, permethrin is converted to 2,2-dichlorovinyl-2,2-dimethylcyclopropane-1-carboxylic acid and 3-phenoxybenzyl alcohol (3PBAlc) by carboxyl-esterase. The latter metabolite is followed by oxidation to form 3-phenoxybenzaldehyde, and finally 3-phenoxybenzoic acid (3PBA) (Figure 2). The metabolites are reported to be endocrine-disrupting agents, but most studies mention permethrin toxicity is derived from itself (Yuan et al., 2010). In general, *trans* isomer is more rapidly metabolized than *cis* isomer, which is related to the lower susceptibility of *cis* isomer to enzymatic hydrolysis of the ester linkage (Soderlund and Casida, 1977; Zhang et al., 2008). Besides hydrolytic pathway by carboxylesterase, oxidative metabolic pathway of both *cis*-and *trans*-permethrin in rat and human hepatic microsomes was recently reported (Scollon et al., 2009). The toxicokinetics of permethrin (with a *cis*:*trans* ratio of 25:75) was investigated after single oral doses to rats (Anadon et al., 1991). The plasma level of permethrin was maximal within 4 hours after dosing, and then was slowly eliminated from plasma with a half-life of 12.4 hours. The bioavailability of permethrin was found to be 60.7%. The maximum permethrin concentrations in the central and peripheral nervous system were higher than plasma concentrations, and declined with half-life similar to those of plasma. Clearance of *trans*- and *cis*-permethrin from the blood was also investigated in a man who drank an emulsifiable concentrate formulation of permethrin (consisting of 43.5% *cis* and 56.5% *trans*) (Gotoh et al., 1998). The serum concentrations of *cis*- and *trans*-permethrin peaked 3-4 hours after ingestion and then declined, with *trans*-permethrin cleared from the blood more quickly than *cis*-permethrin. Levels of the *trans* isomer were below the detectability threshold within 25 hours after exposure, whereas *cis* isomer was still detectable 10 days after exposure. The present study indicated that the differential persistence of *cis* and *trans* isomers in human is consistent with a difference in the metabolic rate of *cis-* and *trans*-permethrin in animal studies (Anadon et al., 1991).

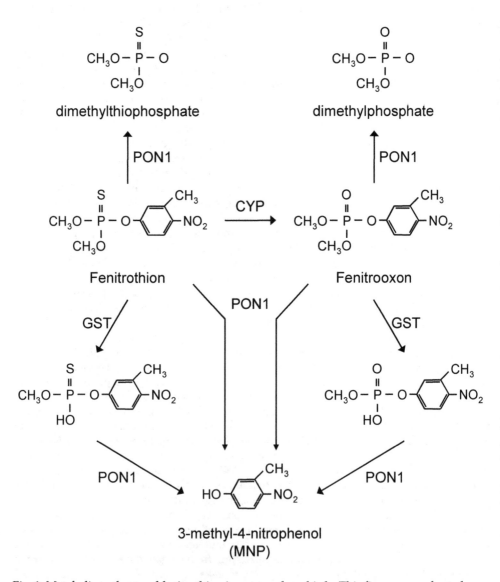

Fig. 1. Metabolic pathway of fenitrothion in mammals or birds. This figure was adopted from WHO (1992) with slight modifications. In mammals, dimethylthiophosphate, dimethylphosphate and MNP were the major urinary metabolites of fenitrothion and fenitrooxon. In birds, MNP was the major urinary metabolite of fenitrothion and fenitrooxon. GST activity was found to be lower than in mammals.

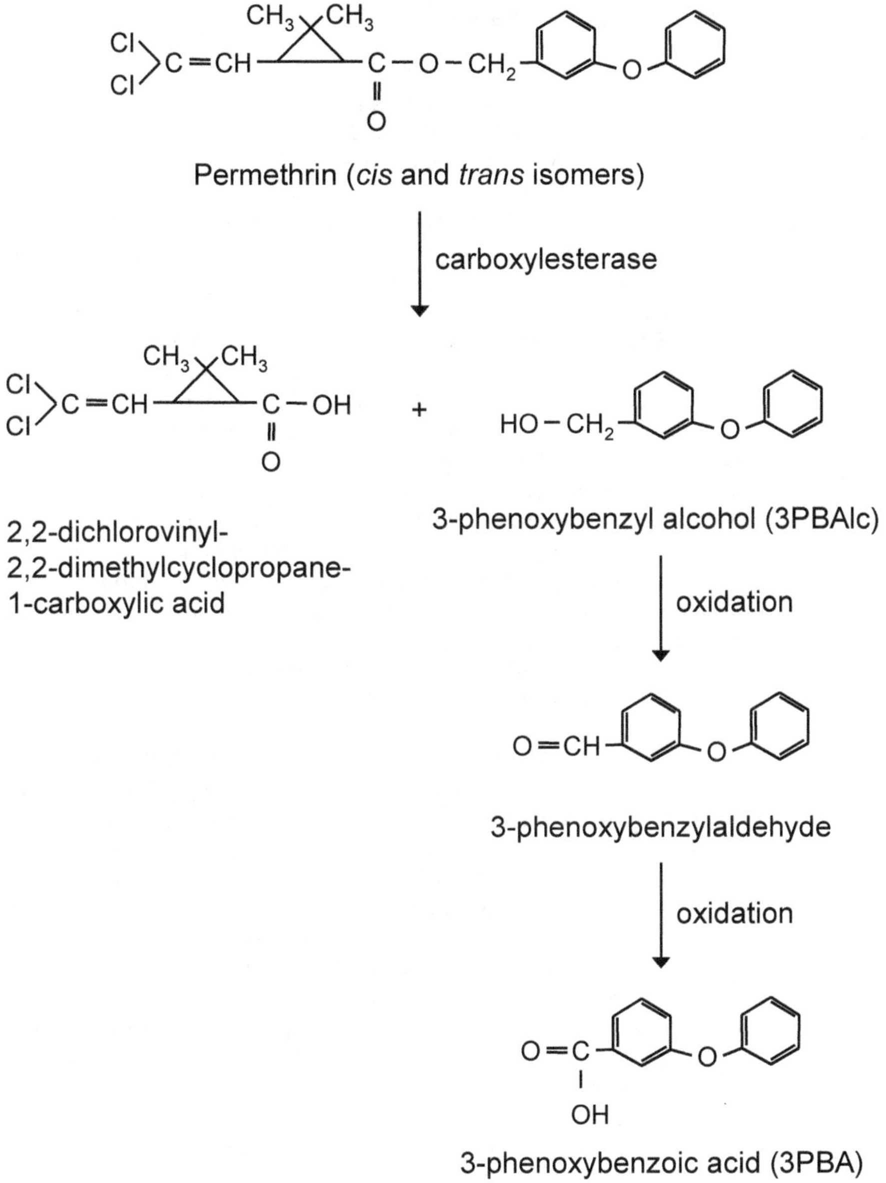

Permethrin (*cis* and *trans* isomers)

carboxylesterase

2,2-dichlorovinyl-
2,2-dimethylcyclopropane-
1-carboxylic acid

+

3-phenoxybenzyl alcohol (3PBAlc)

oxidation

3-phenoxybenzylaldehyde

oxidation

3-phenoxybenzoic acid (3PBA)

Fig. 2. Metabolic pathway of permethrin in mammals. This figure was adopted from WHO (1990) with slight modifications. Carboxylesterase plays an important role in the hydrolytic biotransformation of permethrin. The expression of carboxylesterase is ubiquitous with high levels in various tissues. Amon various animal tissues, the highest hydrolase activity is typically found in the liver and other tissues, such as testis, kidney and plasma. Unlike mouse, rat, rabbit, horse and cat, human plasma contains no carboxylesterase.

3. Toxicity for experimental animals and humans

Fenitrothion

Since it is a cholinesterase (ChE) inhibitor, exposure to fenitrothion causes ChE activity depression in plasma, red blood cells, brain, and liver tissues. The acute toxicity of fenitrothion is considered to be low in mammals, because of the high metabolic rate (Hayes, 1982; Spencer, 1981).

Animal: The no-observed-adverse-effect levels (NOAEL), based on brain-ChE activity, were 10 mg/kg diet in both short- and long-term studies on rats, in long-term studies on mice, and were 50 mg/kg diet in short-term studies on dogs, respectively (WHO, 1992). Fenitrothion was given to female rats by gavage every other day from gestational day 6-15 at doses 3, 15, 30 and 45 mg/kg (Berlińska and Sitarek 1997). The maternal death rates were 39% and 88% at doses of 30 and 45 mg/kg, respectively. At 30 mg/kg, fenitrothion caused a significant decrease in maternal body weight gain, food consumption, hemoglobin and hematocrit values, and absolute weights of liver and kidney, but an increase in relative weights of adrenal and ovary. At 15 mg/kg, fenitrothion significantly decreased maternal relative liver weight. Although fenitrothion at doses of 3-30 mg/kg did not induced teratogenic effects, at 30 mg/kg it showed embryotoxicity, such as a significant increase in the frequency of early resorption per litter, postimplantation loss, and fetuses and litters with dilation of the cerebral ventricles. Furthermore, fenitrothion produced delayed ossification of sternum and cranium, and decreased fetal body weight and length. The frequency of fetuses and litters with dilation of the cerebral ventricles was increased at a dose of 15 mg/kg. Thus, Berlińska and Sitarek concluded that the NOAEL for developmental toxicity in rats was 3 mg/kg per day, and the lowest-observed-adverse-effect level was 15 mg/kg per day.

Recent studies showed the endocrine-disrupting effect of fenitrothion. Berger and Sultatos (1997) demonstrated that fenitrothion treatment caused a dose-dependent decrease in 2-hydroxyestradiol and 4-hydroxyestradiol production in mouse hepatic microsomes even at a dosage as low as 7 mg/kg, and an increase in 16 alpha-hydroxyestrone and estriol production. In another study, 7-week-old castrated Sprague-Dawley rats were subcutaneously treated with testosterone propionate (50 µg/day in 0.2 ml corn oil) and orally with corn oil vehicle or fenitrothion (15 or 30 mg/kg per day) once a day for 7 days (Tamura et al. 2001). Both fenitrothion doses caused significant decreases in the weights of ventral prostate, seminal vesicle, and levator ani plus bulbocavernosus muscles. In contrast, blood acetylcholinesterase activity was only inhibited at the higher dose (30 mg/kg). Tamura et al. also demonstrated in an *in vitro* experiment that fenitrothion blocked dihydrotestosterone-dependent androgen receptor (AR) activity in a concentration-dependent and competitive manner in HepG2 human hepatoma liver cells, which were transiently transfected with human AR and an AR-dependent luciferase reporter gene, suggesting that fenitrothion may be a competitive androgen receptor antagonist.

On the other hand, Okahashi et al. (2005) suggested that lower-dose fenitrothion did not cause disruption of endocrine systems in animals. They administered fenitrothion to Crj:CD(SD)IGS parental rats at concentrations of 10, 20, and 60 (3.81 mg/kg per day) ppm in the diet for 10 weeks prior to mating, and throughout mating, gestation and lactation. Their offspring were exposed from weaning until maturation at the age of 10 weeks. In the parental animals, brain cholinesterase activity was remarkably reduced in males exposed to 60 ppm fenitrothion and in females exposed to 20 and 60 ppm fenitrothion. Reproductive performance, organ weights, histopathology, and sperm analytical parameters were not influenced. In the offspring, no effects on anogenital distance, retention of areolae/nipples, onset of puberty, organ weights, histopathological findings, and sperm parameters were observed. In conclusion, fenitrothion had no effects on the reproductive or endocrine systems of the parental animals and their offspring, even at a toxic dose suppressing brain cholinesterase activity in parental animals. The concentration of 60 ppm (3.81 mg/kg per day) is 750 times higher than the acceptable daily intake (ADI) of fenitrothion (0.005 mg/kg body weight). Therefore, any potential risk of exposure may be negligible, and fenitrothion at in-use levels in the environment may be unlikely to cause disruption of human endocrine systems.

Human: WHO (1990) classified technical grade fenitrothion as "moderately hazardous" (Class II). ADI of fenitrothion was established as 0.005 mg/kg body weight by the Joint FAO/WHO Expert Committee on Pesticide Residues in 2000. Nosal and Hladka (1968) reported that administration of fenitrothion at a single oral dose of 0.042-0.33 mg/kg body weight and repeated administration of 0.04-0.08 mg/kg body weight to human volunteers did not cause inhibition in plasma and erythrocyte ChE, and the urinary MNP was completely excreted within 24 hours. Chronic symptoms of exposure to fenitrothion in humans include general malaise, fatigue, headache, loss of memory and ability to concentrate, anorexia, nausea, thirst, loss of weight, cramps, muscular weakness and tremors.

Permethrin

Permethrin acts on the axons in the peripheral and central nervous systems, causing prolonged opening of sodium channels. The acute toxicity of permethrin in mammalians is relatively low, though the LD50 value varies considerably according to the vehicle used and the cis:trans isomeric ratio (FAO, 1999; U.S. EPA, 2007).

Animal: NOAEL is assigned at 5 mg/kg body weight per day for permethrin with an isomer ratio of cis:trans 40:60 from the viewpoint of the effects on liver weight in 2-year and 26-week studies in rats, and a 3-month study in dogs. NOAEL is not available for respective cis and trans isomers (WHO, 1990). The rat appeared to be the most sensitive species with an oral LD50 of 400 mg/kg body weight for cis:trans 40:60 permethrin administered in corn oil, against 650 mg/kg body weight in mice. The neurotoxicity of intravenous- or orally-administered cis-permethrin is over 10-fold greater than that of trans isomer. Neonatal rats are more sensitive than adult rats to the acute toxic effects of permethrin, which are thought to be related to differences in permethrin metabolism.

In their acute neurotoxicity study, Freeman (1993a) performed a functional observational battery (FOB) approximately 12 hours following administration of 10, 150 or 300 mg/kg of technical grade permethrin (mixture of cis and trans) in corn oil to male and female Sprague-Dawley rats. At doses of 150 and 300 mg/kg, permethrin caused salivation, tremor, splayed hindlimbs, abnormal posture, staggered gait, decreased grip strength, exaggerated reaction to sound, exaggerated hindlimb flexion, convulsions, and mortality. No treatment-related effects were observed at the lowest dose of 10 mg/kg. In a behavioral neurotoxicity study (McDaniel and Moser, 1993), technical grade permethrin was administered by gavage in corn oil to Long Evans hooded rats, at doses of 25, 75 and 150 mg/kg, and the FOB were evaluated at 2 and 4 hours following treatment. Results of the present study are consistent with the acute regulatory study (Freeman, 1993a) including tremor, chromodacryorrhea, decreased grip strength and an exaggerated startle response. However, the absence of salivation, splayed hindlimbs and convulsions and the presence of aggressive sparring in the latter study (McDaniel and Moser, 1993) were inconsistent with the findings of the regulatory acute neurotoxicity study (Freeman, 1993a).

In a subchronic neurotoxicity study (Freeman, 1993b), technical grade permethrin was administered through the diet to male and female Sprague-Dawley rats, at concentrations of 250, 1500 and 2500 ppm (18, 101 and 170 mg/kg per day, respectively). At the 1500 and 2500 ppm dietary levels, permethrin produced tremor, splayed hindlimbs, abnormal posture, a staggered gait, and decreased grip strength. No such effects were observed at the lowest dose of 250 ppm.

Effects of permethrin on endocrine or reproductive function are investigated, but the report is very limited. Castrated rats (5-week-old) were pretreated with testosterone propionate and orally given permethrin (mixture of *cis* and *trans*, 24.8% and 71.8%) at doses of 10, 50 and 100 mg/kg per day for 10 days. A mixture of *cis*- and *trans*-permethrin showed anti-androgen-like effects on male rats such as significant reductions in androgen-dependent sex accessory tissue (ventral prostate, seminal vesicles, levator ani and bulbocavernosus muscles, Cowper's gland and glans penis) weights (Kim et al., 2005). *cis*-Permethrin at 0, 35 and 70 mg/kg was orally administered to IRC mice for 6 weeks, and male reproductive toxicity was investigated. This chemical dose-dependently decreased testicular and plasma testosterone levels, along with a dose-dependent increase in circulating LH and declines in epididymal sperm count and sperm motility (Zhang et al., 2007). Testicular residue concentrations of *cis*-permethrin from the individual animals were also strongly inversely correlated with testicular testosterone levels. The exposure-related reductions in mRNA and protein expression levels of peripheral benzodiazepine receptor, steroidogenic acute regulatory protein and cytochrome P450 side-chain cleavage, which are involved in testosterone synthesis in testis, were observed, as well as structural changes in Leydig cell mitochondria, suggesting that the mitochondrial damage caused by permethrin exposure may result in a reduction of testosterone synthesizing elements and thereby decrease testosterone levels. In a follow-up study in mice, *cis*-permethrin induces reproductive toxicity whereas at the same dose *trans*-permethrin does not because of a faster metabolic rate than *cis* isomer (Zhang et al., 2008). Zhang et al. also reported that *cis*-permethrin caused structural abnormalities in the seminiferous tubules. However, it must be noted that these studies to date have used dose levels much higher than encountered by non-occupationally exposed humans.

Human: Permethrin is a moderately to practically non-toxic pesticide in EPA toxicity class II or III, depending on the formulation. Formulations in the case of possible eye and skin irritation are grouped into class II. Permethrin belongs to the type I group of pyrethroids because it lacks a cyano group, and typically causes tremor (T-syndrome), incoordination, hyperactivity, prostration, and paralysis. An ADI of 0.05 mg/kg body weight for *cis:trans* 40:60 or 25:75 permethrin was established in 1987. Rishikesh et al. (1978) evaluated staff involved with bagging, mixing, or spraying a 5% preparation of permethrin (cis/trans ratio, 25:75) in Nigeria by a questionnaire and urinalysis. Regardless of the protective equipment worn by the sprayers, only 2 mg of permethrin was absorbed after exposure to 6 kg of permethrin, which was excreted in 24 hours.

4. Toxicity for ecosystem

Fenitrothion

In the environment, fenitrothion is degraded by photolysis and hydrolysis. In the presence of ultraviolet radiation or sunlight, the half-life of fenitrothion in water is less than 24 hours.

The presence of micro-flora may accelerate degradation. Miyamoto et al. (1966) studied the degradation of fenitrothion by *B. subtilis*. The major metabolite was aminofenitrothion, and other minor metabolites detected were dimethyl thiophosphoric acid and desmethyl fenitrothion. In the bacteria, aminofenitrothion is further degraded to desmethyl aminofenitrothion, but the rate is slower than the parent compound. No reduction of desmethyl fenitrothion to desmethyl aminofenitrothion was detected, and dimethyl phosphoric acid was not formed from aminofenitrothion (Figure 3). Thus, the degradation of fenitrothine in *B. subtilis*, may be quite different from the metabolic route of experimental animals and humans. In the absence of sunlight or microbial contamination, fenitrothion is stable in water. In soil, biodegradation is the primary route, though photolysis may also play a role. Airborne concentrations of fenitrothion and its levels in water may decrease rapidly by photolysis and hydrolysis. The concentrations of fenitrothion that are likely to be found in the environment do not have any effects on microorganisms in soil or water. In laboratory studies, fenitrothion is highly toxic for aquatic invertebrates in both freshwater and seawater, while fish are less sensitive to fenitrothion than invertebrates, and the most sensitive life stage is early larva. Fenitrothion is highly toxic to bees (LD50, 0.03-0.04 µg/bee) when bees are exposed to direct treatment or to dried residues on foliage (U.S. EPA, 1987).

Furthermore, fenitrothion was found to be highly toxic to upland game birds, but not so toxic to waterfowl. Indeed, the acute oral LD50 values were determined to be 23.6 and 1190 mg/kg body weight for bobwhite quail and mallards, respectively. Even in reproduction studies, NOEL was 10 mg/kg body weight for the quail and 100 mg/kg body weight for the mallard, respectively. There are quantitative differences in the composition of metabolites of fenitrothion between mammalian and avian species. For example, in rats, mice and dogs, demethylated products at the *P-O*-methyl linkage by GST accounted for 30 to 60% of the total urinary metabolites (Hollingworth et al., 1967; Miyamoto et al., 1976), whereas in the birds only 10 to 15%. This may be due to lower GST activity in avian species compared with that in mammalians. Mihara et al. (1979) also revealed that oxidative activities of the *m*-methyl group of fenitrothion and fenitrooxon in livers from hen, quail, pheasant and duck were higher than those of mammalian liver, while *O*-demethylate activity for fenitrothion or fenitrooxon was lower in these birds. In birds, MNP is the major metabolite of fenitrothion by hydrolysis, though a pathway exists with oxidation of the *m*-methyl group of fenitrothion or fenitrooxon. The metabolite MNP is then conjugated with uridine diphosphate glucuronic acid or 3'-phosphoadenosine-5'-phosphosulfate by catalytic action of uridine diphosphate glucuronosyltranferase (UGT) and sulfotransferase (SULT), respectively (Mackenzie et al., 1997). Hepatic UGT and SULT activities investigated in vitro for MNP in Japanese quail, mice and rats revealed lower UGT activity for MNP in quail than rats and mice, but no significant difference in SULT activity (Lee et al., 2007). In addition, the SULT activity was only one-tenth of the UGT activity, suggesting that the latter enzyme plays an important role in MNP elimination in vivo. Li et al. (2008) reported that the birds treated with 100 mg/kg of MNP induced acute toxicological responses such as dyspnea and tremor, and finally death. MNP may cause acute toxicity and death, possibly by a rapid decrease in blood pressure followed by ischemic shock, because the potential vasodilatory action of MNP had been reported in rats (Mori et al., 2003; Taneda et al., 2004). However, none of the rats died after treatment with 100 mg/kg of MNP (Li et al., 2007), suggesting that the sensitivity to MNP is higher in quail than in rats. For these reasons, fenitrothion causes higher toxicity in birds than in mammals.

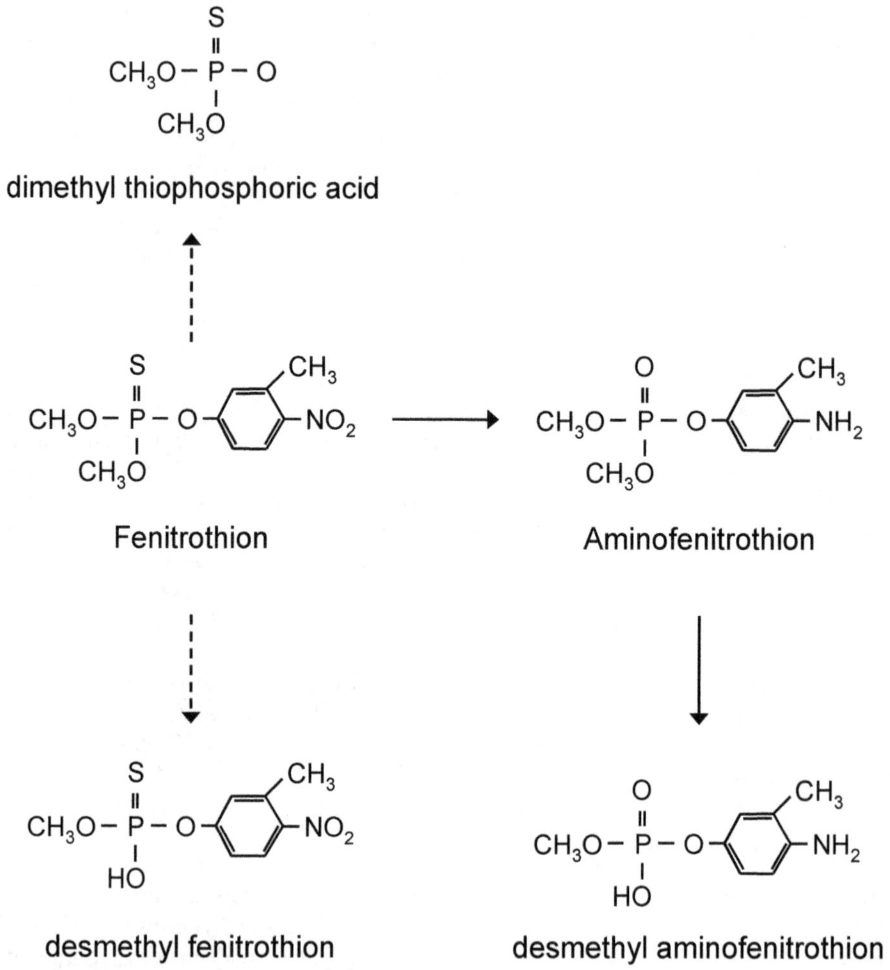

Fig. 3. Pathway of degradation of fenitrothion by *B. subtilis* in environment. This figure was adopted from Miyamoto et al. (1966) with slight modifications. Aminofenitrothion was the major metabolite of fenitrothion in *B. subtilis*.

Permethrin

Permethrin is photodegraded by sunlight in water and on soil surfaces. Under aerobic conditions in soil, permethrin degrades with a half-life of 28 days. Permethrin deposited on plants degrades with a half-life of approximately 10 days. Thus, in the environment, permethrin is hydrolyzed, and the resultant acid and alcohol are conjugated. However, permethrin itself evidences very little movement within the environment, because it binds very strongly to soil particles and is nearly insoluble in water and not expected to leach or contaminate groundwater. Permethrin has been shown to be highly toxic for aquatic arthropods and fish, because they have lower levels of carboxylesterase activity than mammals. However, the extreme susceptibility to permethrin may be ascribed to its high sensitivity to sodium channels rather than low carboxylesterase activity. Permethrin is also highly toxic to honey bees (LD50, 0.11 µg/bee), yet exhibits very low toxicity to birds when given orally or fed in the diet. LD50 is >3000 mg/kg body weight for a single oral dosage and >5000 mg/kg diet for dietary exposure, respectively. One of the reasons for the different toxicity of permethrin among species is negatively correlated its toxic action to their body temperature, thus generally showing more acute effects on cold-blooded animals (insects, fish, etc) than warm-blooded animals (mammals and birds).

5. Interactive toxicity of insecticide mixture

Organophosphorus insecticides are being increasingly used in combination with pyrethroid insecticides. Fenitrothion is used in combination with other pesticides to enhance ChE inhibition by nature. However, fenitrothion inhibits not only ChE but also other esterase activity such as carboxylesterase. Trottier et al. (1980) reported that the oral administration of fenitrothion to male CD rats at a dose of 0, 2.5, 5, 10, or 20 mg/kg per day for 30 consecutive days significantly decreased liver carboxylesterase activity (by 50-80%) on days 8-30 at doses more than 2.5 mg/kg per day but had returned to control values by day 45 (15 days after termination of treatment) at all doses except 20 mg/kg per day, at which a decrease of 25% was still observed. At this dose, the values had returned to normal by day 87 (57 days after termination of treatment). A significant decrease in renal carboxylesterase activity (by 20-70%) was also observed on days 8-30 at doses over 5 mg/kg per day. Recovery of the activity was rapid, and the values were comparable to those of controls by day 38 (8 days after the end of treatment).

As described at the metabolism of permethrin, carboxylesterase plays an important role in detoxication of permethrin. Ortiz et al. (1995) examined the interactions between a commercial formulation of methyl parathion and a commercially formulated product of permethrin in male rats. When rats were treated with the mixture, 380 mg/kg of methyl parathion reduced the LD50 of permethrin by only 9.0%, whereas when rats received methyl parathion at 464 mg/kg, the LD50 of permethrin was reduced by 37% (P < 0.001). Results indicated that methyl parathion modified the acute toxicity of permethrin. Another study examined the effect of organophosphorus insecticide dichlorvos on excretion levels of urinary cis-permethrin-derived 3PBA in rats (Hirosawa et al., 2011). After cis-permethrin injection (20 mg/kg) via the tail vein of rats pretreated intraperitoneally with dichlorvos (low dose, 0.3 mg/kg; high dose, 1.5 mg/kg), the amounts of urinary 3PBA excretion over 48 hours were decreased to 81.1% and 70.3% of dichlorvos non-treated rats in the low- and high-dose dichlorvos groups, respectively. The plasma concentration of cis-permethrin-derived 3PBAlc in high-dose dichlorvos group was significantly lower than that in the dichlorvos non-treated group one hour after cis-permethrin injection. In contrast, no differences were observed in the excretion levels of urinary 3PBA after injection of 3PBAlc between the dichlorvos non-treated group and the high-dose dichlorvos group. These results suggested that dichlorvos may have inhibited the metabolism of the co-exposed cis-permethrin and thereby decreased the amount of urinary 3PBA excretion. In our recent study, we evaluated male reproductive toxicity after co-exposure to diazinon (3 mg/kg) and cis-permethrin (35 mg/kg) in mice. Exposure to diazinon alone and the mixture with cis-permethrin inhibited plasma and liver carboxylesterase activities. In the co-exposed mice, the urinary *cis*-permethrin metabolite decreased compared to that in mice exposed to *cis*-permethrin alone. The co-exposure significantly decreased plasma testosterone levels and increased the number of degenerated germ cells within the seminiferous tubule, whereas exposure to each chemical did not. We concluded that diazinon inhibited the plasma and liver carboxylesterase activities and the metabolic rate of co-exposed *cis*-permethrin, which resulted in accentuating the reproductive toxicity of *cis*-permethrin (Wang et al., unpublished data submitted to the Journal).

Recently, since we could not find any study on the interaction between fenitrothion and permethrin, the toxicity of permethrin may be enhanced by fenitrothion via depression in carboxylesterase activity. Since fenitrothion and permethrin are used in the same place, if not purposefully in mixtures, the two insecticides could conceivably be combined. Until now, the risk assessments of combined toxicity to mammals are still insufficient and further detailed studies are warranted.

6. References

Aardema H, Meertens JH, Ligtenberg JJ: Organophosphorus pesticide poisoning: cases and developments. *Neth J Med* 66:149-153, 2008.

Anadon A, Martinez-Larranaga MR, Diaz MJ, et al: Toxicokinetics of permethrin in the rat. *Toxicol Appl Pharmacol* 110:1-8, 1991.

Berger CW Jr, Sultatos LG: The effects of the phosphorothioate insecticide fenitrothion on mammalian cytochrome P450-dependent metabolism of estradiol. *Fundam Appl Toxicol* 37:150-157, 1997.

Berlińska B, Sitarek K: Disturbances of prenatal development in rats exposed to fenitrothion. *Rocz Panstw Zakl Hig* 48:217-228, 1997.

Clifford NJ, Nies AS: Organophosphate poisoning from wearing a laundered uniform previously contaminated with parathion. *JAMA* 262:3035-3036, 1989.

Elliott M, Farnham AW, Janes NF, et al: A photostable pyrethroid. *Nature* 246:169-170, 1973

Etzel RA, Forthal DN, Hill RH Jr, et al: Fatal parathion poisoning in Sierra Leone. *Bull World Health Organ* 65:645-649, 1987.

Eyer F, Meischner V, Kiderlen D, et al: Human parathion poisoning. A toxicokinetic analysis. *Toxicol Rev* 22:143-163, 2003.

FAO: Pesticide Residues in Food, Toxicological Evaluations; Food and Agriculture Organization of the United Nations and World Health Organization: Rome, 1999.

Freeman C: Permethrin technical: acute neurotoxicity screen in rats. *FMC Corporation* Study No. A92-3646, 1993a.

Freeman C: Permethrin technical: subchronic neurotoxicity screen in rats. *FMC Corporation* Study No. A92-3647, 1993b.

Gotoh Y, Kawakami M, Matsumoto N, et al: Permethrin emulsion ingestion: clinical manifestations and clearance of isomers. *J Toxicol Clin Toxicol* 36:57-61, 1998.

Hayes WJ: Pesticides Studied in Man. *Williams and Wilkins.* Baltimore, London. 1982.

Hayes WJ, Laws ER: Handbook of Pesticide Toxicology, Volume 2: Classes of Pesticides. *Academic Press Inc.* NY, 1990.

Hirosawa N, Ueyama J, Kondo T, et al: Effect of DDVP on urinary excretion levels of pyrethroid metabolite 3-phenoxybenzoic acid in rats. *Toxicol Lett* 203:28-32, 2011.

Hoilingworth RM, Metcalf RL, Fukuto TR: The selectivity of sumithion compared with methylparathion: Metabolism in the white mouse. *J Agric Food Chem* 15:242-249, 1967.

Hoffmann U, Papendorf T: Organophosphate poisonings with parathion and dimethoate. *Intensive Care Med* 32:464-468, 2006.

Kim SS, Lee RD, Lim KJ, et al: Potential estrogenic and antiandrogenic effects of permethrin in rats. *J Reprod Dev* 51:201-210, 2005.

Koyama K, Suzuki R, Kikuno T: Serum fenitrothion concentration and toxic symptom in acute intoxication patients. *Chudoku Kenkyu* 19:41-47, 2006.

Laynez BF, Martínez GL, Tortosa FI, et al: Fatal food poisoning by parathion. *Med Clin (Barc)* 108:224-225, 1997.

Lee CH, Kamijima M, Li C, et al: 3-Methyl-4-nitrophenol metabolism by uridine diphosphate glucuronosyltransferase and sulfotransferase in liver microsomes of mice, rats, and Japanese quail (Coturnix japonica). *Environ Toxicol Chem* 26:1873-1878, 2007.

Li C, Suzuki AK, Takahashi S, et al: Effects of 3-methyl-4-nitrophenol on the reproductive toxicity in female Japanese quail (Coturnix japonica). *Biol Pharm Bull* 31:2158-2161, 2008.

Li C, Taneda S, Suzuki AK, et al: Effects of 3-methyl-4-nitrophenol on the suppression of adrenocortical function in immature male rats. *Biol Pharm Bull* 30:2376-2380, 2007.

Mackenzie PI, Owens IS, Burchell B, et al: The UDP glycosyltransferase gene superfamily: recommended nomenclature update based on evolutionary divergence. *Pharmacogenetics* 7:255-269, 1997.

McDaniel KL, Moser VC: Utility of a neurobehavioral screening battery for differentiating the effects of two pyrethroids, permethrin and cypermethrin. *Neurotoxicol Teratol* 15:71-83, 1993.

Meaklim J, Yang J, Drummer OH: Fenitrothion: toxicokinetics and toxicologic evaluation in human volunteers. *Environ Health Perspect* 111:305-308, 2003.

Mihara K, Misaki Y, Miyamoto J: Metabolism of fenitrothion in birds. *J Pesticide Sci* 4:175-185, 1979.

Miyamoto J, Kitagawa K, Sato Y: Metabolism of organophosphorus insecticides by *Bacillus subtilis*, with special emphasis on Sumithion. *Jpn J Exp Med* 36:211-225, 1966.

Miyamoto J: Mechanism of low toxicity of Sumithion toward mammals. *Residue Rev* 25:251-264, 1969.

Miyamoto J, Mihara K, Hosokawa S: Comparative metabolism of m-methyl-^{14}C-sumithion in several species of mammals *in vivo*. *J Pest Sci* 1:9-21, 1976.

Miyamoto J, Sato Y, Kadota T, et al: Studies on the mode of action of organophosphorus compounds. Part I. Metabolic fate of 32P-labelled Sumithion and methylparathion in guinea-pigs and white rats. *Agric biol Chem* 27:381-389, 1963.

Miyamoto J: Studies on the mode of action of organophosphorus compounds. Part III. Activation and degradation of Sumithion and methylparathion in vivo. Agric biol Chem 28:411-421, 1964.

Mori Y, Kamata K, Toda N, et al: Isolation of nitrophenols from diesel exhaust particles (DEP) as vasodilatation compounds. *Biol Pharm Bull* 26:394-395, 2003.

Nosál M, Hladká A: Determination of the exposure to fenitrothion (O,O-dimethyl-O-3-methyl-4-nitrophenyl-thiophosphate) on the basis of the excretion of p-nitro-m-cresol by the urine of the persons tested. *Int Arch Arbeitsmed* 25:28-38, 1968.

Okahashi N, Sano M, Miyata K, et al: Lack of evidence for endocrine disrupting effects in rats exposed to fenitrothion in utero and from weaning to maturation. *Toxicology* 206:17-31, 2005.

Ortiz D, Yáñez L, Gómez H, et al: Acute toxicological effects in rats treated with a mixture of commercially formulated products containing methyl parathion and permethrin. *Ecotoxicol Environ Saf* 32:154-158, 1995.

Rishikesh N, Clarke JL, Martin HL, et al: Evaluation of Decamethrins and Permethrin Against *Anopheles gambiae* and *Anopheles funestus* in a Village Trial in Nigeria. World Health Organization, Division of Vector Biological Control; 1978.

Scollon EJ, Starr JM, Godin SJ, et al: In vitro metabolism of pyrethroid pesticides by rat and human hepatic microsomes and cytochrome P450 isoforms. *Drug Metab Dispos* 37:221-228, 2009.

Soderlund DM, Casida JE: Effects of pyrethroid structure on rates of hydrolysis and oxidation by mouse liver microsomal enzymes. *Pestic Biochem Physiol* 7:391-401, 1977.

Spencer EY: Guide to the Chemicals Used in Crop Protection. 7th edition. Publication 1093. *Research Branch. Agriculture Canada.* 1981.

Tamura H, Maness SC, Reischmann K, et al: Androgen receptor antagonism by the organophosphate insecticide fenitrothion. *Toxicol Sci* 60:56-62, 2001.

Taneda S, Kamata K, Hayashi H, et al: Investigation of vasodilatory substances in diesel exhaust particles (DEP): isolation and identification of nitrophenol derivatives. *J Health Sci* 50:133-141, 2004.

Trottier B, Fraser AR, Planet G, et al: Subacute toxicity of technical fenitrothion in male rats. *Toxicology* 17:29-38, 1980.

U.S. Environmental Protection Agency: Guidance for the Reregistration of Pesticide Products Containing Fenitrothion. US EPA, Office of Pesticide Programs, Registration Div., Washington, DC. 132 pp. July, 1987.

U.S. Environmental Protection Agency: Reregistration Eligibility Decision (RED) for Permethrin. US EPA, Office of Prevention, Pesticides and Toxic Substance, Office of Pesticide Programs, U.S. Government Printing Office: Washington, DC, 2007.

WHO: Environmental Health Criteria 133: Fenitrothion. World Health Organization, International Programme on Chemical Safety, Geneva, Switzerland, 1992.

WHO: Environmental Health Criteria 94: Permethrin. World Health Organization, International Programme on Chemical Safety, Geneva, Switzerland, 1990.

WHO: The WHO recommended classification of pesticides by hazard and Guidelines to classification 1990-1991, Geneva, World Health Organization (WHO/PCS/90.1), 1990.

Worthing CR, Walker SB: The Pesticide Manual: A World Compendium. 8th edition. Published by The British Crop Protection Council, 1987.

Yuan C, Wang C, Gao SQ, et al: Effects of permethrin, cypermethrin and 3-phenoxybenzoic acid on rat sperm motility in vitro evaluated with computer-assisted sperm analysis. *Toxicol In Vitro* 24:382-386, 2010.

Zhang SY, Ito Y, Yamanoshita O, et al: Permethrin may disrupt testosterone biosynthesis via mitochondrial membrane damage of Leydig cells in adult male mouse. *Endocrinology* 148:3941-3949, 2007.

Zhang SY, Ueyama J, Ito Y, et al: Permethrin may induce adult male mouse reproductive toxicity due to *cis* isomer not *trans* isomer. *Toxicology* 248:136-141, 2008.

3

Chlorfluazuron as Reproductive Inhibitor

Farzana Perveen
Chairperson, Department of Zoology
Hazara University, Garden Campus
Mansehra
Pakistan

1. Introduction

Benzoyl phenyl ureas (BPUs) inhibit chitin synthesis during growth and development in insects and act as moult disruptors, therefore, they have been called insect growth regulators (Wright and Retnakaran, 1987; Binnington and Retnakaran, 1991). IGRs, such as dimilin, are effective against a considerable range of insect larvae and adults in a variety of situations. The compound disrupts the moulting process by interfering with chitin synthesis. Research on the different aspects of dimilin as a chitin synthesis inhibitor, toxicant, ovicide, disrupting adult emergence, and residual effects have been done with various insect species, e.g., Jakob, 1973; James, 1974; Qureshi et al.,1983; Naqvi and Rub, 1985; Ganiev, 1986; Khan and Naqvi, 1988; Gupta et al., 1991; Tahir et al., 1992; Nizam, 1993. Several modes of action have been reported for these pesticides. For example, phagodeterrents and repellents (Abro et al., 1997), chitin synthesis inhibition (Hajjar and Casida, 1979), growth inhibition and abnormal development (Hashizume, 1988), ovicidal action (Hatakoshi, 1992), insecticidal effects on the reproductive system (Chang and Borkovec, 1980) and neurotoxic effects on insect behaviour (Haynes, 1988).

1.1 Chlorfluazuron

Chlorfluazuron (Atabron®) is a benzyl phenyl urea (BPU) chitin-synthesis inhibitor (CSI) and insect growth regulator (IGR) is formed by Ishihara Sangyo Kaisha, Japan. Some important details concerning the insecticide chlorfluazuron are given below (provided by Ishihara Sangyo Kaisha, Japan) (Perveen, 2005):

Common name	:	Chlorfluazuron (proposed to ISO)
Other names	:	Atabron® or Helix® or Aim®
Source	:	Ishihara Sangyo Kaisha Ltd., Tokyo, Japan
Code number	:	IKI-7899, CG-112913, pp-145
Formulation type	:	5% w/w (Emulsifiable concentration: EC)
Chemical name	:	[1-{3,5-dichlor-4-(3-chlor-5-trifluoromethyl-2-pyridyloxy) phenyl}-3-(2,6-ifluorobenzoyl) urea] (IUPAC nomenclature)

1.1.1 Salient physical and chemical properties (Perveen, 2005)

Appearance	:	Crystalline solid at 20 °C
Odor	:	Odorless

Melting point   :   222.0 – 223.9 °C (decomposes after melting)
Vapor pressure   :   <10 – 8 p$_a$, <10-10 Torr at 20 °C
Volatility   :   Relatively non-volatile
Specific gravity   :   1.4977 at 20 °C
Stability   :   No detectable decomposition over at least 3 months at 50 °C

1.2 Spodoptera litura

The *S. litura* is found in most of the Caroline and in the South Pacific Island including American Samoa. It also occurs in the northern two thirds of Australia. The moth is also widespread throughout India and recognized as quarantine pest in EU legislation. It is present in Mediterranean Europe and Africa. It is the most commonly intercepted in the UK, on imported ornamentals and their products. *Spodoptera litura* is also a destructive pest of subtropical and tropical agriculture, and has the potential to be a serious pest of glasshouse crops in northern Europe. It was found as feeding on impatience on Victoria Peak on Hong Kong Island and readily switched to (western) lettuce (Etman and Hooper, 1979). In 1974, Etman and Hooper initiated an investigation into the radiobiology of *S. litura*, and reported that it was a significant pest of cotton in the Ord River region, Australia (Etman and Hooper, 1979). Its larvae are a major cosmopolitan pest of a wide range of crops (Skibbe et al., 1995). Matsuura and Naito (1997) reported that *S. litura* causes serious widespread damage to many agricultural crops in the far southern of the Central Japan every year. They hypothesized that adult *S. litura* immigrate into Japan from overseas every year by long-distance migration. The larvae are destroyed many economically important crops such as *Gossypium hirsutum* L., *Brassica oleracea* L., *Spinacea oleracea* L., *Trifolium alexandrinum* L., *Medicago sativa* L., *Arachis hypogaea* L., *Phaseolus aureus* Roxb., *Phaseolus vulgaria* L. and *Nicotiana tabacum* L. during different seasons throughout the year in Pakistan (Younis, 1973). Their larvae eat nearly all types of herbaceous plants. Some examples of plants are: tobacco, *Nicotiana tabacum* L.; tomatoes, *Lycopersicum esculentum* Mill.; cauliflower, *Brassica botrytis* L.; beetroot, *Beta vulgaris conditiva* L.; silver beet (swiss chard) *Beta vulgaris cicla* (L.); peanuts, *Arachis hypogaea* L.; beans, *Phaseolus vulgaris* L.; banana, *Musa* paradisiaca L.; strawberry, *Fragaria* vesca L.; apple, *Malus pumila* Mill.; lettuce, *Lactuca sativa* L.; zinnia, *Zinnia elegans* Jacq.; dahlia, *Dahlia pinnata* Cav.; ape, *Alocasia macrorrhiza* (L.); geranium, *Pelargoniumx zonale*; St. John's lily, *Crinum asiaticum* L.; mangrove lily, *Crinum pedunculatum* (Fragrant); leek, *Allium porrum* (Leek); horsetail she oak, *Casuarina equisetifolia* L.; *Fuchsia* and many other garden plants (Baloch and Abbasi, 1977). Several common names have been used for *S. litura*, for example defoliator cutworm, oriental leafworm, cluster caterpillar and common cutworm. The larvae are quite polyphagous for example eat all types of herbaceous plants and have been reaching the status of international pest. In 1968, a panel convened by the International Atomic Agency listed species of *S. litura* as a pest on which basic and applied research was needed in order to evaluate the potential of the sterile insect release method for control (Anonymous, 1969).

Eggs of *S. litura* are laid in batches, on plants and other surfaces such as pots, benches or glasshouse structures. Eggs are normally laid in the irregular furry masses covered with orange-brown hairs giving them a "felt-like" appearance on the underside of a leaf of a food plant similar to the egg of the brown locust, *Locustana pardalina* (Walk.) (Matthee, 1951). On hatching, larvae (caterpillars) are 2–3 mm long with white bodies and black heads and are very difficult to detect visually. If they emerge from eggs laid on glasshouse structures or hanging pots, they can reach the plants below by "parachuting" down on silken threads. The overall colour of the later-stages of the larvae can vary from light to dark brown, and the body is strongly speckled with tiny white spots. Initially, when larvae grow become a translucent green with a dark thorax. The young larvae are smooth-skinned with a pattern of red, yellow, and green lines, and with a dark patch on the mesothorax. Larvae initially eat only the flesh of their food leaves, leaving the veins intact. Later, as they grow, they eat whole leaves, and even flowers and fruit (Khuhro et al., 1986).

Many populations are extremely resistant to pesticides and, if they become well established, can be exceptionally difficult to control. In these cases, it is important that a comprehensive treatment programme is implemented, incorporating a range of reliable control methods, including physical destruction of insects. The *S. litura*, as it is the most common to be encountered in a UK nursery, but the larvae and adults of all noctuids are similar in appearance and are difficult to tell apart without laboratory examination (Khuhro et al., 1986). Larvae become brown with three thin yellow lines down the back, one in the middle and one on each side. A row of black dots runs along each side, and a conspicuous row of dark triangles decorates each side of the back. The last-instar larva is very dark, with four prominent yellow triangles on the mesothorax. When disturbed, the larvae curl into a tight spiral with the head protected in the centre. Larvae further develop characteristic markings on their backs. These include: a square of four yellow spots, each on a black patch, located just behind the head; a further pair of black patches just behind these, and another pair of black patches towards the end of the larva; typically, there are three orange-brown lines, punctuated with dashes of black and yellow along the back of the body. Depending on the background colour, these markings may be more evident on some larvae than others; the larvae ultimately grow up to 4.5 cm long; larvae are nocturnal, and during the day can be found at the base of plants or under pots. The feeding activity of young larvae causes "windows" in the leaves, while older larvae can completely defoliate plants if present in large numbers. Stems, buds, flowers and fruits may also be damaged. The larvae burrow into the soil below the plant for several centimeters and pupate there without a cocoon. As they do so, they produce a quantity of fluid, and drown in this if they pupate in captivity in an empty glass jar. They pupate successfully if 0.5 cm of sand is provided in the container. In January in Melbourne, the pupal stage lasts three weeks, but larvae that pupate at the end of summer emerge the following spring. The red-brown pupae are up to 2 cm long. The thoracical, ventral knobs found on covering of pupa (Khuhro et al., 1986). Adult moths with brown colour are up to 2 cm long with a wingspan of approximately 4 cm. The fore-wings are brown, with a large number of pale cream streaks and dashes and, when the adults are newly emerged, there may be a violet tint to the fore-wing. The hind-wings are a translucent white, edged with brown. The hind-wings are silvery white. It has a wingspan of about 4.0 cm. The males but not the females have a blue-grey band from the apex to the inner margin of each forewing. The pheromones of this species (specific sex-attractant scents used by females to attract males) have been elucidated. As the adults are nocturnal, light or pheromone traps should be used for monitoring purposes. Seek assurance from suppliers that plants are free from this pest as part of any commercial contract: carefully inspect new plants and produce on arrival, including any packaging material, to check for eggs and larvae and for signs of damage (Khuhro et al., 1986).

Early notification of the presence of this pest, will allow rapid implementation of a comprehensive treatment programme, and will help eradicate it quickly from nursery. Established outbreaks are very damaging and difficult to eradicate. Various methods of control of *S. litura* have been investigated. Biologically, it has been controlled by the nematode, *Steinernema carpocapsae* (Weiser) and parasitoid fly, *Exorista japonica* (Townsend). A baculovirus has also been used. Resistant species of plants are also grown to save the crops from this pest. Resistant tomatoes are most commonly cultivated (Khuhro et al., 1986).

2. Effects of chlorfluazurn on reproduction of *Spodoptera litura*

Reproductive inhibition induced by BPUs has been reported the most widely when applied to adults or eggs of insect pests rather than to application to larvae or pupae (Fytizas, 1976). When these compounds were applied to females, males or both sexes of insect pests, BPUs induced a variety of effects on reproduction; they caused a decrease in fecundity, fertility and/or hatchability. It has been reported that treatment of adult insect pests with diflubenzuron disrupts the secretion of adult cuticle (Hunter and Vincent 1974; Ker, 1977), and the production of peritrophic membrane in the grasshopper, *Locusta migratoria* (L.) (Clark et al., 1977) and the meal worm, *Tenebrio molitor* L. (Soltani, 1984; Soltani et al., 1987). In addition, topical treatments of male and female adult boll weevils, *Anthonomous grandis* Boheman; stable flies, *Stomoxys calcitrans* (L.) and *M. domestica* with TH-6040 [{N-(4-chlorophenyl)-N-26-difluorobenzoyl}urea] caused significant reduction of egg fertility and hatchability. It also causes inhibition in the fecundity of female adults of several species of insect pests (Holst, 1974; Taft and Hopkins, 1975; Crystal, 1978; Hajjar and Casida, 1979; Otten and Todd, 1979). Diflubenzuron applied to adult females caused a decrease in fecundity in the Mexican bean beetle, *Epilachna varivestis* Mulsant and Colorado potato beetle, *Leptinotarsa decemlineata* Say (Holst, 1974), and adversely affected egg viability in *St. calcitrans* and *M. domestica* (Wright and Spates, 1976). When 2 day-old female adults of the Japanese beetle, *Oryzae japonica* Willemse, were starved for 6 hours, and then allowed to consume 500 μg a.i. of diflubenzuron over another 6 hours, the fecundity of the treated females, in term of number of eggs laid per pod, was significantly decreased from controls. In controls, most pods gave an egg hatch of 82.5% but a hatch of only 8.5% hatched in the treated females (Lim and Lee, 1982). Similarly, treating eggs with diflubenzuron caused reduction in hatching in the mosquitoes, *Culex pipiens* Say, *C. quinquefascialus* Say (Miura et al., 1976), the almond moth, *Ephestia cautella* (Walk.) (Nickle, 1979), the two-spotted lady beetle, *Adalia bipunctata* (L.) and the seven-spotted lady beetle, *Coccinella septempunctata* L. (Olszak, 1994). Reports of inhibition of reproduction when larvae or pupae (instead of adults or eggs) were treated with BPUs are rare and little literature is available (Madore et al., 1983). Brushwein and Granett (1977), working with the spruce budworm, *Choristoneura fumiferana* (Celemens), demonstrated that certain moult-inhibiting IGRs such as EL-494 (Eli Lilly and Co., New York, USA) fed to sixth-instar larvae, caused reproductive failure in adults surviving after the larval treatment. Therefore, in this research newly ecdysed fifth-instar larvae and newly ecdysed pupae of *S. litura* were used as test materials. Chlorfluazuron, a comparatively new IGR and BPU that was discovered by Ishihara Sangyo Kaisha Ltd., Tokyo, Japan, that has been developed and sold commercially as Atabron®, Helix® and Aim® in many countries, including Japan in cooperation with Novelties Co. Ltd, ICI-AGRO and Ciba Geigy. It is a relatively highly active chitin synthesis inhibitor and it is, therefore, an effective treatment for the control of major lepidopteran insect pests in crops such as cotton, fruits, tea, vegetables and where insect resistance to conventional insecticides is becoming a serious problem. Chlorfluazuron exhibits no activity against important beneficial insects (Haga et al., 1992). The highly selective insecticidal activity of chlorfluazuron is particularly suited to integrated pest management programmes. Although chlorfluazuron has contact toxicity at higher rates, the major route of toxicity to insects is ingestion, and it has no root, systemic or foliar translaminar activity. Like other BPUs, chlorfluazuron is believed to disrupt chitin formation and, thus, kills the insects when they moult. This mode of action necessarily means that it is effective only against immature insects and that it is relatively slow actions. When higher dosages of chlorfluazuron were applied to newly ecdysed fifth-instar larvae, it had a devastating effect on the *S. litura* population by killing them during larval, pupal and adult stages (Hashizume, 1988). In insect pest management, the purpose of research is to maintain the pest population below the economic injury level. The mode of action of chlorfluazuron, as a CSI is known to some extent. However, but the knowledge of its effects on reproduction are rare. Insect structure and physiology may vary considerably during growth and development, with certain stages being more susceptible to insecticides than others. For example, the cuticle varies in its composition during larval development and this has been related to changes in IGR susceptibility. The activity of various insecticides detoxifying enzymes, such as MFO, glutathione S-transferase and epoxide hydrase also fluctuate during the life cycle of an insect (Yu, 1983). For this purpose newly ecdysed fifth-instar larvae and newly ecdysed pupae were selected for the treatments for the present research. The main objective of the present research is to determine the effects of sublethal doses (LD$_{10}$: 1.00 ng larva^{-1}; 0.12 ng female pupa^{-1}; 1.23 ng male pupa^{-1} or LD$_{30}$: 3.75 ng larva^{-1}) of chlorfluazuron on the reproduction (e.g., fecundity, fertility and hatchability) when ha been apply to newly ecdysed fifth-instar larvae and newly ecdysed pupae of *S. litura* (Perveen, 2000a).

2.1 Experimental procedures
2.1.1 Insect rearing
Experiments were conducted with *Spodoptera litura* (F.) (Lepidoptera: Noctuidae) taken from a stock that was established from eggs obtained from Aburahi Laboratory of Shionogi Pharmaceutical (Koga-Shiga-Pref., Japan). The larvae of *S. litura* were reared in the laboratory under controlled conditions on the artificial diet Insecta LF® (Nihon Nohsan-kohgyo, Kanagawa, Japan). The rearing temperature was maintained at 25±1 °C, with a L16:D8 hour photoperiod and 50-60% r.h. To facilitate observations, the dark period was set from 06:00 to 14:00 hours. Adults were fed on a 10% sucrose solution soaked in cotton. The eggs, which were laid on Rido® cooking paper (Lion, Tokyo, Japan), were collected every 3rd day and kept in 90 ml plastic cups (4 cm in diameter: 4×4 cm high) for hatching under the same environmental conditions (Perveen, 2000a).

2.1.2 Chlorfluazuron and its application
Sublethal doses, LD$_{10}$ (1.00 ng larva^{-1}; 0.12 ng female pupa^{-1}; 1.23 ng male pupa^{-1}) or LD$_{30}$ (3.75 ng larva^{-1}) were applied to newly ecdysed fifth-instar larvae and newly ecdysed pupae. These LD$_{10}$ and LD$_{30}$ values were calculated based on interpret alone of the results of the toxicity data of larval tests at adult emergence. The treated and untreated insects, at all developmental stages including fifth- and sixth-instar larvae, pupae and adults, were weighed separately, on different developmental days, using an analytical balance (Sartorius Analytical AC-2105, Tokyo, Japan) to a precision of 0.001 mg, to determine the effect of chlorfluazuron on the body weight. The duration of each developemental stage was also strictly recorded (Perveen, 2000a).

2.1.3 Mating

After both larval and pupal treatments, females and males that emerged between 2 and 8 hour (most adults emerged in the dark photoperiod) on the same day were collected at 0200–1000 hour. These adults were considered as 0 day old and paired just before the dark photoperiod (12 hour old) of the next day. Each female and male pair was kept separately in a plastic cup (430 cm³; height: 8.0 cm; diameter: 9.5 cm) for the whole life-span. The cup was padded with Rido cooking paper on its wall and with a disc of 70 mm filter paper on the bottom. The pairs were fed throughout their life by cotton wool soaked in 10% sugar solution in small plastic cups. All *S. litura* were examined daily (Perveen, 2000a).

To determine the effects of sublethal doses of chlorfluazuron on reproductivity, seven different mating combinations of female and male crosses were established. These were: (1) Untreated female mated with untreated male ($_U$♀×$_U$♂); (2) LD_{10}-treated female mated with untreated male ($_{LD10}$♀×$_U$♂); (3) Untreated female mated with LD_{10}-treated male ($_U$♀×$_{LD10}$♂); (4) LD_{10}-treated female mated with LD_{10}-treated male ($_{LD10}$♀×$_{LD10}$♂); (5) LD_{30}-treated female mated with untreated male ($_{LD30}$♀×$_U$♂); (6) Untreated female mated with LD_{30}-treated male ($_U$♀× $_{LD30}$♂); (7) LD_{30}-treated female mated with LD_{30}-treated male ($_{LD30}$♀× $_{LD30}$♂). For the fecundity, fertility and hatchability experiments, 15–30 pairs were used for each cross. Eggs were laid on the cooking paper after 24 hours and were collected during 0800–1000 hour, cut out and kept in cups (90 cm³) for hatching. Eggs laid were hatched within 84 hours. Observation of oviposition continued until the death of female. Four days after each collection of eggs, the fecundity, fertility and hatchability of the laid eggs were assessed. After the natural death of females, the spermatophores were separated from the bursa copulatrix with a fine forceps in 0.9% NaCl (Saline or Ringer's solution: Barbosa, 1974) under the binocular microscope (10×magnification) (Olympus Co. Ltd., Tokyo, Japan) (Perveen, 2000a).

2.1.4 Data analysis

Data for the effects of sublethal doses of chlorfluazuron on reproductivity and viability were analyzed using analysis of variance, one way ANOVA (Concepts, 1989; Minitab, 1997; Walpol and Myers1998) at $P<0.0001$ and Scheffe's F-test (multiple range) (Scheffe, 1953) at 5%. Hatchability percentage values were normalized by arcsin transformation before statistical analysis (Anderson and McLean, 1974).

2.2 Results

When the LD_{10} (1.00 ng larva⁻¹; 0.12 ng female pupa⁻¹; 1.23 ng male pupa⁻¹) and the LD_{30} (3.75 ng larva⁻¹) of chlorfluazuron were applied to newly ecdysed fifth-instar larvae or newly ecdysed pupae, it was observed that the fecundity of the resulting adults and the fertility and hatchability of their eggs, was significantly reduced {$P<0.0001$ (for larval treatment); $P<0.05$ (for pupal treatment)}, compared with untreated adults, but no significant differences were observed between larval and pupal treatments ($P<0.02$) (Tables 1 and 2). When chlorfluazuron was applied to newly ecdysed fifth-instar larvae at sublethal doses, the number of eggs oviposited by a treated females mated with an untreated male ($_T$♀×$_U$♂) was suppressed to the same degree as an untreated female mated with a treated male ($_U$♀×$_T$♂) or a treated female mated with a treated male ($_T$♀×$_T$♂) (Table 1). The mean female fecundity was 2250±198 eggs when both male and female were untreated (control), i.e., ($_U$♀×$_U$♂), (Table 2). When the female was treated either by the LD_{10} or LD_{30} and mated with an untreated male, the fecundity was 1462±353 ($LD10♀×U♂$) and 1266±237 ($LD30♀× U♂$), respectively. When the male was treated with the LD_{10} and mated with an untreated female, the fecundity was 1407±334 ($U♀×LD10♂$). However, in the same cross when the male was treated with LD_{30} instead of the LD_{10} ($U♀×LD30♂$), the fecundity was 1270±215. When both sexes were treated with the LD_{10} ($LD10♀×LD10♂$), it was 1330±295 and when both sexes were treated with the LD_{30} concentration ($LD30♀×LD30♂$), it was 1331 ±295. In all the crosses, the fecundity was significantly reduced when compared with the control cross ($U♀×U♂$) (Table 1) (Perveen, 2000a).

Mating pairs[a] (female×male)	n[a]	Fecundity[b,c] (mean±SD)	Fertility[b,c] (mean±SD)	Hatchability%[c,d] (mean±SD)
$U♀×U♂$	30	2250±198a	1984±208a	88.4±6.6a
$LD10♀×U♂$	30	1462±353b	1010±315b	68.9±11.8b
$U♀×LD10♂$	28	1407±334b	688±317c	48.3±17.1c
$LD10♀×LD10♂$	30	1330±295b	643±265c	48.6±14.2c
$LD30♀×U♂$	29	1266±237b	828±206b	65.8±11.4b
$U♀×LD30♂$	30	1270±215b	36±155d	28.8±11.2d
$LD30♀×LD30♂$	29	1331±295b	33±121d	27.1±9.1d

[a]LD_{10}: 1.00 ng larva[-1]; LD_{30}: 3.75ng larva[-1]; U: untreated (control); ♀: female adults; ♂: male adults; n: number of pairs used
[b]Number of eggs oviposited during the whole life of female adult were counted (fecundity) and from the eggs number of hatched that larvae were counted (fertility).
[c]Data were analyzed using one way ANOVA (Concepts, 1989) at $P<0.0001$). Means within columns followed by different letters indicate significant differences by Scheffe's F-test (Scheffe, 1953) at 5%.
[d]Hatchability % values were normalized by arcsin transformation before statistical analysis (Anderson and McLean 1974).

Table 1. Effects of sublethal doses of chlorfluazuron on fecundity, fertility and hatchability after topical application of newly ecdysed-fifth instar larvae of *Spodoptera litura* (Source: Perveen, 2000a).

When the LD_{10} of chlorfluazuron was applied to newly ecdysed pupae and the resulting adults were paired, the control fecundity was 2170±175. It was not significantly reduced ($P<0.02$) compared with the larval treatment. The fecundity in the treated pupal cross ($LD10♀×U♂$) was 640±83. It was not significantly reduced ($P<0.02$) compared with the same cross with treated larvae. In the pupal treatment, the fecundity was suppressed to a similar degree in the crosses $LD10♀×U♂$, $U♀×LD10♂$, $LD10♀× LD10♂$. This reduction was significant ($P<0.0001$) when compared with the control cross (Table 1) (Perveen, 2000a). There was no significant reduction ($P<0.02$) between the larval and pupal treatments with respect to fecundity (Tables 1 and 2) (Perveen, 2005).
The mean fertility of females was 1984±208 larvae when both the male and female were untreated, i.e. the control ($U♀×U♂$) (Table 1). When the female was treated with either the LD_{10} or LD_{30} and mated with an untreated male, the fertility was ($LD10♀×U♂$: 1010±315; $LD30♀× U♂$: 828±206, respectively), i.e., significantly reduced compared with the control cross ($U♀× U♂$). When the male was treated with the LD_{10} and mated with an untreated female, the fertility ($U♀× LD10♂$: 688±317) was significantly reduced compared with the crosses, i.e.,

LD_{10}♀×U♂ and LD_{30}♀×U♂. However, in the same cross when the male was treated with the LD_{30} instead of LD_{10}, the fertility was (U♀× LD_{30}♂: 368±155) significantly lower than the U♀×LD_{10}♂ cross. When both sexes were treated with the LD_{10}, the fertility was (LD_{10}♀×LD_{10}♂: 643±265) not significantly different from the U♀ ×LD_{10}♂ cross. Similarly, when both sexes were treated with the LD_{30}, the fertility was (LD_{30}♀×LD_{30}♂: 333±121) not significantly different from the U♀×LD_{30}♂ cross (Table 1) (Perveen, 2000a).

Mating pairs[a] (female×male)	n[a]	Fecundity[b,c] (mean±SD)	Fertility[b,c] (mean±SD)	Hatchability %[c,d]
U♀×U♂	15	2170±175a	2123±177a	97.8a
LD_{10}♀×U♂	15	1640±83b	1090±79b	66.5b
U♀×LD_{10}♂	15	1580±75b	827±49c	52.3c
LD_{10}♀×LD_{10}♂	15	1524±76b	751±51c	49.3c

[a]LD_{10}: 0.12 ng female pupa[-1]; 1.23 ng male pupa[-1]; n: number of pairs used
[b]Number of eggs oviposited (fecundity) the during whole life of female adults were counted and from the number of eggs that hatched larvae were counted (fertility).
[c]Data were analyzed using one-way ANOVA (Concepts, 1989) at P < 0.001. Means within a column followed by different letters indicate significant differences according to Scheffe's F-test (Scheffe, 1953) at 5%.
[d]Hatchability% values were normalized by arcsin transformation before statistical analysis (Anderson and McLean 1974).

Table 2. Effects of a sublethal doses of chlorfluazuron on fecundity, fertility and hatchability after topical application of newly ecdysed pupae of *Spodoptera litura* (Source: Perveen, 2005).

When the LD_{10} of chlorfluazuron was applied to newly ecdysed pupae and resulting adults were paired, the fertility of the control cross, U♀×U♂ was 2123±177, which was not significantly reduced (P<0.02) than that in the larval treatment. In the same way, with the LD_{10}♀×U♂ cross, the fertility was 1090±79, which was not significantly reduced (P<0.02) than the same cross in the larval treatment. The fertility was reduced 61 –65% in the U♀×LD_{10}♂ cross, which was significantly reduced (P<0.02) than the LD_{10}♀×U♂ cross with a fertility reduction of 46–49%. The fertility of the U♀×LD_{10}♂ cross was not significantly reduced (P<0.02) than the LD_{10}♀×LD_{10}♂ cross, which was reduced 65–68% for both larval and pupal treatments. There were no significant reductions (P<0.02) between larval and pupal treatments with respect to fertility (Tables 1 and 2). The mean hatchability during a female life-span was 88.4(±6.6)% when both male and female were untreated, i.e. the control (U♀×U♂) cross (Table 1). When the female was treated, either by the LD_{10} or LD_{30} and mated with an untreated male, the hatchabilities were 68.9 (±11.8)% (LD_{10}♀×U♂) and 65.8(±11.4)% (LD_{30}♀×U♂), respectively significantly reduced compared with the control cross. When the male was treated with the LD_{10} and mated with an untreated female, the hatchability, 48.3(±17.1)% (U♀×LD_{10}♂) was significantly reduced than the LD_{10}♀×U♂ and LD_{30}♀×U♂ crosses. However, in the same cross when the male was treated with LD_{30}, instead of LD_{10}, hatchability, 28.8(±11.2)% (U♀×LD_{30}♂), significantly reduced than the U♀×LD_{10}♂ crosses. When both sexes were treated with LD_{10}, the hatchability was 48.6(±14.2)% (LD_{10}♀×LD_{10}♂), not significantly different from the U♀×LD_{10}♂ cross. Similarly, when both sexes were treated with LD_{30}, the hatchability was 27.1(±9.1)% (LD_{30}♀×LD_{30}♂), which was not significantly different from the U♀×LD_{30}♂ cross (Table 3.1). The hatchability for the U♀×U♂ cross was 88.4% and

97.8%, respectively, for the larval and pupal treatments. The larval-treated cross, $LD10♀×U♂$, was 68.9%, which is not significantly reduced (P<0.02) than the same cross of the pupal treatment in which the hatchability was 66.5%. The hatchability was reduced 48.7% and 52.7% in the $U♀×LD10♂$ cross, respectively, for the larval and pupal treatments. In the $LD10♀×LD10♂$ cross, it was reduced to 48.6% and 49.3%, respectively, for these treatments. There was no significant reduction (P<0.02) between larval and pupal treatments with respect to hatchability (Tables 1 and 2) (Perveen, 2000a).

2.3 Discussion
When chlorfluazuron was applied to newly ecdysed fifth instars at sublethal doses, LD_{10} (1.00 ng larva^{-1}) or LD_{30} (3.75 ng larva^{-1}), it was observed that the fecundity of resulting adults as well as the hatching rate of their eggs was suppressed. The hatching rate of eggs oviposited by an untreated female mated with a treated male was suppressed to the same degree as that of eggs oviposited by a treated female mated with a treated male. However, Madore et al. (1983) studied the effects when different concentrations of sublethal doses of the UC-62644 (chlorfluazuron-25) fed to sixth instar larvae of spruce budworm. Homologous crosses between adults of the 0.01, 0.025 and 0.034 ppm treatments showed 0, 69 and 97% reduction, respectively, in the numbers of eggs laid per 30 pairs of moths when compared with control. Emam et al. (1988) reported the fecundity of *S. littoralis* adults decreased significantly from 977.64 eggs in control to 421.75 eggs, a decrease of about 56%, for adults feeding 10% honey solution containing 0.5 p.p.m. chlorflufluazuron. The corresponding fertility inhibition amounted to 32%. In the present case the fertility was significantly different when only the female was treated or only the male was treated. It is obvious from the results that the fertility and hatchability were affected more when the male was treated in comparison with the female, as also reported by Abro et al. (1997), who found that males were more sensitive to insecticides than females, when five concentrations of cyhalothrin and fluvalinate were tested against fourth instar larvae of *S. litura*.

2.4 Conclusion
To clarify the sublethal effects of chlorfiuazuron on reproductivity of common cutworm, *Spodoptera litura*, experiments were conducted under laboratory conditions. Reduction in the body weight was observed in the larvae and pupae when treated with a sublethal dose (LD_{30}: 3.75 ng larva^{-1}) and in the adults when treated with sublethal doses (LD_{10}: 1.00 ng larva^{-1}; LD_{30}: 3.75 ng larva^{-1}) as newly ecdysed fifth instar larvae of *S. litura*, although the number of matings per female and life span of adult females and males remained unaffected by the same treatments. When sublethal doses were applied only to females or only to males, or both sexes, the average fecundity reduction was up to 35–44%. When only females were treated with sublethal doses, fertility was reduced by 49–58%; when only males were treated fertility was reduced by 65–81% and when both sexes were treated, fertility was reduced by 68±83%. Hatchability was reduced by 22–26% when only females were treated, by 44–66% when only males were treated and by 45–72% when both sexes were treated with LD_{10} or LD_{30} doses as newly ecdysed fifth instars. The results from these observations suggest that the fecundity was reduced to a similar degree when only females or only males or both sexes were treated with LD_{10} or LD_{30} doses as newly ecdysed fifth instars. However the fertility and hatchability were affected more when only males were treated with LD_{10} and much more when treated with LD_{30}. Currently, work is in progress to find out the main reasons for the sublethal effects of chlorfluazuron on reproductivity and viability.

3. Effects of chlorfluazurn on female reproductive system of *Spodoptera litura*

In many insects oviposition requires the development of the ovary, egg maturation, mating and, in some insects, feeding of the females. Ovarian development, which includes oöcyte growth and vitellogenesis, is under the hormonal control, of either juvenile hormone or ecdysteroid (Engelmann, 1979). In many insects, juvenile hormone (JH) regulates the biosynthesis and uptake of vitellogenin by the oöcytes. Among Lepidoptera, e.g., the tobacco hawkmoth, *Manduca sexta* L. (Sroka and Gilbert, 1971; Nijhout and Riddiford, 1974) and the large white butterfly *Pieris brassica* L. (Karlinsky, 1963 and 1967; Benz, 1969), juvenile hormone is required for full development of the ovaries in adults, whereas in the silkworm *Bombyx mori* L. (Chatani and Ohnishi, 1976), giant silk moth, *Hyalophora cecropia* (L.) (Williams, 1952; Pan, 1977), ailanthus silkmoth, *Samia cynthia* (Drury) (Takahashi and Mizohata, 1975) and ricemoth, *Corcyra cephalonica* (Stainton) (Deb and Chakarvorty, 1981) ovarian development occurs as part of adult development initiated by ecdysteroid. Juvenile hormone or juvenile hormone analogue (JHA) application at a critical period, however, induces abnormal development of the ovary as well as other tissues, although juvenile hormone analogues can replace natural juvenile hormone in regulating oöcyte maturation (Nomura, 1994). In the normal state, the ovary develops during one day before and after eclosion in the presence of juvenile hormone (as described above) and a haemolymph factor stimulates the ovary to start oviposition. When S-71639 was applied to pupae, it inhibited adult emergence when a relatively a high dose was applied. If adults did emerge, they could not oviposit through inhibition of the haemolymph factor, hatchability was also reduced (Hatakoshi and Hirano, 1990). The effects of diflubenzuron on fecundity resulted from treatment of adult females by contact or ingestion (Leuschner, 1974; Fytizas, 1976). When *O. japonica*, adults females were fed diflubenzuron, it retarded the maturation of oöcytes (Lim and Lee, 1982). In *T. molitor*, diflubenzuron reduced mealworm longevity (Soltani et al., 1987), the number of oöcytes per ovary, the duration of the oviposition period and the fecundity (Soltani, 1984). Diflubenzuron, topically applied (0.5 μg insect^{-1}) to codling moth, *Cydia pomonella* L. on pupal ecdysis, inhibited the growth and development of oöcytes. It delayed adult emergence and caused a decrease in both the thickness of the follicular epithelium and the size of the basal oöcytes during pupal development. On the other hand, the size of basal oöcytes, the protein content per ovary and the number of oöcytes per ovary recorded in newly emerged adults were significantly reduced by the diflubenzuron treatment. These results, together with observations in several other species, indicated that the reduction in fecundity and egg viability was probably due to interference by diflubenzuron with vitellogenesis (Soltani and Mazouni, 1992). Under laboratory conditions, effects of topical application of sublethal doses of chlorfluazuron (LD$_{10}$: 1.00 ng larva^{-1} or LD$_{30}$: 3.75 ng larva^{-1}) on newly ecdysed fifth-instar on fecundity, fertility and hatchability have been investigated. Thus, it is investigated the causes of the decrease in these parameters. To obtain more information, sublethal doses of chlorfluazuron topically have been applied to newly ecdysed fifth-instar larvae of *S. litura* and the effects on female reproductive system during ovarian development and oögenesis have been observed.

3.1 Experimental procedure
3.1.1 Ovary measurement
Experimental *S. litura* were rared in the same way as mentioned in Section 2.1.1. Sublethal doses, LD$_{10}$ (1.00 ng larva^{-1}; 0.12 ng female pupa^{-1}; 1.23 ng male pupa^{-1}) or LD$_{30}$ (3.75 ng

larva[-1]) were applied to newly ecdysed fifth-instar larvae same mentioned in Section 2.1.2. To determine the effects of chlorzuazuron on the ovaries, control and treated batches of insects, collected from the fifth day after pupation to the seventh day after adult emergence, were used, depending on the experiment requirements. Ovaries were dissected from these insects in Ringer's solution under a binocular microscope (10×magnifcation: Nikon, Nippon Kogaku, Tokyo, Japan), and the lengths of pedicle, vitellarium, and germarium of each ovariole were measured. The number of mature oöcytes in each of the ovarioles was also counted. The Ringer's solution was removed, and the freshly dissected ovaries were placed in a small covered container that had been preweighed. The dissected ovaries were weighed on an analytical balance (AC-205, Sartorius Analytical, Tokyo, Japan) and kept in the same container in the oven for 24 h at 62±1 °C for evaporation of water. The dried ovaries were reweighed (Perveen and Miyata, 2000).

3.1.2 Histology
The procedure to stain the nuclei for cell density from the germarium during female adults required age was adapted from the method described by He (1994). First, the germarium of the female were removed and kept on a microscope slide and carefully crushed with micro forceps until it was extended and roughly evenly distributed over the slide. Second, several drops of 3:1 methanol-acetic acid were introduced to the slide to fix the preparation for 15 min and then the excess fixing solution was absorbed with a filter paper after the fixation. Third, several drops of a 2–5% Giemsa solution dissolved by Sorensen-Gomori buffer solution (monobasic and dibasic sodium phosphate, 0.07 M, pH 6.8) were introduced to the slide for 10–30 min to stain the preparation. Then after the staining, the slide was washed gently and carefully and dried in the air. Finally, the air-dried preparation was checked with a phase-contrast microscope at 20×magnification (Perveen and Miyata, 2000).
The length and width of the basal oöcytes were measured from the 5th d after pupation to 0 day after adult emergence in control and treated insects. Oöcyte measurements were made on three to four basal oöcytes per pair of ovaries that were taken from 9-10 insects. Measurements were made with a graduated slide under the phase contrast microscope at 400×magnification (BH₂, Olympus, Tokyo, Japan). The size of basal oöcytes was calculated by the formula used by Loeb et al. (1984) for the size of a prolate spheroid, $4/3\pi(ab^2)$, where a is the radius of the long and b is the short dimension of the same oöcyte (Perveen and Miyata, 2000).
The thickness of the follicular epithelium of basal oöcytes was observed by making a parafilm microtomy conducted according to the method described by Yoshida (1994). Basal oöcytes were fixed in Carnoy's solution for 3 hour, washed in 70% ethanol for 2 hour, and dehydrated in an alcohol series 70, 80, 90, and 95% and twice in 100%, followed by a benzene and ethanol solution (1:1), each for 30 min. Incubation was done three times at 60 °C in benzene and paraffin (1:1), and then in parafin only, each for 30 min. Five-mm microtome sections were cut into a rolling ribbon. It was stained in xylene I (10 min), xylene II (5 min), followed by an ethanol series of 100, 90, 80, and 70%, each for 5 min, Mayer's hematoxylin for 15 min, and washed under running water. Microtome sections were mounted in 1% eosin (10 min), distilled water (2 sec), followed by an alcohol series 50, 70, 80, and 90%, and twice in 100% (each for 1-5 min), xylene and 100% ethanol (1:1; 5 min), xylene I (5 min), xylene II (5 min). Finally, the sections were embedded on microscopic graduated slide in a drop of Canada balsam. The microscopic graduated slide was covered with a glass cover slip. The thickness of follicular epithelium was measured under a phase

contrast microscope (BH₂, Olympus, Tokyo, Japan) at 400×magnifcation (Perveen and Miyata, 2000).

3.1.3 Data analysis

Data were analyzed using analysis of variance, one way ANOVA (Concepts, 1989) at $P<0.01$ and Scheffe's F-test (Scheffe, 1953) at 5%.

3.2 Results

The morphology of the adult female reproductive system of *S. litura* is shown in Figure 1.

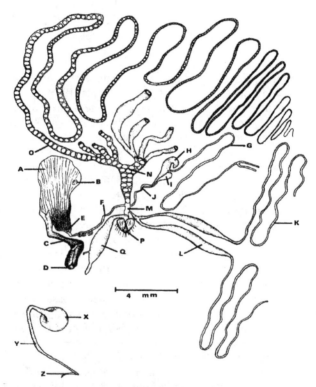

Fig. 1. The morphology of the female reproductive system of *S. litura*: A: corpus bursae; B: signum, C: ductus bursae; D: ostium bursae; and E: diverticulum of bursa copulatrix; F: ductis seminalis; G: spermathecal gland; H: utriculus I: lagena of spermatheca; J: ductus receptaculi; K: accessory gland (paired); L: accessory gland reservoir (paired); M: vestibulum; N: calyx of the unpaired oviductus communis; O: one of four ovarioles of ovary (paired); P: papillae anales; Q: rectum; X: corpus, Y: collum, and Z: frenum of spermatophores; (Source: Etman and Hooper, 1979).

3.2.1 Effects on ovarian development

Sublethal doses of chlorfluazuron (LD₁₀: 1.00 ng larva⁻¹; LD₃₀: 3.75 ng larva⁻¹), applied to newly ecdysed fifth-instar larvae significantly (P<0.0001) reduced the body weight, fresh ovarian weight and dry ovarian weight in newly emerged adults when compared with the controls (Table 3). Significant reductions were not observed in fresh body weight (P=0.0567), fresh ovarian weight (P=0.7788) and dry ovarian weight (P=0.5757), when the LD_{10} and LD_{30} treatments were compared. Similarly, ratios of fresh ovarian/fresh body weight (31.0%), dry ovarian/fresh ovarian weight (28.0%) and dry ovarian/fresh body weight (9.0%), were not significantly different (Table 3) (Perveen and Miyata, 2000).

T^a	n^a	$FBW^{a,b}$ (M±SD) mg	$FOW^{a,b}$ (M±SD) mg	$DOW^{a,b}$ (M±SD) mg	% R=FOW /$FBW^{a,b}$ (M±SD)	% R=DOW /$FBW^{a,b}$ (M±SD)	% R=DOW /$FOW^{a,b}$ (M±SD)
C	30	255±11.6a	81.1±9.2a	23.6±1.4a	31.5±2.9a	28.9±2.4a	9.2±0.7a
LD_{10}	30	229±3.2b	72.0±3.4b	20.0±1.4b	31.2±1.5a	28.1±1.2a	9.0±0.8a
LD_{30}	30	224±5.9b	70.9±4.2b	19.9±1.3b	31.3±2.4a	28.0±0.5a	9.1±0.8a

[a]C: control; T: trearment; LD_{10}: 1.00 ng larva^{-1}; LD_{30}: 3.75 ng larva^{-1}; n: number of insects used; FBW: fresh body weight; FOW: fresh ovarian weight; DOW: dry ovarian weight; % R: percent ratio
[b]Data were analyzed using one-way ANOVA (Concepts, 1989) at P<0.001. Means within a column followed by different letters indicate significant differences according to Scheffe's F-test (Scheffe, 1953) at 5%.

Table 3. Effects of sublethal doses of chlorfluazuron on the ovarian and body weight of newly emerged adults after topical application to newly ecdysed fifth-instar larvae of *Spodoptera litura* (Source: Perveen and Miyata, 2000).

Ovaries are small on the 8th day after pupation. From the 8th day after pupation to the day before adult emergence, ovarian weight slowly increased; after that it increased sharply until the day of adult emergence, and then, increased gradually until the 2nd day after adult emergence, when it reached maximum (120±19.4 mg). Then, in the controls it decreased gradually until the 7th day after adult emergence. The pattern of changes in fresh ovarian weight in the LD_{10}- or LD_{30}-treated females was similar as observed in the controls during various developmental days of pupae and adults. The fresh ovarian weight was significantly reduced on the 8th day after pupation (P<0.0001); on the 9th day after pupation (P<0.0001); on the 1st day after adult emergence (P<0.0003); on the 2nd day after adult emergence (P<0.0001); on the 3rd day after adult emergence (P<0.0001); on the 4th day after adult emergence (P<0.0001); on the 5th day after adult emergence (P<0.0001); on the 6th day after adult emergence (P<0.0001); on the 7th day after adult emergence (P<0.0001) in the LD_{10}- or LD_{30}-treated females compared with the controls, but no significant reduction was observed (P=0.0979–0.970) between the LD_{10}- or LD_{30}-treated females during ovarian development (Figure 2) (Perveen and Miyata, 2000).

In newly emerged the LD_{10}- or LD_{30}-treated adults, the total length of the ovariole was significantly reduced (P<0.0001) compared with the control, but there were no significant reductions (P=0.0508) between the LD_{10}- or LD_{30}-treatments. In the LD_{10}- or LD_{30}-treated insects, the germarium (immature oögonia) was significantly longer (P<0.0001) than that of the pedicle (fully mature ova) and the vitellarium (under developing oöcytes) compared with the controls in which the vitellarium was significantly longer (P<0.0001) than the germarium and pedicle (Figure 3; Table 4) (Perveen and Miyata, 2000; Perveen, 2011).

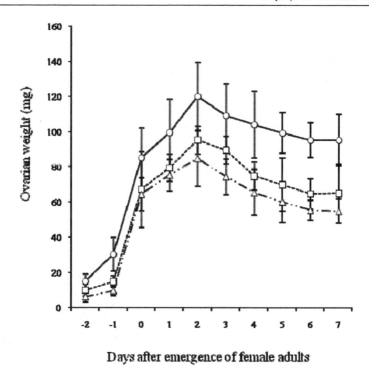

Days after emergence of female adults

Fig. 2. Effect of sublethal doses of chlorfluazuron on ovarian weight during different developmental days (post pupal and 1 to 7 day after adult emergence). For control ($\circ$; n = 10), LD_{10} (1.00 ng larva^{-1}) treated ($\square$; n = 9), and LD_{30} (3.75 ng larva^{-1}) treated ($\triangle$; n = 9) after topical application to newly ecdysed fifth instars of *Spodoptera litura*. Data were analyzed using one-way ANOVA (Concepts, 1989) at $P<0.0001$ and Scheffe's F-test (Concepts, 1989) at 5%. Vertical bars indicate SD; (Source: Perveen and Miyata, 2000).

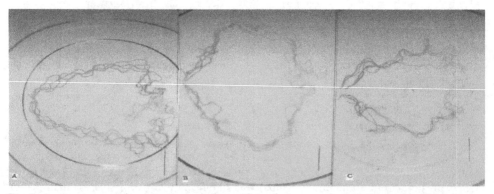

Fig. 3. A comparison of ovarian morphology of newly emerged adult *Spodoptera litura*, in A: untreated (control); B: treated with the LD_{10} dose and C: treated with the LD_{30} dose of chlorfluazuron. Bars in photographs indicate 100 μm (Source: Perveen, 2011).

Treat-ments[a]	n_1	n_2	TLO[bc] (M±SD)mm	LP[bc] (M±SD)mm	LV[bc] (M±SD)mm	LG[bc] (M±SD)mm	R (M±SD)mm	P:V:G
Control	10	80	104.8±5.1a[3]	33.5±3.4a	45.2±4.4a	26.1±4.2a	32:43:25	
LD_{10}	9	72	91.5±5.5b	22.1±5.0b	30.5±2.1b	39.0±2.7b	24:33:43	
LD_{30}	9	72	88.7±9.6b	20.4±4.4	28.6±6.7b	40.0±8.9b	23:32:45	

[a]LD_{10}, 1.00 ng larva^{-1}; LD_{30}, 3.75 ng larva^{-1}; n_1: number of insects used; n_2: number of ovariole measured; TLO: total length of ovariols; LP: length of pedicle; LV: length of vitellarium; LG: length of germarium; RP:V:G: Ratio of pedicle:vitellarium:germanium

[b]In Spodoptera litura, the paired ovaries are composed of 8 ovarioles. Four ovarioles are found on each side of the body cavity, forming several loops. Each ovarioles differentiated into 3 parts: (1) pedicle (fully matured eggs), (2) vitellarium (oöcytes and trophocytes), (3) germanium (oögonia) (Etman and Hooper 1979).

[c]Data were analyzed using one-way ANOVA (Concepts, 1989) at $P<0.0001$. Means within a column followed by different letters indicate significant differences according to Scheffe's F-test (Scheffe, 1953) at 5%.

Table 4. Effects of sublethal doses of chlorfluazuron on ovarian development in newly emerged adults after topical application to newly ecdysed fifth instars of *Spodoptera litura* (Source: Perveen and Miyata, 2000).

When ratios of the length of the pedicle, vitellarium and germarium were compared, they were 32:43:25 for the controls, 24:33:43 for the LD_{10} and 23:32:45 for the LD_{30}. There was a significant reduction when the %ratios of the LD_{10} and LD_{30} were compared with the controls, but there was significant reduction between the LD_{10} and LD_{30} treatments (Table 4). When ovarian maturation was observed untreated females had mature ova with an occasional one or two being absorbed (solid ova) in the ovarioles. In the LD_{10}-treated females, the spacing in the ovarioles and the absorption of ova different from the control. In LD_{30}-treated females, besides the spacing and absorption , sometimes only immature ova (germarium) were found in the ovarioles (data is not presented) (Perveen and Miyata, 2000). Mature ova were not observed in the pupae during the 2nd day before adult emergence, but a few mature ova were found the day before adult emergence. The number of mature ova sharply increased until the 1st day after adult emergence, and then gradually increased until the 2nd day after adult emergence. The maximum number of mature ova 725±2.0 was found on the second day after adult emergence. On the same day, the number of mature eggs was significantly reduced ($P<0.0001$) in the LD_{10}- or LD_{30}-treated females as compared with the controls, but no significant reduction was observed ($P=0.0984$) between the LD_{10} and LD_{30} treatments. From the 2nd day to the 7th day after adult emergence, absorption of mature ova started gradually in the controls, and the LD_{10}- or LD_{30}-treated females. The pattern of maturation of ova in ovaries was similar in the controls, LD_{10} and LD_{30} treatments (Figure 4) (Perveen and Miyata, 2000).

The cell density, expressed as number of nuclei per mm^2, was determined at various days during sexual maturation in the germaria of the controls, LD_{10}- or LD_{30}-treated females (Table 4.3). In the controls, on the 2nd day of adult emergence, the density was 1636±9.17 nuclei mm$^{2-1}$. The cell density increased until the 3rd day, when it was 1829±8.87 nuclei mm$^{2-1}$ and decreased thereafter. On the 4th day of adult emergence, the cell density was 1323±56.20 nuclei mm$^{2-1}$. In the LD_{10}- or LD_{30}-treated insects, the patterns of the cell density change in the germarium were the same as in the controls but the values were significantly ($P<0.01$) lower in the LD_{10}-treated females and more were significantly decreased in the LD_{30}-treated females compared with the controls, during the days the 2nd, 3rd and 4th of adult female development. There was also a significant reduction (P<0.05) in the cell density between the LD_{10}- and LD_{30}-treated females during adult development (Table 5) (Perveen, 2011).

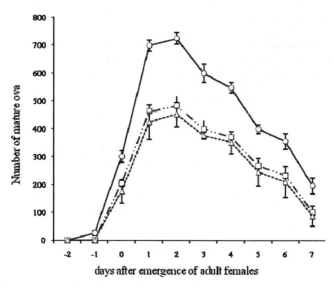

Fig. 4. Effect of sublethal doses of chlorfluazuron on number of mature eggs in the ovaries during different developmental days (post pupal and 1 to 7 day after adult emergence). For control ($\circ$; n = 13), LD_{10} (1.00 ng larva^{-1}) treated ($\square$; n = 11), and LD_{30} (3.75 ng larva^{-1}) treated ($\triangle$; n = 13) after topical application to newly molted fifth instars of *Spodoptera litura*. Data were analyzed using one-way ANOVA (Concepts, 1989) at *P*<0.0001 and Scheffe's *F*-test (Scheffe, 1953) at 5%. Vertical bars indicate SD; (Source: Perveen and Miyata, 2000).

Treatments[a]	n_1	n_2	Cell density number of nuclei $(mm^2)^{-1}$ in the germarium during female adults age (M±SD)[b,c]		
			2 day-old	3 day-old	4 day-old
Control	5	10	1636±9.17a	1829±8.87a	1323±56.20a
LD_{10}	5	10	1570±42.50b	1753±49.91b	1235±9.50b
LD_{30}	5	10	1489±8.60c	1644±7.68c	1089±61.42c

[a]LD_{10}: 1.00 ng larva^{-1}; LD_{30}: 3.75 ng larva^{-1}; n_1: number of insects used; n_2: number of ovariole measured
[b]The age of female adults was taken from the day of adult emergence.
[c]Data were analyzed using one-way ANOVA (Concepts, 1989) at P<0.01. Means within a column followed by different letters indicate significant differences according to Scheffe's F-test (Scheffe, 1953) at 5%.

Table 5. Effects of sublethal doses of chlorfluazuron on the cell density in the germarium after topical application to newly ecdysed fifth-instar larvae of *Spodoptera litura* (Source: Perveen, 2011).

3.2.2 Effects on oöcytes development

In the controls, the basal oöcytes were tiny on the 5th day after pupation, but increased sharply until the 8th day, after which they increased slowly until adult emergence. The maximum size of the basal oöcytes on the day of adult emergence was significantly reduced (P<0.0002) in the LD_{10}- or LD_{30}-treated females, but there was no significant difference (P=0.9976) between LD_{10}- and LD_{30}-treated females (Figure 5) (Perveen and Miyata, 2000).

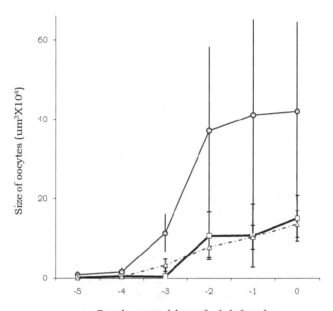

Developmental days of adult females

Fig. 5. Effect of sublethal doses of chlorfluazuron on size of basal oöcytes during different developmental days (5 to 9 day after pupation pupae and newly emerged adults). For control (○), LD_{10} (1.00 ng larva^{-1}) treated (□), and LD_{30} (3.75 ng larva^{-1}) treated (△) after topical application to newly ecdysed 5th instars of *Spodoptera litura*. Data were analyzed using one-way ANOVA (Concepts, 1989) at P<0.0002 and Scheffe's F-test (Scheffe, 1953) at 5%. Verticle bars indicate SD (n = 9-10); (source: Perveen and Miyata, 2000).

The thickness of the follicular epithelium of the basal oöcytes gradually increased and reached a maximum on the 8th day after pupation, after which it sharply declined. On the 8th day after pupation, it was significantly reduced (P<0.005) in the LD_{10}- or LD_{30}-treated females when compared with control females, but there was no significant difference (P=0.8686) between the LD_{10} or LD_{30} treatments. The reduction was approximately 34% in the LD_{10}- and 39% in the LD_{30}-treated females. The patterns of the development of the follicular epithelium of basal oöcytes were not similar in the LD_{10}- or LD_{30}-treated females compared with the controls. However, pattern was similar between the LD_{10}- and LD_{30}-treated females. In the LD_{10}- or LD_{30}-treated females, the follicular epithelium reached maximum on the 9th day after pupation, and then declined. The development of the follicular epithelium was delayed by one day in the LD_{10}- or LD_{30}-treated females compared with the controls. On the 9th day, it was thicker in the LD_{30}-treated females and thickest in

LD$_{10}$-treated females than in control females, but significant differences were not observed (P=0.7611) among these three groups, i.e. controls, LD$_{10}$- and LD$_{30}$-treated females (Figure 6) (Perveen and Miyata, 2000).

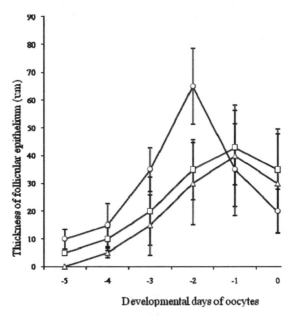

Developmental days of oocytes

Fig. 6. Effect of sublethal doses of chlorfluazuron on thickness of follicular epithelium of basal oöcytes during different developmental days (5 to 9 days after pupation and newly emerged adults). For control (o), LD$_{10}$ (1.00 ng larva^{-1}) treated (□), and LD$_{30}$ (3.75 ng larva^{-1}) treated (Δ) after topical application to newly ecdysed 5th instar larvae of *Spodoptera litura*. Data were analyzed using one-way ANOVA (Concepts, 1989) at P<0.005 and Scheffe's F-test (Scheffe, 1953) at 5%. Vertical bars indicate SD (n = 9-10); (Source: Perveen and Miyata, 2000).

3.3 Discussion

Topical application of sublethal doses of chlorfluazuron (LD$_{10}$: 1.00 ng per larva or LD$_{30}$: 3.75 ng per larva) on newly ecdysed fifth instars had an effect on the reproduction of *S. litura* by reducing fecundity, fertility, and hatchability (Perveen, 2000a). Thus, this study was conducted to establish the causes of the reduction in these parameters. It was found that topical appication of sublethal doses of chlorfuazuron had an affect on ovarian development and oögenesis by decreasing the weight of ovaries during postpupal and adult developmental days of LD$_{10}$ or LD$_{30}$ treated females. The basic factors responsible for the reduction in ovarian weight were reduction in the length of different parts of the ovarioles, decrease in the number of mature ova, reduction in the size of basal oöcytes and thickness of their follicular epithelium, and reduction in protein content of ovarian constituents as compared with the controls. In this study, topical application of sublethal doses of chlorfuazuron signifcantly reduced the ovarian weight to that of the controls in postpupal and adult developmental days (Figure 2; Table 4.3). The ratios of fresh ovarian/fresh body weight, dry ovarian/fresh ovarian weight, and dry ovarian/fresh body weight were the same among the newly emerged control, LD_{10} or LD_{30} treated females, indicating a reduction in the body, fresh, and dry ovarian weights with the same degree of reduction in treated females and controls. However, Soltani and Pickens and DeMilo (1977) reported that 0.5 mg DFB, when topically applied at pupal ecdysis to Cydia pomonella (L.), did not cause the ovarian weight be significantly reduced ($P<0.05$) between control and treated newly emerged adult females. Nor did Hatakoshi (1992) observe any signifcant reduction ($P<0.05$) in the ovarian weight between control and treated last day of pupae to third day after adult emergence of S. litura, when 0.3 ng pyriproxyfen was topically applied at pupal ecdysis. The differences in results observed in these experiments may be related to the pesticides, the kind of insect used and their developmental stages. Spodoptera litura has paired ovaries that branch into four polytrophic meroistic ovarioles located on the ventral side of the body cavity, making several loops of ovarioles, with all basal oöcytes developing simultaneously each ovariole is differentiated into tree portions according to the developmental stages of the oöcytes: (1) the yellowish green pedicle, where fully matured ova are stored; (2) the reddish orange vitellarium, which contains the developing oöcyte and trophocyte follicles which undergo accumulation of yolk proteins, and choriogenesis; and (3) the whitis germarium, which contains oögoia, from which germ cells proliferate and follicles are formed. Similar observations were reported by Riakhel and Dhadialla (1992) and Etman and Hooper (1979), which were confirmed here.

As in other Lepidopterous species, the ovaries of S. litura start to differentiate, and develop at the pupal stage. Indeed, in controls, the thickness of follicular epithelium of basal oöcyte reached its maximum size on the eighth day after pupation. This coincided with the start of follicular epithelium resorption. Histological examination on S. litura showed that topical application of sublethal doses of chlorfuazuron to newly ecdysed ffth-instar larvae affected growth and development of oöcytes during pupal and adult stages by affecting size and thickness of follicular epithelium (Figures 5 and 6). However, in C. pomonella, (in controls) the basal oöcytes reached their maximum size 7 day after pupation. In this insect, this coincided with the start of follicular epithelium resorption. Hence, a 0.5-mg dose of DFB applied topically to newly ecdysed pupae affected the growth and development of oöcytes by causing a decrease in both the thickness of the follicular epithelium and size of basal oöcytes during the pupal development (Soltani and Mazouni, 1992). Lim and Lee (1982) reported that 2-d-old adult females of O. japonica, starved for 6 h and consumed 500 mg (AI) of DFB with two maize discs. The females were found to have retarded ovarian development, caused by a delay of oöcytes development, and an increased percentage of oöcytes resorption. This caused a decrease in fecundity and egg viability of the females. However, significant reduction was not observed either in the number of ovarioles or in the length of basal oöcytes in treated insects. Differences in these results might be a result of the use of different BPUs. Also, the doses used by Lim and Lee (1982) were very high compared with those used in present study or in the Soltani and Mazouni (1992) experiments.

In newly emerged treated adults, the germarium was much longer than the pedicle and vitelarium as compared with the controls in which the vitelarium was longer than the germarium and pedicle (Table 4). This shows that maturation of oöcytes was delayed in treated adult females as compared with the controls. The maximum thickness of the follicular epithelium of basal ooctes was observed on the 9 day after pupation in treated females, whereas it was on the 8 d after pupation in the controls (Figure 6). Subsequent to the 8th or 9th day, resorption of follicular epithelium started in control and treated females, respectively. When ovarian maturation was scored, as depicted in Figure 4.4, a maximum number of matured oöcytes were found in the second day after adult emergence in the controls. From this day, resorption of mature oöcytes started. The chlorfuazuron-treated females showed the same pattern of mature oöcyte resorption up to the seventh day after adult emergence as in the controls. However, Hatakoshi (1992) reported that when 0-day pupae of female S. litura were topically treated with pyriproxyen (0.3 ng per pupa), few or no mature oöcytes were found in newly emerged females, but controls had mature oöcytes with one occasionally being resorbed. The maturation of insect eggs dependent, among other factors, on the materials taken up from the surrounding hemolymph (Telfer et al., 1981), and by materials synthesized by the ovary in situ (Indrasith et al.,1988). These materials include proteins, lipids, and carbohydrates, all of which are required for the embryogenesis (Kunkel and Nordin 1985, Kanost et al., 1990). Difubenzuron also caued a decrease in ovarian protein content in C. pomonella (Soltani and Mazouni, 1992). Decrease in the ovarian protein content suggests an interference of BPUs with vitellogenesis. It has been reported that DFB could affect ecdysteroid secretion from other organs, such as the epidermis, in T. molitor (Soltani, 1984), ovaries in C. pomonella (Soltani et al., 1989a; 1989b), and the concentration of hemolymph constituents in T. molitor (Soltani, 1990). Future studies should clarify the biochemical mechanism. Moreover, this work dose not clarify why signiÞcant differences were not observed ($P<0.0001$) between effects of LD_{10} (1.00 ng larva^{-1}) and LD_{30} (3.75 ng larva^{-1}) treated females, although LD_{30} dose was much higher than LD_{10} dose. Further studies are needed to obtain more knowledge about the effects of chlorfuazuron on oögenesis. Currently, the biochemical mechanism involved has been explored.

3.4 Conclusion

Sublethal doses of chlorfiuazuron (LD_{10}: 1.00 ng larva^{-1} or LD_{30}: 3.75 ng larva^{-1}) topically applied on newly ecdysed fifth instars of S. litura significantly reduced ovarian weight and number of mature eggs in pupae and adults, compared with those of the controls. The ratios of fresh ovarian/fresh body weight, dry ovarian/fresh ovarian weight, and dry ovarian/fresh body weight were the same among controls, LD_{10}, and LD_{30} treated newly emerged adults. In treated adults, the germarium was significantly longer than the pedicle and vitelarium compared with those of the controls, whereas in controls the vitelarium was significantly longer than the germarium and pedicle. This indicates a delayed of maturation of ovarioles in treated cutworms. These doses also disrupt growth and development of oöcytes by significantly affecting the size of basal oöcytes and thickness of follicular epithelium. The maximum size of basal oöcytes recorded on the day of adult emergence was significantly reduced in LD_{10} or LD_{30} treated females, compared with those of the controls. The thickness of the follicular epithelium of basal oöcytes reached to a maximum in the controls on the 8th day and in treated females on the ninth day after pupation. The effects of chlorfiuazuron on ovarian development and oögenesis are presumed to be responsible for the reduction in fecundity caused by sublethal exposure to chlorfiuazuron.

4. Effects of sublethal doses of chlorfiuazurn on male reproductive system of Spodoptera litura

The deep yellow-coloured testes of S. litura are distinctly paired in larvae and they appear as a single round organ in adults. The testes of S. litura resemble those of other lepidopterans, being enclosed in a common membrane called the scrotum. The testes lie dorsally and appear to be held in place by trachea and strands of basement membrane-like material (Amaldoss, 1989). Although reports of several layers surround the testes of lepidopteran have been made, a single capsule and follicular layers are present in *S. litura* (Amaldoss, 1989). Chase and Gilliland (1972) described the *tunica externa* and *interna* as nothing more than basement membranes over the capsule and follicular layers. The intra-follicular layer is divided into eight incomplete compartments in the tobacco leafminer, *Phthorimaea operculella* (Zeller). This layer is similar to that in the larger canna leafroller, *Calpodes ethlius* (Stoll), and tobacco moth, *Ephestia elutella* (Hübner). It also bears the pigments responsible for the bright yellow-coloured testes. It is clear that spermatogenesis persists in the adult testis.

A characteristic feature of the testicular follicles is the presence of large cells or a nucleated mass of protoplasm in the apex of the germarium. This is known as an apical cell or versonian cell. This is the region where there are successive stages of development of the germ cells occur. The upper part contains the primary spermatogonia and is known as the germarium. This is followed by a region called the zone of growth. The region or zone of growth is where spermatogonia multiply and usually become encysted. The maturation zone, where maturation takes place follows. Finally, is the zone of transformation where the spermatocytes develop into spermatids (spermiogenasis) completing spermatogenesis (Amaldoss, 1989). Two distinct types of spermatozoa are produced in the Lepidoptera: eupyrene (nucleated) spermatozoa which can fertilize the egg; and apyrene (anucleated) which are smaller and completely lacking in nuclear material, and do not appear to play any role in activation of the eggs (Doncaster, 1911; Goldschmidt, 1916). The eupyrene sperm can easily be counted in the male tract because they remain in bundles until they are transferred during mating, but the apyrene sperm are dispersed shortly after they leave the testis. Like eupyrene sperm, the apyrene sperm are produced in large numbers, usually contributing over half the sperm complement, and are transferred to the females with the eupyrene sperm during mating. It was thought that the apyrene sperm did not appear to play any role in activation of the eggs (Friedlander and Gitay, 1972). Their function has remained unclear ever since their discovery by Meves (1902), although several hypotheses concerning the function of the apyrene sperm have been proposed (Silberglied et al., 1984). Holt and North (1970 b) proposed that, in the cabbage looper, *Trichoplusia ni* (Hüebner), apyrene sperm might aid the transport of eupyrene sperm from the male reproductive tract to the female reproductive tract. Katsuno (1977 a) reported that the apyrene sperm in the *B. mori*, might facilitate the migration of eupyrene sperm through the cellular barrier, separating the testis from the efferent ducts. Gage and Cook (1994) reported that nutritional stress seriously affected the number and size of eupyrene and apyrene sperm production in the Indian meal moth, *Plodia interpunctella* (Hübner). However, sperm development in Lepidoptera takes place in the larvae (Munson, 1906; Machida, 1929; Garbini and Imberski, 1977). Usually, spermatogenesis starts in the late larval instars and proceeds on a schedule well correlated with the insect's metamorphosis. Studies *in vitro* and *in vivo* indicated that high titre of juvenile hormone inhibits spermatogenesis, and that sperm mitosis and meiosis require sufficient ecdysteroid titre. During the post-embryonic development of eupyrene and apyrene sperm bundles, when the insect is going to pupation, the juvenile hormone titre declines (Leviatan and Friedlander, 1979). Other factors have also been reported to promote spermatogenesis *in vitro* and *in vivo* (Dumser, 1980 a). In adult males *S. litura*, both apyrene and eupyrene sperm appear in bundles in the testis, but in the vas deferens only the eupyrene sperm were still in bundles, as reported for *B. mori* (Katsuno 1977 b), *T. ni* (Holt and North, 1970 a) and the army worm, *Pseudaletia separata* (Walk.) (He, 1994). The testes of early larvae contain a large number of spermatogonial cells near the outer border of the follicles. There is a preponderance of spermatocytes containing primary spermatocytes during the penultimate and early last-instar larvae. Secondary spermatocytes are present in the early last-instar larvae, persisting through to the middle of the last-instar larvae. They then begin to differentiate into spermatids. In the process of elongation and maturation of spermatids, the spermatocytes assume an elliptical shape. The sperm bundles formed as a result of maturation of the spermatids are seen abundantly in adults. Spermiogenesis is, however, not synchronous, and spermatozoa in various stages of differentiation can be detected in the testes of freshly emerged adults (Sridevi et al., 1989a). In the pupa, the apyrene sperm bundles emerge from the testicular follicle into the vas efferens earlier than the eupyrene sperm bundles and the bundles separate when they pass through the basement membrane of the testis (Katsuno, 1977b). The apyrene spermatozoa migrate from the vas efferens into the seminal vesicle through the vas deferens during the pupa (Katsuno, 1977c). In the post-pupal period, however, the eupyrene sperm bundles and apyrene spermatozoa migrate simultaneously through the same way (Katsuno, 1977d). The effects of chlorfluazuron have been examined on male reproductive system during testicular development and spermatogenesis when sublethal doses have been topically applied to newly ecdysed fifth-instar larvae of *S. litura*.

4.1 Experimental procedure
4.1.1 Histology of testis
Testes from newly molted sixth instar larvae to 5-day-old virgin adult males (treated and control), were dissected in 0.9% of NaCl under a binocular microscope. The length and width of each testis were measured by the same procedure as used for the oöcytes measurement. Testis volume was calculated for larval testes using the formula $4/3\pi$ (length×width^2), assuming that the testis is a prolate spheroid (Loeb et al., 1984). For the fused pupal and adult testes, the formula $4/3\pi r^3$ was used, with r as the radius of the globular gonad. The treated and control weight and sheaths thickness of testes were measured by the same procedure used for the ovaries as described above (Perveen, 2000b).

The thickness of treated and control testes sheaths or vas deferenti of untreated and treated relevant stages of insect was observed by making a parafilm microtomy conducted according to the method used by Yoshida (1994) and the procedure to stain the nuclei of sperm was adapted from the method by He (1994) (Perveen, 2000b).

4.1.2 Spermatogenesis
A staining method was used for determining number of the cysts, eupyrene and apyrene sperm for treated and control (He et al., 1995). First, the testis was transferred to a microscopic grid slide (each square= 1mm^2) and crushed until it was evenly distributed on the slide. Secondly, several drops of methanol-acetic acid solution (3 : 1; v/v) were added to the slide to fix the reparation for 15 min, and the excess fixing solution was absorbed with filter paper. Third, several drops of 2-5% Giemsa solution dissolved in Sorensen-Gomori buffer solution (monobasic and dibasic sodium phosphate, 0.07 M, pH 6.8) were added to the slide to stain the preparation for 10-30 min. The slide was washed with water and air dried after staining. Finally, the air-dried preparation was observed for counting of bundles and cysts under a phase contrast microscope at 20×mignification. Cysts were classified into the following six developmental stages as described by Chaudhury and Raun (1966): (1)

spermatogonia; (2) primary spermatocytes; (3) secondary spermatocytes; (4) spermatids; (5) elongated cysts with maturing sperm and (6) bundles with fully matured sperm. The length and width of sperm bundles were measured with a calibrated ocular micrometer a phase contrast microscope at 400×magnification (Perveen, 2000b).

4.1.3 Data analysis
Data were analyzed using analysis of variance, one way ANOVA (Concepts, 1989) at $P<0.0001$ and Scheffe's F-test (Scheffe, 1953) at 5%.

4.2 Results
The structural morphogenesis was seen in sixth-instar larvae of *S. litura* during development. When fifth-instar larvae ecdysed into newly sixth-instar larvae (N: 0 day), larvae remained unchanged for upto 2 hour. Then, they changed to a slender surface during 1st day (S). After that, they changed to being puffy during 2nd day (early-last-instar stage: P). Then, they changed to digging stage during 3rd day (mid-last-instar stage: D). Then, they changed to early burrow during 4th day (pre-late-last-instar stage: B$_1$). After that, it changed to late burrow during 5th day (post-late-last-instar stage: B$_2$). The morphogenesis for phase variations used for convenient of observations during experiments is given in Table 6. The morphology of the adult male reproductive system of *S. litura* is shown in Figure 7 (Perveen, 2000b).

Developmental category (days)	Symbols for phase variations[a]	Name of the phases[a]	Duration (hour day^{-1})
0	N	newly ecdysed	0–2
1	S	slender surface	1st
2	P	puffy (early last instar stage)	2nd
3	D	digging (mid last instar stage)	3rd
4	B$_1$	early burrowed (pre late last instar stage)	4th
5	B$_2$	late burrowed (post late last instar stage)	5th

[a]Symbols for phase variations observed during developmental days of sixth-instar larvae were used for the convenience of observations.

Table 6. Structural morphogenesis during five developmental days of sixth-instar larvae of *Spodoptera litura* (Source: Perveen, 2005).

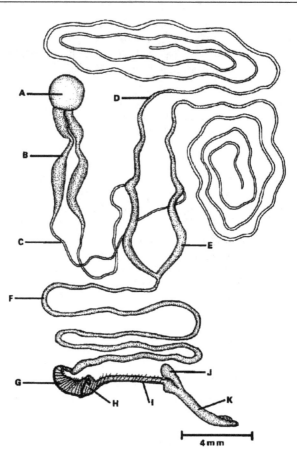

Fig. 7. The morphology of the male reproductive system of *Spodoptera litura*: A: testis; B: seminal vesicle (paired); C: vas deferens (paired); D: accessory glands (paired); E: ductus ejaculatorious duplex; F: primary segment of ductus ejaculatorious simplex; G: muascular area; H: area of frenum formation; and I: area of collum formation of the cuticular secondary segment of the ductus ejaculatorious simplex; J: caecum of aedeagus; K: aedeagus (Source: Etman and Hooper, 1979).

4.2.1 Effects on testicular development

The testes of *S. litura* show the three dimentional measureable structure. Each mature testis consists of four follicles or lobes, each separated by an inner layer of sheath cells. An outer sheath cell layer further surrounds all follicles. In sixth-instar larvae, the volume and weight of the testes gradually then rather sharply increased until the 4th day of sixth-instar larvae. The volume weight decreased when two larval testes fused on the 5th day (last day) after moulting of sixth-instar larvae. They again sharply increased in size, reached a maximum (6.21±1.31 mm^3 and 11.94±0.42 mg, n=30, respectively) on the 0 day of pupation and gradually declined until the 5th day after adult emergence. Sublethal doses of chlorfluazuron rapidly disrupted the development of testes by decreasing the volume and weight of testes compared with the controls. The weight and size of testes were significantly reduced (P<0.001) in the LD_{10}-treated and more significantly reduced (P<0.0001) in LD_{30}-treated males compared with the controls from newly ecdysed sixth-instar larvae to the 5th day after adult emergence. Testes reached their maximum size in treated males (LD_{10}: 4.16±1.54 mm³ and 8.0±0.83 mg; LD_{30}: 2.79±1.00 mm³ and 4.88±1.05 mg; n= 10, respectively) on the same day as the controls. The patterns of development of the testes with respect to the volume and weight were the similar in the controls and the LD_{10}- or LD_{30}-treated males (Figures 8a and b) (Perveen, 2000b).

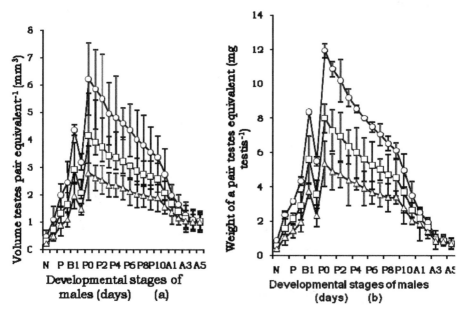

Fig. 8. Effects of sublethal doses of chlorfluazuron (LD_{10}: 1.00 ng larva⁻¹; LD_{30}: 3.75 ng larva⁻¹) on the testis volume (a) and weight (b) of *Spodoptera litura* during newly ecdysed sixth-instar larvae to 5th day after adult emergence; controls: O; LD_{10}: □; LD_{30}: Δ; data analyzed using one-way ANOVA (Concepts, 1989) at P<0.0001 and Scheffe`s F-test (Scheffe, 1953) at 5%; vertical bars: SD; N, S, P and D: larval (Table 5.1), P: pupal and A: adult developmental days; n = 10 for each point; paired larval testes and fused single pupal or adult testis were considered as testes pair equivalent; (Source: Perveen, 2000b).

The thickness of the testis sheath gradually and then sharply increased until the 4th day of moulting of sixth-instar larvae. It decreased when the two larval testes fused on the 5th day (last-day) of the sixth-instar larvae; it again increased and reached to a maximum [(6±0.51)×10⁻² mm; n=10] on the 0 day of pupation and gradually declined in the newly emerged adults. The thickness remained constant until the 2nd day after adult emergence. Sublethal doses rapidly disrupted the development of testis by significantly decreasing (P<0.0001) the thickness of the testes sheath as compared with that of the controls. This reduction occurred from newly ecdysed sixth-instar larvae to the 0 day of pupation in the LD_{10}-treated and the 1st day after pupation in the LD_{30}-treated males. The thickness of the testes sheath in chlorfluazuron-treated males reached a maximum [$(5.9\pm0.67)\times10^{-2}$ mm] in the LD_{10}-treated males on the 1st day and [$(5.9\pm1.1)\times10^{-2}$ mm] in the LD_{30}-treated males on the 2nd day after pupation whereas, in the controls, it was on the 0 day of pupation. This result shows that attainment of the maximum thickness of the testes sheath was delayed by one day in LD_{10}- and by two days in LD_{30}-treated males compared with the controls. However, no significant reduction was observed in the maximum thickness of the testes sheath among the control and LD_{10}- or LD_{30}-treated males. The developmental pattern of the testes sheath in the LD_{10}- or LD_{30}-treated males was similar to that of the controls (Figure 9) (Perveen, 2000b).

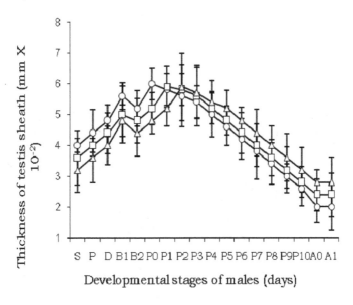

Fig. 9. Effects of sublethal doses of chlorfluazuron (LD_{10}: 1.00 ng larva^{-1}; LD_{30}: 3.75 ng larva^{-1}) on the thickness of testis sheath of *Spodoptera litura* during newly ecdysed sixth-instar larvae to 5th day after adult emergence; controls: O; LD_{10}: □; LD_{30}: Δ; data analyzed using one-way ANOVA (Concepts, 1989) at $P<0.0001$ and Scheffe's F-test (Scheffe, 1953) at 5%; vertical bars: SD; S, P, D and B_1: larval (Table 5.1), P: pupal and A: adult developmental days; n = 10 for each point; paired larval testes and fused single pupal or adult testis were considered as a testes pair equivalent; (Source: Perveen, 2000b).

4.2.2 Effects on spermatogenesis

When the different developmental stages of cysts were observed in testes during spermatogenesis on the 1, 3 and 5 day-old sixth-instar larvae, the number of spermatogonia, primary and secondary spermatocytes was significantly reduced ($P<0.001$) in the LD_{10}- and even more significantly reduced ($P<0.0001$) in the LD_{30}-treated males as compared with those of controls. Spermatids, elongated cysts with mature sperm and eupyrene sperm bundles were not found in the controls and LD_{10}- or LD_{30}-treated larval testes (Table 7) (Perveen, 2000b).

TS[a]	T[a]	n[a]	*Sg[b] (M±SD)	*PS[b] (M±SD)	*SS[b] (M±SD)	St (M±SD)	ECMS (M±SD)	ESB (M±SD)
1st	C	13	2447±18a	4893±35a	1397±18a	0.0±0.0	0.0±0.0	0.0±0.0
	LD_{10}	11	1960±7b	3920±7b	1120±6b	0.0±0.0	0.0±0.0	0.0±0.0
	LD_{30}	10	1320±6c	2641±5b	754±4c	0.0±0.0	0.0±0.0	0.0±0.0
3rd	C	13	1311±8a	4280±9a	3146±8a	0.0±0.0	0.0±0.0	0.0±0.0
	LD_{10}	11	1050±5b	3430±7b	2520±7b	0.0±0.0	0.0±0.0	0.0±0.0
	LD_{30}	10	707±5c	2311±7c	1689±7c	0.0±0.0	0.0±0.0	0.0±0.0
5th	C	13	437±8a	3574±5a	4717±7a	0.0±0.0	0.0±0.0	0.0±0.0
	LD_{10}	11	342±5b	2870±7b	3780±6b	0.0±0.0	0.0±0.0	0.0±0.0
	LD_{30}	10	224±6c	1933±8c	2546±9c	0.0±0.0	0.0±0.0	0.0±0.0

[a]TS: treatment stages; T: treatments; n: number of males used; C: control; LD_{10}: 1.00 ng larva^{-1}; LD_{30}: 3.75 ng larva^{-1}; Sg: spermatogonia; PS: primary spermatocytes; SS: secondary spermatocytes; St: spermatids; ECMS: elongated cysts with mature sperm; ESB: eupyrene sperm bundles; larvae ecdysed usually during 0200 to 0800 hour and collected between 0800 to 1000 hour
[b]Data were analyzed using one-way ANOVA (Concepts, 1989) at $P<0.0001$. Means within columns followed by different letters are significantly different by Scheffe`s F-test (Scheffe, 1953) at 5%.

Table 7. Effect of sublethal doses of chlorfluazuron on spermatogenesis in the testes of sixth-instar larvae during 1st, 3rd and 5th days of development after topical application to newly ecdysed fifth-instar larvae of *Spodoptera litura* (Source: Perveen, 2000b).

When different developmental stages of the cysts were observed in testis during spermatogenesis in newly ecdysed pupae, on the 5th and 10th day after pupation, the number of spermatogonia, primary and secondary spermatocytes, were decreased. However, the spermatids and elongated cysts with mature sperm gradually increased in controls. Eupyrene sperm bundles were not present on the 0-5 day-old pupae. However, they were found on the 10th day after pupation (mean: 1002±3.0 numbers). The pattern of spermatogenesis was the same in the controls, LD_{10}- and LD_{30}-treated male pupae. However, the developmental stages of the sperm were significantly reduced ($P<0.001$) in the LD_{10}-treated and even more significantly reduced ($P<0.0001$) in the LD_{30}-treated males compared with the controls (Table 8) (Perveen, 2000b).

Different developmental stages of cysts were observed in testes during spermatogenesis in newly emerged, 1 and 2 day-old adults. The spermatogonia were not present in newly emerged and 1 day-old adults. Primary spermatocytes were not found in 1 and 2 day-old adults, but they (mean: 174±4.0 number) were found in newly emerged adults. The secondary spermatocytes, spermatids and elongated cysts with mature sperm were present, but gradually decreased in number in the controls. Eupyrene sperm bundles gradually increased in number in the controls. The pattern of spermatogenesis was the same in the controls, LD_{10}- and LD_{30}-treated male pupae. However, the stages of sperm development were significantly reduced ($P<0.001$) in the LD_{10}- and even more significantly reduced ($P<0.0001$) in LD_{30}-treated males as compared with the controls (Table 9) (Perveen, 2000b).

TS[a]	T[a]	n[a]	Sg[b] (M±SD)	PS[b] (M±SD)	SS[b] (M±SD)	St (M±SD)	ECMS (M±SD)	ESB (M±SD)
1st	C	13	262±6a[c]	1747±5a	2710±7a	3670±6a	349±8a	0.0±0.0
	LD_{10}	11	205±6b	1400±4b	2170±6b	2940±7b	280±5b	0.0±0.0
	LD_{30}	10	142±5c	943±6b	1462±5c	1980±4c	189± 5c	0.0±0.0
3rd	C	13	175±7a	1047±8a	2010±4a	4805±5a	697±11a	0.0±0.0
	LD_{10}	11	140±4b	840±5b	1610±3b	3805±5b	560±5b	0.0±0.0
	LD_{30}	10	94±5c	566±6c	1085±3c	2593±7c	377±4c	0.0±0.0
5th	C	13	86±4a	436±5a	698±3a	2969±6a	3581±3a	0.0±0.0
	LD_{10}	11	70±3b	350±4b	560±6b	2660±4b	3339±3b	0.0±0.0
	LD_{30}	10	47±4c	236±3c	377±5c	1792±3c	2263±4c	0.0±0.0

[a]TS: treatment stages; T: treatments; n: number of males used; C: control; LD_{10}: 1.00 ng larva[-1]; LD_{30}: 3.75 ng larva[-1]; Sg: spermatogonia; PS: primary spermatocytes; SS: secondary spermatocytes; St: spermatids; ECMS: elongated cysts with mature sperm; ESB: eupyrene sperm bundles; pupation occurred usually during 0200 to 0800 hour and collected between 0800 to 1000 hour.
[b]Data were analyzed using one-way ANOVA (Concepts, 1989) at $P<0.0001$. Means within columns followed by different letters are significantly different by Scheffe's F-test (Scheffe, 1953) at 5%.

Table 8. Effect of sublethal doses of chlorfluazuron on spermatogenesis in the testes of pupae during 1st, 5th and 10th days of development after topical application to newly ecdysed fifth-instar larvae of *Spodoptera litura* (Source: Perveen, 2000b).

TS[a]	T[a]	n[a]	Sg[b] (M±SD)	PS[b] (M±SD)	SS[b] (M±SD)	St (M±SD)	ECMS (M±SD)	ESB (M± SD)
0	C	13	0±0a	174±4a	524±3a	1048±4a	2097±5a	4893±3a
	LD_{10}	11	35±3b	140±7b	385±3b	840±3b	1680±6b	3920±4b
	LD_{30}	10	24±2c	94±3b	259±2c	566±6c	1132±3c	2641±2c
1st	C	13	0±0a	0±0a	174±6a	611±3a	1047±6a	6902±5a
	LD_{10}	11	14±3b	28±3b	98±1b	490±5b	840±3b	5530±4b
	LD_{30}	10	9±2c	19±2c	66±2c	330±4c	566±2c	3728±5c
2nd	C	13	0.0±0.0	0.0±0.0	43±3a	219±4a	437±6a	8037±6a
	LD_{10}	11	0.0±0.0	0.0±0.0	35±3b	175±3b	350±5b	6440±5b
	LD_{30}	10	0.0±0.0	0.0±0.0	24±2c	118±4c	236±4c	4338±4c

[a]TS: treatment stages; T: treatments; n: number of males used; C: control; LD_{10}: 1.00 ng larva[-1]; LD_{30}: 3.75 ng larva[-1]; M: mean; Sg.: spermatogonia; PS: primary spermatocytes; SS: secondary spermatocytes; St.: spermatids; ECMS: elongated cysts with mature sperm; ESB: eupyrene sperm bundles; adults emerged usually between 2300 to 0200 hour and and between collected 0800 to 1000 hour.
[b]Data were analyzed using 1-way ANOVA (Concepts, 1989) at $P<0.0001$. Means within columns followed by different letters are significantly different by Scheffe's F-test (Scheffe, 1953) at 5%.

Table 9. Effect of sublethal doses of chlorfluazuron on spermatogenesis in the testes of adults during 0, 1st and 2nd day of development after topical application to newly ecdysed fifth-instar larvae of *Spodoptera litura* (Source: Perveen, 2000b).

In the testis of newly emerged LD_{10}- treated adults, the number of eupyrene and apyrene sperm bundles was significantly reduced ($P<0.001$), and even more significantly reduced

(P<0.0001) in LD$_{30}$- treated adults compared with the controls. When males were treated with the LD$_{10}$ or LD$_{30}$, the ratio of eupyrene to apyrene sperm bundles was not significantly changed; apyrene sperm bundle comprised about half of the total sperm complement (Table 10).

T[a]	n[a]	NESB[b] (M±SD)	NASB[b] (M±SD)	Ratios (%)= ESB:ASB
Control	30	4893±546a	4697±520a	51.1:48.9
LD$_{10}$	30	3920±426b	3763±466b	51.0:48.9
LD$_{30}$	30	2641±161c	2386±271c	52.5:47.5

[a]LD$_{10}$: 1.00 ng larva^{-1}; LD$_{30}$: 3.75 ng larva^{-1}; T: treatments; n: number of males used; NESB: number of eupyrene sperm bundle; NASB: number of apyrene sperm bundle; ESB: eupyrene sperm bundle; ASB: apyrene sperm bundle
[b]Data were analyzed using one-way ANOVA (Concepts, 1989) at P<0.0001. Means within columns followed by different letters are significantly different by Scheffe`s F-test (Scheffe, 1953) at 5%.

Table 10. Effect of sublethal doses of chlorfluazuron and comparison of the number of eupyrene and apyrene sperm bundles in the testis of newly emerged unmated adults after topical application to newly ecdysed fifth-instar larvae of *Spodoptera litura* (Source: Perveen, 2000b).

Developmental stages of sperm	T[a]	n[a]	Size (dm in µm)	
			(M±SD)[b]	ranges (min–max)
Spermatogonia (S)	Control	10	7.1±0.7a	6–8
	LD$_{10}$	10	5.9±0.8b	5–7
	LD$_{30}$	10	4.1±0.9c	3–5
Primary spermatocytes (D)	Control	10	15.1±0.7a	14–16
	LD$_{10}$	10	13.0±1.3b	11–14
	LD$_{30}$	10	11.4±1.5c	10–48
Secondry spermatocyte (B$_2$)	Control	10	31.0±1.8a	28–33
	LD$_{10}$	10	29.2±1.6b	27–31
	LD$_{30}$	10	25.0±1.2c	24–27
Spermatids (P$_{10}$)	Control	10	4.1±0.9a	3–5
	LD$_{10}$	10	2.9±0.87b	2–4
	LD$_{30}$	10	1.9±0.7c	1–3

[a]LD$_{10}$: 1.00 ng larva^{-1}; LD$_{30}$: 3.75 ng larva^{-1}; T: treatments; n: number of males used; for S, D and B$_2$ refer to Table 5.1; P$_{10}$: ten day old pupae
[b]Data were analyzed using one-way ANOVA (Concepts, 1989) at $P<0.0001$. Means within columns followed by different letters are significantly different by Scheffe`s F-test (Scheffe, 1953) at 5%.

Table 11. Effect of sublethal doses of chlorfluazuron on the size of various developmental stages of sperm observed in the testes of sixth-instar larvae and puape after topical application to newly ecdysed fifth-instar larvae of *Spodoptera litura* (Source: Perveen, 2000b).

The size of the spermatogonia, primary spermatocytes, secondary spermatocytes and spermatids was significantly reduced (P<0.001) in LD_{10}-treated and more significantly reduced (P<0.0001) in LD_{30}-treated insects compared with the controls (Table 11) in newly emerged adults (Perveen, 2000b).

In newly emerged adults, the width and length of elongated cysts with mature sperm, eupyrene and apyrene sperm bundles were significantly reduced (P<0.001) in LD_{10}- and more significantly reduced (P<0.0001) in LD_{30}- treated insects compared with the controls (Table 12) (Perveen, 2000b).

Developmental stages of sperm	T^a	n^a	sizes of various developmental stages			
			length (μm)[b]		width (μm)[b]	
			M±SD	Ranges (min–max)	M±SD	Ranges (min–max)
Elongated spermatocytes	Control	10	45.5±2.4a	(42–50)	40.0±2.0a	(35–42)
	LD_{10}	10	42.2±1.7b	(40–44)	37.2±1.3b	(35–39)
	LD_{30}	10	40.0±0.8c	(39–41)	35.2±1.6c	(34–38)
Eupyrene sperm bundles	Control	10	98.0±4.4a	(92–105)	33.2±1.9a	(30–36)
	LD_{10}	10	94.7±1.4b	(93–97)	31.3±1.4b	(29–33)
	LD_{30}	10	91.5±1.0c	(90–93)	29.1±1.2c	(28–31)
Apyrene sperm bundles	Control	10	24.9±1.2a	(23–27)	16.1±1.7a	(14–19)
	LD_{10}	10	94.7±1.4b	(21–25)	14.2±1.3b	(12–16)
	LD_{30}	10	20.3±0.9c	(19–22)	12.2±1.2c	(11–14)

[a]LD_{10}: 1.00 ng larva[-1]; LD_{30}: 3.75 ng larva[-1]; T: treatments; n: number of males used
[b]Data were analyzed using one-way ANOVA (Concepts, 1989) at P<0.0001. Means within columns followed by different letters are significantly different by Scheffe`s F-test (Scheffe, 1953) at 5%.

Table 12. Effect of sublethal doses of chlorfluazuron on size of various developmental stages of sperm observed in the testes of newly emerged adults after topical application to newly ecdysed fifth-instar larvae of *Spodoptera litura* (Source: Perveen, 2000b).

T^a	n^a	Eupyrene sperm bundles in vas deferens		
		Pre-adult[b] (M±SD)	Newly emerged adult[b] (M±SD)	One day old adult[b] (M±SD)
Control	30	102±29a	1002±116a	2513±407a
LD_{10}	30	0b	23±4.9b	1621±159b
LD_{30}	30	0b	0c	1080±75c

[a]LD_{10}: 1.00 ng larva[-1]; LD_{30}: 3.75 ng larva[-1]; T: treatments; n: number of males used
[b]Data were analyzed using one-way ANOVA (Concepts, 1989) at P<0.0001. Means within columns followed by different letters are significantly different by Scheffe`s F-test (Scheffe, 1953) at 5%.

Table 13. Effect of sublethal doses of chlorfluazuron on the number of eupyrene sperm bundles in the vas deferens during different developmental days of adults after topical application to newly ecdysed fifth-instar larvae of *Spodoptera litura* (Source: Perveen, 2000b)

In the vas deferens of male pre-adult controls, the mean number of eupyrene sperm bundles was 102 ± 29, but no sperm bundles observed in the LD_{10}- or LD_{30}-treated males of the same age (Table 13) (Perveen, 2000b).

In newly emerged control males, the mean number of eupyrene sperm bundles was 1002 ± 116, and in LD_{10}-treated adult males, 23 ± 4.9. In LD_{30}-treated adults male there was no sperm bundles were observed. Moreover, in 1 day-old LD_{10}-treated adult males, the number of eupyrene sperm bundles was significantly ($P<0.001$) reduced and more significantly ($P<0.0001$) reduced in LD_{30}-treated males compared with the controls (Table 14) (Perveen, 2000b).

In the testis and vas deferens of newly emerged, LD_{10}-treated males, the total number of eupyrene sperm bundles was significantly reduced ($P<0.001$) and more significantly reduced ($P<0.0001$) in LD_{30}-treated males had no sperm bundles in the vas deferens compared with the controls (Table 14) (Perveen, 2000b).

T^a	n^a	NESB in testis[a,b] (M±SD)	NASB in vasa deferens[a,b] (M±SD)	TNESB (testis+ vasa deferens)[a,b] (M±SD)
Control	30	4893±546a	1002±116a	5791±640a
LD_{10}	30	3920±426b	23±4.9b	3943±425b
LD_{30}	30	2641±161c	0±0c	2641±161c

[a]LD_{10}: 1.00 ng larva[-1]; LD_{30}: 3.75 ng larva[-1]; T: treatments; n: number of males used; NESB: number of eupyrene sperm bundle; NASB: number of apyrene sperm bundle; ESB: eupyrene sperm bundle; ASB: apyrene sperm bundle; TNESB: total number of eupyrene sperm bundle
[b]Data were analyzed using one-way ANOVA (Concepts, 1989) at $P<0.0001$. Means within columns followed by different letters are significantly different by Scheffe`s F-test (Scheffe, 1953) at 5%.

Table 14. Effect of sublethal doses of chlorfluazuron on the total number of eupyrene sperm bundles in the testis and vas deferens of newly emerged adults after topical application to newly ecdysed fifth-instar larvae of *Spodoptera litura* (Source: Perveen, 2000b)

4.3 Discussion

Topical application of sublethal doses of chlorfuazuron (LD_{10}: 1.00 ng larva[-1]; LD_{30}: 3.75 ng larva[-1]) has an effect on reproduction of *S. litura* by reducing the fecundity, fertilityand hatchability. Fecundity was reduced to a similar degree (35±44%) when females, males or both sexes were treated with LD_{10} or LD_{30}. Fertility was reduced by 42% or 52% when females were treated with LD_{10} or LD_{30}, respectively, and by 60% or 63%, respectively, when males or both sexes were treated with LD_{10}. Fertility was reduced by 78% or 80% when males or both sexes were treated with LD_{30}. The hatchability was reduced by 20% or 23% when females were treated with LD_{10} or LD_{30}, respectively, and by 37% or 39%, respectively, when males or both sexes were treated with LD_{30}. Hatchability was reduced by 55% or 56% when males or both sexes were treated with LD_{30} (Perveen, 2000a). Thus, this study was conducted to determine the causes of the reduction in these reproductive parameters. Topical application of sublethal doses of chlorfuazuron has an effect on testis development by decreasing the volume and weight of testes and its sheath thickness, and also on spermatogenesis by decreasing the number of cysts, eupyrene and apyrene bundles during different times in development. Reduction in the testes volume and weight might be caused by a reduction in the number of cysts, number and size of sperm bundles, thickness of testes sheath, and/or the reduction in protein content of testes constituents. When females are treated with LD_{10} of chlorfuazuron, no signifiant reduction in the inseminated eupyrene sperm number is observed compared with controls. However, LD_{10} and LD_{30} treatment of males significantly reduces (65.8% and 88.6%) the number of inseminated eupyrene sperm. Moreover, no significant (P<0.0001) differences in the reduction are observed on the inseminated eupyrene sperm number when males or both sexes are treated either with LD_{10} or LD_{30} (Perveen, 2008). Therefore, the main cause of the reduction in the fecundity, fertility and hatchability is a decrease in inseminated eupyrene sperm numbers. In larval *S. litura*, the bright yellow-coloured testes are distinctly paired, reniform and situated between the 5th and 6th abdominal segments. Each of the lateral testicular lobes is made up of four follicles. The testicular lobes are enclosed within two thick, double-layered peritoneal sheaths. Each of these sheaths comprises two layers of epithelial cells. The external sheath is made up of lightly stained cuboidal cells resting on a basement membrane. This forms a common envelope to all the follicles of the testis and is penetrated by tracheal branches. The inner sheath is made up of more darkly stained elliptical cells, within which muscle fibres are distinguisable. Ingrowths from this sheath in the form of double-walled septa penetrate between the follicles to separate them from one another. The testes sheath reaches its maximum thickness at day 0 of pupation in the controls, whereas in LD_{10}- treated males maximum thickness occurs later at 1 day. In LD_{30}-treated males maximum thickness occurs at 2 day after pupation. Difubenzuron affects ecdysteroid secretion from the epidermis in Tenebrio molitor (Soltani, 1984), ovaries of *Cydia pomonella* (Soltani et al., 1987; Soltani et al., 1989b) and also the concentrations of haemolymph constituents in *T. molitor* (Soltani, 1992). Ecdysteroids have been reported to stimulate spermatogenesis in many insect species (Dumser, 1980b; Gelman and Hayes, 1982). In Mamestra brassica (Shimizu et al., 1985), *Heliothis virescens* (Loeb, 1986) and *L. dispar* (Loeb et al., 1988) testes synthesize ecdysteroids. The treatment with chlorfuazuron effects ecdysteroid production by the testis sheet remains to be investigated. Gelman and Hayes (1982) and Gelman et al. (1988) observed in *Ostrinia nubilalis*, the size and weight of separate testes paired. Topical application of the sublethal dose, LD_{10} of chlorfuazuron significantly reduces (P<0.0001) the weight and size of testes, and this is even greater in LD_{30}-treated males. However, the topical application of similar sublethal doses of chlorfuazuron significantly reduced (P<0.0001) ovarian weight in post-pupal and developing pharate adult females as compared with that of the controls (Perveen and Miyata, 2000). However, significant differences are not observed in ovarian weight when adult females are treated with LD_{10} or LD_{30} doses. The maturation of insect testes depends, among other factors, upon the materials that are taken up from the surrounding haemolymph and by materials synthesized by the testes *in situ*. These materials include protein, lipid and carbohydrate, all of which are required for development of the genital tract (Kunkel and Nordin, 1985; Kanost et al., 1990). In newly emerged males from LD_{10}- or LD_{30}-treated larvae the eupyrene and apyrene sperm bundles are significantly (P<0.0001) smaller in size and number compared with those of controls. Spermatozoa descend regularly from the testis through the vas deferens into the seminal vesicles, which fill with eupyrene sperm bundles with cysts and individual apyrene sperm. Yoshida (1994) also reported that the number of eupyrene sperm bundles in the testis and vas deferens of newly emerged treated (LD_{10}) males is reduced by 36%, and the initiation of sperm movement from testis to seminal vesicle was delayed. The present results show that initiation of sperm movement from testis to vas deferens is delayed after chlorofuazuron treatment and this is caused by the delay of spermatogenesis. When males or both sexes are treated on the first day of pairing, few or none mate. However, seven to nine pairs mate in control or female treated crosses. During the next day, seven to 10 pairs mate in 10±13 pairs of seven combinations of crosses (Perveen, 2008). These results suggest that the delay of the first mating is caused by the delayed spermatogenesis. The effect of chlorfuazuron on testicular development and spermatogenesis is one of the factors responsible for the reduction in fecundity, fertility and hatchability caused by the sublethal doses of chlorfuazuron. More work is in progress to determine the biochemical mechanism of these effects in *S. litura*.

4.4 Conclusion
The physiological mechanism of action of chlorfluazuron describes here on testicular development and spermatogenesis when sublethal doses (LD_{10}: 1.00 ng larva^{-1} or LD_{30}: 3.75 ng larva^{-1}) are applied topically to the cuticle of newly ecdysed fifth instars of *Spodoptera litura*. These doses disrupt the growth and development of testes by decreasing the volume and weight of testes and thickness of testes sheath as compared with that of the controls. Additionally, such doses disrupt spermatogenesis by reducing the number and size of eupyrene and apyrene sperm bundles in the testis. Very few or no eupyrene sperm bundles are observed in vas deferens of pre- and newly ecdysed adults compared with controls. This result shows that the transfer of sperm bundles from testes to vas deferens is delayed in treated males. The effects of chlorfluazuron on testicular development and spermatogenesis are thought to be one of the factors responsible for the reduction in fecundity, fertility and hatchability caused by sublethal doses of chlorfluazuron.

5. References

Abro, G. H., Memon, A. J. and Syed, T. S. (1997). Sublethal effects of cyhalothrin and fluvalinate on biology of *Spodoptera litura* F. Pakistan J. Zool. 29: 181–184.

Anderson, V. L. and McLean, R. L. (1974). Design of experiments: a realistic approach. Dekker, New York. p. 70.

Amaldoss, G. (1989). Ultrastructure and physiology of the ductus ejaculatorius duplex and seminal vesicle of male *Spodoptera litura* (Fabricius) (Lepidoptera: Noctuidae). Proc. Indian Acad. Sci. (Anim. Sci.) 98 (6): 405–418.

Anonymous, (1969). Summary of panel meeting in "Sterile male technique for eradication or control of harmful insect". Intl. Atom. Ener. Com. Vienna. pp. 80

Baloch, S. H. and Abbasi, Q. D. (1977). Investigation of the ecology and biology of the cutworm (Noctuidae: Lepidoptera) in the Hyderabad region. First Annu. Res. Report, Pakistan Sci. Found. Islamabad. p. 80.

Barbosa, P. (1974). Dissecting fluids and saline solutions. In: "Manual of basic techniques in insect' histology". (Mass, A. D. ed.), Oxford Publisher, London. p. 243.

Bargers, A. and Raw, F. (1967). Soil biology. Academic Press. New York. pp. 53. Beeman, R. W. (1982). Advances in mode of action of insecticides. Ann. Rev. Entomol. 27: 253–281.

Benz, G. (1969). Influence of mating, insemination and other factors on oÖgenesis and oviposition in the moth *Zeiraphera diniana*. J. Ins. Physiol. 15: 55–71.

Binnington, K. and Retnakaran, A. (1991). Epidermis: A biologically Active target for metabolic inhibitors. *In*: "Physiology of the insect epidermis" (Binnington, K. and Retnakaran, A. eds.) CSIRO Publications, Australia. p. 307–334.

Brushwein, J. R. and Granett, J. (1977). Laboratory tests of two moulting disruptants, El-494 and dimilin, against the spruce budworm (Lepidoptera: Tortricidae): mortality and reproductive effects. Me. Life Sci. Agr. Exp. St. Mis. Report. p. 192.

Chase, J. A. and Gilliland, F. R. J. (1972). Testicular development in Thet tobacco budworm. Ann. Entomol. Soc. Amer. 65: 901– 906.

Chatani, F. and Ohnishi, E. (1976). Effects of ecdysone on theovarian development of *Bombyx* silkworm. Develop. Gro. Diff. 18. 48–484.

Chaudhury, M. F. B. and Raun, A. S. (1966). Spermatogenesis And testicular development of the European corn borer, *Ostrinia nubilalis* (Lepidoptera: Pyraustidae). Ann. Entomol. Soc. Amer. 59: 1157–1159.

Clark, L., Temple, G. H. R. and Vincent, J. F. V. (1977). The effects of a chitin inhibitor, dimilin on the reproduction of the peritrophic membrane in the locust, *Locusta migratoria*. J. Insect Physiol. 24: 241–246.

Concepts, A. (1989). Super ANOVA. Abacus Concepts, Berkeley, CA. pp. 120.

Crystal, M. M. (1978). Diflubenzuron-induced decrease of egg hatch of screw worms (Diptera: Calliphoridae). J. Med. Entomol. 15: 52–56.

Deb, D. C. and Chakaravorty, S. (1981). Effects of a juvenoid on the growth and differentiation of the ovary of *Corcyra cephalonica* (Lepidoptera). J. Insect Physiol. 27: 103–111.

Doncaster, L. (1911). Some stages in the spermatogenesis of Abraxas *grossulariata* and its variety *lacticolor*. J. Gent. 1: 179–184.

Dumser, J. B. (1980 a). The regulation of spermatogenesis in insects. Ann. Rev. Entomol. 25: 341–369.

Dumser, J. B. (1980 b). *In vitro* effects of ecdysterone on the spermatogonial cell in *Locusta*. Intl. J. Invert. Rep. 2: 165–174.

Emam, A. K., El-Refal, S. A. and Degheele, D. (1988). Effect of Sublethal dosages of four chitin synthesis inhibitors on the reproduction potential and F-1 generation of the Egyptian cotton leafworm *Spodoptera litoralis* (boisd.) (Lepidoptera: Noctuidae). Med. Van de Facland. Rij. Gent. 53: 249–254.

Engelman, F. (1979). Insect vitellogenin: identification, bio-synthesis, and role in vitellogenesis. Adv. Insect Physiol. 27: 49–108.

Etman, A. A. M. and Hooper, G. H. S. (1979). Developmental and reproductive biology of *Spodoptera litura* (F.) (Lepidoptera: Noctuidae). J. Aus. Entomol. Soc. 18: 363–372.

Friedlander, M. and Gitay, H. (1972). The fate of the normal-a-nucleated spermatozoa in inseminated females of the silkworm, Bombyx mori. J. Morphol. 138: 121–129.

Fytizas, E. (1976). L'action du TH6040 sur la metamorphose de *Dacus oleae* Gemel. (Diptera, Trypetidae). *Z. Angew. Entomol. 31*: 440–444. (Abstract in English read).

Gage, M. J. G. and Cook, P. A. (1994). Sperm size or number? Effects of nutritional stress upon eupyrene and apyrene sperm production strategies in the moth, *Plodia interpunctella* (Lepidoptera: Pyralidae). Func. Ecol. 8: 594–599.

Ganiev, S. G. (1986). Residual amount of dimilin in the plantand soil of paistaccio stands in SW Kirgizia. *Lesn. Khoz. 10*: 63–70.

Gelman, D. B. and Hayes, D. K. (1982). Methods and markers for synchronizing the maturation of fifth-stage diapause-bound and non diapause-bound larvae, pupae and pharate adults of the European cornborer, *Ostrinia nubilalis* (Lepidoptera: Pyralidae). Ann. Entomol. Soc. Amer. 75: 485–493.

Gelman, D. B., Woods, C. W. and Borvokovec, A. B. (1988). Ecdysteroid profile for haemolymph and testes from larvae, pupae, and pharate adults of the European cornborer, *Ostrinia nubilalis*. Arch. Ins. Biochem. Physiol. 7: 267–279.

Goldschmidt, R. (1916). The function of the apyrene spermatozoa. Science 44: 544–546.

Gupta, H. C., Bareth, S. S. and Sharma, S. K. (1991). Bioefficacy of edible and non-edible oils against pulse, *Callosobrochus chinensis* L. on stored pulses and their effect on germination. Agri. Biol. Res. 7: 101–107.

Haga, T., Toki, T., Tsujii, Y. and Nishiyama, R. (1992). Development of an insect growth regulator, chlorfluazuron. J. Pestic. Sci.17: S103–S113.

Hajjar, N. P. and Casida, J. E. (1979). Structure-activity relationships of benzoyl phenyl urea as toxicants and chitin synthesis inhibitors in *Oncopeltus fasciatus*. Pestic. Biochem. Physiol. 11: 33–45.

Hashizume, B. (1988). Atabron® 5E, a new IGR insecticide (chlorfluazuron). Jpn. Pestic. Inform. 58: 32–34.

Hatakoshi, M. (1992). An inhibitory mechanism over oviposition in the tobacco cutworm, *Spodoptera litura* by juvenile hormone analogue pyriproxyfen. J. Insect. Physiol. 38: 793–801.

Hatakoshi, M. and Hirano, M. (1990). Effects of S-71639 on the reproduction of *Spodoptera litura*. Adv. Invert. Rep. 5: 429–434.

Haines, L. C. (1981). Changes in colour of a secretion in the reproductive tract of adult males of *Spodoptera littoralis* (Boisduval) (Lepidoptera: Notuidae) with age and mated status. Bull. Entomol. Res. 71: 501–598.

Haynes, K. F. (1988). Sublethal effect of neurotoxic insecticides on insect behaviour. *Ann. Rev. Entomol. 33*: 149–168.

He, Y. (1994). Reproductive strategies in the armyworm, *Pseudaletia separata* (Lepidoptera: Noctuidae): with special reference to rearing density. Ph. D. Thesis. Graduate School of Agricultural Sciences, Nagoya University, Japan. pp. 122.

He, Y., Tanaka, T. and Miyata, T. (1995). Eupyrene and apyrene sperm and their numerical fluctuations inside the female reproductive tract of the armyworm, *Pseudaletia separata*. J. Insect. Physiol. 41: 689–694.

Holst, H. Z. (1974). Die fertility ätsbeeinflussende Wirkung des neuen insektizids P D 60-40 bei *Epilachna varivestis* Muls (Col. occinellidae) und *Leptinotarsa decemlineata* Say. (Col. Chrysomelidae). Z. Pflkranzk. Pflschutz. 81: 1–7. (Abstract in English read)

Holst, H. Z. (1974). Die fertility ätsbeeinflussende Wirkung des neuen insektizids P D 60-40 bei *Epilachna varivestis* Muls (Col. occinellidae) und *Leptinotarsa decemlineata* Say. (Col. Chrysomelidae). Z. Pflkranzk. Pflschutz. 82: 1–7. (Abstract in English read)

Holt, G. G. and North, D. T. (1970a). Spermatogenesis in the cabbage looper, *Trichoplusia ni*. (Lepidoptera: Noctuidae). Ann. Entomol. Soc. Amer. 63: 501–507.

Holt, G. G. and North, D. T. (1970b). Effects of gamma Irradiation on the mechanisms of sperm transfer in *Trichoplusia ni*. J. Insect. Physiol. 16: 2211–2222.

Hunter, E. and Vincent, J. F. V. (1974). The effects of a novel insecticide on insect cuticle. Experientia 4: 484–490.

Indrasith, L. S., Sasaki, T., Yaginuma, T. and Yamashita, O. (1988). The occurrence of a premature form of egg-specific protein in vitellogenic follicles of Bombyx mori. J. Comp. Physiol. 158: 1–7.

Jakob, W. L. (1973). Development inhibition of mosquitoes and house fly by urea analogues. J. Med. Entomol. 10: 540–543.

James, E. W. (1974). Laboratory and field evaluation of TH-6040, against House fly and stable fly. J. Econ. Entomol. 67: 746–747.

Kanost, M. R., Kawooya, J. K., Law, J. H., Ryan, R. O., Heusden, V. M. C. and Ziegler, R. (1990). Insect haemolymph proteins. In: "Advances Insect physiology" (Evans, P. D. and Wigglesworth, V.B. eds.), Academic Press, London. v. 22, p. 299–396.

Karlinsky, A. (1963). Effects de l'ablation des corpora allata imaginaux sur le developement ovarien de Pieris brassicae L. (Lepidoptera). C. R. Acad. Sci. Paris. 256: 4101–1103. (Abstract in English read)

Kasuga, H., Osani, M., Yonezawa, Y. and Aigaki, T. (1985). An energy supply device for spermatozoa in the spermatophore of the silkworm. Develop. Gro. Diff. 27: 514–522.

Katsuno, S. (1977a). Studies on eupyrene and apyrene spermatozoa in the silkworm Bombyx mori L. (Lepidoptera: Bombycidae). I. The intra testicular behaviour of the spermatozoa at various stages from the 5th instar to the adult. Appl. Entomol. Zool. 12: 142–147.

Katsuno, S. (1977 b). Studies on eupyrene and apyrene spermatozoa in the silkworm, Bombyx mori L. (Lepidoptera: Bombycidae). II. The intra testicular behaviour of the spermatozoa after emergence. Appl. Entomol. Zool. 12: 148–153.

Katsuno, S. (1977c). Studies on eupyrene and apyrene spermatozoa in the silkworm, Bombyx mori L. (Lepidoptera: Bombycidae). III. The post testicular behaviour of spermatozoa at various stages from the pupa to adult. Appl. Entomol. Zool. 12: 142–247.

Katsuno, S. (1977d). Studies on eupyrene and apyrene spermatozoa in the silkworm Bombyx mori (Lepidoptera: Bombycidae). IV. The behaviour of the spermatozoa in the internal reproductive organs of female moths. Appl. Entomol. Zool. 12: 352–359.

Ker, F. (1977). Investigation of locust cuticle using the insecticide diflubenzuron. J. Insect Physiol. 23: 39–48.

Khan, M. F. and Naqvi, S. N. H. (1988). Toxicological studies of diflubenzuron and nicotine cyanurate on house flies larvae. Pakisrtan J. Entomol. 3: 95–106.

Khuhro, R. D., Abbasi, Q. D., Abro, G. H., Baloch, S. H. and Soomor, A. H. (1986). Population dynamics of Spodoptera litura (F.) as influenced by ecological conditions and its host plant preference at Tandojam. Pakisrtan J. Entomol. 18: 351–357.

Kunkel, J. G. and Nordin, J. H. (1985). Yolk proteins. In: "Comprehensive Insect Biochemitry, Physiology and Pharmacology" (Kerkut, G. A. and Gilbert, L. I. eds.), Pergamon Press, Oxford. v. 1, p. 83–111.

Leuschner, K. (1974). Effects of the new insecticide PDD 60-40 on nymphs and adults of the cabbage bug, Eurydema oleraceum L. (Heteroptera, Pentatomidae). Z. Pflkrankh. Pflschutz. 81: 8–12.

Leviatan, R. and Friedlander, M. (1979). The eupyrene-apyrene dichotomous spermatogenesis of Lepidoptera. I. The relationship with post embryonic development and the role of the decline in juvenile hormone titre towards pupation. Develop. Biol. 68: 515-525.

Lim, S. J. and Lee, S. S. (1982). Toxicity of diflubenzuron to the grass hopper, Oxy japonica; effect on reproduction. Entomol. Exp. Appl. 31: 154-158.

Loeb, M. J. (1986). Ecdysteroids in testis sheaths of Heliothis Virescens larvae: An immunological study. Arch. Insect Biochem. Physiol. 3: 173-180.

Loeb, M., Brand, E. P. and Birnbaum, M. J. (1984). Ecdysteroid production by testes of budworm, Heliothis virescens, from last-larval instar to adult. J. Insect Physiol. 30: 375-381.

Loeb, M. J., Brandt, E. P., Woods, C. W. and Bell, R. A. (1988). Secretion of ecdysteroids by sheaths of testis of the gypsy moth, Lymantria dispar, and its regulation by testis ecdysiotropin. J. Exp. Zool. 248: 94-100.

Machida, J. (1929). Feine experimentelle untersuchung über die apyrene spermatozoen des seidenspinners Bombyx mori L. Zur. Beitru. Wissen. 9: 466-510. (Abstract in English read)

Madore, C. D., Drion, G. B. and John, B. D. (1983). Reduction of reproductive potential in spruce budworm (Lepidoptera: Tortricidae) by a chitin-inhibiting insect growth regulator. J. Econ. Entomol. 76: 708-710.

Matsuura, H. and Naito, A. (1997). Studies on the cold-hardiness and overwintering of Spodoptera litura F. (Lepidoptera: Noctuidae) VI. possible overwintering areas predicted from meteorological data in Japan. Appl. Entomol. Zool. 32: 167-177.

Matthee, J. J. (1951). The structure and physiology of the egg of Locustana pardalina (Walk.). Bull. Develop. Agri. For. Un. S. Afr. no 316: p. 83.

Mellanby, K. (1976). Pesticides, the environment and the balance of nature.In: "Pesticides and Human Welfare" (Gunn, D. I. and Steven, J. G. R. eds) Oxford University Press, p. 217-227.

Meves, F. (1902). Über oligpyrene und apyrene spermien und über ihre entstehung, nach beobachtugen an Paludina und Pygaera. Arch. Mikro. Anatom. 61: 1-84. (Abstract in English read)

Minitab, Inc. (1979). Statistics software. Release 12. Minitab Inc., Stat College, USA. p. 16801.

Mitsuhashi, J. (1995). A continuous cell line from pupal ovaries of the common cutworm, Spodoptera litura (Lepidoptera: Noctuidae). Appl. Entomol. Zool. 30: 75-82.

Munson, J. P. (1906). Spermatogenesis of the butterfly, Papilio rutulus. Proc. Boston Soc. Natu. His. 33: 43-124.

Naqvi, S. N. H. and Rub, A. (1985). Effect of dimilin on the activity of melanization enzyme of Musca domestica L. (PCSIR strain). J. Sci. Karachi Univ. 13: 171-177.

Nickle, D. A. (1979). Insect growth regulators: New protectants against the almond moth in stored inshell peanuts. J. Econ. Entomol. 72: 816-819.

Nijahout, M. M. and Riddiford, L. M. (1974). The control of egg maturation by juvenile hormone in the tobacco hornworm, Manduca sexta. Biol. Bull. 146: 377-392.

Nizam, S. (1993). Effects of allochemicals against 3rd instar larvae of Musca domestica L. (Malir strain). Ph. D. Thesis, Department of Zoology, University of Karachi. pp. 250.

Nomura, M. (1994). Physiological effects of pyriproxyfen against the common cutworm, *Spodoptera litura*. M. Sc. Thesis, Laboratory of Applied Entomology, Nagoya University. pp. 125.

Olszak, R. W., Pawlik, B. and Zajac, R. Z. Z. (1994). The influence of some insect growth regulators on mortality and fecundity of the aphidophagous coccinellids, *Adalia bipunctata* L. and *Coccinella septempunctata* L. (Col., Coccinellidae). J. Appl. Entomol. 117: 58–63.

Ottens, R. J. and Todd, J. W. (1979). Effect of diflubenzuron on reproduction and development of *Graphognathus peregrinus* and *G. leucoloma*. J. Econ. Entomol. 72: 743–746.

Pan, M. L. (1977). Juvenile hormone and vitellogenin synthesis in the *Cecropia* silk moth. Biol. Bull. 153: 336–345.

Perveen, F. (2011) Effects of Sublethal Doses of Chlorfluazuron on Ovarioles in the Common Cutworm, *Spodoptera litura* (F.) (Lepidoptera: Noctuidae). Journal of Life Science, USA (Accepted).

Perveen, F. (2000a). Sublethal effects of chlorfluazuron on reproductivity and viability of *Spodoptera litura* (F.) (Lep., Noctuidae). J. Appl. Entomol. 124: 223–231.

Perveen, F. (2000b). Effects of sublethal dose of chlorfluazuron on testicular development and spermatogesis in the common cutworm, *Spodoptera litura*. J. Physiol. Entomol. 25: 315–323.

Perveen, F. (2005). Effects of chlorfluazuron on the growth of common cutworm, *Spodoptera litura* (Lepidoptera: Noctuidae) with special reference to the morphology and histology of the reproductive system. Ph.D. Thesis, Department of Zoology, University of Karachi.

Perveen, F. (2008). Effects of sublethal doses of chlorfluazuron on the number of inseminated sperm in *Spodoptera litura* (Lepidoptera: Nocutidae). J. Entomol. Sci. 11: 111–121.

Perveen, F. and Miyata, T. (2000). Effects of sublethal dose of chlorfluazuron on ovarian development and oögenesis in the common cutworm, *Spodoptera litura* (F.) Lepidoptera: Noctuidae). Ann. Entomol. Soc. Ame. 93: 1131–1137.

Qureshi, R. A., Qadri, S., Anwarullah, M. and Naqvi, S. N. H. (1983). Effect of neopesticides (JHA) on the morphology, emergence and fertility of *Musca domestica* L. (PCSIR strain) and its relation to phosphatases. Zool. Ang. Entomol. 95: 304–309.

Riakhel, A. S. and Dhadialla, T. S. (1992). Accumulation of yolk proteins in insect oöcytes. Ann. Rev. Entomol. 37: 217–251.

Ross, D. C. and Brown, T. M. (1982). Inhibition of larval growth in *Spodoptera frugipera* by sublethal dietary concentrations of insecticides. J. Agr. Food Chem. 30: 193–196.

Scheffe, H. (1953). A method for judging all contrasts in the analysis of variance. Biometrika 40: 87–104.

Shimizu, T., Moribayashi, A. and Agui, N. (1985). *In vitro* analysis of spermatogenesis and testicular ecdysteroids in the cabbage armyworm, *Mamestra brassicae* L. (Lepidoptera: Noctuidae). Appl. Entomol. Zool. 20: 56–61.

Silberglied, R. E., Shepherd, J. G. and Dickinson, J. L. (1984). Eunuchs: the role of apyrene sperm in Lepidoptera? Amer. Natur. 123: 255–265.

Skibbe, J. T., Challaghan, P. T., Eccles, C. D. and Laing, W. A. (1995).Visulization of pH in the larval midgut of *Spodoptera litura* using ^{31}P-NMR Microscopy. J. Insect Physiol. 42: 777-790.

Soltani, N. (1984). Effects of ingested diflubenzuron on the longevity and peritrophic membrane of adult mealworms (*Tenebrio molitor* L.). Pestic. Sci. 15: 221-225.

Soltani, N., Quennedey, A., Delbecque, J. P. and Delchambre, J. (1987). Diflubenzuron-induced alterations during *in vitro* development of *Tenebrio molitor* pupal integument. Arch. Insect Biochem. Physiol. 5: 201-209.

Soltani, N., Delachambre, J. and Delbecque, J. P. (1989a).

Stage-specific effects of diflubenzuron on the ecdysteroid titres during the developmental of *Tenebrio molitor*: Evidence for a change in hormonal source. Gen. Comp. Endocrinol. 76: 350-356.

Soltani, N., Mauchamp, B. and Delbeque, J. P. (1989b). Effects of diflubenzuron on the ecdysteroid titres in two insect species. Akad Landwirtsch Wiss. Tag. Ber., DDR, Berlin. 274: 171-177.

Soltani, N. and Mazouni, N. S. (1992). OÖgenesis in mealworms: cell density of germarium, thickness of chorion and ecdysteroid production. Effects of regulators. Med. Fac. Landbouww. Univ. Gent. 62: 565-571.

Sridevi, R., Gupta, A. D. and Ramamurty, P. S. (1989a). Spermatogenesis in *Spodoptera litura* (Lepidoptera: Noctuidae). Entomon 14: 01-09.

Sroka, P. and Gilbert, L. I. (1971). Studies on the endocrine control of post emergence ovarian maturation in *Manduca sexta*. J. Insect Physiol. 71: 2409-2419.

Taft, H. M. and Hopkins, R. A. (1975). Boll weevils: field populations controlled by sterilizing emerging over winter females with a TH-6040 sprayable bait. J. Econ. Entomol. 68: 551-554.

Tahir, S., Anwer, T. and Naqvi, S. N. H. (1992). Toxicity and Residual effects of novel pesticides against rice weevil, *Sitophilus oryzae* (L.) (Coleoptera: Curculionidae). Pakistan J. Zool. 24: 111-114.

Takahashi, S. and Mizohata, H. (1975). Critical period necessary for the imaginal moulting and egg maturation of the eri-silkworm, *Samia cynthia* Ricini. Zool. Mag. 84: 283-295.

Telfer, W. H., Rubenstein, E. and Pan, M. L. (1981). How the ovary makes yolk in *Hylophora*. In: "*Regulation of Insect Development and Behaviour*" (Wroclaw, M. K. ed.) Technical University Press, p. 637-654.

Walpole, R. E. and Myers, R. H. (1998). Probability, statistics for engineer scientist, 6th eds. Prentice-Hall, USA.

Williams, C. M. (1952). Physiology of insect diapause. IV. The brain and prothoracic glands as an endocrine system in the *Cecropia* silk. Biol. Bull. 103: 120-138.

Wright, J. E. and Spates, G. E. (1976). Reproductive inhibition activity of the insect growth regulator TH-6040 against the stable and house fly: Effect on hatchability. J. Econ. Entomol. 69: 365-368.

Wright, J. E. and Retnakaran, A. (1987). "Chitin and Bezoyl phenyl Ureas" (Junk, K. eds.) The Hague. v. 1, pp. 225.

Yoshida, K. (1994). Studies on the physiological effects of chlorfluazuron: with special reference to reproductive effects of chlorfluazuron on *Spodoptera litura*. M. Sc. Thesis, Graduate School of Agricultural Sciences, Nagoya University. p. 69.

Younis, M. (1973). Losses of cotton crop due to insect pests. *Proc. Crop. Produc. Sem. Organ. ESSO (Pak.) Ltd*. p. 110.

Yu, S. J. (1983). Age variation in insecticide susceptibility and detoxification capacity of fall armyworm (Lepidoptera: Noctuidae) larvae. J. Econ. Entomol. 76:

4

Organophosphorus Insecticides and Glucose Homeostasis

Apurva Kumar R. Joshi and P.S. Rajini

Food Protectants and Infestation Control Department, Central Food Technological Research Institute (CSIR lab), Mysore, India

1. Introduction

The modern world has heavily thrived on the revolution in agricultural practices that have culminated in tremendous boost in agricultural productivity. Pesticides are perhaps one of the most important and effective strategies of the green revolution. Pesticides are the only class of toxic substances that are intentionally released into the environment for achieving greater good, a decision that far outweighs their toxicological concerns. Organophosphorus insecticides (OPI) are one of the most extensively used classes of insecticides. Chemically they are derivatives of phosphoric (H_3PO_4), phosphorous (H_3PO_3) or phosphinic acid (H_3PO_2) (Abou-Donia, 2003). The OPI were initially introduced as replacements for the much persistent organochlorine insecticides (Galloway & Handy, 2003). With systemic, contact and fumigant action, OPI find use as pest control agents in various situations. OPI are extensively used in agricultural practices for protecting food and commercial crops from various types of insects. In addition, OPI are also used in household situations for mitigating menacing pest varieties. They are not very stable chemically or biochemically and are degraded in soil, sediments and in surface water. Perhaps, it is this instability of these agents that has lead to their widespread and indiscriminate use, which has exposed animal and human life to various forms of health hazard. The increase in their use has lead to wide range of ecotoxicological problems and exposure to OPI is believed to be major cause of morbidity and mortality in many countries.

Huge scientific body of evidence suggests that OPI exposure is a major toxicological threat that may affect human and animal health because of their various toxicities such as neurotoxicity, endocrine toxicity, immunotoxicity, reproductive toxicity, genotoxicity and ability to induce organ damage, alterations in cellular oxidative balance and disrupt glucose homeostasis. Indeed, the data on residue levels of OPI in various sources reported from India does create a huge cause for concern regarding their toxic effects. Samples of raw and bottled water were reported to be contaminated with various OPI residues, some of which were much higher than recommended levels (Mathur et al., 2003). Sanghi et al. (2003) have reported OPI residue levels in breast milk samples in India. Based on the levels of OPI residues, it has been speculated that infants may consume 4.1 times higher levels of malathion than the average daily intake levels recommended by the World Health Organisation. Similarly, human blood samples were reported to be contaminated with residues of monocrotophos, chlorpyrifos, malathion and phosphamidon (Mathur et al.,

2005). Thus, OPI present a realistic environmental threat that could affect various facets of human health.

2. Toxicity of organophosphorus insecticides

The toxicity of active OPI is attributed to their ability to inhibit acetylcholinesterase (AChE, choline hydrolase, EC 3.1.1.7), an enzyme that catalyses the hydrolysis of the neurotransmitter acetylcholine (ACh), leading to cholinergic stress as a result of stimulation of muscarinic and nicotinic ACh receptors (Fukuto, 1990; Sogorb & Vilanova, 2003; Abou-Donia, 2003). The inhibition of AChE by an OPI takes place via a chemical reaction in which the serine hydroxyl moiety (of the active site) is phosphorylated. The phosphorylated enzyme is highly stable and, depending on the groups attached to the central 'P' atom of the OPI molecule, may be irreversibly inhibited.

There are several factors that determine the toxicity of OPI. Important of these are route and levels of exposure, structure of the substance and its interaction with the biotransformation/detoxification system of the body. The metabolic fate of OPI is basically the same in insects and animals. Following absorption, the distribution of OPI is variable. Blood half-lives are usually short, although plasma levels are in some cases maintained for several days. OPI undergo extensive biotransformation, which is complex and involves several metabolic systems in different organs, with simultaneous oxidative biotransformation at a number of points in the molecule, utilizing the cytochrome P-450 isoenzyme system. Metabolism occurs principally by oxidation, hydrolysis by esterases, and by transfer of portions of the molecule to glutathione. Oxidation of OPI may result in more or less toxic products. Most mammals have more efficient hydrolytic enzymes than insects and, therefore, are often more efficient in their detoxification processes. Numerous conjugation reactions follow the primary metabolic processes, and elimination of the phosphorus-containing residue may be via the urine or faeces. Some bound residues remain in exposed animals. Binding seems to be to proteins, principally, since there are limited data showing that incorporation of residues into DNA (Eto, 1974).

2.1 Neurotoxicity

Based on structure-function relationships, OPI are essentially neurotoxicants. Most important of their neurotoxicities is their 'cholinergic toxicity', which is a consequence of acetylcholinesterase (AChE) inhibition by OPI leading to accumulation of ACh and cholinergic stress. Signs of cholinergic toxicity include miosis, muscle fasciculation, excessive glandular secretions, nausea and vomiting (Namba, 1971). In addition, OPI are known to exert two other forms of neurotoxicities- Organophosphorus ester-induced delayed neurotoxicity (OPIDN) and Organophosphorus ester-induced chronic neurotoxicity (OPICN). OPIDN is a neurodegenerative disorder characterized by delayed onset of prolonged ataxia and upper motor neuron spasticity as a result of single or multiple exposures. OPICN refers to other forms of neurotoxicity that is distinct from both cholinergic toxicity and OPIDN. OPICN is characterized by neuronal degeneration and subsequent neurobehavioral and neuropsychological consequences (Abou-Donia, 2003).

2.2 Oxidative stress

Numerous studies provide evidence for the propensity of OPI to disrupt oxidative balance leading to oxidative stress (Soltaninejad & Abdollahi, 2009). Increased lipid peroxidation, protein carbonylation, depletion of cellular antioxidant pools and alterations in enzymatic antioxidant status appear to be chief mechanisms of OPI-induced oxidative stress that often results in pathophysiological changes and organ damage. Several studies have demonstrated usefulness of antioxidant intervention in alleviating oxidative stress and pathophysiological changes induced by OPI (Kamath et al., 2008, Soltaninejad & Abdollahi, 2009). These studies lend unequivocal support to view that oxidative stress mediates as one of the chief mechanisms of OPI toxicity.

3. Organophosphorus insecticides and glucose homeostasis: mechanistic insights

In addition to neurotoxicity and oxidative stress, alterations in glucose homeostasis often culminating hyperglycemia is increasingly being reported as characteristic outcome of OPI toxicity. Meller et al., (1981) have described two cases of human subjects who were hospitalized with many complications including hyperglycemia. With no pseudocholinesterase detected, patients were given pralidoxime (AChE activator), which improved their condition and normalized hyperglycemia. Investigations revealed that they may have been exposed to malathion sprayed in their area. This case presents a classic case of hyperglycemic outcome following exposure to OPI as patients also exhibited miosis and muscle twitching. Numerous experiments have been conducted with experimental animals that reveal hyperglycemia as a characteristic outcome of OPI poisoning. A recent review by Rahimi & Abdollahi (2007) provides an exhaustive account of investigations revealing hyperglycemia in cases of OPI exposure.

There are certain characteristic features of alterations induced by OPI in glucose homeostasis. In cases of exposure to single dose of OPI, hyperglycemia appears to set in rapidly and peak changes are often followed by a trend of normalization. High dose of diazinon has been reported to cause hyperglycemia in mice that follows a trend of normalization (Seifert, 2001). Acute exposure of rats to malathion resulted in hyperglycemia with peak increase occurring at 2.2h after administration followed by decrease after 4h (Rodrigues et al., 1986). A similar case of reversible hyperglycemia has been reported by Lasram et al., (2008) following administration of a single dose of malathion to rats.

Biochemical changes associated with hyperglycemia serve as useful tools to understand etiology of OPI-induced hyperglycemia. Malathion has been reported to cause hyperglycemia in fasted rats. Interestingly, these hyperglycemic responses were not associated with hepatic glycogen depletion. The reversible phase of hyperglycemia was associated with increased glycogen deposition in liver, indicating that glucose may have come from gluconeogenesis (Gupta, 1974). Malathion induced hyperglycemia was associated with AChE inhibition in pancreas. More importantly, the trend of reversibility coincided with spontaneous reactivation of inhibited AChE (Lasram et al., 2008), indicating involvement of AChE-inhibition in hyperglycemia. Increase in blood glucose induced by sub chronic exposure of rats to malathion has been reported to be associated with increased glycogen phosphorylase and phosphoenolpyruvate carboxykinase activities, indicating involvement of both glycogenolytic and gluconeogenic processes. Increase in blood glucose levels induced by sub chronic exposure of rats to acephate has been reported to be associated with decrease in hepatic glycogen content (Deotare & Chakrabarti, 1981).

3.1 Pancreatic dysfunctions

Acute pancreatitis is also a well known complication of OP poisoning (Dressel et al., 1979; Frick et al., 1987; Hsiao et al., 1996), and epidemiological findings indicate that the incidence of pancreatitis is high in OPI intoxication based on various pathophysiological reports (Gokalp et al., 2005). The precise mechanisms underlying OPI-induced pancreatitis are still undefined, although it is believed to involve obstruction of pancreatic ducts and /or enhanced reactive oxygen species (Dressel et al., 1982; Sevillano et al., 2003, Sultatos, 1994). Involvement of oxidative stress following acute exposure to OPI has been reported recently (Banerjee et al., 2001) and it has been demonstrated unequivocally that lipid peroxidation is one of the molecular mechanisms involved in OPI-induced cytotoxicity (Akhgari et al., 2003; Ranjbar et al., 2002; Abdollahi et al., 2004b).

In view of the above, we attempted to understand the potential of repeated oral doses of dimethoate (DM) (at 20 and 40mg/kg b.w/d for 30days; doses corresponding to 1/20 and 1/10LD50 values) to cause alterations in glucose homeostasis and the associated biochemical alterations in pancreas of rats. We observed distinct signs of glucose intolerance among rats administered DM (**Fig. 1**) at time points at which un-treated rats showed normal glucose tolerance after an oral glucose load (3g/kg b.w.). We also observed that DM at both doses caused significant increase in blood glucose levels with concomitant inhibition of acetylcholinesterase activity and depletion of reduced glutathione contents in pancreas (**Table 1**) (Kamath & Rajini, 2007).

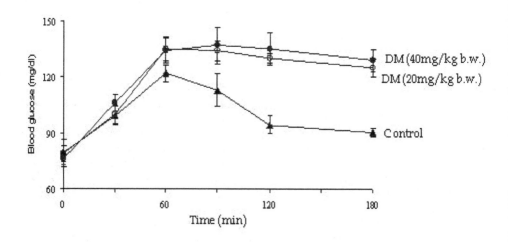

Fig. 1. Oral glucose tolerance at the end of 30 days in control (CTR) and Dimethoate (DM) treated rats.

Treatment	Blood glucose (mg/dl)		AChE (nmoles substrate hydrolyzed /min/mg protein)	GSH (mg/g tissue)
Dimethoate (mg/kg b.w.)	Initial	Final		
0	85.33 ± 3.85	91.33 ± 2.41	4.96 ± 1.47	1.11 ± 0.02
20	87.34 ± 5.23	105.28 ± 3.57[a]	2.94 ± 1.75	0.99 ± 0.05[a]
40	85.00 ± 5.30	138.67 ± 5.70[,b]	0.43 ± 0.21[a,b]	0.91 ± 0.07[a,b]

Values are mean ± SEM (n=6);
[a] Comparison of control and other groups;
[b] Comparison of DM (20mg /kg b.w.) group with DM (40 mg/kg b.w.) group

Table 1. Blood glucose, acetylcholinesterase (AChE) and reduced glutathione (GSH) levels in pancreas of rats administered oral doses of Dimethoate (DM) for 30 days.

Further, DM also caused significant pancreatic damage as reflected by increased amylase (2-3 folds) and lipase (20 & 38%) activities in serum (**Fig 2**). These changes were sharply paralleled by significant damage in pancreatic milieu. There was a dose-related elevation in ROS levels in pancreas of treated rats. While the increase at the lower dose was 66%, a dramatic (150%) increase was evident at the higher dose. Concomitantly, a dose-related increase in TBARS (lipid peroxidation index) levels was observed in the pancreas of DM treated rats. There was 2.5 and 3.7 fold increase in TBARS level at lower and higher doses of DM respectively (**Fig. 3**). Activities of selected antioxidant enzymes were significantly elevated in the pancreas of treated rats compared to that of control rats. (**Table 2**) (Kamath & Rajini, 2007). These results are in accordance with the study of Hagar et al., (2002) who had earlier reported increased blood glucose levels and hyerinsulinemia with concomitant histochemical and ultramicrostructural changes in pancreas of rats following chronic exposure to dimethoate.

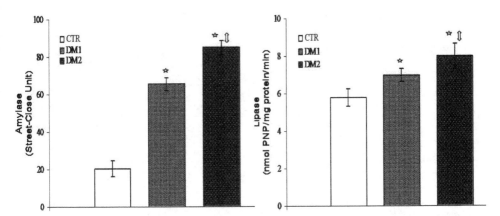

Fig. 2. Changes in pancreatic damage markers in rats induced by Dimethoate after 30 days (DM1: 20 mg/kg b.w/d; DM2: 40 mg/kg b.w/d). Values are mean ± SEM (n=6); * Comparison of control and other groups ($P < 0.01$), ⇕ Comparison of DM1 with DM2 ($P < 0.01$)

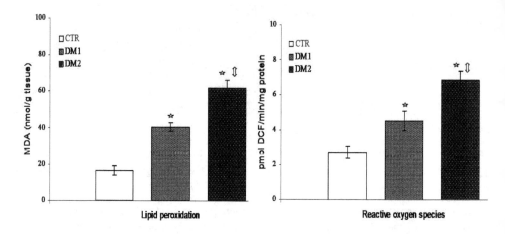

Fig. 3. Extent of lipid peroxidation and ROS levels in pancreas of control (CTR) and Dimethoate treated rats (DM1: 20 mg/kg b.w/d; DM2: 40 mg/kg b.w/d). Values are mean ± SEM (n=6); *Comparison of control and other groups ($P < 0.01$), ⇕ Comparison of DM1 with DM2 ($P < 0.01$).

Several studies have demonstrated pancreatitis after exposure to OPI (Dressel et al., 1979; Moore & James, 1988; Hsiao et al., 1996). Increase in the serum lipase and amylase activities reported by us clearly indicates that DM results in a state of pancreatic damage. Increased serum lipase activity has also been reported after administration of methidathion, an OPI (Mollaoglu et al., 2003). These results agree with earlier reports of acute pancreatitis in humans after accidental cutaneous exposure to DM (Marsh et al., 1988) and increase in amylase activity reported in dogs after diazinon administration (Dressel et al., 1982). Together, these studies clearly indicate that OPI possess propensity to elicit structural and functional alterations in pancreatic milieu that may be associated with disruptions in euglycemic conditions. From these studies, it may be argued that OPI may present a great threat to pancreatic functions in human beings and such threats may have far-reaching consequences on gluco-regulation in human beings.

Group	Enzyme Activity				
	SOD[1]	CAT[2]	GPX[3]	GR[3]	GST[4]
CTR	26.42 ± 2.2	9.38 ± 0.31	27.18 ± 5.24	17.50 ± 1.60	0.03 ± 0.004
DM1	42.72 ± 0.38[a]	10.24 ± 0.32	25.23 ± 3.89	19.72 ± 2.03	0.04 ± 0.003[a]
DM2	56.23 ±1.18[a,b]	15.44 ± 0.51[a,b]	13.85 ± 2.20[a,b]	25.30 ± 1.30[a,b]	0.06 ± 0.003[a,b]

[1]units/mg protein; [2]μmol/min/mg protein; [3]nmol/ min/ mg protein; [4]μmol/ min / mg protein
Values are mean ± SEM (n=6)
[a] Comparison of control (CTR) and other groups;
[b] Comparison of DM1 (DM: 20mg /kg b.w/d) group with DM2 (DM: 40 mg/kg b.w/d) group

Table 2. Antioxidant enzyme activities in pancreas of rats administered oral doses of Dimethoate for 30 days.

3.2 Adrenal involvement

Studies undertaken by several researchers to investigate the mechanisms mediating hyperglycemic effects of OPI have mainly focused on the involvement of cholinergic stress and adrenal functions. We have extensively studied the mechanistic involvement of adrenals in glucotoxicity of OPI in rats mainly under acute and short-term exposure regimes. The rationale for studying adrenal involvement emerged from the typical hyperglycemic behaviour of single dose (oral) of two OPI-acephate and monocrotophos. Single dose of acephate and monocrotophos elicited rapid and transient hyperglycemia after administration. Both OPI were administered to overnight-fasted rats at 1/10 doses of their LD_{50} (LD50; acephate-1400mg/kg b.w., monocrotophos-18mg/kg b.w.). As depicted in **Fig. 4**, both acephate and monocrotophos induced reversible hyperglycemia with peak occurring at 2h after exposure. Acephate induced peak hyperglycemia at 2h (87%), which tended to normalize thereafter and attained near-control values at 8h after administration (Joshi & Rajini, 2009). Similarly, monocrotophos also induced rapid hyperglycemia with peak occurring at 2h (103%). Interestingly, monocrotophos induced hyperglycemia exhibited steep reversibility compared to acephate, with normalization occurring at 6h (Joshi & Rajini, 2010). This trend observed in the present study is consistent with other reports, which demonstrated reversible hyperglycemia in experimental animals following OPI administration. While Malathion has been reported to cause reversible hyperglycemia in rats (Gupta, 1974; Rodrigues et al., 1986; Seifert, 2001; Lasram et al., 2008), acute exposure to diazinon induced reversible hyperglycemia in mice (Seifert, 2001).

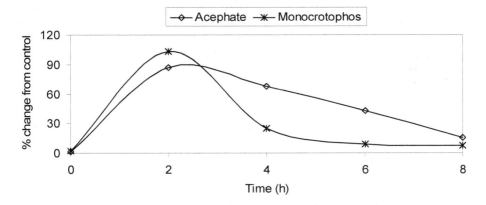

Fig. 4. Time-course of blood glucose levels in rats administered a single oral dose of acephate (140mg/kg b.w.) and monocrotophos (1.8mg/kg b.w.).

Based on the above results, we reasoned that the reversible hyperglycemia could be triggered by transient changes in the hormonal milieu of glucose homeostasis. Adrenals are an important part of the endocrine system and play a key role in glucose homeostasis by secreting glucocorticoid and amine hormones. Glucocorticoid hormones (GCs) (mainly cortisol in man and corticosterone in rodents) are secreted by the adrenal cortex under the control of hypothalamic-pituitary-adrenal axis. Glucocorticoid hormones, along with other key hormones, act to maintain blood glucose levels within narrow limits (Andrews & Walker, 1999). GCs, glucagon and epinephrine raise blood glucose by inhibiting glucose uptake in the periphery and stimulating hepatic glucose release. Hepatic gluconeogenesis serves as the main source of hepatic glucose production during state of prolonged fasting and contributes significantly to development of diabetes mellitus (Pilkis & Granner, 1992). GCs facilitate gluconeogenesis as they exert permissive effect on the process by transcriptional activation of key enzymes of gluconeogenesis viz., glucose-6-phosphatase (G6Pase) (Argaud et al., 1996), phosphoenolpyruvate carboxykinase (PEPCK) (O'Brien et al., 1990) and tyrosine aminotransferase (TAT) (Ganss et al., 1994). Increased glycogenolysis and gluconeogenesis appear to be the two chief mechanisms underlying OPI-induced hyperglycemia. Fenitrothion-induced increase in blood glucose in *S. mossambicus* was associated with decreased hepatic glycogen (Koundinya & Ramamuthi, 1979) and sub chronic exposure of rats to acephate, which caused slight increase in blood glucose also caused depletion of liver glycogen in rats (Deotare & Chakrabarthi, 1981). Abdollahi et al. (2004a) reported increased activity of GP and phosphoenolpyruvate carboxykinase (PEPCK) following sub chronic exposure to Malathion. Acute exposure to diazinon has been shown to cause depletion of liver glycogen with increased activity of glycogen phosphorylase, and also increased activities of gluconeogenesis enzymes in liver (Matin et al., 1989). Valexon is reported to have increased the activity of G6Pase in liver of rats (Kuz'minskaia et al., 1978).

OPI and other AChE inhibiting organophosphate compounds exert strong influences on functioning of hypothalamic-pituitary-adrenal (HPA) axis, leading to increased circulating levels of corticosteroid hormones in vivo. This is particularly true in the case of acute exposure to AChE inhibiting compounds. Studies have shown elevated corticosteroid hormones levels in response to AChE-inhibiting compounds and role of AChE inhibition in the phenomenon. Single dose of Chlorfenvinphos, acephate and methamidophos have been demonstrated to elevate circulating levels of corticosterone and aldostserone after administration of a single dose (Osicka-Kaprowska et al., 1984; Spassova et al., 2000). Soman has been reported to increase plasma corticosterone levels in rodent models (Hudon & Clement, 1986; Fletcher et al., 1998). More importantly, the stressogenic potential (hypercorticosteronemia and induction of liver tyrosine aminotransferase activity) of soman was effectively abrogated by reactivators of inhibited acetylcholinesterase (Kassa, 1995 & 1997). Similarly, stressogenic potential of Cyclohexylmethyl phosphonofluoridate (AChE inhibitor) has been reported to be eliminated by HI-6 (AChE reactivator) (Kassa & Bajgar, 1995). Thus, it is clearly evident that AChE-inhibiting OPI elicit hyper stimulation of adrenal functions, leading to induction of gluconeogenesis enzymes in liver.

Based on the time-course of reversible hyperglycemia induced by acephate and monocrotophos, further experiments were carried out to investigate the adrenal effects of OPI and its role in the ensuing hyperglycemia. We assessed the effects of 2 or 6h exposure to either acephate (oral) or 2 or 4h exposure to monocrotophos (oral) on plasma corticosterone, adrenal cholesterol, blood glucose, key liver gluconeogenesis enzymes (G6Pase and TAT) and hepatic glycogen content in rats. Interestingly, we observed that both acephate and monocrotophos induced strong hypercorticosteronemia with concomitant hyperglycemia and induction of liver gluconeogenesis enzyme activities. Further, hypercorticosteronemia was associated with decrease in adrenal cholesterol pools (effect of monocrotophos on adrenal pools described in the section on 'comparison between single and repeated dose effects'), which is the precursor for corticosterone synthesis (**Table 3 & 4**). Depletion in adrenal cholesterol pools may therefore be attributable to increased synthesis and secretion of corticosterone. Interestingly, both OPI did not cause depletion in hepatic glycogen content. At time points that represented normalization of blood glucose levels, there was phenomenal increase in liver glycogen levels. The data presented above clearly demonstrates co-existence of hypercorticosteronemia and induction of liver gluconeogenesis enzyme activities with hyperglycemia in OPI treated rats, indicating that OPI may trigger induction of liver gluconeogenesis machinery as result of hypercorticosteronemia, leading to hyperglycemia.

	At time interval after administration		
	0h	2h	6h
Plasma corticosterone *	30.9±3.4[a]	55.0±2.5[b]	44.0±2.7[b]
Adrenal cholesterol**	26.5±1.4[a]	15.6±0.56[b]	12.5±0.47[b]
Blood glucose ***	101.6±4.6[a]	182.4±5.2[b]	142.7±5.2[c]
Liver G6Pase#	90.14±4.38[a]	171.93±5.61[b]	112.84±4.18[c]
Liver TAT ##	14.28±1.34[a]	26.31±0.87[b]	23.7±0.48[b]
Hepatic glycogen$	316.2±34.90[a]	325.3±29.12[a]	1145.0±27.92[b]

(Joshi and Rajini, 2009)

Data analyzed by ANOVA followed by Tukey Test (n=6)
* μg/dl; ** mg/g tissue; *** mg/dl
# glucose-6-phosphatase (nmol/min/ mg protein);
tyrosine aminotranferase (nmol/min/mg protein); $ μg/g tissue

Table 3. Biochemical effects of acephate (140mg/kg b.w.) in rats

	At time interval after administration		
	0h	2h	4h
Plasma corticosterone *	36.62±1.2[a]	73.82±3.8[b]	45.65±1.8[a]
Blood glucose **	95.2±1.8[a]	194.8±3.7[b]	121.3±1.9[c]
Liver TAT #	15.86±0.8[a]	32.27±1.2[b]	26.87±1.8[c]
Hepatic glycogen##	213.8±49.2[a]	216.4±21.1[a]	925.7±27.6[b]

(Joshi and Rajini, 2010)

Data analyzed by ANOVA followed by Tukey Test (n=6)
* μg/dl; ** mg/g tissue ; # tyrosine aminotranferase (nmol/min/mg protein); ## μg/g tissue

Table 4. Biochemical effects of monocrotophos (1.8mg/kg b.w.) in rats

Indeed, role of adrenals in glucotoxicity of OPI has been explored earlier. Matin et al., (1989) earlier demonstrated that single dose diazinon (OPI) caused hyperglycemia and induction of liver gluconeogenesis enzymes in normal rats while these changes did not manifest in adrenalectomized rats, indicating the involvement of adrenals in the glucotoxicity of diazinon. Our attempts to study the adrenal and glycemic effects of acephate and monocrotophos revealed that two compounds, which exhibit anticholinesterase property, elicited similar effects. Thus, the effects raised question whether the adrenal and glycemic effects are mediated through the anticholinesterase property of OPI. To address the question, we studied the extent of AChE inhibition elicited by monocrotophos at 2 and 4h after administration. Influence of cholinergic antagonists was investigated at 2h after administration on stressogenic (hypercorticosteronemia and induction of liver TAT activity) and hyperglycemic potential of monocrotophos. For the purpose of mechanistic investigations, we employed two muscarinic cholinergic antagonists- atropine sulphate, a general ACh receptor antagonist that can pass through blood brain barrier (BBB) (Guarini et al., 2004) and methyl atropine nitrate, which is a peripherally active antagonist that does not pass through blood BBB (Pavlov et al., 2006). Both antagonists were administered at 30μmols/kg b.w 3-5 min before monocrotophos (1/10 LD50).

Monocrotophos elicited significant inhibition of AChE activity (>50%) in brain, adrenals and liver at both 2 and 4 h after exposure (**Fig. 5A**). Of the organs studied, maximum inhibition of AChE activity was evident in brain (84 and 78% at 2 and 4 h respectively) while the enzyme activity in adrenals was inhibited to 32 and 34% of control activity at 2 and 4 h after exposure respectively. Similarly, monocrotophos administration reduced liver AChE activities to 47 and 46% of control at 2 and 4 h after exposure respectively. More importantly, we did not observe any spontaneous reactivation of inhibited AChE activity at 4h after administration, which is an important feature of the enzymes' behavior (Reiner and Aldridge, 1967; Reiner, 1971). This indicates that, while hyperglycemic potential of monocrotophos in rats may be a result of its anticholinesterase potency, the reversibility of hyperglycemia is not a consequence of spontaneous reactivation of the enzyme. Reversibility of hyperglycemia may hence be a consequence of counter-regulatory mechanism as reflected by glycogen deposition at 4h after administration. Increase in glycogen content upon 4h exposure is a clear indication of mobilization of glucose into glycogen synthesis pathway as a measure to overcome hyperglycemia.

We also observed that both cholinergic antagonists were potent in offering protection against stressogenic and hyperglycemic potential of monocrotophos. Administration of monocrotophos elicited significant hyperglycemia (103%) (**Fig. 5B**). Pre- treatment of rats with atropine sulfate (106.04 ±1.83 compared to 191.82 ±7.59 mg/dl of monocrotophos alone) and atropine methyl nitrate (123.49 ±4.12 compared to 191.82 ±7.59 mg/dl of monocrotophos alone) offered significant protection against hyperglycemia induced by monocrotophos. It has been earlier demonstrated that diazinon-induced hyperglycemia was mediated by AChE inhibition, as revealed by protective effects of pralidoxime (AChE reactivator) (Seifert, 2001). Monocrotophos-induced hypercorticosteronemia (112%) was effectively prevented by cholinergic antagonists (**Fig. 5C**). Pre-treatment of rats with atropine sulfate (33.98 ±2.89 compared to 76.63 ±1.76 μg/dl of monocrotophos alone) and atropine methyl nitrate (44.67 ±1.64 compared to 76.63 ±1.76 μg/dl of monocrotophos alone) offered significant protection against hypercorticosteronemia induced by monocrotophos. Monocrotophos induced a marked increase in the TAT activity in liver (107%) (**Fig. 5D**). Pre-treatment of rats with atropine sulfate (20.42 ±1.70 compared to 33.38 ±1.09 nmol/min/mg protein) and atropine methyl nitrate (22.39 ±0.79 compared to 33.38 ±1.09 nmol/min/mg protein) offered significant protection against induction of TAT activity. These results clearly indicated that both physiological stress (hypercorticosteronemia and induction of liver TAT activity) and hyperglycemia manifest as a consequence of peripheral muscarinic cholinergic stimulation. Corticosterone exerts hyperglycemic action by up-regulation of gluconeogenesis machinery. Hence, hypercorticosteronemia and induction of liver TAT (gluconeogenesis enzyme) activity accompanying hyperglycemia raises a question whether hypercorticosteronemia is responsible hyperglycemia in monocrotophos-treated rats.

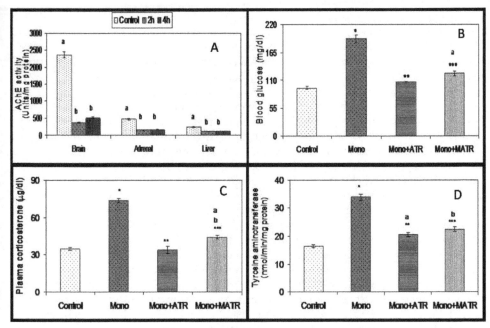

Fig. 5. Protective effects of atropine (ATR) and methyl atropine (MATR) against stressogenic and hyperglycemic potential of monocrotophos (Mono) (Joshi and Rajini, 2010).

Acetylcholine exerts strong influence on functioning of hypothalamus-pituitary-adrenal (HPA) axis. Acetylcholine has been found to increase corticotrophin releasing hormone (CRH) activity of hypothalamus *in vitro* as measured by effect on corticosteroidogenesis, an effect that was antagonized by atropine (Bradbury et al., 1974). ACh has also been shown to increase secretion of immunoreactive CRH from hypothalamus *in vitro* (Calogero et al., 1988), an effect that was antagonized by ACh receptor antagonists, atropine (muscarinic) and hexamethonium (nicotinic). Given the importance of ACh in excitation of HPA axis, assessment of cholinergic stress in activation of HPA axis in monocrotophos treated rats becomes important. The importance of ACh in functioning of HPA axis is further exemplified by the fact that muscarinic receptor agonists such as carbachol (Bugajski et al., 2002) and arecoline (Calogero et al., 1989) were found to increase ACTH and corticosterone *in vivo*. More importantly, the agonist induced increase in ACTH and corticosterone was antagonized by atropine (Bugajski et al., 2002), suggesting role for muscarinic ACh receptor in regulation of HPA axis. Role of anticholinesterase properties of organophosphate compounds in activation of HPA axis is demonstrated by studies showing elimination of stressogenic activity of cyclohexyl methyl phosphonofluoridate (as measured by plasma corticosterone and liver tyrosine aminotransferase activity) by HI-6, a cholinesterase reactivator that sufficiently reactivated inhibited AChE in brain and diaphragm (Kassa & Bajgar, 1995) and protection offered by atropine against diisopropylfluorophosphate induced increase in corticosterone levels (Smallridge et al., 1991). These studies clearly show the influence of ACh and involvement of muscarinic receptors in functioning of HPA axis.

From our data on influence of cholinergic antagonists on stressogenic and hyperglycemic potential of monocrotophos, it could be hypothesized that muscarinic cholinergic stress triggers hypercorticosteronemia, which may lead to induction of liver gluconeogenesis and hyperglycemia. However, experiments conducted with glucocorticoid receptor and adrenergic receptor antagonists revealed that hyperglycemia in mediated by adrenergic mechanisms while hypercorticosteronemia leads to only induction of liver TAT activity (data not shown). Further, we observed that monocrotophos-induced hyperglycemia was completely abolished by a gluconeogenesis inhibitor (data not shown). This establishes that physiological stress and hyperglycemia manifest in monocrotophos treated rats as independent consequence of peripheral cholinergic stress.

We further compared the effects of monocrotophos on adrenal functions and glycemic control in rats following single and repeated doses. Comparison was made between the effects of a single dose (measured 2h after administration) and that of 5 or 10 doses (one dose per day, measured 2h after last dose). In both cases, the oral dose of 1.8mg/kg b.w. was employed for the purpose of comparison. Interestingly, we observed that effects single dose of monocrotophos on adrenal functions and glycemic control was more severe than that of repeated doses. Single dose of monocrotophos elicited hypercorticosteronemia (114%) with concomitant decrease in adrenal cholesterol (33%). These adrenal effects of single dose were accompanied with hyperglycemia (109%) and induction of liver tyrosine aminotransferase activity (113%). However, repeated administration of monocrotophos for 5 or 10days resulted in blunting of responses. In case of repeated exposure, increase in corticosterone was 76 and 67% respectively in 5 and 10d exposure groups with 18 and 13% decrease in adrenal cholesterol. Similarly repeated administration elicited marginal increase in blood glucose (39 and 32%) and induction of liver TAT activity (56 and 61%) (**Fig. 6**).

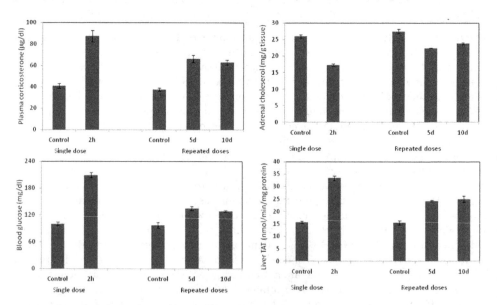

Fig. 6. Adrenal and glycemic effects of monocrotophos.

The above data clearly shows that repeated administration results in blunting of responses. This indicates that multiple administrations are associated with onset of some sort of resistance to the action of OPI. Development of tolerance to cholinesterase inhibitors during multiple administrations is a well documented phenomenon (Brodeur and DuBois, 1964; McPhillips, 1969; Sterri et al., 1980). Tolerance to the elevation of plasma corticosterone by DFP was reported to develop during repeated administration (Kokka et al., 1985). Several studies suggest that cholinergic receptors could be involved in onset of tolerance to OPI, which may be mediated by events such as down regulation of muscarinic receptors (Costa et al., 1982a&b). Tolerance to the toxic effects of dilsulfoton during multiple exposures has been attributed to reduced muscarinic receptor binding in tissues of rats tolerant to the insecticide (Schwab et al., 1981). Blunted responses observed by us in case of repeated administration of monocrotophos may be attributed to tolerance mechanisms such as down regulation of muscarinic receptors. One mechanism that may be responsible for development of resistance is increased blood insulin levels. Comparison of effects of acute and repeated doses of monocrotophos on plasma insulin levels, however, needs to be done. Such a hyperinsulinemic response has been reported in case of exposure to malathion. While malathion caused increase in blood glucose and insulin levels after single exposure and continued dietary administration for 4 weeks, the degree of hyperinsulinemia was markedly greater in dietary group (Panahi et al., 2006). Thus, repeated administration of organophosphorus insecticides leads to blunting of responses. Although blunted, these responses still represent a great threat to euglycemic balance. This is particularly true in the case of constant state of hypercorticosteronemia. This has propensity to affect skeletal muscle glucose metabolism and long-term impairments in such mechanisms may lead to long lasting dysregulation in glucose homeostasis.

3.3 OPI act as pre-disposing factors for onset of diabetes?

Based on our comprehensive studies described above, we have proposed a scheme on the mechanism/s through which OPI might regulate/ disrupt glucose homeostasis (**Fig. 7**). Oxidative stress in pancreatic milieu and glucose intolerance, up regulated gluconeogenesis machinery and hyperglycemia are critical factors in diabetes etiology. With the ability to induce the above-mentioned dysregulations, OPI may have far reaching consequences on diabetic outcomes. This may be a more pertinent issue in the present times since diabetes is fast emerging as a major wide spread disorder that threatens human life. With this realization, our laboratory has also committed to investigate if OPI act as predisposing/aggravating factors for onset or progression of diabetic condition.

We observed that dichlorvos (DDVP) treated rats showed higher (22%) levels of blood glucose compared to normal control rats while as expected, rats injected with the diabetogenic agent, Streptozotocin (STZ) alone also showed elevated (37%) level of blood glucose. However, blood glucose levels of DDVP pre-treated rats administered STZ showed relatively higher blood glucose level compared to all the groups. Liver glycogen levels were significantly lower in rats administered either DDVP (18%) or STZ (19%) alone while, rats administered DDVP followed by STZ revealed further lower levels of glycogen (46 %) **(Table 5)**. Further, we also observed that DDVP pre-treatment resulted in more sever oxidative stress in STZ treated rats. ROS levels were significantly elevated in STZ (40%) and DDVP (55%) groups compared to 'untreated control' group. However, ROS levels were markedly higher (81.23 ± 6.52 pmole DCF/min/mg protein) in 'DDVP+STZ' group of rats. Pancreas of rats administered with either DDVP or STZ alone showed marginally higher levels of the lipid peroxides compared to that in 'untreated controls' while, the levels of lipid peroxides generated in pancreas of 'DDVP+STZ' rats showed significant increase (110%) compared to all other groups. Pancreatic reduced glutathione level in 'DDVP+STZ' rats was significantly lower (37%) while, rats administered with DDVP or STZ alone also had significantly lower levels of GSH, although to a lesser extent (**Table 6**). These results clearly demonstrate that OPI act as pre-disposing factor for diabetes as reflected by higher degree of glucotoxicity of STZ (subdiabetogenic dose) in DDVP treated rats.

Fig. 7. Proposed scheme for OPI-induced alterations in glucose homeostasis.

Generally, an acute intraperitoneal dose of 40–60 mg/kg b.w is employed to induce significant hyperglycemia in rats. For the present study, we employed a lower dose of 25 mg/kg b.w ('sub-diabetogenic dose') in order to examine if pre-treatment with DDVP renders these rats more susceptible to hyperglycemia. Experimental regime began with two groups with 12 rats each-control and DDVP-treated group. The DDVP-treated group animals were orally administered daily DDVP at 20mg/kg b.w/d (corresponding to 1/5 of LD50 value: 100 mg/kg b.w, determined in a preliminary study) for 10 d. After 10 days, rats of the control group were further divided into two sub groups of six animals each ; the first sub group served as control ('untreated control'), while the second sub group of rats was intraperitoneally injected streptozotocin (STZ, 25 mg/kg b.w.) ('STZ'). The group of rats administered with DDVP was also divided into two sub groups; the first sub group of rats served as DDVP control ('DDVP'), while the second sub group of rats was injected with streptozotocin (i.p, 25mg/kg b.w.) ('DDVP+STZ').

Group	Blood glucose[1]	Liver glycogen[2]
CONTROL	113.53[a] ± 2.31	41.55[c] ± 2.01
DDVP	138.37[b] ± 4.17	34.20[b] ± 1.42
STZ	155.03[c] ± 5.09	33.34[b] ± 2.23
DDVP+STZ	188.99[d] ± 4.44	22.62[a] ± 3.52

[1]mg/dl; [2]mg/g tissue; Values are mean ± SEM (n=6); Mean in the same column with different superscript differ significantly ($p<0.05$)

Table 5. Blood glucose and liver glycogen levels in rats treated with DDVP ± STZ (i.p , 25 mg/kg b.w).

Group	ROS[1]	TBARS[2]	GSH[3]
Untreated control	38.76[a] ± 4.04	242.76[a] ± 19.18	1.07[c] ± 0.03
DDVP	60.18[b] ± 4.59	294.94[a] ± 10.65	0.83[b] ± 0.01
STZ	54.27[b] ± 2.89	283.63[a] ± 7.27	0.78[b] ± 0.02
DDVP+STZ	81.23[c] ± 6.52	389.38[b] ± 38.47	0.67[a] ± 0.02

[1]pmol DCF/min/mg protein; [2]nmol/g tissue; [3]mg/g tissue
Values are mean ± SEM (n=6); Mean in the same column with different superscript differ significantly ($p<0.05$)

Table 6. Oxidative stress parameters in pancreas of rats treated with DDVP ± STZ (i.p , 25 mg/kg b.w).

3.4 Do OPI aggravate diabetic outcomes?

To address this question, we investigated the diabetic outcomes in rats experimentally rendered diabetic and post-treated with monocrotophos. Rats were rendered diabetic with an acute dose (60 mg/kg b.w, i.p) of streptozotocin. Monocrotophos was orally administered at a sublethal dose (1/20 LD_{50}, 0.9 mg /kg b.w./d, 5 days) to both normal and diabetic rats. We observed that monocrotophos *per se* moderately increased (25%) the blood glucose levels in normal rats, but significantly aggravated the hyperglycemic outcome in diabetic rats (56% above diabetic rats). We observed the typical lipid profile alterations among diabetic rats characterized by increase in total cholesterol and triglycerides (TG) in serum. While monocrotophos did not impact lipid profile *per se*, its interaction with diabetic component resulted in severe alterations lipid profile that reflected in phenomenal increase in serum triglyceride. Such augmented impairments may have high bearing on cardiovascular health since the cardiovascular risk index was alarmingly high among diabetic rats treated with monocrotophos (**Fig. 8**). Further, monocrotophos resulted in higher degree of hepatic and renal toxicity as reflected by alterations in serum transaminase activities and blood urea nitrogen values respectively (**Table 7**).

Further, it is now established that oxidative stress in an important consequence of diabetic condition and plays as an important pathophysiological factor in progression of diabetic complications (Maritim, 2003). We observed that STZ-induced diabetes was associated with increased lipid peroxidation (122%), depletion of reduced glutathione (10%) and alterations in activities of two important antioxidant enzymes (superoxide dismutase and catalase) in kidney. Monocrotophos further deteriorated oxidative impairments in kidney as evidenced by further increase in lipid peroxidation (170%) and depletion of reduced glutathione (18%) content. Diabetes was associated with marginal decrease in superoxide dismutase activity in kidney, which was further reduced by monocrotophos treatment (35%) (**Fig. 9**).

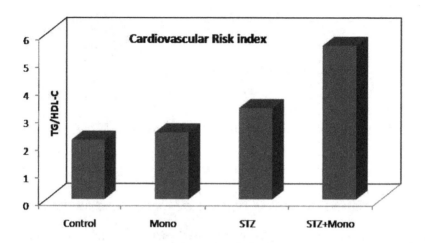

Fig. 8. Effect of repeated oral doses of monocrotophos at $1/20$ LD$_{50}$ (0.9 mg/kg b.w) on cardiovascular index in control and diabetic rats (Begum and Rajini, 2011a).

	Control	Mono	STZ	STZ+Mono
Blood glucose (mg/dl)	101.58 ± 1.4	126.91± 8.9	382.71 ± 14.0[ab]	597.94 ± 12.5[cde]
TC (mg/dl)	38.00 ± 2.1	41.92 ± 1.9	50.45 ± 1.6[a]	50.42 ± 1.2[a]
HDL-C (mg/dl)	31.09 ± 1.2	32.68 ± 1.7	37.84 ± 1.4	35.48 ± 1.3
TG (mg/dl)	66.74 ± 3.5	78.12 ± 6.9	125.44 ± 9.2[a]	193.52 ± 19.4[bcd]
BUN (mg/dl)	33.08 ± 5.6	51.88 ± 8.2	71.18 ± 10.1[a]	78.05 ± 5.2[b]
SC (mg/dl)	0.63 ± 0.06	0.72 ± 0.1	0.81 ± 0.1	0.84 ± 0.04
Serum ALT (U/L)	66.19 ± 2.0	72.24 ± 0.7	80.03 ± 2.4	120.77 ± 9.2[abc]
Serum AST (U/L)	126.57 ± 0.2	156.00 ± 18.7	199.78 ± 14.2[de]	341.55 ± 8.5[abc]
(Begum and Rajini, 2011a)				

Data analyzed by Tukey's HSD test; Mean ± SEM (n=4)
TC: Total cholesterol; HDL-C: High-density lipoprotein ; TG: Triglyceride; BUN: Blood Urea Nitrogen; SC: Serum creatinine

Table 7. Effect of repeated oral doses of monocrotophos at $1/20$ LD$_{50}$ (0.9 mg/kg b.w) on blood glucose, lipid profile and hepatic and renal damage markers in serum in control and diabetic rat.

Our work on interaction of OPI with diabetic component clearly shows that OPI can act as both predisposing and aggravating factors for diabetes. The inference becomes an important consideration to be made as the modern world is facing an escalating situation of alarming increase in the incidence of diabetes. Our study employed a low dose of monocrotophos, which *per se* did not interfere with lipid profile in rats, yet causing augmentation of alteration in lipid profile in diabetic rats. However, other studies have clearly demonstrated that several OPI cause alterations in lipid profile, particularly hypertriglyceridemia (Ryhänen et al, 1984; Ibrahim & El-Gamal, 2003; Rezg et al., 2010). Dyslipidemia or lipid abnormalities play an important role in the progression of diabetes (Goldberg, 2001) and these are characterized by lipid derangements including hypertriglyceridemia, low high-density cholesterol (HDL-C), and a high concentration of small dense low-density lipoprotein (LDL) particles. Further, a state of elevated hypertriglyceridemia is commonly associated with insulin resistance and represents a valuable clinical marker of the metabolic syndrome (Grundy et al., 1999). Propensity of OPI to induce hypertriglyceridemia coupled with their permissive effects of gluconeogenesis in liver creates a serious threat to glucose homeostasis.

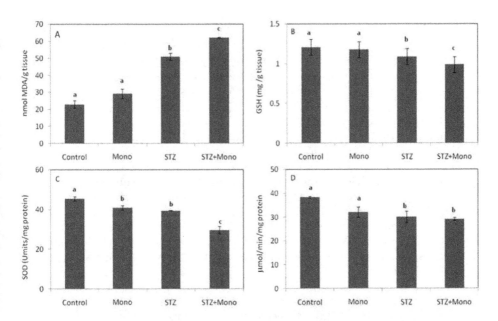

Fig. 9. Effect of repeated oral doses of monocrotophos at $1/20$ LD_{50} (0.9 mg/kg b.w) on oxidative balance in kidney of control and diabetic rats (Begum and Rajini, 2011b).

4. Conclusion

Given the status of OPI as environmental pollutant with residues being detected in biosphere around, which are now being shown to make it into human body, it is almost certain that OPI will interact with etiological factors of diabetes at toxicologically significant levels. Interaction of living system with OPI may have severe two-way impact on glycemic control. As documented facilitators of hepatic glucose output via glycogenolysis and gluconeogenesis, OPI are most likely to elicit hyperglycaemic responses in humans during exposure. Further, OPI may also affect the responsiveness to human system to insulin via multiple mechanisms, causing predisposition to diabetes. From our studies, it is also clear that OPI may also act to augment diabetic outcomes. In most societies, large sections of populations are subject to diabetes risk factors such as unhealthy diet patterns, lack of physical exercise and obesity. With such high odds of risk factors, the burden of constant exposure to OPI (as environmental pollutants) could be a silent aggravating factor that is causing increase in incidence of diabetes.

5. Acknowledgments

The authors we wish to thank the Director, CFTRI for extending support for this research. Indian Council of Medical Research (New Delhi) is gratefully acknowledged for funding the research programme described herein. The first author (AKRJ) thanks the Council of Industrial and Scientific Research (New Delhi) for award of Research Fellowship.

6. References

Abdollahi, M., Donyavi, M., Pournourmohammadi, S., & Saadat M. (2004a). Hyperglycemia associated with increased hepatic glycogen phosphorylase and phosphoenol pyruvate carboxykinase in rats following sub-chronic exposure to malathion. *Comparative Biochemistry & Physiology (C)*, 137, 343-347.

Abdollahi, M., Ranjbar, A., Shadnia, S., et al. (2004b). Pesticides and oxidative stress; review. *Medical Science Monitor*, 10, 141-147.

Abou-Donia, M. (2003). Organophosphorus Ester-Induced Chronic Neurotoxicity. *Archives of Environmental Health*, 58, 484-497.

Akhgari, M., Abdollahi, M., Kebryaeezadeh, A., et al. (2003). Biochemical evidence for free radical-induced lipid peroxidation as a mechanism for sub chronic toxicity of malathion in blood and liver of rats. *Human & Experimental Toxicology*, 22, 205-211.

Andrews, RC., & Walker, BR. (1999). Glucocorticoids and insulin resistance: old hormones, new targets. *Clinical Science*, 96, 513-523.

Argaud. D., Zhang, Q., Malatra, S., et al. (1996). Regulation of rat liver glucose-6-phosphatase gene expression in different nutritional and hormonal states. *Diabetes*, 45, 1563-1571.

Banerjee, BD., Seth, V., & Ahmed, RS. (2001). Pesticide-induced oxidative stress: perspective and trends. *Reviews on Environmental Health*, 16, 1-40.

Begum, K, & Rajini, PS. (2011a). Monocrotophos augments the early alterations in lipid profile and organ toxicity associated with experimental diabetes in rats. Pesticide Biochemistry & Physiology, 99, 33-38.

Begum, K & Rajini, PS. (2011b). Augmentation of hepatic and renal oxidative stress and disrupted glucose homeostasis by monocrotophos in streptozotocin-induced diabetic rats. *Chemico Biological Interactions*, (In press).

Bradbury, MWB., Burden, J., Hillhouse, EW., & Jones, MT. (1974). Stimulation electrically and by acetylcholine of the rat hypothalamus in vitro. *Journal of Physiology*, 239, 269–83.

Brodeur, J., & DuBois, KP. (1964). Studies on the mechanism of acquired tolerance by rats O, O-diethyl S-2- (ehylthio) ethyl phosphorodithioate (Di-syston). *Archives of International Pharmacodynamics*, 149, 560-570.

Bugajski, J., Gadek-Michalska, A., & Bugajski, AJ. (2002). Effect of constitutive- and inducible-cyclooxygenase in the carbachol-induced pituitary-adrenocortical response during social stress. *Journal of Physiology & Pharmacology*, 53, 453-462.

Calogero AE, Kamilaris TC, Gomez MT, et al. (1989). The muscarinic cholinergic agonist arecoline stimulates the rat hypothalamic-pituitary-adrenal axis through a centrally-mediated corticotropin-releasing hormone-dependent mechanism. *Endocrinology*, 125, 2445-2453.

Calogero, AE., Gallucci, WT., Bernardini, R., et al. (1988). Effect of cholinergic agonists and antagonists on rat hypothalamic corticotropin-releasing hormone secretion in vitro. *Neuroendocrinology*. 47, 303-308.

Costa, LG., Schwab, BW., & Murphy, SD. (1982a). Differential alterations of cholinergic muscarinic receptors during chronic and acute tolerance to organophosphorus insecticides. *Biochemical Pharmacology*, 31, 3407-3413.

Costa, LG., Schwab, BW., & Murphy, SD. (1982b). Tolerance to anticholinesterase compounds in mammals. *Toxicology*, 25, 79-97.

Deotare, ST., & Chakrabarti, CH. 1981. Effect of acephate (orthene) on tissue levels of thiamine, pyruvic acid, lactic acid, glycogen and blood sugar. *Indian Journal of Physiology & Pharmacology*, 25, 259-264.

Dressel, TD., Goodale, RL Jr, Zweber B, & Borner JW. 1982. The effect of atropine and duct decompression on evolution of diazinon-induced canine pancreatitis. *Annals of Surgery*, 195: 424-434.

Dressel, TD., Goodale, RL Jr., Ameson, MA., & Borner, JW. (1979). Pancreatitis as a complication of anticholinesterase insecticide intoxication. *Annals of Surgery*, 189, 199-204.

Eto, M. (1974). *Organophosphorus Pesticides: Organic and Biological Chemistry*, 387 pp, CRC Press, Cleveland, OH.

Fletcher, HP, Akbar, WJ, Peoples, RW, & Spratto, GR. (1998). Effect of acute soman on selected endocrine parameters and blood glucose in rats. *Fundamental & Applied Toxicology*, 11, 580-586.

Frick, TW, Dalo, S, O'Leary, JF, et al. (1987). Effects of insecticide, diazinon, on pancreas of dog, cat and guinea pig. *Journal of Environmental Pathology, Toxicology and Oncology*, 7, 1-11.

Fukuto, TR. (1990). Mechanism of action of organophosphorus and carbamate insecticides. *Environmental Health Perspectives*, 87, 245-254.

Galloway, T, & Handy R. (2003). Immunotoxicity of organophosphorus pesticides. *Ecotoxicology*, 12, 345-363.

Ganss, R, Weih, F, & Schutz G. (1994). The Cyclic Adenosine 3'5'-Monophosphate and the glucocorticoids-dependent enhancers are targets for insulin repression of tyrosine aminotransferase gene transcription. *Molecular Endocrinology*, 8, 895-903.

Gokalp, O, Buyukvanh, B, & Cicek E, et al. (2005). The effects of diazinon on pancreatic damage and ameliorating role of vitamin E and vitamin C. *Pesticide Biochemistry & Physiology*, 81, 123-128.

Goldberg, JI. (2001). Diabetic dyslipidemia: causes and consequences. *The Journal of Clinical Endocrinology & Metabolism*, 86, 965–971.

Grundy, SM. (1999). Hypertriglyceridemia, insulin resistance, and the metabolic syndrome. *American Journal of Cardiology*, 83, 25F-29F.

Guarini, S, Cainizzo, MM, Giuliani, D, et al. (2004). Adrenocorticotropin reverses hemorrhagic shock in anesthetized rats through the rapid activation of a vagal anti-inflammatory pathway. *Cardiovascular Research*, 63, 357-365.

Gupta, PK. (1974). Malathion induced biochemical changes in rats. *Acta Pharmacology et. Toxicology*, 35, 191-194.

Hsiao, CT, Yang, CC, Deng, JF, et al. (1996). Acute pancreatitis following organophosphate intoxication. *Journal of Toxicology Clinical Toxicology*, 34, 343-347.

Hudon, M, & Clement, JG. (1986). Effect of soman (pinacolyl methylphosphonofluoridate) on the blood levels of corticosterone and adrenocorticotropin in mice. *Candian Journal of Physiology & Pharmacology*, 64, 1339-1342.

Ibrahim, NA, El-Gamal, BA. (2003). Effects of diazinon, an organophosphate insecticide, on plasma lipid consitutents in experimental animals. *Journal of Biochemistry and Molecular Biology*, 36, 499–504.

Joshi AK, Rajini PS. 2010. Hyperglycemic and stressogenic effects of monocrotophos in rats: Evidence for the involvement of acetylcholinesterase inhibition. *Experimental & Toxicologic Pathology*, (In press).

Joshi, AK, & Rajini, PS. (2009). Reversible hyperglycemia in rats following acute exposure to acephate, an organophosphorus insecticide: role of gluconeogenesis. *Toxicology*, 257, 40-45

Kamath, V, Joshi, AKR, & Rajini, PS. (2008). Dimethoate induced biochemical perturbations in rat pancreas and its attenuation by cashew nut skin extract. *Pesticide Biochemistry & Physiology*, 90, 58-65.

Kamath, V, Rajini, PS. (2007). Altered glucose homeostasis and oxidative impairment in pancreas of rats subjected to dimethoate intoxication. *Toxicology*. 231, 137-146.

Kassa, J, & Bajgar, J. (1995). Comparison of the efficacy of HI-6 and obidoxime against cyclohexyl methyl phosphonofluoridate (GF) in rats. *Human & Experimental Toxicology*, 11, 923-928.

Kassa, J. (1995). Comparison of efficacy of two oximes (HI-6 and obidoxime) in soman poisoning in rats. *Toxicology*, 101, 167-174.

Kassa, J. (1997). Importance of cholinolytic drug selection for the efficacy of HI-6 against soman in rats. *Toxicology*, 116, 147-152.

Kokka, N, Lee, R, & Lomax P. (1985). Effects of organophosphae cholinesterase inhibitors on pituitary adrenal activity and thermoregulation. *In Fifth Annual Chemical Defense Bioscience Review*. 60.

Koundinya, PR, Ramamurthi, R. (1979). Effect of organophosphate pesticide (fenitrothion) on some aspects of carbohydrate metabolism in a freshwater fish, Saotherodon (Tilapia) mossamicus (Peters). *Experientia*, 35, 1632-1633.

Kuz'minskaia, UA, Bersan, LV, & Veremenko, LM. (1978). Activity of the indicator enzymes of liver subcellular structures with the prolonged administration of Valexon. *Vopr. Pitan*, 5, 48-51.

Lasram, MM, Annabi, AB, Rezg R, et al. (2008). Effect of short-time Malathion administration on glucose homeostasis in Wistar rat. *Pesticide Biochemistry & Physiology*, 92, 114-119.

Maritim, AC, Sanders, RA, Watkins (III), JB. (2003). Diabetes, Oxidative Stress, and Antioxidants: A Review. *Journal of Biochemical & Molecular Toxicology*, 17, 24-38.

Marsh, WH, Viekov, GA, & Conradi, EC. (1988). Acute pancreatitis after cutaneous exposure to an organophosphate insecticide. *American Journal of Gastroenterology*, 83, 1158-1160.

Mathur HB, Johnson S, Mishra R., et al. (2003). Analysis of pesticide residues in bottled water (Delhi Region). CSE Report.
http://www.cseindia.org/userfiles/Delhi_uploadfinal_sn.pdf

Mathur, HB, Agarwal, HC, Johnson, S, & Saikia, N. (2005). Analysis Of Pesticide Residues In Blood Samples From Villages Of Punjab. CSE Report.
http://www.cseindia.org/userfiles/Punjab_blood_report.pdf

Matin, MA, Khan, SN, Hussain K, & Sattar S. 1989. Effects of adrenalectomy on diazinon-induced changes in carbohydrate metabolism. *Archives of Toxicology*, 63, 376-380.

McPhillips, JJ. (1969). Altered sensitivity to drugs following repeated injections of a cholinesterase inhibitor to rats. *Toxicology and Applied Pharmacology*, 14, 67-73.

Meller, D, Fraser, I, & Kryger, M. (1981). Hyperglycemia in anticholinesterase poisoning. *Candian Medical Association Journal*, 124, 745-748.

Mollaoglu, H, Yilmaz, HR, Gokalp, O, & Altuntas I. (2003). Methidathion un pankreas uzerine erkileri: Vitamin E ve C nin rolu. *Van Tip Dergisi*, 10, 98-100.

Moore, PG, & James, OF. (1988). Acute pancreatitis induced by acute organophosphate poisoning.
Postgraduate Medical Journal, 57, 660-662.

Namba, T. (1971). Cholinesterase inhibition by organophosphorus compounds and its clinical effects. *Bulletin - World Health Organization*, 44, 289-307.

O'Brien, RM, Lucas, PC, Forest CD, et al. (1990). Identification of a sequence in the PEPCK gene that mediates a negative effect of insulin on transcription. *Science*. 249, 533-537.

Osicka-Kaprowska, A, Lipska M, Wysocka-Paruzewska, B. (1984). Effect of chlorfenvinphos on plasma corticosterone and aldosterone levels in rats. *Archives of Toxicolicology*, 55, 68-69.

Panahi,P, Vosough-Ghanbari, S, Pournourmohammadi, S, et al. (2006). Stimulatory Effects of Malathion on the Key Enzymes Activities of Insulin Secretion in Langerhans Islets, Glutamate Dehydrogenase and Glucokinase. *Toxicology Mechanisms & Methods*, 16, 161–167.

Pavlov, VA, Ochani M, Gallowitsch-Puetra, M, et al. (2006). Central muscarinic cholinergic regulation of the systemic inflammatory response during endotoxemia. *Proceedings of the National Academy of Sciences, USA*, 103, 5219-5223.

Pilkis, SJ, & Granner, DK. (1992). Molecular physiology of the regulation of hepatic gluconeogenesis and glycolysis. *Annual Review of Physiology*, 54, 885-909.

Rahimi, R, & Abdollahi, M. (2007). A review on the mechanisms involved in hyperglycemia induced by organophosphorus pesticides. *Pesticide Biochemistry & Physiology*, , 115-121.

Ranjbar, A, Pasalar, P, & Abdollahi, M. (2002). Induction of oxidative stress and acetylcholinesterase inhibition in organophosphorous pesticide manufacturing workers. *Human & Experimental Toxicology*, 21, 179-182.

Reiner, E, & Aldridge, WN. (1967). Effect of pH on inhibition and spontaneous reactivation of acetylcholinesterase treated with esters of phosphorus acids and carbamic acids. *Biochemical Journal*, 105: 171-179.

Reiner, E. (1971). Spontaneous reactivation of phosphorylated and carbamylated cholinesterase. *Bulletin of World Health Organization,* 44, 109-112.

Rezg, R, Mornagui, B, Benahmed, M, etal. (2010). Malathion exposure modulates hypothalamic gene expression and induces dyslipedemia in Wistar rats. *Food & Chemical Toxicology.* 48, 1473-1477.

Rodrigues, MA, Puga, FR, Chenker, E, & Mazanti, MT. (1986). Short-term effect of malathion on rats' blood glucose and on glucose utilization by mammalian cells in vitro. *Ecotoxicology & Environnemental Safety,* 12 , 110-113.

Ryhänen, R, Herranen, J, Korhonen, K, et al. (1984). Relationship between serum lipids, lipoproteins and pseudocholinesterase during organophosphate poisoning in rabbits. *International Journal Biochemistry.* 16: 687-690.

Sanghi, R, Pillai, MK, Jayalekshmi, TR, & Nair A. (2003). Organochlorine and organophosphorus pesticide residues in breast milk from Bhopal, Madhya Pradesh, India. *Human & Experimental Toxicology.* 22, 73-6.

Schwab, BW, Hand, H, Costa, LG, & Murphy, SD. (1981). Reduced muscarinic receptor binding in tissues of rats tolerant to the insecticide disulfoton. *Neurotoxicology.* 2, 635-47.

Seifert, J. (2001). Toxicological significance of the hyperglycemia caused by organophosphorous insecticides. *Bulletin of Environmental Contamination & Toxicology,* 67, 463-469.

Sevillano, S, de la Mano, AM, Manso, MA, et al. (2003). N-acetylcysteine prevents intra-acinar oxygen free radical production in pancreatic duct obstruction-induced acute pancreatitis. *Biochimica et Biophysica Acta,* 20, 177-184.

Smallridge, RC, Carrb, FE, & Feina, HG. (1991). Diisopropylfluorophosphate (DFP) reduces serum prolactin, thyrotropin, luteinizing hormone, and growth hormone and increases adrenocorticotropin and corticosterone in rats: Involvement of dopaminergic and somatostatinergic as well as cholinergic pathways. *Toxicology & Applied Pharmacology,* 180, 284-295.

Sogorb, MA, & Vilanova, E. (2002). Enzymes involved in detoxification of organophosphorus, carbamate and pyrethroid insecticides through hydrolysis. *Toxicology Letters,* 128, 215-228.

Soltaninejad, K, & Abdollahi, M. (2009). Current opinion on the science of organophosphate pesticide and toxic stress: A systematic review. *Medical Science Monitor,* 15, RA75-RA90.

Spassova, D, White, T, & Singh, AK. (2000). Accute effects of acephate and methamidophos on acetylcholinesterase activity, endocrine system, and amino acid concentrations in rats. *Comparative Biochemistry & Physiology Part (C),* 12, 79-89.

Sterri, SH, Lyngaas, S, & Fonnum, F. (1980). Toxicity of soman after repetitive injection of sublethal doses in rat. *Acta Pharmacology et Toxicology (Copenh).* 46, 1-7.

Sultatos, LG. (1994). Mammalian toxicology of organophosphorous pesticides. *Journal of Toxicology & Environmental Health,* 43, 271-289.

5

DDT and Its Metabolites in Mexico

Iván Nelinho Pérez Maldonado, Jorge Alejandro Alegría-Torres,
Octavio Gaspar-Ramírez, Francisco Javier Pérez Vázquez,
Sandra Teresa Orta-Garcia and Lucia Guadalupe Pruneda Álvarez
Laboratorio de Toxicología, Facultad de Medicina,
Universidad Autónoma de San Luis Potosí,
San Luis Potosí, SLP,
México

1. Introduction

DDT (dichlorodiphenyltrichloroethane) was first synthesized in 1874, and its insecticidal properties were discovered in 1939 by Paul Hermann Müller (Stapleton 1998). The U.S. military began using DDT extensively for mosquito control in 1944, particularly in the Pacific, where much of the action of World War II took place in highly malarious areas (Stapleton 1998). In 1955, the World Health Organization (WHO) started a global malaria control program with DDT; by 1958, 75 countries had joined and, at the peak of the campaign, 69,500 tons of pesticides mainly DDT [1,1,1-trichloro-2,2-bis(p-chlorophenyl)ethane] were applied to 100 million dwellings each year (Wijeyaratne, 1993). For the control of malaria, houses were sprayed twice a year with DDT wettable powder to kill resting mature Anopheles mosquito. Later, the Stockholm Convention on Persistent Organic Pollutants, which came into force on 17 May 2004, outlawed the use of 12 chemicals including DDT (UNEP, 2004). However, one exemption clause allows malaria-endemic nations to use DDT, strictly for disease vector control. The United Nations Environment Program estimates that about 25 countries will use DDT under exemptions from the DDT pesticide ban (POPs, 2009). Thus, in regard to presence of DDT around the world can be divided into three scenarios: Sites where DDT is still in use; sites where the presence is due to DDT sprayed several years ago, and sites where the presence of DDT is the result of a long-range transport of the insecticide to areas where it was never used like the Antarctic. In Mesoamerica (Mexico, Costa Rica, El Salvador, Guatemala, Honduras, Mexico, Nicaragua and Panama) DDT was used until the year 2000, Mexico and Nicaragua being the last nations that applied the insecticide in agriculture and for the control of malaria. Table 1 lists the period and the total amount of DDT used in each Mesoamerican country by the malaria control programs. The amount used (approximately 85,000 tons between 1946 and 1999) together with the high environmental persistence of DDT and its metabolites, provide the necessary conditions for DDT to become a contaminant of concern for this region of the world (ISAT, 2002). Taking into account the environmental persistence and the toxicity of DDT, a program for the control of malaria without using insecticides in Mesoamerica was developed between 2004 and 2007, with assistance from the Pan American Health Organization [PAHO; (Chanon et al., 2003; PAHO, 2008)]. The phase-out of DDT in Costa

Rica, El Salvador, Guatemala, Honduras, Mexico, Nicaragua and Panama was part of a regional proposal supported by the Global Environment Facility (GEF) and the United Nations Environmental Program with the participation of the North American Commission for Environmental Cooperation (CEC).

2. DDT in Mexico

In 1944, and for the first time, houses were sprayed with DDT in Temixco, Morelos, Mexico (Stapleton 1998). The spray was applied to the walls and ceilings of residences. Studies done two months after the spraying, showed that there was a 99% reduction in the incidence of Anopheles (Stapleton 1998). In 1947–48, the spraying of DDT began in other Mexican regions, such as Veracruz, Mexico City and Baja California (Stapleton 1998). By 1948, the first clear evidence of malaria control appeared in the areas first sprayed with DDT; the overall parasite rate in the state of Morelos was found to be 10%, and the rate in the sprayed towns was found to be 1% (Stapleton 1998). In 1936 it was estimated that half of the Mexican population lived in endemic regions and was subject to a malaria mortality rate of 0.5%, or about 36,000 deaths per year (Stapleton 1998).During the 1930s and 1940s, malaria became the third cause of death in the country. However, the antimalaria campaign was not generalized until 1956 (CCE 1998). The success of DDT was outstanding, malaria cases decreased from 41,000 in 1955 to 4,000 in 1960 (Fernández de Castro 1998); in 1970 the campaign was relaxed and the cases increased to 57,000 (Fernández de Castro 1998). However, this was also the time in which DDT production peaked in Mexico, with more than 80 thousand tonnes produced annually (CCE 1998). In recent years, the incidences of malaria have declined significantly, to less than 5,000 cases. Since 1982 there have been no deaths from this disease. As, showed, Malaria is a long-standing public health problem that has inhibited development in large areas of the country. Approximately 60% of the Mexican territory, representing an area inhabited by close to 45 million people, provides an environment suitable for malaria transmission. This includes the Pacific coast, the Gulf of Mexico slopes, the Yucatan peninsula and interior basins of the high plateau. (CCE 1998). In actuality, Mexico operates a malaria control program that has substantially reduced the incidence of this disease. In 1995, Mexico initiated an integrated pest management approach for malaria to reduce the heavy dependence on pesticides. Much of the success of Mexico's malaria control program (there have been no recorded deaths from malaria since 1982) is due to improvements in sanitation, increased disease surveillance, and integrated pest management schemes that focus pesticide applications on critical habitats and stages in the mosquito's life cycle (Government of Mexico 1998). Since 1998, DDT was substituted with pyrethroids in the malaria control program. In other hand, In the area of agriculture, as much as 1,000 tonnes per year were used (CCE 1998).Application rates in the north of Mexico, were among the highest in the world (CCE 1998). However, the growing concern about DDT persistence has had a significant impact on agricultural practices in Mexico. During the early 1970s the US Food and Drug Administration (USFDA) began rejecting the importation of commodities due to high residue levels, especially of DDT (CCE 1998). Therefore, some agricultural areas changed to newer pesticides in order to comply with the USFDA regulations. By 1990, DDT was limited to campaigns addressing public sanitation (CCE 1998). In recognition of DDT's environmental and human health effects, Mexico shifted the emphasis of its anti-malarial campaigns away from DDT beginning in the 1980s and 1990s, and the use of the pesticide was gradually reduced. In 1997, the Intergovernmental Forum on Chemical Safety agreed there was sufficient evidence to take international action to restrict and reduce the use of DDT.

Country	Period of use	Total tons
México	1957-2000	69,545.00
Nicaragua	1959-1991	2,172.00
Costa Rica	1957-1985	1,387.00
Guatemala	1958-1979	4,790.00
Honduras	1950-1978	2,640.00
El Salvador	1946-1973	4,271.00
Panamá	1967-1971	189.00

Table 1. History of DDT use in Mesoamerica countries (ISAT, 2002).

3. Environmental pathways of exposure to DDT

The physicochemical properties of DDTs (Table 2) show the extent of their volatility and the high KOW/KOA value shows that they are more likely to partition into environmental sectors which exhibit greater organic phases (biota, soil, and sediment). The concentration of DDTs in the water samples may be limited due to characteristically low water solubility. In other hand, the exposure pathways are the processes by which DDT may be transported from the pollution source to living organisms. In the malaria areas, the source of DDT was the household-spraying of the insecticide. Since the beginning of the control program of malaria, DDT was sprayed on the ceilings and walls, both indoors and outdoors. Therefore, after spraying, indoor dust (or indoor soil in some cases), and the external surface soil in those areas next to the dwellings, were the media first to become contaminated with DDT. From these points, the insecticide could be transported from one medium to another by different processes.

Compound	Molecular weight	Vapor pressure (Pa)	Aqueous solubility (mol/m³)	Henry's law constant (Pa m³/mol)	Log Kow*	Log Koa
p′p-DDT	354.5	0.00048	0.00042	1.1	6.39	9.73
p′p DDE	319.0	0.00340	0.00079	4.2	6.93	9.70
p′p-DDD	321.0	0.00120	0.00230	0.5	6.33	10.03

Table 2. Physicochemical properties of DDT and its metabolites at 25⁰ C. (Sahsuvar et al. 2003; Shen and Wania 2005).

Soil and Dust

Several studies have identified indoor house dust as an important pathway of toxicant exposure. Often levels of pollutants found in house dust, including compounds banned long ago such as DDT, are significant sources of exposure for the general population, especially children (Butte and Heinzow 2002; Hwang et al. 2008; Rudel et al. 2003). Moreover, analyses of compounds in house dust are a measure of indoor contamination, but may also provide valuable information on the assessment of human indoor exposure (Butte and Heinzow 2002). Also, outdoor soil is considered an important exposure pathway for the general population and children to compounds banned long ago (Herrera-Portugal et al. 2005). However, it is important to note that longer residence times and elevated contaminant concentrations in the indoor environment may increase the chance of exposure to these contaminants by 1,000-fold compared to outdoor exposure (Hwang et al. 2008).

Tables 3 and 4 show DDT levels in outdoor and indoor surface soils, respectively. Taking into account the guideline for DDT in residential soil: 0.7 mg/kg from Canada (Environment Canada, 2007) different scenarios have been observed in Mexico. Regarding outdoor levels, in general lower levels were found in household outdoor samples (Table 3). With exception of levels found in Chiapas and Oaxaca, that have levels lower than Canadian guide (Table 3). In other hand, high levels are recorded in indoor levels in different regions of Mexico, generally higher than Canadian guideline (Table 4). Also, we can note that the higher levels of DDT in those environment media were found in indoor dust samples, generally with levels above the Canadian guideline (Table 5). Moreover, the data in Tables 3, 4 and 5 indicate high levels of total DDT in soil and dust in all regions studied in Mexico when compared with studies around the world.

Water

DDT, DDD and DDE (DDTs) are only slightly soluble in water, with solubilities of 3.4 ppb, 160 ppb and 120 ppb, respectively (ATSDR 2010). In this regard, sedimentation is the most important factor for the disappearance of DDT from water. However, it has also been suggested that contaminated sediments are a main source of DDT inputs to the water column (Zeng 1999). In order to study the degree of pollution in water bodies located in tropical areas, DDTs were quantified in a relatively small stream. The levels of total DDT found in the tropical area was 280 pg/L (Carvalho et al. 2009). In other hand, levels of total DDT found in the tropical area in United states of America was 10300 pg/L (California 1999)

Sediments

As stated above, sediments act as the primary reservoir for excess quantities of DDT. Therefore, it is very important to analyze the concentrations in this medium. In Table 6 it is shown that DDT concentrations in Mexican samples are lower than those detected in other countries, where DDT was used either for the control of malaria or for agricultural practices. Whether this difference can be explained by an increased degradation or by a DDT mass reduction caused by water currents carrying suspended DDTs out of the contaminated area, are issues that deserve further research. However, we cannot exclude another explanation. The Mexican studies, results of which are shown in Table 6, were not designed to assess the amount of DDT in sediments due to spraying. In fact, a sediment sample collected in a river near an area where the insecticide was used intensively for vector control, had DDT concentrations of up to 70.0 mg/kg (Gonzalez-Mille et al. 2010). Discvutir disminución.

Location	Total DDT (mg/Kg)	Region	Reference
Chiapas	0.95	Southeastern	Martínez-Salinas et al. 2011
Chiapas	8.20	Southeastern	Yañez et al. 2002
Oaxaca	0.90	Southeastern	Yañez et al. 2002
Tabasco	0.04	Southeastern	Torres-Dosal et al. 2011
Chihuahua	0.45	North	Díaz-Barriga et al. 2011
Veracruz	0.01	Southeastern	Espinosa-Reyes et al. 2010
Puebla and Mexico	0.07	Central	Waliszewski et al. 2008

Table 3. Total DDT levels in outdoor surface soil (mg/Kg) in different Mexican Regions.

Mexican state	Total DDT (mg/Kg)	Region	Reference
Chiapas	6.8	Southeastern	Martínez-Salinas et al. 2011
Chiapas	7.1	Southeastern	Yañez et al. 2002
Oaxaca	0.15	Southeastern	Yañez et al. 2002
Chihuahua	0.95	North	Díaz-Barriga et al. 2011

Table 4. Total DDT levels in indoor soil (mg/Kg) in different Mexican Regions.

Food and Biota

Due to their lipophilic attributes and high persistence, the DDTs may bioaccumulate significantly in animal species (Fisher 1995). Furthermore, biomagnification has been observed; for example, DDT concentration increased with each successive trophic level in a food chain (Fisher 1995).Taking into account these properties, food ingestion can be considered a pathway of exposure. In Mexico, studies have been done in different food items, such as fish, hen's egg, butter and cow's milk and muscle. In Table 7, total DDT levels in different food items are presented. Considering fish, the concentrations of DDT, in organisms collected in Mexico, are above normal values. As is shown in Table 8, where DDTs levels in Fish are depicted for different countries. We can note that the food item with high levels of DDT are food rich in fat as butter and cow's milk (Table 7).

Location	Total DDT (mg/Kg)	Region	Reference
Chiapas	6.9	Southeastern	Martínez-Salinas et al. 2011
Chihuahua	1.0	North	Díaz-Barriga et al. 2011

Table 5. Total DDT levels in dust (mg/Kg) in different Mexican Regions.

Location	Total DDT (µg/Kg)	Region	Reference
Estado, Mexico (bay)	1.5	Southeastern	Noreña-Barroso et al. 1998
Mexico (bay)	0.6	Southeastern	Noreña-Barroso et al. 2007
Mexico (river)	74.0	Southeastern	Gonzalez-Mille et al. 2010
Mexico (lagoon)	4.9	Southeastern	Botello et al. 2000
China (Bay)	7.8		Liu et al. 2011
China (river)	3.8		Tan et al. 2009
Korea (bay)	3.4		Khim et al. 2001
Japan (bay)	1.2		Kim et al. 2007

Table 6. Total DDT levels in sediment (mg/Kg) in different Mexican Regions.

Air

Because DDTs have a Henry's Law constant value of 10-4/10-5 atm m3 mol, they are considered moderate volatile compounds [5]. Therefore, these compounds can be transported by air, either in the gaseous phase or adsorbed to atmospheric particles [5].Photodegradation of DDT occurs slowly; thus, residues of these pesticides are ubiquitous in the atmosphere, although at lower concentrations. Information on the atmospheric levels of OCs in Mexico is scarce. Previous studies in southern Mexico found that DDT and toxaphene concentrations in air were 1-2 orders of magnitude above levels in the Laurentian Great Lakes and arctic regions (24-26). Atmospheric levels in southern Mexico were generally higher than those in central Mexico (27), Costa Rica (28, 29) and Cuba (30), and comparable to those in Belize (31).

Recently, two important studies regarding DDT and its metabolites in air has been developed. Passive air samplers (PAS) were deployed at four sampling sites at the southern Mexico in 2002-2004 and eleven sampling sites across Mexico during 2005-2006 (referencia). The total DDT levels ranged from 239 to 2360 pg/m3 in 2002-2004 (referencia) and from 15 to 1975 pg/m3 in 2005-2006 (refernecia). Table ? shows the Total DDT air levels in the sampling sites in both studies. Total DDT tended to be higher in the south (poner siglas) and some central sites (sigles). While, the other central and northern sites had lower total DDT levels (poner siglas). It is important to note that the higher levels were found in tropical sites where DDT was used for health campagnes or for agriculture as MT, CEL and VC (Table ?).

Location	Food	Total DDT (mg/Kg)	Region	Reference
Mexico country	Butter	88.0		Waliszewski et al. 2003
Veracruz	Cow´s milk	39.0	Southeastern	Pardio et al. 2003
Campeche	Oysters	5.9	Southeastern	Carvalho et al. 2009
Campeche	Oysters	1.5	Southeastern	Gold-Bouchot et al. 1995
Tabasco	Oysters	6.2	Southeastern	Botello et al. 1994
Campeche	Shrimps	0.25	Southeastern	Gold-Bouchot et al. 1995
Campeche	Mussels	1.44	Southeastern	Gold-Bouchot et al. 1995
Baja California	Mussels	9.16	North	Gutierrez-Galindo et al. 1988

Table 7. Total DDT levels in food items (mg/Kg) in different Mexican Regions.

Location	Total DDT (ng/g lipid)	Region	Reference
Veracruz, Mexico	25.0	Southeastern	Gonzalez-Mille et al. 2010
Chiapas, Mexico	4.7	Southeastern	Pérez-Maldonado et al. 2010
Hidalgo, Mexico	7.3	Central	Fernandez-Bringas et al. 2008
Costa Rica	0.6		Pérez-Maldonado et al. 2010
Honduras	< LOD (below detection limit)		Pérez-Maldonado et al. 2010
Nicaragua	3.9		Pérez-Maldonado et al. 2010
El Salvador	3.8		Pérez-Maldonado et al. 2010
Guatemala	< LOD		Pérez-Maldonado et al. 2010

Table 8. Total DDT levels in fish (mg/Kg) in different Mexican Regions.

Community	Total DDT (pg/m3)	Region	Reference
Baja California	338	North	Wong et al. 2009
Chihuahua	34	North	Wong et al. 2009
Yucatan	1975	Southeastern	Wong et al. 2009
Colima	750	North	Wong et al. 2009
Veracruz	129	Southeastern	Wong et al. 2009

Community	Total DDT (pg/m3)	Region	Reference
Morelos	500	Central	Wong et al. 2009
Sinaloa	76	North	Wong et al. 2009
Mexico DF	55	Central	Wong et al. 2009
Nuevo Leon	15	North	Wong et al. 2009
San Luis Potosi	21	Central	Wong et al. 2009
Veracruz	50	Southeastern	Wong et al. 2009
Tabasco	239	Southeastern	Alegría et al. 2008
Chiapas	2360	Southeastern	Alegría et al. 2008
Chiapas	547	Southeastern	Alegría et al. 2008
Veracruz	1200	Southeastern	Alegría et al. 2008

Table 9. Total DDT levels in air (pg/m3) in different Mexican Regions.

4. Human exposure to DDT

Biomonitoring studies are a useful instrument in formulating environmental health policies. For example, in Mexico during the last decade, studies in children led to the reduction or elimination of different chemicals such as lead, lindane and inclusive DDT. Furthermore, the biomonitoring of susceptible populations is a valuable method for the identification of critical contaminants, as has been shown in the United States with the National Health and Nutritional Examination Survey (NHANES III; Needham et al., 2005). Information about human exposure to chemicals is very limited; besides, in relation to children, the information is even more scarce. In this regard, in this text, we shall present data for breast milk, blood serum and adipose tisssue. Results have been done in adults and children.

In other hand, a biomarker of exposure is a xenobiotic substance or its metabolite (s), that is measured within a compartment of an organism. The preferred biomarkers of exposure are generally the substance itself or substance-specific metabolites in readily obtainable body fluid(s) or excreta. DDT and its metabolites DDD,DDE, DDA, and MeSO2-DDE (3-methylsulphonyl-DDE), can be measured in adipose tissue, blood serum, urine, feces, semen, or breast milk .

Breast Milk

Psychological and medical studies have underlined the benefits of nursing which raises immunological defenses and provides a healthier development of the baby. Parallel findings have increased concern about the excretion of drugs and environmental contaminants contained in breast milk, since it is considered the main route for eliminating deposited organochlorine pesticides from a mother's body (Jensen and Slorach 1991; Sonawane 1995; Cupul-Uicab et al. 2008).

Because of their lipophilic nature and high persistence, DDT and its metabolites accumulate in lipophilic human body parts, particularly in lipid-rich tissues such as adipose tissue and subsequently translocated and excreted through milk fat. A major concern is that milk is the first (and in some areas the only) food for the newborn child.

Concentrations of DDTs (DDT, DDD and DDE) in human milk have been shown to be higher in communities exposed to this insecticide, than in non-exposed populations (Table 10). For example, levels from a cotton area where DDT was used for agricultural purposes (Coahuila) were higher or similar to those obtained in samples collected in a malarious area (Veracruz, Yucatan) where DDT was extensively used (Table 10). And both were higher than the concentrations quantified in urban areas (Mexico DF), where DDT has never been used.

The World Health Organization's Acceptable Daily Intake (ADI) for DDTs is 20 mg/kg/day (Lu 1995). In this regard, considering a body mass of 5 kg, a milk intake of 0.85 kg/day, a proportion of fat in milk of 0.035, and a DDT concentration in milk of 10.4 mg/kg (total DDT concentration in samples collected during 1996 1997 in a suburban malarious area; Albert et al. 1980), the estimated daily intake is three times higher than the ADI. Furthermore, if the maximum range concentration found in some studies (36.5 mg/kg; Albert et al. 1980) is taken into account, the ADI is surpassed 11 times. However, when calculated the ADI with DDT concentration in milk of 2.4 mg/kg (Waliszewski et al. 2009), the ADI calculated is lower than 20 mg/kg/day. It is important to remark that the chronological levels obtained by Waliszewski et al. (1996, 1999, 2001, 2002, 2009) have a decreasing tendency. That result coincides with the restriction and prohibition for DDT use in Mexico.

Serum

In this document, we presented data regarding DDT and its metabolites levels in children (Table 11) and adults (Table 12)

Children appear to be particularly suitable for a monitoring program, as they are not directly exposed to occupational pollution; thus, children normally reflect present trends of environmental exposure more accurately than do adults (Link et al., 2005). Moreover, it is well established that children are potentially at a higher risk than adults for adverse health effects from exposure to many environmental chemicals (Guzelian et al., 1992; Bearer, 1995; Carlson, 1998; Galson et al., 1998; Aprea et al., 2000; Needham and Sexton, 2000; Adgate and Sexton, 2001; Brent et al., 2004; IPCS, 2006).

Location	Total DDT	Region	Reference
Coahuila	10400.0	North	Albert et al. 1980
Mexico DF	900.0	Central	Torres-Arreola et al. 1999
Veracruz	7815.0	Southeastern	Pardio et al. 1998
Veracruz	6280.0	Southeastern	Waliszewski et al. 1996
Veracruz	4700.0	Southeastern	Waliszewski et al. 1999
Veracruz	4700.0	Southeastern	Waliszewski et al. 2001
Veracruz	3740.0	Southeastern	Waliszewski et al. 2002
Morelos	4320.0	Central	Lara et al. 2000
Yucatan	3065.0	Southeastern	Rodas-.Ortíz et al. 2008
Veracruz	2335.0	Southeastern	Waliszewski et al. 2009

Table 10. Total DDT levels in human milk (ng/g lipid) in different Mexican Regions.

Adipose tissue

Adipose tissue biopsy has been used in epidemiological studies to assess chronic exposure to DDT. This is a logical choice because the DDTs are accumulated in adipose tissue due to its lipid solubility. The half-life of DDT in human adipose tissue is approximately seven years (Woodruff et al. 1994).

As in serum, DDTs in adipose tissue are a good biomarker of exposure for communities exposed to DDT. When compared to an urban non-exposed community [46], the levels of DDTs (especially those of DDE), were higher in the exposed population (Table 9). In the same table it can be observed that the concentrations of DDT in adipose tissue from workers of the malaria program were higher than the levels found in people living in an agricultural area or in malarious areas. In the workers, a linear model that included an index of chronic exposure, the use of protective gear,and recent weight loss explained 55% of the variation of p,p'-DDE conncentrations in adipose tissue. The index of chronic exposure was constructed according to worker position and based on the historical duration and intensity of DDT application [48].

When the concentrations of DDTs in adipose tissue were expressed by age group, two groups were identified as the most exposed. Those groups were children and elderly people [49]. The levels in elderly people can be explained by the accumulation of DDT in a chronic exposure scenario,whereas the concentration in children may be the result of an exposure to multiple pathways (soil, household dust, air, water, food, etc.). It is interesting that the group less exposed to DDT was the 0–2 years, a group that may be exposed to DDT through lactation [49].

Location	Total DDT (ng/g lipid)	Region	Reference
Chiapas	22280.0	Southeastern	Trejo-Acevedo et al. 2009
Oaxaca	7500.00	Southeastern	Perez-Maldonado et al. 2006
Quintana Roo	11300.0	Southeastern	Perez-Maldonado et al. 2006
Chihuahua	35000.00	North	Díaz-Barriga et al. 2011
Queretaro	2170.0	Central	Trejo-Acevedo et al. 2009
Durango	2270.0	North	Trejo-Acevedo et al. 2009
San Luis Potosí	1990.0	Central	Trejo-Acevedo et al. 2009
Guanajuato	940.0	Central	Trejo-Acevedo et al. 2009
Veracruz	1910.0	Southeastern	Trejo-Acevedo et al. 2009
Michoacan	550.0	Central	Trejo-Acevedo et al. 2009
Zacatecas	700.0	Central	Trejo Acevedo et al. 2009

Table 11. Total DDT levels in serum (ng/g lipid) of children living in different Mexican Regions.

A monitoring program of DDTs in adipose tissue is needed in order to assess the body burden, now that in Mexico this insecticide has been eliminated from the malaria program. However, due to ethical constraints, it is not always possible to obtain adipose tissue samples form healthy individuals. Therefore, alternative matrices are needed; for example, a good correlation between adipose tissue concentration and levels in human milk [50] or human serum [51] has been reported .When the geometric DDE levels in lipid bases are used for the estimation of the adipose tissue/serum DDE ratio, a value near unity is obtained [51].

5. Health effects

DDT and its metabolites have been associated with neurological effects (Dorner and Plagemann 2002; Fenster et al. 2007; Torres-Sánchez et al. 2007; Rocha-Amador et al. 2009), asthma (Sunyer et al. 2006), immunodeficiency (Dewailly et al. 2000; Vine et al. 2000; Vine et al. 2001; Belles-Isles et al. 2002; Bilrha et al. 2003; Cooper et al. 2004; Dallaireet al. 2004), apoptosis (Pérez-Maldonado et al. 2004) and DNA damage in immune cells in children (Yáñez et al. 2004; Herrera-Portugal et al. 2005b).

Location	Total DDT (ng/g lipid)	Region	Reference
Chiapas	12750.0	Southeastern	Yañez et al. 2002
Oaxaca	8050.0	Southeastern	Yañez et al. 2002
Tabasco	8700.0	Southeastern	Torres-Dosal et al. 2011
Mexico, DF	20.0	Central	Lopez-Carrillo et al. 1997
Veracruz	4500.00	Southeastern	Waliszewski et al. 2000
Morelos	20.0	Central	Lopez-Carrillo et al. 2001
San Luis Potosí	1715.0	Central	Yañez et al. 2004

Table 12. Total DDT levels in serum (ng/g lipid) of adults living in different Mexican Regions.

Location	Total DDT	Region	Reference
Coahuila	18400.0	Central	Albert et al. 1980
Mexico DF	6100.0	Central	Albert et al. 1980
Puebla	2700.0	Central	Albert et al. 1980
Veracruz	10000.0	Southeastern	Waliszewski et al. 1996
Veracruz	61000.0	Southeastern	Rivero-Rodriguez et al. 1997
Veracruz	5700.0	Southeastern	Waliszewski et al. 2001
Veracruz	2600.0	Southeastern	Waliszewski et al. 2010
Puebla	800.0	Central	Waliszewski et al. 2010
Veracruz	1900.0	Southeastern	Waliszewski et al. 2011
Veracruz	1400.0	Southeastern	Herrero-Mercado et al. 2010
Veracruz	900.0	Southeastern	Herrero-Mercado et al. 2011

Table 13. Total DDT levels in Adipose Tissue (ng/g lipid) of adults living in different Mexican Regions.

5.1 Cancer

Although it has been suggested that the estrogenic activity of DDE may be a contributing factor for development of breast cancer in women, levels of these compounds are not consistently elevated in breast cancer patients. It was initially reported that levels of $p,p2$-DDE were elevated in breast cancer patients (serum or tissue) versus controls [52].More recent studies and analysis of organochlorine levels in breast cancer patients versus controls show that these contaminants are not elevated in the latter group [53–56]. The study of occupationally exposed workers has not found clear increased risks for other cancers [57].

Two case-control studies of breast cancer have been carried out in Mexico City, with conflicting results. The first study, conducted by Lopez-Carrillo et al. [58] in Mexico City, compared 141 cases of breast cancer with 141 age-matched controls. All subjects were identified at three referral hospitals between March 1994 and April 1996. The arithmetic mean of serum DDE in lipid basis was 562 ppb±676 for the cases and 505 ppb±567 for the controls.The age-adjusted odds ratios for breast cancer regarding the serum level of DDE were 0.69 (95% confidence interval, 0.38–1.24) and 0.97 (CI, 0.55–1.70) for the contrasts between tertile 1 (lowest level) and tertiles 2 and 3, respectively. These estimates were unaffected by adjustment for body mass, accumulated time of breast-feeding and menopause, and other breast cancer risk factors. These results do not lend support to the hypothesis that DDT is causally related to breast cancer. The second study conducted by Romieu et al. [59] compared 120 cases and 126 controls, selected from six hospitals in Mexico City, from 1989 to 1995. Serum DDE levels in lipid basis were higher among cases (mean=3840 ppb±5980) than among controls (mean=2510 ppb±1970).After adjusting for age, age at menarche,duration of lactation, Quetelet index, and serum DDT levels, serum DDE levels were positively r elated to t he risk for b reast cancer (adjusted O R_{Q1-Q2}=1.24), 382 F. Díaz-Barriga et al. (CI, 0.50–3.06; OR$_{Q1}$ -$_{Q3}$=2.31, 95 percent, CI, 0.92–5.86; OR$_{Q1}$ -$_{Q4}$=3.81, CI, 1.14–12.80). The increased risk associated with higher serum DDE levels was more apparent among postmenopausal women (OR$_{Q1}$ -$_{Q4}$=5.26, 95%, CI, 0.80–34.30). Serum DDT level was not related to the risk for breast cancer. In addition to the differences in the comparison of cases and controls, the difference in the serum DDE levels among the women studied is remarkable. Participants from both studies came from similar hospitals, and there were no apparent differences between case and control selection that could explain this divergence. Differences in laboratory procedures is the most feasible explanation.

5.2 Endocrine disruption
DDT is known to have adverse effects on wildlife via endocrine disruption.Clear effects include thinning of the eggshell, feminization, reproduction impairment and development effects [60]. In Mexico two studies in humans have reported findings in this area. Gladen and Rogan [61] found that DDE might affect women's ability to lactate in a study conducted in an agricultural town in northern Mexico. Two hundred and twenty-nine women were followed from childbirth until weaning or until the child reached 18 months of age.DDE was measured in breast milk samples taken at birth, and women were followed to see how long they lactated.Median duration was 7.5 months in the lowest DDE group and 3 months in the highest. The effect was confined to those who had lactated previously – but not for first pregnancies – and it persisted after statistical adjustment for other factors. Rodriguez et al. [62] conducted a study aimed at determining the capability of long-term exposure to DDT of altering the normal endocrine function of the hypothalamus-hypophysis-gonads axis in humans. This included 70 workers dedicated to control malaria in the State of Guerrero,Mexico. The main activities of these workers were the application of pesticides,detection of malaria cases and promotion of preventive measures for control vectors. The average time of exposure to technical grade DDT was 25 years (range: 4–35), their last exposure being 5 months before sampling. An interview gathered information on the occupational history, reproductive performance, life styles and other relevant factors. Blood and urine samples were collected to measure serum levels of DDT and metabolites as

well as levels of LH, FSH, prolactine, and testosterone. Participants ranged in age from 22 to 69 years, and had been employed in the sanitation campaign from 4 to 37 years. Ninety-seven percent of the participants were sprayers of DDT at some time in their occupational history, and 15% are current sprayers. Average levels of DDT and metabolites expressed as µg/g of extractable lipids were: total DDT, 60.1; $p,p2$-DDE, 37.41; $p,p2$-DDT, 21.52; $p,p2$-DDD, 1.07 and $o2p$-DDT 0.11. Results show a positive association of LH and FSH with DDT metabolites. An increase of 10 mg/g of $p,p2$-DDE was associated with an increase of 1.95 UI/L in LH (p=0.01); and 1.10 UI/L per each 10 mg/g of $p,p2$-DDT (p=0.02). FSH increased 1.09 UI/L per each 10 mg/g of $p,p2$- DDT (p=0.03). There was a negative association of DDE with testosterone, DDT in Mexico 383 especially for those participants under 55 years of age. These associations suggest direct toxicity to the testicles, especially the Leydig cells, as observed with antineoplasic drugs.

5.3 Genotoxicity

Some studies have reported genotoxic effects in humans heavily exposed to DDT [8]. Therefore, this area has been studied in Mexico. Studies were done in workers from the control program of malaria, and in women living in malarious areas. Herrera et al. [63] evaluated chromosomal translocations in a sample of the above-mentioned workers.Nineteen male sprayers (median age 46 years; range: 22–64),working in campaigns to control malaria vectors in the state of Guerrero, Mexico were included in this study.DDT data obtained in a previous study from eleven individuals, 5 women and 6 men, living in Mexico City, and occupationally unexposed to DDT were used as reference group. Chromosomal aberrations in lymphocytes were analyzed with a chromosome painting technique, a highsensitivity technique for detecting complex chromosomal translocations. $o,p2$- DDT was the only isomer significantly associated with the frequency of chromosomal translocations (p=0.003). Individuals presenting the higher levels of $o,p2$-DDT in serum (>0.79 mg/g fat; n=4) had a mean frequency of chromosomal translocations (5.1¥1000 metaphases), two times higher than that observed in workers occupationally exposed to 0.5 Gy of radiation.A positive relationship between the duration of exposure to DDT, measured as years working for the vector control program, and chromosomal translocations was observed (Fig.2). These results suggest an increased risk for diseases with a genetic component, such as cancer.

Yañez et al. [64] evaluated the association of blood DDT levels and DNA damage using the single cell gel electrophoresis assay. A group of 53 postpartum women were selected from two different areas in San Luis Potosí to assure different exposure levels, one with antecedents of malaria and DDT spraying and the other without malaria. Mean and range levels of DDT, DDE and DDD in whole blood were 5.57 ppm (0.02–20.69), 6.24 ppm (0.04–39.16) and 1.16 ppm (0.01–5.63), respectively. The significant correlation of DNA damage, measured as DNA migration with the logarithm of DDT,DDE and DDD was 0.60, 0.62 and 0.43, respectively. Fig. 3 shows the shape of the association of DNA migration with DDE concentration in whole blood, as obtained by the regression: DNA migration= 71.58+7.62(logDDE), this association was not modified by age, smoking habits, nutrition or occupation. This observational finding in the epidemiological study with postpartum women was reevaluated in an in vitro study.Human blood cells were exposed to three doses of DDT, DDD and DDE.DNA damage was assessed by two different techniques: single-cell electrophoresis and flow cytometry. Results obtained by either technique showed that DNA damage was induced by the three organochlorides and a dose-response was observed with

DDT. The data suggest a DDT-induced DNA fragmentation, and this outcome was also observed with DDE and DDD.

6. References

[1] Stapleton DH (1998) Parassitologia 40 :149
[2] North American Commission for Environmental Cooperation (1998) North American Regional Action Plan on DDT. In: The Sound Management of Chemicals Initiative. Regional Commitments and Action Plans.Montreal, Canada, p 67
[3] Fernández de Castro J (1988) Panorama histórico y epidemiológico del paludismo en México. Secretaría de Salud,México
[4] Implementation Plan to Control Malaria with Alternatives Methods to DDT in Mexico 1999–2001.Government of Mexico, January 1999
[5] ISAT, 2002. Diagnóstico situacional del uso de DDT y el control de la malaria. Informe regional para México y Centroamérica. Instituto Salud Ambiente y Trabajo, México.

[6]ATSDR (2010) Toxicological Profile for DDT,DDE, and DDD.Agency for Toxic Substances and Disease Registry,Atlanta,Georgia
[7] Zeng EY,Yu CC, Tran K (1999) Environ Sci Technol 33:39
[8] Fisher SW (1995) Rev Environ Contam Toxicol 142:87
[9] Albert LA (1996) Rev Environ Contam Toxicol 147:1
[10] Lu FC (1995) Reg Toxicol Pharmacol 21 :352
[11] Woodruff T,Wolff MS, Davis DL,Hayward D (1994) Environ Res 65:132

6

Presence of Dichlorodiphenyltrichloroethane (DDT) in Croatia and Evaluation of Its Genotoxicity

Goran Gajski[1], Marko Gerić[1], Sanda Ravlić[2],
Željka Capuder[3] and Vera Garaj-Vrhovac[1]
[1]Institute for Medical Research and Occupational Health
[2]Ruđer Bošković Institute, Center for Marine Research
[3]Dr. Andrija Štampar Institute of Public Health
Croatia

1. Introduction

Pesticide is any substance or mixture of substances used for preventing, destroying or controlling any pest. According to the Food and Agriculture Organization (FAO) pest is defined as an organism that: (1) is a vector of human and animal disease; (2) can interfere with or cause harm to any step in food, wood and agricultural goods production, transport and storage; (3) can lower the quality of animal food or drugs. Extended definition of pesticide includes all substances: (1) used as plant growth regulator, defoliant and desiccant or agent for prevention of premature fruit fall; (2) applied to crops in order to protect them from deterioration during storage and transport (Food and Agriculture Organization [FAO], 2005). Every pesticide consists of active ingredient whose main function is to cause harm on the pest and inert carrier substance to improve storage, handling, application, efficiency and safety of the pesticide (World Health Organization [WHO], 1990).

For easier dealing with large numbers of pesticides, there are several classifications which divide pesticides in some classes. According to the chemical structure pesticides can be classified as: pyrethroids, carbamates, thiocarbamates, dithiocarbamates, organophosphates, organochlorides, phenoxy and benzoic acid herbicides, triazines, triazoles, ureas etc. (Kamir, 2000). Another classification is done considering the target organism, so pesticides can be classified as: insecticides, fungicides, herbicides, virucides, avicides, molluscicides, nematocides, rodenticides etc. (WHO, 1990). It is important to notice that pesticide's mode of action is not restricted just to the target organism; it also has negative effects on other organisms. That is why the World Health Organization (WHO) recommends the classification of pesticides by hazard. The base of this classification is determining the lethal dose (LD_{50}) for 50% of rat population after acute oral and dermal pesticide exposure. Depending of the LD_{50} value, as presented in Table 1, pesticides are classified as extremely, highly, moderately or slightly hazardous (WHO, 2009a).

It can be assumed that fighting against the pests started when human civilization was introduced to agriculture which ensured enough food for further development. The first known historical use of pesticide was observed in Mesopotamia 2500 BC. People in ancient Sumer used sulphur to treat crops thus killing pests (Miller, 2004). Around 500 BC ancient Greek historian Herodotus described the use of castor-oil plant as a mosquito repellent. People in Egypt who lived in swampy areas lit the castor-oil laps which had an unpleasant smell. Also they fixed fishing nets around their beds before going to sleep. This information suggests that ancient Egyptians used similar techniques to deal with mosquitos as today's insecticide-treated nets (Charlwood, 2003; WHO, 2007). After long period of natural pesticides use (e.g. arsenic, mercury, nicotine sulphate, pyrethrum, etc.), in 20th century the revolution of synthesized pesticides started.

Class		LD$_{50}$ for rats (mg/kg of body mass)			
		Oral		Dermal	
		Solids	Liquids	Solids	Liquids
I a	Extremely hazardous	5 or less	20 or less	10 or less	40 or less
I b	Highly hazardous	5 – 50	20 – 200	10 – 100	40 – 400
II	Moderately hazardous	50 – 500	200 – 2000	100 – 1000	400 – 4000
III	Slightly hazardous	over 500	over 2000	over 1000	over 4000

Table 1. The WHO pesticide classification based on LD$_{50}$ (adapted from WHO, 2005).

DDT (1,1,1-trichloro-2,2-di(4-chlorophenyl)ethane) was one of the most used pesticides in the mid 20th century. This organochloride was synthesized in 1874 by Othmar Zeidler and its pesticides properties were discovered in 1939 by Paul Hermann Müller who was awarded Nobel Prize in Physiology or Medicine for this discovery (Agency for Toxic Substances and Disease Service [ATSDR], 2002; Turusov et al., 2002). Commercially used DDT is a mixture of several isomeric forms: p,p'-DDT (85%), o,p'-DDT (15%), o,o'-DDT (trace amounts) and its breakdown products DDE (1,1-dichloro-2,2-bis(p-chlorophenyl)ethylene) and DDD (1,1-dichloro-2,2-bis(p-chlorophenyl)ethane). The basic chemical properties of p,p'-DDT, p,p'-DDE and p,p'-DDD are shown in Table 2. According to WHO, DDT is moderately hazardous pesticide with LD$_{50}$ of 113 mg/kg (WHO, 2009a).

In the period of almost 35 years, 2 million tons of DDT was used to control malaria and typhus, thus contaminating water, soil and air. When DDT and its metabolites enter the environment they have the potential to stay adsorbed in the sediment for more than 100 years. The best example of DDT's persistency is that it has been found in the areas 1000 km off the spraying spot and even in the Arctic animals and ice where it has never been used (ATSDR, 2002). One of the most popular acts that contributed global DDT ban in 1973, was Rachel Carson's book *Silent Spring* written in 1962 where she emphasized the problem of the DDT use (Carson, 2002). Today DDT is still legally produced (China, India, and North Korea) and used in the countries of the third world (e.g. India, Ethiopia, South Africa etc.) for malaria control (WHO, 2009b).

There are four possible ways for DDT to enter organisms: (1) ingestion, (2) inhalation, (3) dermal exposure, and (4) placental transport. Once in the body DDT is metabolized mainly in the liver and partly in the kidneys to its most familiar metabolites DDE and DDD. Detailed metabolic pathways are shown in Figure 1. Conjugated forms of 2,2-bis(4-chlorophernyl)-acetic acid (DDA) are then excreted through urine and faeces. The most dangerous property of DDT is that it possesses high bioaccumulation and biomagnification potential. As the trophic level rises, the concentration of DDT and/or its metabolites increases. All three compounds are reported to be harmful to either human or animals

(Brooks, 1986; Gold & Brunk, 1982, 1983; Morgan & Roan, 1974; Peterson & Robinson, 1964). As mentioned in the book *Silent spring*, DDT (especially metabolite DDE) has negative effects on bird's reproductive system by reducing Ca2+ transport which results with eggshell thinning thus increasing lethality (Carson, 2002).

	p,p'-DDT	*p,p'*-DDE	*p,p'*-DDD
Chemical formula	$C_{14}H_9Cl_5$	$C_{14}H_8Cl_4$	$C_{14}H_{10}Cl_4$
Chemical structure			
Molecular mass	354.49	318.03	320.05
Physical state	Solid[a]	Cristalline solid[a]	Solid[a]
Colour	Colourless crystals, white powder	White	Colourless crystals, white powder
Melting point	109 °C	89 °C	109 – 110 °C
Boiling point	Decomposes	336 °C	350 °C
Solubility in water	0.025 mg/L[b]	0.12 mg/L[b]	0.09 mg/L[b]
Solubility in organic solvents	Slightly soluble in ethanol, very soluble in ethyl ether and acetone	Lipids and most organic solvents	NA*

Table 2. Chemicals properties of *p,p'*-DDT, *p,p'*-DDE, and *p,p'*-DDD (adapted from ATSDR, 2002), [a] room temperature, [b] at 25°C, *NA – not available.

Not only birds are affected by this persistent pollutant, animals with higher amounts of fatty tissue have high levels of DDT and its metabolites (e.g. polar bears, orcas, belugas etc. (Crinnion, 2009; Galassi et al., 2008; Glynn et al., 2011; Okonkwo et al., 2008)). Although the acute DDT poisoning is rare, there are reports of negative effects on human health when exposed to low concentrations of DDT and its metabolites for longer periods. As shown in animals, the major damage was done on reproductive system where DDT can interfere with reproductive hormones. Also, there have been some reports regarding neurotoxicity, hepatotoxicity, imunotoxicity and genotoxicity (ATSDR, 2002). According to International Agency for Research on Cancer (IARC) DDT is classified as possible carcinogen to humans (IARC, 2009) and the increase in frequency of some cancers was detected (e.g. breast, testicular cancer etc.; Aubé et al., 2008; McGlynn et al., 2008).

As mentioned above, DDT introduced revolution in control of vector-borne diseases. Today, there are reports that some *Anopheles* species are resistant to DDT (Morgan et al., 2010) which implies that new methods of fighting vectors should be introduced. According to WHO one can observe the decrease in both number of countries using pesticides to control vectors (from 72 countries in 2006 to 61 countries in 2007) and in total amount of used DDT for vector control (from 5.1 million kg in 2006 to 3.7 million kg in 2007). Also WHO stimulates the import of new technologies to improve efficiency of insecticide-treated nets and long-lasting insecticidal nets (WHO, 2007). The importance of introducing new technologies for non harmful and efficient global pest fighting can be seen in estimation that only in United States, environmental and social costs of pesticide use are 9645 million dollars per year (Pimentel, 2005).

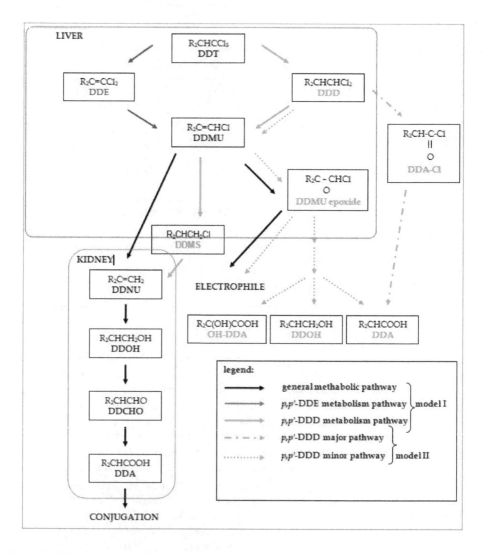

Fig. 1. Models of DDT metabolic pathways (adapted from Gold & Brunk, 1982 and Peterson & Robinson, 1964).

2. Presence of DDT in Croatia

2.1 Production and use in Croatia

Croatia, as a Mediterranean country, has been affected by malaria since the ancient times, especially in the coastal area. Only areas along the Velebit and Biokovo and the districts Hvar and Supetar have been spared from the disease. Most exposed to malaria have been the cities Nin, Benkovac, Obrovac, Skradin, Knin, Drniš, Šibenik, Imotski, Vrgorac and Metković area, the valley of the Neretva River, where malaria was called *Morbus naronianus* - Neretva disease. In 1820 French pharmacists Pierre Joseph Pelletier & Jean Biename Caventou isolated the alkaloid quinine, while doctor Lujo Adam from the island Lošinj was probably the first in the world who injected quinine sulphate into the veins of people sick from malaria. In 1902 doctor Rudolf Battara conducted the first controlled study of drug quinine prophylaxis on the overall population in Nin. The success of the study encouraged implementation of this prophylaxis in other places in Dalmatia in the upcoming years (Dugački, 2005).

DDT was first used in Croatia in 1941, when spraying was carried out against the lice of people as part of educational campaigns in the School of Public Health. The first mass use of DDT was carried out on the front of Srijem in 1945 by spraying soldiers as prevention of typhoid fever with the preparation "neocid", which was donated by the International Red Cross (Bakić, 2011). Companies Chromos from Zagreb and Zorka from Šabac, with their preparations Pantakan and Pepein, were the first producers of DDT in Croatia and the former Yugoslavia. Since 1946 DDT was produced in the form of dust, and since 1949 in the form of concentrated emulsions (Table 3). Production of pesticides in Croatia was small, but large quantities of pesticides were imported. Data on quantities of imported pesticides does not exist, except for the year 1957, when active substances for plant protection products in quantities of 1435 tons were imported. The reason for the continuing production decline of DDT was the lack of foreign exchange quotas for the purchase of organic substances necessary for the production of pesticides. Production and use of DDT in Croatia was until 1972, when its use was banned in agriculture, while in forestry DDT was still used until 1984 (Table 4; National Implementation Plan [NIP], 2004).

Year and purpose	Type of pesticide	Amount (t)	Note
1958. production	DDT	2150	1600 t sprayer concentrate
1959. plan	DDT	3000	550 t emulsifiable concentrate
1959. consumption	DDT	3122	
	DDT + Lindane	687	
1963. production	DDT	2327	
	DDT + Lindane	1854	4.6-6.6 % DDT + 0.3-0.7 lindane content of active substance in preparation

Table 3. The planned and produced quantities of DDT, according to the available data for Yugoslavia in some years (adapted from Hamel, 2003).

Year	Consumption of DDT in forestry (kg/year)	Consumption of DDT in agriculture (kg/year)
1963.	2312	-
1964.	-	1784
1965.	53428	1196
1966.	280	14051
1967.	-	16323
1968.	600	4183
1969.	-	6051
1970.	-	5450
1971.	2363	4296
1972.	4912	1078
1973.	884	0
1974.	8437	0
1975.	6907	0
1976.	8437	0
1979. – 1987.	18658	0

Table 4. Consumption of DDT in kg per year in agriculture and forestry in Republic of Croatia (adapted from NIP, 2004).

Today, list of active substances permitted for use in the Republic of Croatia is synchronized with the official list of active substances permitted in the means of the European Union. The regulations require from pesticide manufacturers to provide data on all possible risks to human health and the environment, as well as data on the effectiveness of pesticides and information on possible contamination in order to obtain licenses for the production and use.

2.2 Methodology
Residues of DDT and its metabolites were analyzed in samples of surface water, soil and food. One of the most sensitive techniques for measuring the rest of DDT and its metabolites in samples from the environment is gas chromatography with electron capture detector (GC/ECD). DDT and its metabolites were determined by applying analytical methods or modifications of the method: International Organization for Standardisation [ISO] 6468 (2002), ISO 10382 (2002), Reference Methods for Marine Pollution (United Nations Environment Programme/International Atomic Energy Agency [UNEP/IAEA], 1982) and EN 1528 1-4 (1996), according to the scheme; extraction, purification of the extracts and quantitative analysis.

2.2.1 Sample extraction and cleanup
Water samples, two to four litres, were extracted with methylene chloride (HPLC grade for spectroscopy, Merck, Darmstadt, Germany) as solvent using an Ultra Turrax system. Extracts were dried over granular anhydrous sodium sulphate and concentrated using rotary evaporator (ISO 6468, 2002). The soil sample was sieved (<2 mm) and stored at room temperature for two days before the experiments. 10 g soil sample (dry matter) was extracted in an ultrasonic bath (ISO 10382, 2002) and the extract was dried by passing them through anhydrous sodium sulphate. Determination of DDT in food samples was based on

method EN 1528 1-4 (1996). Approximately 10-20 g of food were homogenized and extracted with hexane, the extract was evaporated in a rotary evaporator and in nitrogen stream to ensure dryness. Contents of fat were determined weighing a dry sample. Milk fat portions were extracted from each individual sample of milk, cheese, cream and yoghurt according to the method by Sannino et al. (1996). Butter does not normally require extraction procedures. An automated gel permeation chromatographic (GPC) procedure was used to determine DDT residues in fatty foods. About three grams of the fat or less was dissolved into methylene chloride and cleaned up by GPC with a Biobeads SX3 column (OI Analytical, College Station, TX, USA) and a methylene chloride as eluant. About 10 g of fish tissue was weighed, homogenized with anhydrous sodium sulphate and extracted with pesticide grade hexane (Merck, UNEP/IAEA, 1982). The extract was condensed in a rotary flask vaporator to a specific aliquot (5 ml). The aliquot was then subjected to acid treatment by adding concentrated sulphuric acid (Merck). All samples were cleaned up with florisil, commercially available cartridges, 3-6 mL, 500-1000 mg (Kodba & Voncina, 2007). Extracts were concentrated using evaporator in nitrogen stream. An aliquot of each extract was transferred to vials for the quantitative analysis.

2.2.2 Gas chromatograph analysis

The samples were analyzed by gas chromatograph (GC) Shimadzu (Models GC 2010 and GC 17A series, Tokyo, Japan) equipped with autosampler and electron capture detector (ECD) on two fused silica capillary columns of different polarity. Nitrogen was used as the carrier gas with a flow rate of about 1.4 mL/min and as the makeup gas. The injection was set at splitless mode. The injection port and detector temperature were 250 and 300 ^{0}C, respectively. The compounds p,p'-DDE, p,p'-DDD, o,p'-DDT and p,p'-DDT were identified by comparing peak retention times between samples and known standards. The standard samples of the 18 pesticides were obtained from Dr. Ehrenstorfer GmbH (Augsburg, Germany, and AccuStandard Inc., New Haven, CT, USA) with the purities of 97-99%. Calibration standard curves were created and DDT residues were quantitatively determined by comparison of the retention times and peak areas of the sample chromatogram with those of standard solutions run under the same operating conditions. Peak confirmation was done by running the samples and the standard on another column and comparing. The concentrations of DDT residues in each sample were reported as ng/L or µg/kg. Limit of quantification for every matrix is presented in Table 5. The laboratory has participated (annually) in intercalibration study proficiency test.

Limit of quantification					
		p,p'-DDE	p,p'-DDD	o,p'-DDT	p,p'-DDT
Water	µg/L	0.0005	0.001	0.001	0.001
Soil (dry mater)	µg/g	0.002	0.003	0.004	0.003
Milk (liquid) and milk products (milk fat)	µg/g	0.0001	0.0004	0.0004	0.0004
Fish and fish products (wet weight)	µg/g	0.0005	0.001	0.0006	0.001
Food (fat)	µg/g	0.001	0.002	0.004	0.002

Table 5. Limit of quantification for p,p'-DDE, p,p'-DDD, o,p'-DDT and p,p'-DDT.

2.3 Results

During his lifetime a human being has been exposed to DDT and other pesticides through food, water and the environment. A level of these exposures can be established by analysing of samples from the environment. All food samples come from the market in Croatia while samples of surface water and soil were from different areas of Croatia. Food, particularly dairy products, meat and fish, has been identified as the primary immediate intake route of DDT and other organochlorine pesticides for the population (Johansen et al., 2004).

Samples of surface water of the rivers Sava, Krka, Mrežnica, Kupa, Zrmanja and Cetina were analyzed. The concentration of total DDT in surface waters in Croatia for the period 2009-2010 amounted from below detectable limit to 0.021 µg/L, with median value 0.0016 µg/L (Table 6). The average concentrations of total DDT in surface waters were 0.0026 µg/L. These results comply with those reported by Drevenkar et al. (1994), Drevenkar & Fingler (2000), and Fingler et al. (1992) in their papers. DDT may reach surface waters primarily by runoff, atmospheric transport and drift, or by direct application (e.g. to control mosquito-borne malaria). DDT is practically insoluble in water; but some DDT may be adsorbed onto the small amount of particulate matter present in water (ATSDR, 2002).

The concentration of DDT and metabolites in analyzed samples of soil were below 0.5 µg/g of dry matter (legally prescribed limit values). These analyses were conducted on the samples collected over the last two years and only from a few locations and therefore cannot be found sufficient to make any kind of general conclusions about the current situation. There is no systematic monitoring of DDT in soil in Croatia, resulting with little relevant data. Studies such as Picer et al. (2004) are rare examples of organized research in this area in Croatia, conducted as a part of post war damages in areas were there were concerns that soil was contaminated.

Data on levels and distribution of DDT and other persistent organic pollutant (POPs) in surface waters and soils in Croatia is insufficient, despite these hydrophobic substances being extremely important for assessing environmental contamination. DDT compound in sample of fish tissue were present in very low concentrations, although it is well known that DDT bioaccumulates in marine species. The mean value of DDT in fish tissue, analyzed in year 2007, was 3.8 µg/kg wet weights, with median value of 4.7 µg/kg . Similar results are reported by Krautchaker & Reiner (2001), and Bošnir et al. (2007). The five dairy products have been examined for the rest of DDT: milk, butter, cheese, cream and yoghurt. The mean values of the residual concentrations of total DDT in the examined dairy products were 29.6 µg/kg fat, respectively, with median 25.2 µg/kg (Table 6). The presence of DDT in milk and milk products has also been reported by Krautchaker & Reiner (2001) and Bošnir et al. (2010) as well as in other countries (Nevein et al., 2009; Dawood et al., 2004; Bulut et al., 2010). These studies have found that DDT complex were the most frequent contaminants in dairy products. Heck et al. (2006) concluded there is no difference in DDT in raw and pasteurized milk. The concentration of DDT in meat products sampled at food markets was 1.2-740.0 µg/kg fat, with median value 16.2 µg/kg fat (Table 6). These results should be taken with caution since the origin of meat is unknown; whether the meat is from domestic production or imported from other countries of the world. Meat is imported primarily from developing countries where there is limited or no control over the use and/or control of pesticide residues in foods. These results comply with those reported by Krautchaker & Reiner (2001), Covaci et al. (2004), and Tompić et al. (2011).

	Value range	Mean	Median
Surface and ground water (ng/L)	0.6 – 20.5	2.6	1.6
Soil (µg/g; dry matter)	0 – 0.005	0.002	0.002
Fish and fish products (µg/kg; wet weight)	0.2 – 31.0	3.8	4.7
Meat and meat products (µg/kg; fat)	1.2 – 74.0	51.7	16.2
Milk and milk products (µg/kg; milk fat)	11.3 – 79.9	29.6	25.2

Table 6. Concentrations of DDT compounds in different samples.

Lamb meat results (Table 7) demand special attention, since the concentrations of p,p'-DDE were generally high. Similar results were found by Tompić et al. (2011) in samples of lamb imported from Bulgaria. p,p'-DDE are found in every examined sample of lamb meat, which indicates the need for continuous monitoring of concentrations of this metabolites in samples of lamb meat.

	p,p'-DDE (µg/kg)	p,p'-DDD (ng/g)	p,p'-DDT (ng/g)	Total DDT (ng/g)	Total DDT (µg/kg)
Lamb meat	739.0	-	0.4	74.3	743.0
Lamb meat	173.0	-	0.5	17.8	178.0
Lamb meat	609.0	0.6	-	1.5	615.0
Lamb meat	964.0	6.1	1.2	103.7	1037.0
Lamb meat	354.0	1.4	0.5	37.3	373.0

Table 7. Concentrations of DDT compounds in samples of lamb meat.

Generally, it was observed that the total DDT or its metabolites residues were bellow acceptable and legally prescribed boundaries. These results highlight the need for regular analyzing of a larger number of samples from the environment to DDT residues and other chemicals of POPs, especially in imported food.

3. Cytogenetic methods for detection of pesticide genotoxicity

Pesticide exposure is ubiquitous, due not only to agricultural pesticide use and contamination of foods, but also to the extensive use of these products in and around residences. Because of their biological activity, the use of pesticides may cause undesired effects to human health. Pesticides tend to be very reactive compounds that can form covalent bonds with various nucleophilic centers of cellular biomolecules, including DNA (Crosby, 1982). For instance, the induction of DNA damage can potentially lead to adverse reproductive outcomes, the induction of cancer and many other chronic diseases (Ribas et al., 1996; Lander et al., 2000; Meinert et al., 2000; Ji et al., 2001). A great variety of tests and test systems based on microbes, plant and animals have been developed in order to asses the genotoxic effects of xenobiotic agents, including pesticides. Biomonitoring studies on human populations exposed to pesticides are employing circulating lymphocytes as biomarkers of exposure (and perhaps of effect). Those studies have essentially focused on cytogenetic end-points such as chromosomal aberrations (CA), sister-chromatid exchanges (SCE) and micronuclei (MN) frequency. Genetic damage at the chromosomal level entails an alternation in either chromosome number or chromosome structure, and such alternations can be measured as CA or MN frequency. The SCE analysis was also adopted as an indicator of genotoxicity, although the exact mechanism that leads to an increased exchange of segments between sister chromatids is not known in detail at present (Palani-Kumur & Panneerselvam, 2008). Recent studies revealed the nucleotide pool imbalance can have severe consequences on DNA metabolism and it is critical in SCE formation. The modulation of SCE by DNA precursors raises the possibility that DNA changes are responsible for the induction of SCE and mutations in mammalian cells (Popescu, 1999; Ashman & Davidson, 1981). While increased levels of CA have been associated with increased cancer risk (Hagmar et al, 1994, 1998), a similar conclusion has not been reached for SCE or MN. However, high levels of SCE and MN frequency have been observed in persons at higher cancer risk due to occupational or environmental exposure to a wide variety of carcinogens (Fučić et al, 2000; Vaglenov et al, 1999; Fenech et al, 1997). Evidence of CA increases, mainly as structural chromosomal aberrations in occupationally exposed populations. The sensitivity of SCE is lower than that of the CA test in detecting genotoxic effects related to pesticide exposure and fewer data are therefore, available for MN than for the other cytogenetic endpoints (Bolognesi, 2003). Exposure to potential mutagens or carcinogens can provide an early detection system for the initiation of cell disregulation. Biomarkers of effect are generally pre-clinical indicators of abnormalities and the most frequently used in genotoxicity assessment are comet assay and cytokinesis-block micronucleus test that are being proposed as a useful biomarkers for early effects. The cytogenetic endpoints can give indication of genetic damage; hence they are used as effective biomarkers of exposure *in vivo* and *in vitro*. In recent years, the comet (single-cell gel) assay has been established as a useful technique for studying DNA damage and repair (Tice, 1995). The comet assay combines the single-cell approach typical of cytogenetic assays with the simplicity of biochemical techniques for detecting DNA single strand breaks. The advantages of the comet assay include its simple and rapid performance, its sensitivity for detecting DNA damage, and the use of extremely small cell samples (Hartmann et al., 1998). The advantage of micronucleus assay is its simplicity and speed over the assay of chromosomal aberration. Both techniques have become an important tool for genotoxicity testing because of their simplicity of scoring and wide applicability in different cell types. These techniques became the methods of choice for studies of environmental and occupational exposure to air pollutants, metals, radiation, pesticides, and other xenobiotics.

3.1 Comet assay
The comet assay, also known as the single-cell gel electrophoresis assay (SCGE), is a method for detecting DNA strand breakage (single-strand DNA breaks, alkali-labile sites, double-strand DNA breaks, incomplete repair sites, and inter-strand cross-links) in virtually any nucleated cell (Collins et al., 2004, 2008; Shaposhnikov et al., 2008).
First quantification of DNA damage in individual cells was done by Rydberg & Johanson (1978). After gamma–irradiation they embedded cells in agarose on microscopic slides and lysed under mild alkali conditions. Upon neutralization, the cells were stained with acridin orange and the extent of DNA damage was measured by the ratio of green (indicating double–stranded DNA) to red (indicating single–strand DNA) fluorescence. To enhance the sensitivity for the DNA damage detection, Östling & Johanson (1984) proposed that strand breaks would enable DNA loops to stretch out upon electrophoresis, so the microelectrophoretic procedure under pH of 9.5 was developed. As reported by Singh et al. (1988) this pH of 9.5 is below the limit for DNA unwinding, and was notify to detect only double strand breaks (DSB), with more strongly alkaline conditions (pH 10 or above) needed for unwinding and detection of single strand breaks (SSB). It has been shown that

neutral or mildly alkaline comet assay has the same limit of detection of DNA damage (SSB) as the alkaline comet assay, although the use of neutral pH does effect the comet image obtained (Collins, 2004). The comet tails are less pronounced at netural pH, and this can be an advantage when a less sensitive method is needed, for example when investigating cells that have large amount of background, or induced damage is high (Angelis et al., 1999).

Alkaline version of the comet assay was presented by Singh et al. (1988) in which DNA is allowed to unwind at pH>13. In their paper DNA damage was measured as the migrating distance of DNA from the nucleoid. In 1990, Olive et al. (1990) also under alkaline conditions developed the concept of the tail "moment", a combination of tail length and DNA content, as a measure of DNA damage. Also, in 1990, the name "Comet assay" was introduced and the application of the first image analysis program was described (Olive, 1989; Olive et al., 1990; Sviežená et al., 2004). Image analysis has become essential for objective measurement of low-dose effects, or for distinguishing small differences among sub-populations of cells. Strong alkaline conditions enabled clearer images, and besides SSBs other types of DNA damage could be detected, such as alkaline labile sites (Tice et al., 2000). Olive et al. (1990) revealed that employing milder alkaline (pH 12.3) conditions prevents conversion of alkaline labile sites into breaks. Under the pH>13 alkali labile sites are formed into SSBs, thus revealing otherwise hidden damage. Therefore, by modifying the pH of lysis and/or electrophoresis over the range of 9.5-13.5, one can apply a comet assay of different sensitivity, but of similar limits of detection (Collins et al., 1997; Angelis et al., 1999; Wong et al., 2005).

In its basic form, comet assay gives limited information on the type of DNA damage being measured. Single strand breaks detected by standard alkaline method are not the most interesting of lesions, because they are quickly repaired, and are not regarded as a significant lethal or mutagenic lesion. Many genotoxic agents do not induce strand breaks directly. They may create apurinic/pyrimidinic (AP) sites, which are alkali labile and are probably converted to breaks while DNA is in the electrophoresis solution at high pH. Furthermore, it is not possible to determine whether the high level of breaks in the comet assay is the indicator of high damage or efficient repair, due to temporary presence of breaks in the lesions repair via base excision or nucleotide excision (Collins et al., 1997).

More recently, the assay was modified further to enable the detection of specific kinds of DNA damage by combining the assay with the use of a purified DNA repair enzymes, which recognize the lesions along the DNA and convert them into the breaks expressed as an increase in comet DNA migration. Briefly, the DNA in the gel, following lysis, is digested with a lesion-specific repair endonuclease, which introduces breaks at sites of damage. In principle, any lesion for which a repair endonuclease exists can be detected in this way. To date, endonucleases most commonly used in the modified comet assay are the bacterial enzymes which recognize different types of oxidative damage. The first enzyme to be used was endonuclease III, a glycosylase which recognizes a variety of oxidized pyrimidines in DNA and removes them, leaving an AP site (Doetsch et al., 1987). Formamidopyrimidine DNA glycosylase (FPG) has the ability to convert altered purines, including 8-oxoguanine, into DNA breaks (Collins, 2007; Dušinská & Collins, 1996). When using these enzymes to measure oxidative DNA damage the usual practice is to incubate a slide with buffer alone in parallel with the enzyme slide. Slide with buffer would be a valid control slide, due to small increase in strand breaks on incubation without enzyme (Collins, 2009; Gajski et al., 2008). Recently, a mammalian analogue of FPG, 8-oxoguanidine DNA glycosylase, or OGG1, has been applied in the comet assay (Smith et al., 2006). OGG1 is the major base extension repair enzyme that initiates the repair of oxidative base lesion. It is a bifunctional DNA glycosylase possessing both DNA glycosylase and AP lyase activities (Boiteux & Radicella, 2000). In human it is named hOGG1. hOGG1 recognizes both 8-oxoguanine (8-oxodG) and 8-oxoadenine (8-oxodA) and removes these oxidized bases from double-stranded DNA, initiating the base lesion repair process (Smith et al., 2006). 8-hidroxy-2-deoxyguanosine (8-OHdG) lesion causes G→T and A→C transversions (Moriya, 1993) that have been reported as the sites of spontaneous oncogene expression and ultimately cancer manifestation (Valko et al., 2004; Bartsch, 1996; Shinmura & Yokota, 2001). Deletion of the hOGG1 gene was shown to be associated with accumulation of 8-OHdG lesion and increase in mutational risk (Hansen & Kelley, 2000; El-Zein et al., 2010).

For evaluation of DNA specific damage the comet assay has also been coupled with the method of fluorescent in situ hybridization (Comet–FISH). Since its initial development, Comet–FISH has been used to handle a number of quite different questiones. First, it was used to identify chromosome-specific areas on electrostretched DNA fibres and to determine their special distribution (Santos et al., 1997). Further applications were then to detect region–specific repair activities (Horvathova et al., 2004; McKenna et al., 2003; Mellon et al., 1986), genotoxic effects in total DNA and in telomeres (Arutyunyan et al., 2005), or in tumor relevant genes, like TP53 (Schaeferhenrich et al., 2003). Also specific chromosomal alternations (Harreus et al., 2004) were studied as were genetic instabilities (Tirukalikundram et al., 2005). Comet–FISH has also been used to discriminate between DNA double-and single-strand breaks (Fernandez et al., 2001). Whereas results from the Comet assay alone reflect only the level of overall DNA damage, the combination with the FISH-technique allows the assignment of the probed sequences to the damaged or undamaged part of the comet (tail or head, respectively). If two fluorescence signals are obtained with a probe for a particular gene in the head of a comet, this indicates that the gene is in an undamaged region of DNA, whereas the appearance of a spot or several spots in the tail of a comet indicates that a break or breaks has/have occurred in the proximity of the probed gene.

3.2 Micronucleus assay

Human exposure to environmental mutagens can be monitored using cytokinesis–block micronucleus (CBMN) assay (Natarajan et al., 1996) which is an efficient biomarker for diagnosing genetic damage and/or genome instability at the chromosome/molecular level in animal and/or human cells. It provides a comprehensive measure of chromosome breakage, chromosome loss, chromosome rearrangments, non-disjuction, gene amplification, necrosis and apoptosis (Fenech, 2000, 2006; Kirsch-Volders et al., 2000).

In the classical cytogenetic techniques, chromosomes are studied directly by observing and counting aberrations in methaphases (Natarajan & Obe, 1982). The complexity and laboriousness of enumerating aberrations in methaphase and the confounding effect of artefactual loss of chromosomes from methaphase preparations has stimulated the development of a simpler system of measuring chromosome damage. More than a century ago micronuclei were described in the cytoplasm of erythrocytes and were called "fragment of nuclear material" by Howell or "corpuscules intraglobulaires" in the terminology of Jolly in the late 1800s and early 1900s. To the hematologists these structures are known as "Howell-Jolly bodies". Similar structures were described in mouse and rat embryos and in *Vicia faba* (Thoday, 1951) and called "fragment nuclei" or "micronuclei". In the early 1970s the term micronucleus test was suggested for the first time by Boller & Schmidt (1970) and

Heddle (1973) who showed that this assay provided a simple method to detect the genotoxic potential of mutagens after *in vivo* exposure of animals using dividing cell population such as bone marrow erythrocytes. A few years later it was shown by Countryman & Heddle (1976) that peripheral blood lymphocytes could also be used for the *in vivo* micronucleus approach and they recommended using micronuclei as a biomarker in testing schemes. As only dividing cells could express micronuclei, for the *in vitro* micronucleus studies it was necessary to establish cell proliferation and micronucleus induction at the same time. The decisive breakthrough of micronuclei as assay for *in vitro* genotoxicity testing came with work of Fenech & Morley (1986) that developed the CBMN assay. In the CBMN assay, cells that have completed nuclear division are blocked from performing cytokinesis using cytochalasin–B and are consequently readily identified by their binucleated appearance (Fenech & Morley, 1985, 1986). Whereas micronuclei originate from chromosome fragments or whole chromosomes that leg behind at anaphase during nuclear division, their occurrence is proven in binucleated cells. As a consequence, the CBMN assay has been shown to be more accurate and more sensitive than the conventional methods that do not distinguish between dividing and nondividing cells (Fenech & Morley, 1986; Fenech, 1991; Kirsch-Volders & Fenech, 2001).

Baseline or spontaneous micronucleus frequencies in culture human lymphocytes provide an indicator of accumulated genetic damage occurred during the lifespan of circulating lymphocytes. The half-life and mean lifespan of T-lymphocytes has been estimated to be three to four years, respectively (Natarajan & Obe, 1982; Buckton et al., 1967). The observed genetic instability may also reflect accumulated mutations in the stem cell lineage from which the mature lymphocytes originate. The type of mutations that could contribute to spontaneous micronuclei include: 1) mutations to kinetochore proteins, centromeres, and spindle apparatus that could lead to unequal chromosome distribution or whole chromosome loss at anaphase, and 2) unrepaired DNA strand breaks induced endogenously or as a result of environmental mutagens, which may result in acentric chromosome fragments. Studies using kinetochore antibodies to identify whole chromosomes suggest that approximately 50% of spontaneously occurring micronuclei are the consequence of whole chromosome loss and the rest are presumably derived from acentric chromosome fragments (Thompson & Perry, 1988; Fenech & Morley, 1989; Eastmond & Tucker, 1989). The spontaneous micronucleus frequency refers to the incidence of micronucleus observed in the absence of the environmental risk or exposure that is being assessed. The spontaneous micronucleus frequency of a population has to be established to determine acceptable normal values as well as providing baseline data for those situations when spontaneous micronucleus frequencies for individuals is not known before exposure. Micronuclei harbouring whole chromosomes are primarily formed from failure of the mitotic spindle, kinetochore, or other parts of the mitotic apparatus or by damage to chromosomal sub-structures, alterations in cellular physiology, and mechanical disruption. An increased number of micronucleated cells is a biomarker of genotoxic effects and can reflect exposure to agents with clastogenic modes of action (chromosome breaking; DNA as target) or aneugenic ones (aneuplodigenic; effect on chromosome number; mostly non-DNA target) (Albertini, 2000). The advantage of the CBMN assay is its ability to detect both clastogenic and aneugenic events, leading to structural and numerical chromosomal aberrations, respectively (Kirsch-Volders et al., 2002; Mateuca et al., 2006). Micronuclei observed in cultured lymphocytes are believed to arise primarily *in vitro* from: 1) chromatid-type chromosomal aberrations formed during DNA replication on a damaged template, 2) chromosome-type aberrations initiated before the mitosis and duplicated at replication, or 3) disturbances of mitotic apparatus leading to chromosome lagging. Micronuclei arising *in vivo*, inducible by both clastogenic and aneugenic mechanisms, can be scored in exfoliated epithelial cells (Salama et al., 1999) sampled, e.g., from buccal or nasal mucosa or urine, or in peripheral blood mononuclear cells e.g., isolated lymphocytes (Surrallés et al., 1996; Albertini, 2000). The CBMN assay is the preferred method for measuring micronuclei (MNi) in cultured human and/or mammalian cells because scoring is specifically restricted to once–divided cells. For scoring MNi uniform criteria should be used. Only MNi not exceeding 1/3 of the main nucleus diameter, clearly separable from the main nucleus and with distinct borders and of the same color as the nucleus, should be scored. In practice, 1000–2000 cells are often scored per subject in lymphocyte studies utilising the CBMN technique, while more cells (3000–5000 per subject) are evaluated in epithelial cells due to the lower baseline MNi frequency. In the CBMN method, only binucleate cells should be analysed for MNi; further divisions of a binucleate cell, usually resulting in cells with 3–4 nuclei, are highly irregular and show high MNi rates (Fenech et al., 2003).

The discovery of kinetochore-specific antibodies in the serum of scleroderma CREST (*Calcinosis, Raynaud's phenomenon, Esophageal dysmotility, Sclerodactyly and Telangiectasia*) patients (Moroy et al., 1980) has made it possible to determine the contents of micronuclei. Kinetochore immunofluorescence has been rapidly developed for the *in situ* detection of aneuploidy and chromosome breakage in human micronuclei (Degrassi & Tanzarella, 1988; Hennig et al., 1988; Eastmond & Tucker, 1989). Also, the FISH technique using chromosome-specific DNA probes has improved the detection and evaluation of structural chromosomal aberrations. The combination of the micronucleus assay with FISH using a DNA probe specific to the centromeric regions, or with antibodies that specifically stain kinetochore proteins, provides the methodology to distinguish between micronuclei containing either one or several whole chromosomes, which are positively labeled (centromere positive micronucleus), or acentric chromosome fragments, which are unlabeled due to the absence of centromere (centromere negative micronucleus) (Natarajan et al., 1996; Mateuca et al., 2006; Benameur et al., 2011). Except for cytogenetic damage measured by the number and distribution of micronuclei, according to the new criteria for micronuclei scoring, the CBMN assay also detects the nucleoplasmic bridges (NPBs), as well as nuclear buds (NBUDs). Current evidence suggests that NPBs derive from dicentric chromosomes which the centromeres have been pulled to the opposite poles of the cell during the anaphase stage, and are therefore indicative of the DNA mis-repair, chromosome rearrangement or telomere end-fusion. According to the new criteria applicable to the CBMN assay, NBUDs arise from the elimination of the amplified DNA and possibly from the elimination of the DNA-repair complexes, which therefore, may be considered a marker of gene amplification and altered gene dosage (Fenech, 2006; Fenech & Crott, 2002; Fenech et al., 2011; Garaj- Vrhovac et al., 2008; Lindberg et al., 2006; Thomas et al., 2003).

The significance of the CBMN assay lies in the fact that every cell in the system studied is scored cytologically for its viability status (necrosis, apoptosis), its mitotic status (mononucleated, metaphase, anaphase, binucleated, multinucleated) and its chromosomal instability or damage status (presence of MNi, NPBs, NBUDs and number of centromere probe signals amongst nuclei of binucleated cell if such molecular tools are used in combination with the assay). In this respect, the micronucleus assay has evolved into a comprehensive method employed in measuring chromosomal instability of the phenotype and altered cell viability and represents an effective tool to be used in research of cellular and nuclear dysfunctions caused by *in vitro* or *in vivo* exposure to toxic substances (Fenech, 2006; Garaj- Vrhovac et al., 2008; Thomas & Fenech, 2011).

4. Genotoxicity of *p,p'*-DDT in human peripheral blood lymphocytes

4.1 Methodology
4.1.1 Chemicals
Chromosome kit P was from Euroclone, Milan, Italy; RPMI 1640, was from Invitrogen, Carlsbad, CA, USA; cytochalasin B, histopaque-1119, ethidium bromide, low melting point (LMP) and normal melting point (NMP) agaroses were from Sigma, St Louis, MO, USA; heparinised vacutainer tubes from Becton Dickinson, Franklin Lakes, NJ, USA; acridine orange form Heidelberg, Germany; Giemsa from Merk, Darmastad, Germany; All other reagents used were laboratory-grade chemicals from Kemika, Zagreb, Croatia. DDT was administered as *p,p'*-DDT (Sulpeco, Bellefonte, PA, USA) in final concentration of 0.025 mg/L at different time points.

4.1.2 Blood sampling and treatment
Whole blood samples were taken from a healthy female donor who had not been exposed to ionizing radiation, vaccinated or treated with drugs for a year before blood sampling. Whole venous blood was collected under sterile conditions in heparinised vacutainer tubes containing lithium heparin as anticoagulant. The comet assay and the micronucleus assay were conducted on whole blood cultivated at 37 °C in an atmosphere with 5% CO_2 (Heraeus Heracell 240 incubator, Langenselbold, Germany). The whole blood was treated with 0.025 mg/L *p,p'*-DDT for 1, 2, 4, 8, 24 and 48 h for the cytotoxicity assay and alkaline comet assay and 24 and 48 h for the CBMN assay. In each experiment, a non treated control was included.

4.1.3 Cell viability (cytotoxicity) assay
The indices of cell viability and necrosis were established by differential staining of human peripheral blood lymphocytes (HPBLs) with acridine orange and ethidium bromide using fluorescence microscopy (Duke & Cohen, 1992). Lymphocytes were isolated using a modified Ficoll-Histopaque centrifugation method (Singh, 2000). The slides were prepared using 200 μL of HPBLs and 2 μL of stain (acridine orange and ethidium bromide). The suspension mixed with dye was covered with a cover slip and analyzed under the epifluorescence microscope (Olympus AX 70, Tokyo, Japan) using a 60× objective and fluorescence filters of 515–560 nm. A total of 100 cells per repetition were examined. The nuclei of vital cells emitted a green fluorescence and necrotic cells emitted red fluorescence.

4.1.4 The alkaline comet (SCGE) assay
The alkaline comet assay was carried out as described by Singh et al. (1988). Briefly, after the exposure to *p,p'*-DDT, 5 μL of whole blood was mixed with 100 μL of 0.5% LMP agarose and added to fully frosted slides pre-coated with 0.6% NMP agarose. After solidifiying, the slides were covered with 0.5% LMP agarose, and the cells were lysed (2.5 M NaCl, 100 mM EDTANa2, 10mM Tris, 1% sodium sarcosinate, 1% Triton X-100, 10% dimethyl sulfoxide, pH 10) overnight at 4°C. After the lysis the slides were placed into alkaline solution (300 mM NaOH, 1 mM EDTANa2, pH 13) for 20 min at 4°C to allow DNA unwinding and subsequently electrophoresed for 20 min at 1 V/cm. Finally, the slides were neutralized in

0.4 M Tris buffer (pH 7.5) for 5 minutes 3 times, stained with EtBr (20 μg/mL) and analyzed at 250× magnification using an epifluorescence microscope (Zeiss, Göttingen, Germany) connected through camera to an image analysis system (Comet Assay II; Perceptive Instruments Ltd., Haverhill, Suffolk, UK). The tail length parameter was used to measure the level of DNA damage and a total of 100 randomly captured nuclei were examined from each slide. Tail length (i.e. the length of DNA migration) is related directly to the DNA fragment size and is presented in micrometers (μm). It was calculated from the centre of the nucleus.

4.1.5 Cytokinesis-blocked micronucleus (CBMN) assay
The micronucleus assay was performed in agreement with guidelines by Fenech & Morley, (1985). After the exposure to p,p'-DDT the whole blood (500 μL) was incubated in a Euroclone medium at 37 °C in an atmosphere of 5% CO_2 in air. Cytochalasin-B was added at a final concentration of 3 μg/mL 44 h after the culture was started. The cultures were harvested at 72 h. The lymphocytes were fixed in methanol-acetic acid solution (3:1), air-dried and stained with 5% Giemsa solution. All slides were randomised and coded prior to analysis. Binuclear lymphocytes were analyzed under a light microscope (Olympus CX41, Tokyo, Japan) at 400× magnification. Micronuclei, nucleoplasmic bridges and nuclear buds were counted in 1000 binucleated cells and were scored according to the HUMN project criteria published by Fenech et al. (2003).

4.1.6 Statistics
For the results of the comet assay measured after treatment with p,p'-DDT statistical evaluation was performed using Statistica 5.0 package (StaSoft, Tulsa, OK, USA). Multiple comparisons between groups were done by means of ANOVA on log-transformed data. Post hoc analyses of differences were done by using the Scheffé test. Differences in the frequency of micronuclei, nucleoplasmic bridges, and nuclear buds were assessed using the chi-square test. $P<0.05$ was considered significant.

4.2 Results
4.2.1 Vital staining using ethidium bromide and acridine orange
The viability of HPBLs exposed to aqueous p,p'-DDT (0.025 mg/L) for different lengths of time as determined by acridine orange and ethidium bromide staining, using fluorescence microscopy was not significantly affected (data not shown). Changes were determined according to the different staining of the nucleus (Figure 2).

 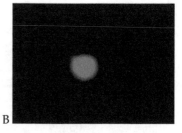

Fig. 2. Cell viability microphotographs represent viable lymphocyte from the un-exposed sample (A; green), and dead lymphocyte (B; red) from sample treated with aqueous solution of p,p'-DDT.

4.2.2 Induction of DNA strand breaks

The whole blood was exposed to aqueous solution of p,p'-DDT (0.025 mg/L) and the DNA damage in HPBLs was determined with the alkaline comet assay. Figure 3 represents different levels of DNA fragmentation between non-exposed sample and sample exposed to aqueous solution of p,p'-DDT. Statistically significant ($P<0.05$) increase in the amount of DNA strand breaks was observed after all exposure times to p,p'-DDT (Figure 4).

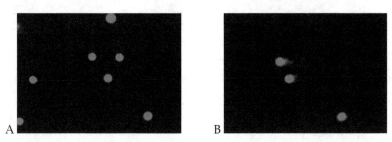

Fig. 3. Comet assay microphotographs represent undamaged lymphocytes from the un-exposed sample (A). Image (B) represents damaged lymphocytes that have comet appearance after the treatment with aqueous solution of p,p'-DDT.

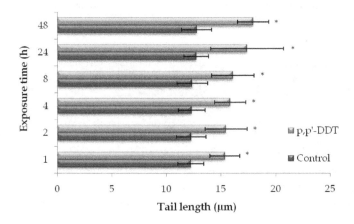

Fig. 4. Tail length (length of DNA migration) as comet assay parameter in human peripheral blood lymphocytes after exposure to low concentration of aqueous p,p'-DDT. * Statistically significant compared to corresponding control ($P<0.05$).

4.2.3 Induction of micronuclei, nucleoplasmic bridges and nuclear buds

The genotoxic activity of p,p'-DDT (0.025 mg/L) was further evaluated using the CBMN assay. Figure 5 represents binucleated lymphocytes from the non-exposed sample and samples exposed to aqueous solution of p,p'-DDT. Following p,p'-DDT treatment for 24 and 48 h, increase in the frequency of MNi was detected for both exposure times (Table 8). Additionally, statistically significant induction ($P<0.05$) of NBPs and NBUDs was also observed following p,p'-DDT treatment for 24 and 48 h (Table 9).

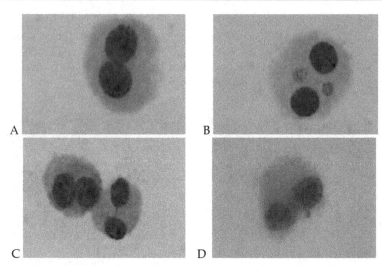

Fig. 5. Cytokinesis-block miconucleus assay microphotographs represent binucleated lymphocyte from the un-exposed sample (A). Image (B) represents binucleated lymphocyte with two micronuclei, image (C) binucleated lymphocyte with nucleoplasmic bridge (right), and image (D) binucleated lymphocyte with nucler bud after the treatment with aqueous solution of p,p'-DDT.

Exposure time (h)	Sample	1 MN	2 MNi	3 MNi	Total no. of MNi
24	Control	0.00±0.00	0.00±0.00	0.00±0.00	0.00±0.00
	p,p'-DDT	28.50±13.44*	1.50±2.12	0.50±0.71	33.00±19.80*
48	Control	6.00±1.41	0.00±0.00	0.00±0.00	6.00±1.41
	p,p'-DDT	30.50±9.19*	1.50±2.12	0.50±0.71	35.00±15.56*

Table 8. Incidence of micronuclei (MNi) as cytokinesis-block micrunucleus assay parameter in human peripheral blood lymphocytes after exposure to low concentration of aqueous p,p'-DDT, * Statistically significant compared to corresponding control ($P<0.05$).

Exposure time (h)	Sample	Total no. of NPBs	Total no. of NBUDs
24	Control	0.00±0.00	0.00±0.00
	p,p'-DDT	10.50±0.71*	10.50±4.95*
48	Control	0.00±0.00	2.50±0.71
	p,p'-DDT	14.00±11.31*	20.50±3.54*

Table 9. Incidence of nucleoplasmic bridges (NPBs) and nuclear buds (NBUDs) as cytokinesis-block micrunucleus assay parameters in human peripheral blood lymphocytes after exposure to low concentration of aqueous p,p'-DDT, * Statistically significant compared to corresponding control ($P<0.05$).

5. Discussion

Due to uncontrolled use for several decades, many pesticides, among them DDT, probably the best known and the most useful insecticide in the world has damaged wild life and might have adverse effects on human health. Because of its current use in countries of the Third World, DDT still enters environment and in that way it can still represent health risk for human population, even in countries that have banned its use almost 40 years ago (ATSDR, 2002; Crinnion, 2009; Ecobichon, 2000, 2001; Eskenazi et al., 2009; Gajski et al., 2007; Turusov et al., 2002).

Large number of epidemiological studies regarding health risk of DDT confirmed that it represents major threat for wild life and human health (Aronson et al., 2010; Demers et al., 2000; Donato & Zani, 2010; Ecobichon, 1995; Martin et al., 2002; van Wendel de Joode et al., 2001; Wojtowicz et al., 2004, 2007; Woolcoot et al., 2001). Although tested in several genotoxicity studies on bacterial and animal models (Amer et al., 1996; Binneli et al., 2007, 2008a, 2008b; Canales-Aquirre, et al., 2011; Donnato et al., 1997a, 1997b; Galindo Reyes et al., 2002; Gauthier et al., 1999; Uppala et al., 2005) there is still necessity for conducting cytogenetic research regarding its genotoxicity employing sensitive methods to reveal the exact mechanisms of action of this chemical.

Combining different cytogenetic methods may play an important role in assessing genotoxic damage from different environmental chemical or physical agents. With these methods it is possible to evaluate the level of primary DNA damage or the dynamics of its repair even after short-term exposure to these agents (Garcia-Sagredo, 2008; Gajski & Garaj-Vrhovac, 2008). The comet assay is a sensitive method for measuring and analyzing DNA damage at the single cell level, and can be used both in *in vivo* and *in vitro* (Collins, 2004; Collins et al., 2008; Dušinská & Collins, 2008). The comet assay detects single and double stranded breaks at the level of DNA molecule, sites of incomplete repair, alkali labile sites, and DNA-DNA and DNA-protein cross-links (Piperakis, 2009). Furthermore, micronucleus assay can indicate cellular and nuclear dysfunction caused by *in vitro* or *in vivo* exposure to toxic substances. It is a reliable method for measuring chromosomal instability and altered cellular viability (Fenech, 2009; Fenech et al., 2003). It includes micronuclei, which are biomarkers of chromosome breakage and whole chromosome loss, nucleoplasmic bridges, which are biomarkers of DNA misrepair and telomere end-fusions, and nuclear buds, which are biomarkers of elimination of amplified DNA and DNA repair complexes (Fenech, 2007; Garaj-Vrhovac et al., 2008).

Considering the lack of data on the effect of DDT on the cellular genome, and taking into account its usage in some countries of the Third World and its environmental persistence, the aim of this study was to evaluate the genotoxic potential of a low concentration of aqueous *p,p'*-DDT upon *in vitro* exposure of HPBLs of different duration, by using alkaline comet assay and CBMN assay. Our results showed that exposure of HPBLs to aqueous *p,p'*-DDT increased DNA damage in time dependent fashion as measured by the comet assay. In addition, CBMN assay parameters revealed a wider scale of chromosomal alterations after *p,p'*-DDT treatment.

Cytogenetic studies of DDT are mainly based on *in vitro* research on the animal models and its genotoxicity was evaluated in a variety of test systems. Results obtained by studying cytogenetic effects of DDT on DNA of shrimp larvae (*Litopenaeus stylirostris*) indicated that DDT causes DNA adducts and/or breaks (Galindo Reyes et al., 2002). DDT and its metabolites DDE and DDD showed a clear genotoxic effect on haemocytes of zebra mussel (*Dreissena polymorpha*) specimens in different concentrations that have been found in several aquatic ecosystems worldwide, with a greater genotoxic potential of the DDE in respect to the other two chemicals (Binneli et al., 2008a, 2008b). DDT has also the ability to induce chromosomal aberrations in mouse spleen indicating its genotoxicity (Amer at al., 1996). In addition, DDT was genotoxic towards lymphocytes and mammary epithelial cells of female rats showing an increase in lipid peroxidation, the outcome of the growth level of free oxygen radicals, which lead to an oxidative stress (Canales-Aquirre et al., 2011). DDT also induces cellular and chromosomal alterations in the rat mammary gland, which is consistent with the hypothesis that it can induce early events in mammary carcinogenesis (Uppala et al., 2005). Additionally, beluga whales (*Delphinapterus leucas*) inhabiting the St. Lawrence estuary are highly contaminated with environmental pollutants including DDT which can induced significant increases of micronucleated cells in skin fibroblasts of an Arctic beluga whale (Gauthier et al., 1999).

Regarding human test system, the cytogenetic effect of DDT was investigated both *in vitro* and *in vivo*. *In vitro*, certain DDT concentrations have the effects on human leukocyte functions (Lee et al., 1979), are causing chromosomal aberrations (Lessa et al., 1976), DNA strand breaks (Yáñez et al., 2004), and apoptosis induction which is preceded by an increase in the levels of reactive oxygen species (Pérez-Maldonado et al., 2004, 2005). *In vivo*, DDT is able to induce chromatid lesions (Rabello et al., 1975), increase in chromosomal aberrations and sister chromatid exchanges (Rupa et al., 1989, 1991), DNA strand breaks (Yáñez et al., 2004; Pérez-Maldonado et al., 2006), apoptosis (Pérez-Maldonado et al., 2004) as well as cell cycle delay and decrease in mitotic index (Rupa et al., 1991).

Before 1973 when it was banned, DDT entered the air, water and soil during its production and use as an insecticide. DDT is present at many waste sites and from these sites it might continue to contaminate the environment. DDT still enters the environment because of its current use in other areas of the world. DDT may be released into atmosphere in countries where it is still manufactured and used; it can also enter the air by evaporation from contaminated water and soil and than it can be deposited on land or surface water. This cycle of evaporation and deposition may be repeated many times and as a result, DDT can be carried long distances in the atmosphere (ATSDR, 2002; Crinnion, 2009; Donato & Zani, 2010; Gajski et al., 2007; Torres-Sánchez & López-Carrillo 2007). These chemicals have been found in bogs, snow and animals even in the Artic and Antarctic regions, far from where they were ever used. DDT can last in the soil for a very long time, potentially for hundreds of years. Most DDT breaks down slowly into DDE and DDD, generally by the action of microorganisms and can be deposited in other places like in the surface layers of soil; it may get into rivers and lakes in runoff or get into groundwater. In surface waters, DDT will bind to particles in the water, settle and be deposited in the sediment. DDT is than taken up by small organisms and fish in the water. It accumulates to high levels in adipose tissue of fish and marine mammals, reaching levels many thousands of times higher than in water. DDT can also be absorbed by some plants and by the animals that can directly impact human population and like that represent a major health threat (ATSDR, 2002; Beard, 2006; Gajski et al., 2007; Gauthier et al., 1999).

All of these findings suggests that DDT is still present not only in poorly developed countries of the Third World but it can still be found in other countries that have banned its use almost 40 years ago due to its stability and long persistence in the environment.

6. Conclusion

Significant levels of DDT and its metabolites can still be found in biological samples of serum, adipose tissue and maternal milk of populations that are not occupationally exposed. People are usually exposed to DDT through food, inhalation or dermal contact. Also, there are evidences on damages to the health, specially related to the reproductive area, and more recently damages at cellular level, as well as, alteration in the psychomotor development of children exposed in uterus. Although there are studies dealing with adverse effects of pesticide exposure there is still great need for elucidating the exact mechanism and health consequences related to DDT exposure and its metabolites. Our data in conjunction with other available data regarding pesticide genotoxicity have identified that DDT induces DNA strand breaks in human peripheral blood lymphocytes *in vitro* as well *in vivo*. In our study, this effect was noted even after the treatment with very low concentration of aqueous DDT. These results also confirms previous findings that DDT induces alterations of the ultrastructure of cells and DNA damage by causing single strand breakage and adducts in DNA molecule. Present study also confirms that combinations of sensitive techniques like alkaline comet assay and cytokinesys-block micronucleus assay are useful for the assessment of cellular and DNA alterations after exposure to mutagens and carcinogens from the environment. Results obtained in this research indicate the need for further environmental and food monitoring, and cytogenetic research using sensitive methods in detection of primary genome damage after exposure to DDT to establish the impact of such chemicals on human genome and health.

7. Acknowledgment

The study is a part of a research project supported by the Ministry of Science, Education and Sports of the Republic of Croatia (Grant no. 022-0222148-2125).

8. References

Agency for Toxic Substances and Disease Service. (2002). *Toxicological profile for DDT, DDE and DDD*, U.S. Department of Health and Human Services, Public Health Service, ATSDR, Atlanta, USA

Albertini, R.J.; Anderson, D.; Douglas, G.R.; Hagmar, L.; Hemminki, K.; Merlo, F.; Natarajan, A.T.; Norppa, H.; Shuker, D.E.G.; Tice, R.; Waters, M.D. & Aitio, A. (2000). IPCS guidelines for the monitoring of genotoxic effects of carcinogens in humans. *Mutation Research*, Vol.463, No.2, pp. 111–172, ISSN 1383-5742

Amer, S.M.; Fahmy, M.A. & Donya, S.M. (1996). Cytogenetic effect of some insecticides in mouse spleen. *Journal of Applied Toxicology*, Vol.16, No.1, pp. 1-3, 1099-1263, ISSN 0260-437X

Angelis, K.J.; Dušinská, M. & Collins, A.R. (1999). Single cell gel electrophoresis: detection of DNA damage at different levels of sensitivity. *Electrophoresis*, Vol.20, No.10, pp. 2133-2138, ISSN 0173-0835

Aronson, K.J.; Wilson, J.W.; Hamel, M.; Diarsvitri, W.; Fan, W.; Woolcott, C.; Heaton, J.P.; Nickel, J.C.; Macneily, A. & Morales, A. (2010). Plasma organochlorine levels and prostate cancer risk. *Journal of Exposure Science and Environmental Epidemiology*, Vol.20, No.5, pp. 434-445, ISSN 1559-0631

Arutyunyan, R.; Rapp, A.; Greulich, K.O.; Hovhannisyan, G.; Haroutiunian, S. & Gebhart, E. (2005). Fragility of telomeres after bleomycin and cisplatin combined treatment measured in human leukocytes with the Comet–FISH technique. *Experimental Oncology*, Vol.27, No.1, pp. 38-42, ISSN 1812–9269

Ashman, C.R. & Davidson, R.L. (1981). Bromodeoxyuridine mutagenesis in mammalian cells is related to deoxyribonucleotide pool imbalance, *Molecular and cellular biology*, Vol.1, No.3, pp.254–260, ISSN 0270-7306

Aubé, M., Larochelle, C. & Ayotte, P. (2008). 1,1-dichloro-2,2-bis(p-chlorophenyl)ethylene (*p,p'*-DDE) disrupts the estrogen-androgen balance regulating the growth of hormone-dependent breast cancer cells. *Breast Cancer Research*, Vol.10, No.1, R16, ISSN 1465-5411

Bakić, J. (2011) Seven centuries of the prevention of spreading of the infectious diseases into Croatia - Overview of the 60th anniversary of establishment of modern pest control in Croatia. *Hrvatski Časopis za Javno Zdravstvo*, Vol.7, No.25, ISSN 1845-3082

Bartsch, H. (1996). DNA adducts in human carcinogenesis: Etiological relevance and structure-activity relationship. *Mutation Research*, Vol.340, No.2, pp. 67-79, ISSN 1383-5742

Beard, J, & Australian Rural Health Research Collaboration. (2006). DDT and human health. *Science of the Total Environment*, Vol.355, No.1-3, pp. 78-89, ISSN 0048-9697

Benameur, L.; Orsière, T.; Rose, J. & Botta, A. (2011). Detection of environmental clastogens and aneugens in human fibroblasts by cytokinesis-blocked micronucleus assay associated with imunofluorescent staining of CENP-A in micronuclei. *Chemosphere*, Vol.84, No.5, pp, 676-680, ISSN 0045-6535

Binelli, A.; Riva, C.; Cogni, D. & Provini, A. (2008a). Genotoxic effects of p,p'-DDT (1,1,1-trichloro-2,2-bis-(chlorophenyl)ethane) and its metabolites in Zebra mussel (D. polymorpha) by SCGE assay and micronucleus test. *Environmental and Molecular Mutagenesis*, Vol.49, No.5, pp. 406-415, ISSN 1098-2280

Binelli, A.; Riva, C.; Cogni, D. & Provini A. (2008b). Assessment of the genotoxic potential of benzo(a)pyrene and pp'-dichlorodiphenyldichloroethylene in Zebra mussel (Dreissena polymorpha). *Mutation Research*, Vol.649, No.1-2, pp. 135-145, ISSN 1383-5718

Binelli, A.; Riva, C. & Provini, A. (2007). Biomarkers in Zebra mussel for monitoring and quality assessment of Lake Maggiore (Italy). *Biomarkers*, Vol.12, No.4, pp. 349-368, ISSN 1354-750X

Boiteux, S. & Radicella, P. (2000). The human OGG1 gene: structure, functions, and its implication in the process of carcinogenesis. *Archives of Biochemistry and Biophysics*, Vol. 377, No. 1, pp. 1– 8, ISSN 0003–9861

Boller, K. & Schmid, W. (1970). Chemical mutagenesis in mammals. The Chinese hamster bone marrow as an in vivo test system. Hematological findings after treatment with trenimon. *Humangenetik*, Vol.11, No.1, pp. 35–54, ISSN 0018–7348

Bolognesi, C. (2003). Genotoxicity of pesticides: a review of human biomonitoring studies. *Mutation Research*, Vol.543, No.3, pp. 251–272, ISSN 1383-5742

Bošnir, J.; Puntarić, D.; Smit, Z.; Klarić, M.; Grgić, M. & Kosanović, L.M. (2007). Organochlorine pesticides in freshwater fish from the Zagreb area. *Arhiv za Higijenu Rada i Toksikologiju*, Vol.58, No.2, pp. 187-193, ISSN 0004-1254

Bošnir, J.; Novoselac, V.; Puntarić, D.; Klarić, I.; Miškulin, M. (2010). Organochlorine pesticide residues in cow's milk from Karlovac County, Croatia. *Acta Alimentaria*. Vol.39, No.3, pp. 317-326, ISSN 0139-3006

Brooks, G.T. (1986). Insecticide metabolism and selective toxicity. *Xenobiotica*, Vol.16, No.10-11, pp. 989-1002, ISSN 0049-8254

Buckton, K.E.; Court Brown, W.M. & Smith, P.G. (1967). Lymphocyte survival in men treated with X-rays for ankylosing spondylitis. *Nature*, Vol.241, No.5087, pp. 470-473, ISSN 0028-0836

Bulut, S.; Akkaya, L.; Gok, V.; & Konuk, M. (2010). Organochlorine pesticide residues in butter and kaymak in afyonkarahisar, Turkey. *Journal of Animal and Veterinary Advances*, Vol.9, No.22, pp. 2797-2801, ISSN 1680-5593

Canales-Aguirre, A.; Padilla-Camberos, E.; Gómez-Pinedo, U.; Salado-Ponce, H.; Feria-Velasco, A. & De Celis, R. (2011). Genotoxic effect of chronic exposure to DDT on lymphocytes, oral mucosa and breast cells of female rats. *International Journal of Environmental Research and Public Health*, Vol.8, No.2, pp. 540-553, ISSN 1661-7827

Carson, R. (2002). *Silent Spring*, Mariner Books, ISBN 0618249060, New York, USA

Charlwood, D. (2003). Did Herodotus describe the first airborne use of mosquito repellents. *Trends in Parasitology*, Vol. 19, No. 12, pp. 555-556, 1471-4922

Collins, A.R. (2004). The comet assay for DNA damage and repair: principles, applications, and limitations. *Molecular Biotechnology*, Vol.26, No.3, pp. 249-261, ISSN 1073–6085

Collins, A.R. (2009). Investigating oxidative DNA damage and its repair using the comet assay. *Mutation Research*, Vol.681, No.1, pp. 24-32, ISSN 1383-5742

Collins, A.R.; Dobson, V.L.; Dušinská, M.; Kennedy, G. & Stetina, R. (1997). The comet assay: what can it really tell us? *Mutation Research*, Vol.375, No.2, pp. 183-193, ISSN 0027-5107

Collins, A.R.; Oscoz, A.A.; Brunborg, G.; Gaivão, I.; Giovannelli, L.; Kruszewski, M.; Smith, C.C. & Stetina, R. (2008). The comet assay: topical issues. *Mutagenesis*, Vol.23, No.3, pp. 143-151, ISSN 1464-3804

Countryman, P.I. & Heddle, J.A. (1976). The production of micronuclei from chromosome aberrations in irradiated cultures of human lymphocytes. *Mutation Research*, Vol.41, No.2-3, pp. 321–332, ISSN 0027–5107

Covaci, A.; Gheorghe, A. & Schepens, P. (2004). Distribution of organochlorine pesticides, polychlorinated biphenyls and a-HCH enantiomers in pork tissues. *Chemosphere*, Vol.56, No.8, pp. 757–766, ISSN 0045-6535

Crinnion, W.J. (2009). Chlorinated pesticides: threats to health and importance of detection. *Alternative Medicine Review*, Vol. 14, No. 4, pp. 347-359, ISSN 1089-5159.

Crosby, D.G. (1982). Pesticides as environmental mutagens, In: *Genetic Toxicology: An agricultural Perspective*, R.A. Fleck & A. Hollander, (Eds.), 201-218, Plenum Press, ISBN 9780306411359, New York, USA

Dawood, A.W.A.; Abd El-Maaboud, R. M.; Helal, M.A.; Mohamed, S.A. & Waleed, H.A. (2004). Detection of organochlorine pesticide residues in samples of cow milk collected from Sohag and Qena overnorates. *Assiut University Bulletin for Environmental Researches*, Vol.7, No.2, pp. 105-116, ISSN 1110-6107

Degrassi, F. & Tanzarella, C. (1988). Immunofluorescent staining of kinetochores in micronuclei: a new assay for the detection of aneuploidy. *Mutation Research*, Vol.203, No.5, pp. 339–345, ISSN 0165-7992

Demers, A.; Ayotte, P.; Brisson, J.; Dodin, S.; Robert, J. & Dewailly, E. (2000). Risk and aggressiveness of breast cancer in relation to plasma organochlorine concentrations. *Cancer Epidemiology, Biomarkers and Prevention*, Vol.9, No.2, pp. 161-166, ISSN 1055-9965

Doetsch, P.W.; Henner, W.D.; Cunningham, R.P.; Toney, J.II. & Helland, D.E. (1987). A highly conserved endonuclease activity present in *Escherichia coli*, bovine, and human cells recognizes oxidative DNA damage at sites of pyrimidines. *Molecular and Cellular Biology*, Vol.7, No.1, pp. 26–32, ISSN 0270-7306

Donato, M.M.; Jurado, A.S.; Antunes-Madeira, M.C. & Madeira, V.M. (1997a). Bacillus stearothermophilus as a model to evaluate membrane toxicity of a lipophilic environmental pollutant (DDT). *Archives of Environmental Contamination and Toxicology*, Vol.33, No.2, pp. 109-116, ISSN 0090-4341

Donato, M.M.; Jurado, A.S.; Antunes-Madeira, M.C. & Madeira, V.M. (1997b) Effects of a lipophilic environmental pollutant (DDT) on the phospholipid and fatty acid contents of Bacillus stearothermophilus. *Archives of Environmental Contamination and Toxicology*, Vol.33, No.4, pp. 341-349, ISSN 0090-4341

Donato, F. & Zani, C. (2010). Chronic exposure to organochlorine compounds and health effects in adults: diabetes and thyroid diseases. *Annali di Igiene: Medicina Preventiva e di Comunita*, Vol.22, No.3, pp. 185-198, ISSN 1120-9135

Drevenkar, V. & Fingler, S. (2000). Persistent organochlorine compounds in Croatia. *Arhiv za Higijenu Rada i Toksikologiju*, Vol.51, No.Suppl, pp. 59–73, ISSN 0004-1254

Drevenkar, V.; Fingler, S. & Fröbe, Z. (1994). Some organochlorine pollutants in the water environment and their influence on drinking water quality, In: *Chemical Safety, International Reference Manual*, Richardson, M.L. (Ed.), 297-310, Wiley-VCH, ISBN 9783527286300, Weinheim, Germany

Dugački, V. (2005). Dr. Rudolf Battara operation in Nin in 1902, the first systematic battle attempt against malaria in Croatia. *Medica Jadertina*, Vol.35, No.Suppl, pp. 33-40, ISSN 0351-0093

Duke, R.C. & Cohen, J.J. (1992). Morphological and biochemical assays of apoptosis, In: *Current Protocols in Immunology*, J.E. Coligan & A.M. Kruisbeal, (Eds.), 1-3, John Willey & Sons, ISBN 9780471522768, New York, USA

Dušinská, M. & Collins, A. (1996). Detection of oxidised purines and UV-induced photoproducts in DNA of single cells, by inclusion of lesion-specific enzymes in the comet assay. *Alternatives to Laboratory Animals*, Vol.24, No.3, pp. 405–411, ISSN 0261-1929

Dušinská, M. & Collins, A.R. (2008). The comet assay in human biomonitoring: gene-environment interactions. *Mutagenesis*, Vol.23, No.3, pp. 191-205, ISSN 1464-3804

Eastmond, D. A. & Tucker, J. D. (1989). Identification of aneuploidyinducing agents using cytokinesis-blocked human lymphocytes and an anti-kinetochore antibody. *Environmental and Molecular Mutagenesis*, Vol.13, No.1, pp. 34-43, ISSN 0893-6692

Ecobichon, D.J. (1995). Toxic effects of pesticides, In: Casarett and Doull's toxicology: the Basic Science of Poisons, C.D. Klaassen, M.O. Amdur & J. Doull, (Eds.), 643-689, Macmillan Co., ISBN 0071054766, New York, USA

Ecobichon, D.J. (2000). Our changing perspectives on benefits and risks of pesticides: a historical overview. *Neurotoxicology*, Vol.21, No.1-2, pp. 211-218, ISSN 0161-813X

Ecobichon, D.J. (2001). Pesticide use in developing countries. *Toxicology*, Vol.160, No.1-3, pp. 27-33, ISSN 0300-483X

El-Zein, R.A.; Monroy, C.M.; Cortes, A.; Spitz, M.R.; Greisinger, A. & Etzel, C.J. (2010). Rapid method for determination of DNA repair capacity in human peripheral blood lymphocytes amongst smokers. *Bio Med Central Cancer*, Vol.10, No.1, pp. 439, ISSN 1471-2407

Eskenazi, B.; Chevrier, J.; Rosas, L.G.; Anderson, H.A.; Bornman, M.S.; Bouwman, H.; Chen, A.; Cohn, B.A.; de Jager, C.; Henshel, D.S.; Leipzig, F.; Leipzig, J.S.; Lorenz, E.C.; Snedeker, S.M. & Stapleton, D. (2009). The Pine River statement: human health consequences of DDT use. *Environmental Health Perspectives*, Vol.117, No.9, pp. 1359-1367, ISSN 0091-6765

Fenech, M. (1991). Optimisation of micronucleus assays for biological dosimetry. *Progress in Clinical and Biological Research*, Vol.372, pp. 373-386, ISSN 0361-7742

Fenech, M. (2000) The in vitro micronucleus technique. *Mutation Research*, Vol.455, No.1-2, pp. 81-95, ISSN 0027-5107

Fenech, M. (2006). Cytokinesis-block micronucleus assay evolves into a "cytome" assay of chromosomal instability, mitotic dysfunction and cell death. *Mutation Research*, Vol.600, No.1-2, pp. 58-66, ISSN 0027-5107

Fenech, M. (2007). Cytokinesis-block micronucleus cytome assay. *Nature Protocols*, Vol.2, No.5, pp. 1084-1104, ISSN 1750-2779

Fenech, M. (2009). A lifetime passion for micronucleus cytome assays - reflections from Down Under. *Mutation Research*, Vol.681, No.2-3, pp. 111-117, ISSN 1383-5742

Fenech, M.; Perepetskaya, G. & Mikhalevic, L. (1997). A more comprehensive application of the micronucleus technique for biomonitoring of genetic damage rates in human populations. Experiences from the Chernobyl catastrophe, *Environmental and Molecular Mutagenesis*, Vol.30, No.2, pp.112–118, ISSN 0893-6692

Fenech. M.; Chang, W.P.; Kirsch-Volders, M.; Holland, N.; Bonassi, S.; Zeiger, E. & HUman MicronNucleus project. (2003). HUMN project: detailed description of the scoring criteria for the cytokinesis-block micronucleus assay using isolated human lymphocyte cultures. *Mutation Research*, Vol.534, No.1-2, pp. 65-75, ISSN 1383-5718

Fenech, M. & Crott, J.W. (2002). Micronuclei, nucleoplasmic bridges and nuclear buds induced in folic acid deficient human lymphocytes-evidence for breakage-fusion-bridge cycles in the cytokinesis-block micronucleus assay. *Mutation Research*, Vol.504, No.1-2, pp. 131-136, ISSN 0027-5107

Fenech, M.; Kirsch-Volders, M.; Natarajan, A.T.; Surralles, J.; Crott, J.W.; Parry, J.; Norppa, H.; Eastmond, D.A.; Tucker, J.D. & Thomas, P. (2011). Molecular mechanisms of micronucleus, nucleoplasmic bridge and nuclear bud formation in mammalian and human cells. *Mutagenesis*, Vol.26, No.1, pp. 125-132, ISSN 0267-8357

Fenech, M. & Morley, A. A. (1985). Measurement of micronuclei in lymphocytes. *Mutation Research*, Vol.147, No.1-2, pp. 29-36, ISSN 0027-5107

Fenech, M. & Morley, A.A. (1986). Cytokinesis-block micronucleus method in human lymphocytes: effect of in vivo ageing and low dose X- irradiation. *Mutation Research*, Vol.161, No.2, pp. 193–198, ISSN 0027–5107

Fenech, M. & Morley, A.A. (1989). Kinetochore detection in micronuclei: an alternative method for measuring chromosome loss. *Mutagenesis*, Vol.4, No.2, pp. 98-104, ISSN 0267-8357

Fernández, J.L.; Vázquez-Gundín, F.; Rivero, M.T.; Genescá, A.; Gosálvez, J. & Goyanes, V. (2001). DBD-fish on neutral comets: simultaneous analysis of DNA single- and double-strand breaks in individual cells. *Experimental Cell Research*, Vol.270, No.1, pp. 102-109, ISSN 0014-4827

Fingler, S.; Drevenkar, V.; Tkalčević, B. & Šmit, Z. (1992). Levels of polychlorinated biphenyls, organochlorine pesticides, and chlorophenols in the Kupa river water and in drinking waters from different areas in Croatia, *Bulletin of Environmental Contamination and Toxicology*, Vol. 49, No. 6, pp. 805-812 ISSN 1432-0800

Food and Agriculture Organization of United Nations. (2005). *International Code of Conduct on the Distribution and Use of Pesticides*, FAO, ISBN 9251054118, Rome, Italy

Fučić, A.; Markučič, D.; Mijić, A. & Jazbec, A.M. (2000). Estimation of genome damage after exposure to ionising radiation and ultrasound used in industry. *Environmental and molecular mutagenesis*, Vol.36, No.1, pp. 47–51, ISSN 0893-6692

Gajski, G. & Garaj-Vrhovac, V. Application of cytogenetic endpoints and comet assay on human lymphocytes treated with atorvastatin in vitro. *Journal of Environmental Science and Health Part A Toxic/Hazardous Substances & Environmental Engineering*, Vo.43, No.1, pp. 78-85, ISSN 1093-4529

Gajski, G.; Garaj-Vrhovac, V. & Oreščanin V. (2008). Cytogenetic status and oxidative DNA-damage induced by atorvastatin in human peripheral blood lymphocytes: standard and Fpg-modified comet assay. *Toxicology and Applied Pharmacology*, Vol.231, No.1, pp. 85-93, ISSN 0041-008X

Gajski, G.; Ravlić, S.; Capuder, Ž. & Garaj-Vrhovac, V. (2007). Use of sensitive methods for detection of DNA damage on human lymphocytes exposed to p,p'-DDT: Comet assay and new criteria for scoring micronucleus test. *Journal of Environmental Science and Health Part B*, Vol.42, No.6, pp. 607-613, ISSN 0360-1234

Galassi, S.; Bettinetti, R.; Neri, M.C.; Falandysz, J.; Kotecka, W.; King, I.; Lo, S.; Klingmueller, D. & Schulte-Oehlmann, U. (2008). p,p'-DDE contamination of the blood and diet in central European populations. *Science of the Total Environment*, Vol. 390, No.1, pp. 45-52, ISSN 0048-9697

Galindo Reyes, J.G.; Leyva, N.R.; Millan, O.A. & Lazcano, G.A. (2002). Effects of pesticides on DNA and protein of shrimp larvae Litopenaeus stylirostris of the California Gulf. *Ecotoxicology and Environmental Safety*, Vol.53, No.2, pp. 191-195, ISSN 0147-6513

Garaj-Vrhovac, V.; Gajski, G. & Ravlić, S. (2008). Efficacy of HUMN criteria for scoring the micronucleus assay in human lymphocytes exposed to a low concentration of p,p'-DDT. *Brazilian Journal of Medical and Biological Research*, Vol.41, No.6, pp. 473-476, ISSN 1414-431X

Garcia-Sagredo, J.M. (2008). Fifty years of cytogenetics: a parallel view of the evolution of cytogenetics and genotoxicology. *Biochimica et Biophysica Acta*, Vol.1779, No.6-7, pp. 363-375, ISSN 0006-3002

Gauthier, J.M.; Dubeau, H. & Rassart, E. (1999). Induction of micronuclei in vitro by organochlorine compounds in beluga whale skin fibroblasts. *Mutation Research*, Vol.439, No.1, pp. 87-95, ISSN 1383-5718

Glynn, A.; Lignell, S.; Darnerud, P.O.; Aune, M.; Ankarberg, E.H.; Bergdahl, I.A.; Barregård, L. & Bensryd, I. (2011). Regional differences in levels of chlorinated an brominated pollutants in mother's milk from primiparous women in Sweden. *Environment International*, Vol. 37, No. 1, pp. 71-79, ISSN 0160-4120

Gold, B. & Brunk, G. (1982). Metabolism of 1,1,1-trichloro-2,2-bis(p-chlorophenyl)-ethane and 1,1-dichloro-2,2-bis(p-chlorophenyl)ethane in the mouse. *Chemico-Biological Interactions*, Vol.41, No.3, pp. 327-339, ISSN 0009-2797

Gold, B. & Brunk, G. (1983). Metabolism of 1,1,1-trichloro-2,2-bis(p-chlorophenyl)ethane (DDT), 1,1-dichloro-2,2-bis(p-chlorophenyl)ethane, and 1-chloro-2,2-bis(p-chlorophenyl)ethene in the hamster. *Cancer Research*, Vol.43, No.6, pp. 2644-2647, ISSN 0008-5472

Hagmar, L.; Brogger, A.; Hansteen, I.L.; Heim, S.; Hogstedt, B.; Knudsen, L.; Lambert, B.; Linnainmaa, K.; Mitelman, F.; Nordenson, I.; Reuterwall, C.; Salomaa, S.I.; Skerfving, S. & Sorsa, M. (1994). Cancer risk in human predicted by increased levels of chromosomal aberrations in lymphocytes: Nordic Study Group on the Health Risk of Chromosome Damage. *Cancer research*, Vol.54, No.11, pp. 2919–2922, ISSN 0008-5472

Hagmar, L.; Bonassi, S.; Stromberg, U.; Brogger, A.; Knudsen, L.; Norppa, H. & Reuterwall, C. (1998). The European Study Group on Cytogenetic Biomarkers and Health, Chromosomal aberrations in lymphocytes predict human cancer. A report from the European Study Group on Cytogenetic Biomarkers and Health (ESCH), *Cancer Research*, Vol.58, No.18, pp. 4117–4121, ISSN 0008-5472

Hamel, D. (2003). Inventarizacija perzistentnih organskih onečišćavala – pesticidi, Available from http://www.cro-cpc.hr/projekti/pops/Pest_Izvjestaj.pdf

Hansen, W.K. & Kelley, M.R. (2000). Review of mammalian DNA repair and translational implications. *The Journal of Pharmacology and Experimental Therapeutics*, Vol.295, No.1, pp. 1-9, ISSN 0022-3565

Harreus, U.A.; Kleinsasser, N.H.; Zieger, S.; Wallner, B.; Reiter, M.; Schuller, P. & Berghaus, A. (2004). Sensitivity to DNA-damage induction and chromosomal alterations in mucosa cells from patients with and without cancer of the oropharynx detected by a combination of Comet assay and fluorescence in situ hybridization. *Mutation Research*, Vol.563, No.2, pp. 131-138, ISSN 1383-5718

Hartmann, A.; Fender, H. & Speit, G. (1998). Comparative biomonitoring study of workers at a waste disposal site using cytogenetic tests and the Comet (Single-Cell Gel)

assay. *Environmental and Molecular Mutagenesis*, Vol.32, No.1, pp. 17-24, ISSN 0893-6692

Heck, M.C.; Sifuentes dos Santos, J.; Bogusz Junior, S.; Costabeber, I & Emanuelli, T. (2006). Estimation of children exposure to organochlorine compounds through milk in Rio Grande do Sul, Brazil. *Food Chemistry*, Vol.102, No.1, pp. 288-294, ISSN 0308-8146

Heddle, J.A. (1973). A rapid in vivo test for chromosome damage. *Mutation Research*, Vol.18, No.2, pp. 187–190, ISSN 0027-5107

Hennig, U.G.G.; Rudd, N.L. & Hoar, D.I. (1988). Kinetochore immunofluorescence in micronuclei: a rapid method for the in situ detection of aneuploidy and chromosome breakage in human fibroblasts. *Mutation Research*, Vol.203, No.6, pp. 405-414, ISSN 0165-7992

Horváthová, E.; Dusinská, M.; Shaposhnikov, S. & Collins, A.R. (2004). DNA damage and repair measured in different genomic regions using the comet assay with fluorescent in situ hybridization. *Mutagenesis*, Vol.19, No.4, pp. 269-276, ISSN 0267-8357

International Agency for Research on Cancer. (2011). *Agents reviewed by the IARC monographs*, Volumes 1-101, IARC, Lyon, France

International standard HRN EN ISO 6468. (2002). *Water quality-Determination of certain organochlorine insecticides, policlorinated biphenyls and chlorbenzenes-Gas chromatographic method after liqid-liquid extraction*

International standard ISO 10382. (2002). *Soil quality: Determination of organochlorine pesticides and polychlorinated biphenyls; Gaschromatographic method with electron capture detection*

Ji, B.T.; Silverman, D.T. & Stewart, P.A.; Blair, A.; Swanson, G.M.; Greenberg, R.S.; Hayes, R.B.; Brown, L.M.; Lillemoe, K.D.; Schoenberg, J.B.; Pottern, L.M.; Schwartz, A.G. & Hoover, R.N. (2001). Occupational exposure to pesticides and pancreatic cancer. *American Journal of Industrial Medicine*, Vol.39, No.1, pp. 92–99, ISSN 0271-3586

Johansen, P.; Muir, D.; Asmund, G. & Riget, F. (2004). Human exposure to contaminants in the traditional Greenland diet. *Science of the Total Environment*, Vol.331, No. 1-3, pp. 189–206, ISSN 0048-9697

Kamrin, M.A. (2000). *Pesticide Profiles, Toxicity, Environmental Impact, and Fate*, Lewis Publisher, ISBN 0849321794, New York, USA

Kirsch-Volders, M. & Fenech, M. (2001). Inclusion of micronuclei in non-divided mononuclear lymphocytes and necrosis/apoptosis may provide a more comprehensive cytokinesis block micronucleus assay for biomonitoring purposes. *Mutagenesis*, Vol.16, No.1, pp. 51-58, ISSN 0267-8357

Kirsch-Volders, M.; Sofuni, T.; Aardema, M.; Albertini, S.; Eastmond, D.; Fenech, M.; Ishidate, M.Jr.; Lorge, E.; Norppa, H.; Surrallés, J.; von der Hude, W. & Wakata, A. (2000). Report from the In Vitro Micronucleus Assay Working Group. *Environmental and Molecular Mutagenesis*, Vol.35, No.3, pp. 167-172, ISSN 0893-6692

Kirsch-Volders, M.; Vanhauwaert, A.; De Boeck, M. & Decordier, I. (2002). Importance of detecting numerical versus structural chromosome aberrations. *Mutation Research*, Vol.504, No.1-2, pp. 137–148, ISSN 0027-5107

Kodba, Z.C. & Voncina, D.B. (2007). A rapid method for the determination of organochlorine, pyrethroid pesticides and polychlorobiphenyls in fatty foods using

GC with electron capture detection. *Chromatographia*, Vol.66, No.7-8, pp. 619–624, ISSN 0009-5893

Krauthacker, B. & Reiner, E. (2001). Organochlorine compounds in human milk and food of animal origin in samples from Croatia. *Arhiv za Higijenu Rada i Toksikologiju*, Vol.52, Vol.2, pp. 217–227, ISSN 0004-154

Lander, B.F.; Knudsen, L.E.; Gamborg, M.O.; Jarventaus, H. & Norppa, H. (2000). Chromosome aberrations in pesticide-exposed greenhouse workers. *Scandinavian Journal of Work, Environment & Health*, Vol. 26, No.5, pp. 436–442, ISSN 0355-3140

Lee, T.P.; Moscati, R. & Park, B.H. (1979). Effects of pesticides on human leukocyte functions. *Research Communication in Chemical and Pathological Pharmacology*, Vol.23, No.3, pp. 597-609, ISSN 0034-5164.

Lessa, J.M.; Beçak, W.; Nazareth Rabello, M.; Pereira, C.A. & Ungaro, M.T. (1976). Cytogenetic study of DDT on human lymphocytes in vitro. *Mutation Research*, Vol.40, No.2, pp. 131-138, ISSN 0027-5107

Lindberg, H.K.; Wang, X.; Järventaus, H.; Falck, G.C.; Norppa, H. & Fenech, M. (2006). Origin of nuclear buds and micronuclei in normal and folate-deprived human lymphocytes. *Mutation Research*, Vol.617, No.1-2, pp. 33-45, ISSN 0027-5107

Martin, S.A.Jr.; Harlow, S.D.; Sowers, M.F.; Longnecker, M.P.; Garabrant, D.; Shore, D.L. & Sandler, D.P. (2002). DDT metabolite and androgens in African-American farmers. *Epidemiology*, Vol.13, No.4, pp. 454-458 ISSN 1044-3983

Mateuca, R.; Lombaert, N.; Aka, P.V.; Decordier, I. & Kirsch-Volders, M. (2006). Chromosomal changes: induction, detection methods and applicability in human biomonitoring. *Biochimie*, Vol.88, No.11, pp. 1515–1531, ISSN 0300-9084

McGlynn, K.A.; Quraishi, S.M.; Graubard, B.I.; Weber, J.P.; Rubertone, M.V. & Erickson, R.L. (2008). Persistent organochlorine pesticides and risk of testicular germ cell tumors. *Journal of the National Cancer Institute*, Vol. 100, No. 9, pp. 663-671, ISSN 0027-8874.

McKenna, D.J.; Gallus, M.; McKeown, S.R.; Downes, C.S. & McKelvey- Martin, V.J. (2003). Modification of the alkaline Comet assay to allow simultaneous evaluation of mitomycin C-induced DNA cross-link damage and repair of specific DNA sequences in RT4 cells. *DNA Repair*, Vol. 2, No.8, pp. 879–890, ISSN 1568-7864

Meinert, R.; Schuz, J.; Kaatsch, P. & Michaelis, J. (2000). Leukemia and non-Hodgkin's lymphoma in childhood and exposure to pesticides: results of a register-based cite-control study in Germany. *American Journal of Epidemiology*, Vol.151, No.7, pp. 639–646, ISSN 0002-9262

Mellon, I.; Bohr, V.A.; Smith, C.A. & Hanawalt, P.C. (1986). Preferential DNA repair of an active gene in human cells, *Proceedings of the National Academy of Sciences of the United States of America*, Vol. 83, No.23, pp. 8878–8882, ISSN 0027-8424

Miller, GT. (2004). *Living in the environment*, Brooks/Cole Publishing, ISBN 978-0534997298, Pacific Grove, USA

Morgan, J.C.; Irving, H.; Okedi, L.M.; Steven, A. & Wondji, C.S. (2010). Pyrethroid resistance in an *Anopheles funestus* population from Uganda. *PLoS ONE*, Vol. 5, No. 7, e11872, ISSN 1932-6203

Morgan, D. & Roan C. (1974). The metabolism of DDT in man, In: *Essays in Toxicology*, W.Jr. Hayes, (Ed.), 39-97, Academic Press, ISBN 978-0121076054, New York, USA

Moriya, M. (1993). Single-stranded shuttle phagemid for mutagenesis studies in mammalian cells: 8-Oxoguanine in DNA induces targeted G.C->T.A transversions in simian kidney cells. *Proceedings of the National Academy of Sciences of the United States of America*, Vol.90, No. 3, pp. 1122–1126, ISSN 0027-8424

Moroy, Y.; Peebles, C.; Fritzler,M.J.; Steigerwald, J. & Tan,E.M. (1980). Autoantibody to centromere (kinetochore) in scleroderma sera. *Proceedings of the National Academy of Sciences of the United States of America*, Vol.77, No.3, pp. 1627–1631, ISSN 0027-8424

Natarajan, A.T.; Boei, J.J.; Darroudi, F.; Van Diemen, P.C.; Dulout, F.; Hande, M.P. & Ramalho, A.T. (1996). Current cytogenetic methods for detecting and effects of mutagens and carcinogens. *Environmental Health Perspectives*, Vol.104, No.3, pp. 445–448, ISSN 0091-6765

Natarajan, A.T. & Obe, G. (1982). Mutagenicity testing with cultured mammalian cells: cytogenetic assays, In: *Mutagenicity: New Horizons in Genetic Toxicology*, J.A. Heddle, (Ed.), 171-213, Academic Press, ISBN 978-0123361806, New York, USA

National Implementation Plan for the Stocholm Convention, Republic of Croatia. (2004). Available from http://www.cro-cpc.hr/projekti/pops/NIP_eng.pdf

Nevein, S.A. & Eman, M.S.Z. (2009). Detection of some organochlorine pesticides in raw milk in Giza Governorate. *Journal of Applied Sciences Research*, Vol.5, No.12, pp. 2520-2523, ISSN 1816-157X

Okonkwo, J.O.; Mutshatshi, T.N.; Botha, B. & Agyei, N. (2008). DDT, DDE and DDD in human milk from South Africa. *Bulletin of Environmental Contamination and Toxicology*, Vol. 81, No. 4, pp. 348-354, ISSN 0007-4861.

Olive, P.L. (1989). Cell proliferation as a requirement for development of the contact effect in Chinese hamster V79 spheroids. *Radiation Research*, Vol.117, No.1, pp. 79–92, ISSN 0033-7587

Olive, P.L.; Banáth, J.P. & Durand, R.E. (1990). Heterogeneity in radiation-induced DNA damage and repair in tumor and normal cells measured using the "comet" assay. *Radiation Research*, Vol.122, No.1, pp. 69-72, ISSN 0033-7587

Östling, O.; Johanson, K.J. (1984). Microelectrophoretic study of radiationinduced DNA damages in individual mammalian cells. *Biochemical and Biophysical Research Communications*, Vol.123, No.1, pp. 291–298, ISSN 0006–291X

Pérez-Maldonado, I.N.; Athanasiadou, M.; Yáñez, L.; González-Amaro, R.; Bergman, A. & Díaz-Barriga, F. (2006). DDE-induced apoptosis in children exposed to the DDT metabolite. *Science of the Total Environment*, Vol.370, No.2-3, pp. 343-351, ISSN 0048-9697

Pérez-Maldonado, I.N.; Díaz-Barriga, F.; de la Fuente, H.; González-Amaro, R.; Calderón, J. & Yáñez, L. (2004). DDT induces apoptosis in human mononuclear cells in vitro and is associated with increased apoptosis in exposed children. *Environmental Research*, Vol.94, No.1, pp. 38-46, ISSN 0013-9351

Pérez-Maldonado, I.N.; Herrera, C.; Batres, L.E.; González-Amaro, R.; Díaz-Barriga, F. & Yáñez, L. (2005). DDT-induced oxidative damage in human blood mononuclear cells. *Environmental Research*, Vol.98, No.2, pp. 177-184, ISSN 0013-9351

Peterson, J. & Robison, W. (1964). Metabolic products of p,p'-DDT in the rat. *Toxicology and Applied Pharmacology*, Vol.6, pp. 321-327, ISSN 0041-008X

Picer, M.; Picer, N.; Holoubek, I.; Klanova, J. & Hodak Kobasić, V. (2004). Chlorinate hydrocarbons in the atmosphere and surface soil in the areas of the city Zadar and Mt. Velebit Croatia. *Fresenius Environmental Bulletin*, Vol.13, No.8, pp. 712–718, ISSN 1018-4619

Pimentel, D. (2005). Environmental and economic costs of the application of pesticides primarily in the United States. *Environment, Development and Sustainability*, Vol.7, No.2, pp. 229-252, ISSN 1387-585X

Piperakis, S.M. (2009). Comet assay: a brief history. *Cell Biology and Toxicology*, Vol.25, No.1, pp. 1-3, ISSN 1573-6822

Popescu, N.C. (1999). Sister chromatid exchange formation in mammalian cells is modulated by deoxyribonucleotide pool imbalance. *Somatic Cell and Molecular Genetics*, Vol.25, No.2, pp. 101–108, ISSN 0740-7750

Rabello, M.N.; Dealmeida, W.F.; Pigati, P.; Ungaro, M.T.; Murata, T.; Perira, C.A. & Beçak, W. (1975). Cytogenetic study on individuals occupationally exposed to DDT. *Mutation Research*, Vol.28, No.3, pp. 449-454, ISSN 0027-5107

Ribas, G.; Surrales, J.; Carbonell, E.; Xamena, N.; Creus, A. & Marcos, R. (1996). Genotoxicity of the herbicides alachlor and maleic hydrazide in cultured human lymphocytes. *Mutagenesis*, Vol.11, No.3, pp. 221–227, ISSN 0267-8357

Rignell-Hydbom, A.; Lidfeldt, J.; Kiviranta, H.; Rantakokko, P.; Samsioe, G.; Agardh, C.D. & Rylander, L. (2009). Exposure to p,p'-DDE: a risk factor for type 2 diabetes. *PLoS ONE*, Vol.4, No.10, e7503, ISSN 1932-6203

Rupa, D.S.; Reddy, P.P. & Reddi, O.S. (1989). Frequencies of chromosomal aberrations in smokers exposed to pesticides in cotton fields. *Mutation Research*, Vol.222, No.1, pp. 37-41, ISSN 0027-5107

Rupa, D.S.; Reddy, P.P.; Sreemannarayana, K. & Reddi, O.S. (1991). Frequency of sister chromatid exchange in peripheral lymphocytes of male pesticide applicators. *Environmental and Molecular Mutagenesis*, Vol.18, No.2, pp. 136-138, ISSN 1098-2280

Rydberg, B.; Johanson K.J. (1978). Estimation of DNA strand breaks in single mammalian cells, In: DNA Repair Mechanisms, P.C. Hanawalt, E.C. Friedberg & C.F. Fox, (Eds.), 465-468, Academic Press, ISBN 9781555813192, New York, USA

Salama, S.A.; Serrana, M. & Au, W.W. (1999). Biomonitoring using accessible human cells for exposure and health risk assessment. *Mutation Research*, Vol.436, No.1, pp. 99-112, ISSN 1383-5742

Sannino, A.; Mambriani, P.; Bandini, M. & Bolzoni, L. (1996). Multiresidue method for determination of organochlorine insecticide residues in fatty processed foods by gel permeation chromatography. *Journal of AOAC International*, Vo.79, No.6, pp. 1434-1446, ISSN 1060-3271

Santos, S.J.; Singh, N.P. & Natarajan, A.T. (1997). Fluorescence in situ hybridization with comets. *Experimental Cell Research*, Vol.232, No.2, pp. 407–411, ISSN 0014-4827

Schaeferhenrich, A.; Sendt, W.; Scheele, J.; Kuechler, A.; Liehr, T.; Claussen, U.; Rapp, A.; Greulich, K.O. & Pool-Zobel, B.L. (2003). Putative colon cancer risk factors damage global DNA and TP53 in primary human colon cells isolated from surgical samples, *Food and Chemical Toxicology*, Vol.41, No.5, pp. 655–664, ISSN 0278-6915

Shaposhnikov, S.A.; Salenko, V.B.; Brunborg, G.; Nygren, J. & Collins, A.R. (2008) Single-cell gel electrophoresis (the comet assay): loops or fragments? *Electrophoresis*, Vol.29, No.14, pp. 3005-3012, ISSN 0173-0835

Shinmura, K. & Yokota, J. (2001). The OGG1 gene encodes a repair enzyme for oxidatively damaged DNA and is involved in human carcinogenesis. *Antioxidants & Redox Signaling*, Vol.3, No.4, pp. 597–609, ISSN 1523-0864

Singh N.P. (2000). A simple method for accurate estimation of apoptotic cells. *Experimental Cell Research*, Vol.256, No.1, pp. 328-337, ISSN 0014-4827

Singh, N.P.; McCoy, M.T.; Tice, R.R. & Schneider, E.L. (1988). A simple technique for quantitation of low levels of DNA damage in individual cells. *Experimental Cell Research*, Vol.175, No.1, pp. 184-191, ISSN 0014-4827

Smith, C.C.; O'Donovan, M.R. & Martin, E.A. (2006). hOGG1 recognizes oxidative damage using the comet assay with greater specificity than FPG or ENDOIII. *Mutagenesis*, Vol.21, No.3, pp. 185–190, ISSN 0267-8357

Surrallés, J.; Falck, G. & Norppa, H. (1996). In vivo cytogenetic damage revealed by FISH analysis of micronuclei in uncultured human T lymphocytes. *Cytogenetics and Cell Genetics*, Vol.75, No.2-3, pp.151–154, ISSN 0301-0171

Sviežená, B.; Gálová, E.; Kusenda, B.; Slaninová, M.; Vlček, D.; Dušinská, M. (2004). The Comet assay and the troubles with its application in the green alga *Chlamydomonas reinhardtii*. *Czech Phycology*, Vol.4, No.1, pp. 163-174, ISSN 1802-5439

The European Standard EN 1258/1-4. (1996). *Fatty food: Determination of pesticides and polychlorinated biphenyls (PCBs), CEN/TC 275 N0360 Food analysis – Horizontal methods*

Thoday, J.M. (1951). The effect of ionizing radiation on the broad bean root. Part IX. Chromosome breakage and the lethality of ionizing radiations to the root meristem, *The British Journal of Radiology*, Vol.24, No.286, pp.276–572, ISSN 0007-1285

Thomas, P. & Fenech, M. (2011). Cytokinesis-block micronucleus cytome assay in lymphocytes. *Methods in Molecular Biology*, Vol.682, No.3, pp. 217-234, ISSN 1940-6029

Thomas, P.; Umegaki, K. & Fenech, M. (2003). Nucleoplasmic bridges are a sensitive measure of chromosome rearrangement in the cytokinesis-block micronucleus assay. *Mutagenesis*, Vol.18, No.2, pp. 187-194, ISSN 0267-8357

Thompson, E. J. & Perry, P. E. (1988). The identification of micronucleated chromosomes: a possible assay for aneuploidy. *Mutagenesis*, Vol.3, No.5, pp. 415-418, ISSN 0267-8357

Tice, R.R. (1995). Applications of the single cell gel assay to environmental biomonitoring for genotoxic pollutants, In: *Biomonitors and Biomarkers as Indicators of Environmental Change*, F.M. Butterworth, (Ed.), 314-327, Plenum Press, ISBN 9780306451904, New York, USA

Tice, R.R.; Agurell, E.; Anderson, D.; Burlinson, B.; Hartmann, A.; Kobayashi, H.; Miyamae, Y.; Rojas, E.; Ryu, J.C. & Sasaki, Y.F. (2000). Single cell gel/comet assay: guidelines for in vitro and in vivo genetic toxicology testing. *Environmental and Molecular Mutagenesis*, Vol.35, No.3, pp. 206-221, ISSN 0893–6692

Tirukalikundram, S.; Kumaravel, T.S. & Bristow, R.G. (2005). Detection of genetic instability at HER-2/neu and p53 loci in breast cancer cells using Comet- FISH. *Breast Cancer Research and Treatment*, Vol.91, No.1, pp. 89–93, ISSN 0167–6806

Tompić, T.; Šimunić-Mežnarić, V.; Posedi, M. & Božidar, B. (2011). Pesticides in food, *Hrvatski Časopis za Javno Zdravstvo*, Vol. 7, No. 25, ISSN 1845-3082

Torres-Sánchez, L. & López-Carrillo, L. (2007). Human health effects and p,p'-DDE and p,p'-DDT exposure: the case of Mexico. *Ciência & Saúde Coletiva*, Vol.12, No.1, pp. 51-60, ISSN 1413-8123

Turusov, V.; Rakitsky, V. & Tomatis, L. (2002). Dichlorodiphenyltrichloroethane (DDT): ubiquity, persistence, and risks. *Environmental Health Perspectives*, Vol.110, No.2, pp. 125-128, ISSN 0091-6765

United Nations Environment Programme/International Atomic Energy Agency. (1982). *Determination of DDTs, PCBs in selected marine organisms by gas-liquid chromatography*, Reference Methods for Marine Pollution Studies, No.17, UNEP

Uppala, P.T.; Roy, S.K.; Tousson, A.; Barnes, S.; Uppala, G.R. & Eastmond, D.A. (2005). Induction of cell proliferation, micronuclei and hyperdiploidy/polyploidy in the mammary cells of DDT- and DMBA-treated pubertal rats. *Environmental and Molecular Mutagenesis*, Vol.46, No.1, pp. 43-52, ISSN 1098-2280

Vaglenov, A.; Nosko, M.; Georgieva, R.; Carbonell, E.; Creus, A. & Marcos, R. (1999). Genotoxicity and radioresistance in electroplating workers exposed to chromium. *Mutation Research*, Vol.446, No.1, pp. 23–34, ISSN 1383-5718

Valko, M.; Izakovic, M.; Mazur, M.; Rhodes, C.I. & Telser, J. (2004). Role of oxygen radicals in DNA damage and cancer incidence. *Molecular and Cellular Biochemistry*, Vol.266, No.1-2, pp. 37-56, ISSN 0300-8177

van Wendel de Joode, B.; Wesseling, C.; Kromhout, H.; Monge, P.; García, M. & Mergler, D. (2001). Chronic nervous-system effects of long-term occupational exposure to DDT. *Lancet*, Vol.357, No.9261, pp. 1014-1016 ISSN 0140-6736

Wójtowicz, A.K.; Gregoraszczuk, E.L.; Ptak, A. & Falandysz, J. (2004). Effect of single and repeated in vitro exposure of ovarian follicles to o,p'-DDT and p,p'-DDT and their metabolites. *Polish Journa of Pharmacology*, Vol.56, No.4, pp. 465-472 ISSN 1230-6002

Wójtowicz, A.K.; Milewicz, T. & Gregoraszczuk, E.L. (2007). DDT and its metabolite DDE alter steroid hormone secretion in human term placental explants by regulation of aromatase activity. *Toxicology Letters*, Vol.173, No.1, pp. 24-30, ISSN 0378-4274

Wong, W.C.; Szeto, Y.T.; Collins, A.R. & Benzie, I.F.F. (2005). The comet assay: a biomonitoring tool for nutraceutical research. *Current Topics in Nutraceutical Research*, Vol. 3, No.1, pp. 1–14, ISSN 1540-7535

Woolcott, C.G.; Aronson, K.J.; Hanna, W.M.; SenGupta, S.K.; McCready, D.R.; Sterns, E.E. & Miller, A.B. (2001). Organochlorines and breast cancer risk by receptor status, tumor size, and grade (Canada). *Cancer Causes and Control*, Vol.12, No.5, pp. 395-404, ISSN 0957-5243

World Health Organization. (1990). *Public health impact of pesticides used in agriculture*, WHO, ISBN 9241561394, Geneva, Switzerland

World Health Organization. (2007). *Insecticide-treated mosquito nets: a WHO Position Statement*, WHO, Available from http://www.who.int/malaria/publications/ atoz/itnspospaperfinal/en/index.html

World Health Organization. (2009a). *The WHO recommended classification of pesticides by hazard and guidlines to classification*, WHO, ISBN 9789241547963, Geneva Switzerland

World Health Organization. (2009b). *Global insecticide use for vector - borne disease control* (4th ed), WHO, ISBN 9789241598781, Lyon, France

Yáñez, L.; Borja-Aburto, V.H.; Rojas, E.; de la Fuente, H.; González-Amaro, R.; Gómez, H.; Jongitud, A.A. & Díaz-Barriga, F. (2004). DDT induces DNA damage in blood cells. Studies in vitro and in women chronically exposed to this insecticide. *Environmental Research*, Vol.94, No.1, pp. 18-24, ISSN 0013-9351

Part 2

Vector Management

7

Vector Control Using Insecticides

Alhaji Aliyu
Department of Community Medicine,
Ahmadu Bello University, Zaria
Nigeria

1. Introduction

At the end of the 19th century, it was discovered that certain species of insects, other arthropods and fresh water snails were responsible for the transmission of some diseases of pubic health importance. Since effective vaccines or drugs were not always available for the prevention or treatment of these diseases, control of transmission then had to rely mainly on control of vectors. The control programmes included among others, use of mosquito nets, drainage of gutters, filling of potholes and other water bodies used by insects for breeding. The 1940s sawed the discovery of DDT insecticide (dichlorodiphenyl tricloroethene) which was a major breakthrough in the control of vector-borne diseases. DDT also appeared to be effective and economical in the control of other biting flies (tsetse fly, simulium, and sand fly) and midges and of infestations with fleas, lice, bedbugs and triatomine bugs.

The initial large scale success achieved in the control programme was short-lived as the vectors developed resistance to the insecticides in use, thereby creating a need for new more expensive chemicals. Interest in alternatives to the use of insecticides such as environmental management (source reduction) and biological control, has been revived because of increasing resistance to the commonly used insecticides among important vector species e.g. (malaria) and also because of concerns about the effects of DDT and certain other insecticides on the environment. For many vector species, environmental sanitation[1] through source reduction and health education is the fundamental means of control; other methods should serve as a supplement, not as a substitute. Thus, in recent years, the practices of vector control have evolved, and environmental management and modification have come to the fore, both for disease control and for agricultural and other economic purposes[2], this is a complex and multi-disciplinary field[3]. Effective application of any control measure must be based on a fundamental understanding of the ecology, bionomics and behaviour of the target vector species and its relation to its host and the environment. Effective vector control also requires careful training and supervision of pest control operations and periodic evaluation of the impacts of the control measures.

In more recent years, less reliance has been placed on the use of a single method of chemical control; there is a shift towards more integrated vector control involving several types of environmental management supplemented by more than one method of chemical control and the use of drugs. More attention has been paid to community participation (a key component of primary health care (PHC) in eliminating breeding sites of vectors (including clearing of weeds/bushes near residential houses) and reducing vector densities.

Finally, there is also need to provide on continuous basis information on single, effective and acceptable methods for vector source reduction and personal protection to individuals and families in the community at a reasonable cost.

In a chapter of this nature only a few, but more important species of vectors could be discussed and even such a discussion could only be brief, emphasizing only the important features of the arthropod which serve to illustrate how the arthropod affects public health.

2. Malaria and its vectors

Human malaria is a number one public enemy and an illness caused by the bite of an infective female anopheles mosquito which transfers parasites called plasmodium from person to person. Four plasmodium parasites exists (P. falciparum, P. ovale, P. malariae and P. vivax) but only one (P. falciparum) is of vital importance in disease transmission in Nigeria. The important vectors in Nigeria are anopheles gambiae and A. funestus. These mosquitoes breed readily in ditches and collections of water in empty receptacles around the houses. The disease is endemic in Nigeria and over 90% of the population is at risk. Fifty percent of the population will have at least one attack per year, 300, 000 children and 11% of pregnant women die of malaria each year respectively and millions of dollars is lost each year in treatment of malaria.

The table below shows some of these vectors and diseases transmitted by them.

Type	Designation	Disease	Vectors	Mode of transmission
Protozoa	Plasmodium spp Trypanosome spp	Malaria African sleeping sickness	Mosquitoes Tsetse fly	Bite (sg) Bite (sg)
Filarial nematodes	Wucheria Bancrofti Onchocerca Volvula	Filariasis Onchocerciasis	Mosquitoes Black flies	Bite Bite
Viruses	YF	Yellow Fever	Aedes	Bite (sg)
Bacteria	Cholera/diarrhea Typhoid fever Bacillary dysentery	Cholera Typhoid fever Dysentery	Cockroaches	Mechanical carriers

Table 1. Pathogens and vectors of some human diseases (sg) indicates definite involvement of salivary gland.

The results of several surveys showed that the most prevalent species of malaria parasites is p. falciparum with a prevalent rate of between 80% and 100% of all positive blood films. P. vivax is conspicuously absent in West Africans who because of their Duffy negative erythrocyte membrane are immune to P. vivax[4, 5].

It is estimated that there are more than 400 species of anopheles in the world, but only 40 of these are important vectors (as transmitters or carriers) of malaria. Only the female anopheles mosquito transmits malaria to humans. Some anopheles mosquitoes can also carry other human diseases like Filariasis and some viruses. There are two other genera of mosquitoes which are important carriers of other mosquito-borne diseases: Aedes is a vector of viral diseases- Yellow fever and Dengue while culex mosquito is the vector of Filariasis and Japanese encephalitis. For practical purposes and entomologically, it is useful to be able to differentiate between these three types of mosquitoes; so the main distinguishing features in each stage of the life cycle are shown below in the diagram

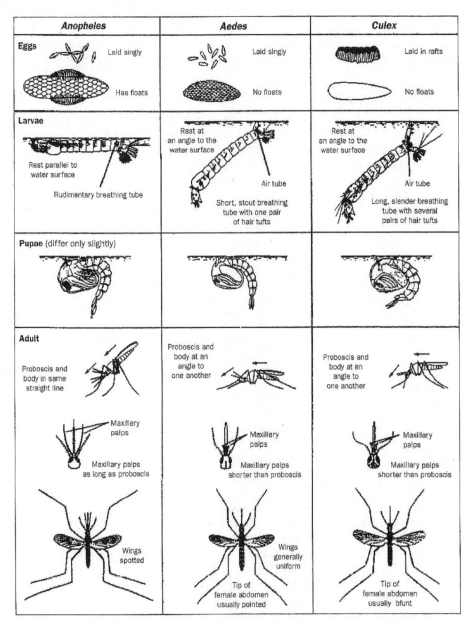

Fig. 1. Life cycles of different mosquito genera (online, 2011).

Basically, the mosquito has four stages in its life cycle: eggs → larva→ pupa → adult. It is useful to understand the life cycle and natural history of the anopheles mosquito not only for epidemiological reasons but most importantly for the fact that all the four stages are targets to control mosquito vector. There are several features of the behaviour of anopheles mosquitoes which are important in understanding malaria epidemiology and also for planning mosquito control.

Malaria is holo-endemic in most regions of Nigeria; ideal climatic conditions for the propagation and transmission of the infection are a temperature between 26°C – 30°C and relative humidity of over 60 percent.

3. Disease control

The general methods of malaria control can be grouped into three measures directed against the parasite in man, measures directed against the vector and measures designed to prevent mosquito-man contact, these are summarized in table 2

Principal goal	Interventions
Treatment	Outpatient treatment of uncomplicated malaria
	Inpatient treatment of severe and complicated malaria
	Home treatment
Prevention	
(a) Inhibit mosque ito breeding	i) Source reduction (drainage, filing in ditches)
	ii) Chemical larviciding
	iii) Management of agricultural, industrial and urban development to avoid breeding sites
(b)Kill adult mosquitoes	i) IRS (indoor residual spraying)
	ii) ITMs (insecticide treated materials – bed nets, curtains, etc)
(c) Prevent mosquito contact	i) ITMs
	ii) Repellants, sprays, coils, etc
(d) Reduce malaria infection and morbidity in humans	i) IPT (intermittent preventive treatment of pregnant women)
	ii) Chemoprophylaxis

Table 2. Interventions to control malaria[5].

The first involves the use of appropriate anti-malarial agents to treat clinical malaria and the use of chemoprophylaxis among the vulnerable groups. The problem facing most tropical countries is serious as majority of patients with clinical malaria are either untreated or treated inadequately by self-medication. Also, their governments cannot afford to buy sufficient anti-malarial drugs for their needs and most people cannot afford to purchase effective treatments[7]. The annual per capita expenditure on anti-malarial drugs in most of sub-Saharan Africa is still <US$10. An adult (60kg) course of Chloroquine costs US$0.08 but the new artemisinin based combinations (ACTs) cost more than five times as shown in table 3.

Drug	US$
Chloroquine	0.13
SP	0.14
Amodiaquine	0.20
Artemisinin-based combinations (ACTs)	1-3

Table 3. Average cost of a full course of adult outpatient treatment[8].

This is currently unaffordable to most patients surviving barely on less than a US$1 per day. The treatment of malaria has become quite challenging, and the emergence of resistant strains of the parasite. In Nigeria where at least 80%[9,10] of the people live in rural areas and are supperstitious illiterates, early recognition and the right treatment are not adhered to. This encourages drug resistance.

The Abuja declaration on Roll Back Malaria on 25 April 2000 and agreed to by African Heads of State sets an ambitious goals to reduce the burden of malaria (insecticide-treated nets, prompt access to treatment and prevention of malaria in pregnancy) by the year 2010[8]. Achieving high coverage in both IPT and use of ITNs among the vulnerable groups and the general population has remained elusive for many countries in sub-Saharan Africa[11]. A major barrier to net ownership is poverty as the price of a net represents a large proportion of the income of a poor household, this has been reported in various studies[12, 13, 14].

Environmental control (source reduction) offers the best practical and easy measure to control disease vector as it eliminates the breeding places. The filling of mosquito breeding sites with soil, ash or rubbish and is most suitable for reducing breeding in small depressions, water holes or pools, which does not require much filling material. On the other hand, drainage of water can be achieved by constructing open ditches; however, the drainage systems used in agriculture or for the transportation of sewage and rainwater in cities often promote breeding because of poor design and maintenance.

4. The use of insecticides

The first house-spraying campaigns after the Second World War, showed the capacity of this interventions to produce profound reductions in malaria transmission in a wide variety of circumstances. In Africa, the intervention was used in 1960s and 1970s but later abandoned except in some countries in southern and eastern Africa where residual insecticide spraying (IRS) remained the cornerstone of malaria control strategy. Since 2005, however, there has been renewed interest in large scale IRS programmes. To date, 25 out of the 42 malaria endemic countries in the WHO Africa region have included IRS in their control strategy for malaria control. Of these, 17 countries routinely implement IRS as a major malaria control intervention, six including Nigeria are piloting the IRS in a few districts while 2 are planning to pilot. At the end of 2006, the National Malaria Control Programme (NMCP) and its partners initiated a pilot IRS project in 3 local government areas/districts in 3 states (Lagos, Plateau, Borno states) using the WHOPES approved insecticides:

a. Lambdacyhalothrin CS 10% ($0.030g/m^2$)
b. Alphacypermethrin WP10% ($0.030g/m^2$)
c. Bifenthrin WP 10% ($0.050g/m^2$)

The evaluation of the local context of the 3 chemicals confirmed that residual effectiveness of the insecticides lasted for at least 4 months.

Residual spraying of houses involves the treating of interior walls and ceilings using a handheld compression spraying and is effective against mosquitoes that favour indoor resting before or after feeding. For a more detailed discourse of notable insecticide formulations, spray pumps, spraying techniques and maintenance of equipment, see Jan A. Rozendaal[15].

The figure 2, below summarizes and describes the IRS management principles;

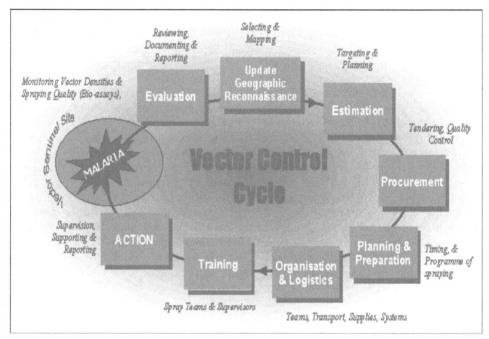

Fig. 2. IRS Management cycle.

Mosquito nets are an old technology and most people in sub-Saharan Africa are aware of the existence of nets. In some countries such as Nigeria and the Gambia nets have been in regular use for over hundred years. Similarly, in some parts of Africa, net ownership is a well established social norm and nets are widely available. For instance, recently, it was found that 70% of households (HHs) in Dar es Salaam (Tanzania) owned a mosquito net[16] and around 35% of HHs were found to own a net in urban Burkina Faso[17].

Insecticide treated mosquito nets have had significant impact in reducing morbidity and mortality among children under-five years old and pregnant women where ITNs have been appropriately and extensively used in malaria endemic areas. The potential epidemiological advantages and public health benefits of treating nets with insecticide for protection against malaria were recognized in the mid-1980s. Specifically, the efficacy of insecticide treated nets (ITNs) for the control of malaria in children under-5 years of age has recently been demonstrated by several large scale studies[18, 19, 20, 21] which find reductions in all cause – mortality, ranging from 16% to 63%. These insecticides which have been approved by WHOPES (World Health Organization Pesticides Evaluation Scheme) are safe. They have the following properties; provide personal protection from mosquito bites, effective against other insects: bedbugs, flies and cockroaches, community and household mass effective which may be more important in some contexts[22] and mosquito nuisance effect. There have been reports also of dead insects on the nets and on the floor, less mosquito noise and that ITNs providing a better night's sleep than a net alone[16].

Most nets (and other materials) need to be treated and retreated with insecticides to increase their effectiveness. There are several insecticides that can be used which have proved to be safe. The insecticides recommended by the National Malaria and Vector Control Programme for treatment/retreatment of nets are those approved by WHOPES and registered by NAFDAC (National Foods and Drugs Administration Commission) are shown in table 4 below:

Generic name	Trade name	Dose per 1 net	Manufacturer
Alphacypermethrin 10%.SC[a]	Fendona	6ml	BASF
Deltamelthrin 1%.SC	K-Othrine, Ko-Tab	40ml	Aventis
Etofenprox 10%.EW[b]	Vectron	30ml	Mitsui
Permethrin 10%.EC[c]	Peripel, Imperator	75ml	Aventis
Cyfluthrin 5% EW	Solfac EW	15ml	Bayer
Lambdacyhalothrin 2.5% CS[d]	Icon	10ml	Syngenta

Key:
SC[a]   =   Suspension concentrate
EW[b]   =   Emulsion oil in water
EC[c]   =   Emulsifiable concentrate
CS[d]   =   Capsule suspension

Table 4.

For a detailed description of preparation of insecticide mixtures and treatment methods, the reader should consult guidelines on the use of insecticide-treated mosquito nets[23].

A single impregnation of a cotton or nylon mosquito net will provide protection for 1 year[24,25]. Nylon tends to retain permethrin and deltamethrin better than cotton. The impregnated nets can be washed and can tolerate small tears/holes without markedly reducing the protective effect. Recently, long lasting nets (LLNs) have been developed and have the advantage of retaining insecticidal activity for years (so the nets will not lose its potency with repeated washings). Despite various government policies, cost of the nets remains a significant barrier and a long obstacle to the Roll Back Malaria goal of universal coverage – defined as one long lasting insecticide-treated nets for every two people in the household with 80% usage. ITN development is a public good. The development of insecticide resistance in the 1950s and recently by vectors has been a cause for concern. There have been reports of resistance to DDT in a wide range of sub-Saharan African countries[26], but this has not reached an operationally significant level. With the exception of the Gezira region of Sudan[27], widespread loss of vector susceptibility is not yet a big problem in Nigeria. A worrying new development is the emergence of knockdown resistance to pyrethroid insecticides in natural populations of anopheles mosquitoes in Cote d'Ivoire and Burkina Faso[28, 29, 30] where insecticides are widely used in cotton production.

Currently, pyrethroids are the only insecticides used for net treatment and are also increasingly used for spraying, so there is threat that if widespread resistance develops, the interventions will gradually become less cost-effective over time.

Finally, there is need to describe importantly the WHO integrated vector management (IVM) concept. Vector control is well suited for integrated approaches because vectors are responsible for multiple diseases and since interventions are effective against several vectors (use of insecticides) the concept of IVM was developed as a result of lessons learnt from integrated pest management which was used in Agriculture.

Integrated vector management (IVM) is a major component of the global campaign against malaria. The revised strategic plan for RBM recommended that from 2006 – 2010, 80% of the population at risk need to be protected using effective vector control measures. IVM creates synergies between various vector-borne disease control programmes. Utilization of single method could be optimized to control more than one vector-borne disease, eg ITNs can control malaria, lymphatic filariasis and to some extend leishmaniasis. IVM operates in the context of inter-sectoral collaboration. The application of IVM principles to vector control will contribute to the judicious use of insecticides and extend their useful life.

5. References

[1] World Health Organization. Chemical methods for the control of arthropod vectors and pests of Public Health importance. WHO, Geneva; 1984.

[2] Najera J A. Malaria control: achievements, problems and strategies. Parasitologica 2001, 43:1-89

[3] Molineaux I, Grammiccia G: The Garki project; WHO 1980

[4] Garnham P.c.C. (1971).Progress in parasitology. The Atlone Press, London

[5] Kara Hanson, Catherine Goodman, Jolines, Sylvia Meek, David Bradley and Anne Mills. Global Forum for health Research: The Economics of Malaria Control Interventions, 2004.

[6] Stages of life cycle of vector parasites, available at
   www.who.int/docstore/water_control_htlm (cited 11/05/2011)

[7] Miller, L.H., Mason, S J, Clyde, DF and McGinnis M H. Resistance factors to plasmodium vivax: Duffy genotype Fyfy. New England Journal of Medicine: 1976; 255:302-304

[8] Denis MB, Tsuyuoka R, Poravuth Y et al. Surveillance of the efficacy artersunate and mefloquine combination for the treatment of uncomplicated malaria in Cambodia. Tropical Medicine International health 2006; 11:1160-1366

[9] WHO. The Africa Malaria Report 2003, WHO/CDS/MAL/2003/1903:31-37

[10] Salako A A. Environmental health education in schools and communities in Nigeria: The essence. The Nigerian Clinical Review Journal (2008): 14; 12:17-25

[11] The Challenges of Malaria Infection in Nigeria: The Nigerian clinical Review. March/April 2011; 89:3-6.

[12] Anna Maria van Eijk, Jenny Hill, Victor A. Alegana et al. Coverage of Malaria protection in pregnant women in sub-Saharan Africa: A synthesis and analysis of National survey data. Medi-Link Journal. April 2011: 12:4:20-42

[13] Aliyu A A, Alti-Mu'azu M, Insecticide-treated net usage and malaria episodes among boarding students in Zaria, Northern Nigeria. Annals of African Medicine, 2009; 8:2:85-89.

[14] Aniefiok M. Acceptance and use of insecticide treated bed nets (ITN in the Roll Back Malaria Programme in Kuje Area Council, Abuja. Journal of Environmental health 2005; 2:26-33

[15] Insecticide-treated nets. New findings from Kenya, published by Centers for Disease Control and Prevention (CDC), Atlanta, USA, Confirm effectiveness in areas of intense transmission. Africa Health (Mera) July 2003; Issue 6

[16] Jan A. Rozendaal: Vector control (Methods used by individuals and communities) 1997.

[17] Jones C. Bed nets and malaria. Postgraduate Doctor. May 2002; vol. 24 No. 2:23-25.

[18] Guiguemde T R et al. Household expenditure on Malaria prevention and treatment of families in the town of Bobo-Diolasso, Burkina Faso. Trans. of the Royal Society of Tropical Medicine and Hygiene 1994: 88: (3)285- 287

[19] Alonso P L, Lindsay SW, Schellenberg J. et al. A malaria control net trial using insecticide-treated bed nets and targeted chemoprophylaxis in a rural area of the Gambia, West Africa. The impact of the interventions on mortality and morbidity from malaria. Transactions of the Royal Society of Tropical Medicine and Hygiene, 1993; 87(S):37-44.

[20] D'Alessandro U, Olaleye BO, McGuire W et al. Mortality and Morbidity from malaria in Gambian children after introduction of an impregnated bed net programme. Lancet 1995;345 (8948).479-83

[21] Binka F N, Kubaje A, Adjuik M et al. Impact of permethrin impregnated bed nets on child mortality in Kassena-Nankana district, Ghana: A randomized controlled trial. Tropical Medicine and International health 1996; 1(2):147-54.

[22] Nevill CG, Same ES, Mung'ala VO et al. Insecticide-treated bed nets reduce mortality and severe morbidity from malaria among children on the Kenyan Coast. Tropical Medicine and International Health 1996: 1(2) 139-46

[23] Nicholas J. White. Malaria Gordon C. Cook and Alimuddin I Zumla (Eds). Manson's Tropical Diseases! 22nd edition, China, Saunders Elsevier: 2009; 1201-1300

[24] WHO Guidelines on the use of insecticide-treated nets for the prevention and control of malaria in Africa.

[25] Lindsay SW, Gibson MF. Bed nets revisited – old idea, new angle. Parasitology Today 1988:4:270-272

[26] Alonso PI, Lindsay SW, Armstrong IRM, et al. The effects of insecticide-treated bed nets on mortality of Gambian children. Lancet 1991;337:1499-1502

[27] WHO Vector resistance to pesticides, Fifteenth Report of the WHO Expert Committee of Vector Biology and Control, Geneva; 1992

[28] El Gaddal A A, Haridi A A M, Hassan F T, Hussein H. Malaria Control in Gezira-Managil Irrigated Scheme of the Sudan. Journal of Tropical Medicine and Hygiene 1985; 88:153-157

[29] Martinez Torres D, Chandre F, Willamson M S et al. Molecular characterization of pyrethroid knockdown resistance (kdr) in the major malaria vector anopheles Gambiae SS. Insect Molecular Biology 1998; 7(2):174-184

[30] Elissa N, Morchet J, Riviere F, Mevnier J-Y, Yaok. Resistance of Anopheles gambiae SS to pyrethroids in Cote d'Ivoire. Annales de la Societe Belge de Medecine Tropicale 1993; 73:291-294.

8

Behavioral Responses of Mosquitoes to Insecticides

Theeraphap Chareonviriyaphap
Department of Entomology, Faculty of Agriculture,
Kasetsart University, Bangkok,
Thailand

1. Introduction

Many people living in areas of the tropical and subtropical world are at serious risk of infection from a wide variety of vector-borne diseases, most notably malaria and dengue. Globally, approximately 50-100 million people are estimated to be at risk of infection with dengue viruses (the cause of dengue fever/dengue haemorrhagic fever) and between 100-300 million live in malaria endemic areas (World Health Organization [WHO], 2009). The viruses responsible for dengue are transmitted primarily by *Aedes aegypti*, a predominately urban, day-biting mosquito that often resides in and around human dwellings and preferentially feeds on humans (Gubler, 1997); whereas the 4 human malaria parasites (*Plasmodium*) are transmitted by a wide variety of *Anopheles* species (Service & Townson, 2002). Dengue vector has proven extremely resilient to control measures because of its close association and exploitation of domestic and peridomestic human settings (Reiter & Gubler, 1997). On the other hand, malaria vectors display a more diverse array of host seeking behaviors and preference, biting patterns and larval breeding habitats (Pates & Curtis, 2005; Sinka et al., 2011). Despite decades of extensive research, efficacious and commercially viable vaccines for these 2 important vector-borne diseases are not yet available. Therefore, the prevention and control of dengue and malaria remains dependent on various vector control strategies to reduce risk of transmission; in some instances this requires the use of various chemical insecticides as larvicides, space spray and indoor residual spray (IRS) applications, and use of insecticide-impregnated bed nets to control adult mosquito blood feeding (Roberts & Andre, 1994; WHO, 1999; Reiter & Gubler, 1997; Grieco et al., 2007).

Chemical insecticides, including organochlorines, organophosphates, carbamates, and synthetic pyrethroids, have long been used with great effect in public health vector control programs worldwide (Reiter & Gubler, 1997; Roberts & Andre, 1994; WHO, 1992). Although DDT used ceased in many countries several decades ago, the chemical has returned for use in malaria control IRS programs in Africa because of some its superior attributes (Roberts & Tren, 2010). The dramatic impact of DDT on mosquito populations in terms of both toxicity and behavior suppressing disease transmission is well known but in some instances the actual mechanisms at work remain unclear and poorly understood. Most studies on insecticides have placed attention exclusively on the direct toxicological (knockdown and killing) effects on mosquito populations; whereas far less research has focused on the behavioral responses and outcomes as a result of chemical exposure (Roberts, 1993; Roberts et al., 1997; Chareonviriyaphap et al., 1997). Studies have shown that most chemical compounds influence insect locomotor (movement) behavior, often resulting in profound excitation and pre-mature movement away from treated surfaces or areas, one or more kinetic mechanisms resulting in so-called "avoidance behavior" (Kongmee et al., 2004; Miller et al., 2009). The term "excito-repellency" is often used to describe behavior that is stimulated by direct contact with a chemical that results in abnormal excitation (sometimes termed 'irritancy') and spatial repellency that results without the insect making physical contact with the chemical (Roberts et al., 1997). This chapter describes the various behavioral responses to insecticides in some important mosquito vectors of malaria (*Anopheles* species) (Table 1) and dengue (*Aedes aegypti*) (Table 2) and the tropical house mosquito, *Culex quinquefasciatus* (Table 3) using two types of laboratory systems and field assays using experimental huts.

2. History of test systems to study mosquito behaviors

Behavioral responses of mosquitoes to chemical compounds can be directly demonstrated by using various laboratory devices and field assay systems. For laboratory assays, many of the variations have been reviewed (Roberts et al., 1997). The WHO developed the first test box using plywoood in its construction attempting to access the excitation ("irritability") of exposed mosquitoes following physical contact with insecticides (WHO, 1970). This system was subsequently referred to as an "excito-repellency" test box (Rachou et al., 1963). Subsequently, the test system was further modified by other investigators interested in behavioral avoidance responses exposed to DDT and some of the early synthetic pyrethroids (Charlwood & Paraluppi, 1978; Roberts et al., 1984; Bondareva et al., 1986; Rozendaal et al., 1989; Quinones & Suarez, 1989; Ree & Loong, 1989). Years later, a lightproof test chamber was designed to study the irritant response of *Anopheles gambiae*, an importent malaria vector in Africa, to several chemical compounds (Evans, 1993). One key concern with all these test systems was associated with the technical difficulties of the test boxes for introducing and removing test specimens. Other concerns were controling for the various physiological conditions of wild-caught mosquitoes, and selecting the ideal range of concentrations for chemical compounds. In addition, no single or set of statistical methods for analysis of data has been fully accepted nor has any test system been specifically designed to truely discriminate between contact excitation and noncontact repellency responses (Roberts et al., 1997). An improved excito-repellency test device that was able to clearly differentiate between excitation and spatial repellency was developed and initially tested against several field populations of *Anopheles albimanus* from Central America (Roberts et al., 1997; Chareonviriyaphap et al., 1997). Unfortunately, this fixed prototype was cumbersome to handle and required considerable time for attaching the chemcial-treated test papers. Several years on, a more field-friendly test system was designed that was both collapsible and easily transportable (Chareonviriyaphap & Aum-Aong, 2000; Chareonviriyaphap et al., 2002) and was used to investigate the behavioral responses of various mosquito species and geographical populations from Thailand and a few populations from elsewhere in Asia (Chareonviriyaphap et al., 2001; Sungvornyothin et al., 2001; Kongmee et al., 2004; Pothikasikorn et al., 2005). Recently, a novel modular assay system was developed for mass screening of chemical actions; including contact irritancy, spatial repellency, or toxicity responses, on adult mosquitoes (Grieco et al., 2005). This modular system is substantially reduced in size compared to the previous excito-repellency box and minimizes the treated surface area and therefore the amount of chemical required for testing. In field observations, numerous attempts have been made to determine behavioal responses of mosquitoes using specially constructed experimental huts (Smith, 1965; Roberts et al., 1984, 1987; Roberts & Alecrim, 1991; Rozendaal et al., 1989; Bangs, 1999; Grieco et al., 2000; Grieco et al., 2007; Polsomboon et al., 2008; Malaithong et al., 2010). Most experimental hut studies have been conducted to observe the behavior of *Anopheles* mosquitoes with far fewer studies on other genera. Grieco et al., (2007) successfully demonstrated all 3 chemical actions using *Ae. aegypti* as a model system. The results obtained from both laboratory and field studies can help faciliate the choice of the most effective chemical measures to control house-frequenting adult mosquitoes.

Two standard systems used in the laboratory, the excito-repellency box (ERB) and the high throughput screening system (HITSS), and the experimental hut in the field to help characterize the behavioral responses of mosquitoes to chemical compounds are discussed herein.

2.1 Excito-Repellency Box (ERB) test system

Given the complexities of insect behavioral research, the excito-repellency testing design and testing methodology along with methods of analyzing and interpreting test data, have yet to be universally accepted. However, tremendous improvement has occurred in recent years that effectively addressed a number of the previous drawbacks, opening up significance progress in behavioral studies involving innate responses to insecticides. The ERB test system (Fig.1) was developed to distinguish irritancy (Fig. 2) and repellency (Fig. 3) (Roberts, et al., 1997) and was first used to study the avoidance behavior of *Anopheles albimanus* to DDT and synthetic pyrethroids in Belize, Central America in which consistent and reliable results were generated measuring the escape behavior of female mosquitoes (Chareonviriyaphap et al., 1997). Over time, the initial system was modified into a collapsible test chamber with identical dimensions and operational attributes to help alleviate some of the previous handicaps (Chareonviriyaphap et al., 2002). This improved version has been used extensively in the evaluation of behavioral responses by laboratory and field mosquito populations in Thailand.

In 2006, a more field compatible device was designed with a substantial reduction in chamber size to minimize the amount of chemical and treated paper required, provide greater ease of test preparation and allow use of a small number of test specimens (15 vs. 25 females) compared to the larger version (Tanasinchayakul et al., 2006). Using this system, consistent and reliable results have been produced (Muenworn et al., 2006; Polsomboon et al., 2008; Thanispong et al., 2009; Mongkalagoon et al., 2009).

The system comprises of 4 outer stainless steel sheet metal walls (Fig. 1). Each wall is constructed with an aluminum sliding rib on each end and socket providing a surface for the test paper holder in the middle. The test paper holder consists of 2 sides: a sheet of fine mesh iron screen attached on one side and a panel to hold the test papers in place on top on the opposite side. There is a 0.8 cm gap between the test paper and screen barrier to prevent mosquitoes from making physical contact ('noncontact') on the treated paper surface during the repellency assay. The paper holder simply has to be inverted to provide the proper conditions to expose the paper in the contact test. On one end of the chamber a portal is made of overlapping sheets of dental dam and is used for placing mosquitoes inside the chamber and later for removing them after the exposure test period. A Plexiglas door serves to seal the chamber and allow the investigator to view the exposure chamber before and after the actual test period. A stainless steel outer rear door cover is used to shut off all external light inside the chamber when the experiment is being performed.

Each tests consists of enclosing 15 female mosquitoes in each of 4 chambers lined with either insecticide-treated or untreated (control) papers. On the opposite side of the Plexiglas portal is a single exit portal for mosquitoes to escape to a receiving cage. At the beginning of a test, a 3-min rest period allows mosquitoes to adjust to test chamber conditions before the exit portal is opened to initiate the observation period (30 or 60 min depending on experiment). The numbers of mosquitoes escaping from the chamber into receiving cage are recorded at 1 min intervals until test completion.

Fig. 1. Excito-repellency test chamber showing side of exit portal (Roberts et al., 1997).

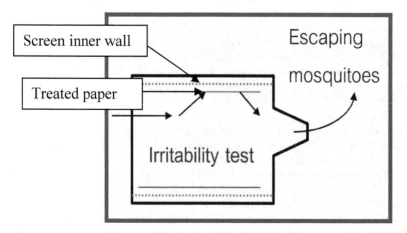

Fig. 2. Excito-repellency test chamber: Irritability test design (Roberts et al., 1997).

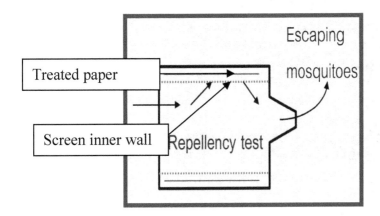

Fig. 3. Excito-repellency test chamber: Repellency test design (Roberts et al., 1997)

2.2 High Throughput Screening System (HITSS)

Even though an ERB test system has been found valuable to evaluate the innate behavioral responses of mosquito to chemical compounds, this system is relatively resource intensive requiring a comparatively large amount of chemical to be used for treating papers and having to use a large number of test mosquitoes. Moreover, a current ERB design is not conducive for mass screening the candidate chemical compounds. With the development of the HITSS assay (Grieco et al., 2005), a smaller amount of chemical and a lower number of mosquitoes is required per test. In addition, the HITSS can also be configured to test each of the three actions of insecticide compounds, namely contact irritancy assay (CIA) (Fig. 4), spatial repellency assay (SRA) (Fig. 5) and the toxicity assay (TOX) (Fig. 6). This modular system is made from a variety of durable materials including a thick, clear plastic cylinder, metal chambers, hard plastic end caps and a butterfly valve to control the opening and closing of the door. For the CIA, a single clear plastic cylinder is connected to a metal chamber lined with either treated or untreated netting material using a butterfly value placed in the open position (Fig. 4). For the SRA, an assay comprises a single clear plastic cylinder and 2 metal chambers on either end containing either treated or untreated (control) netting. The plastic chamber is positioned between the 2 metal chambers using a butterfly value as a linking system (Fig. 5). For the TOX assay, a single metal chamber is equipped with plastic end caps (solid cap on one side and tunnel cap on the other end). Netting material treated with either chemical active ingredient (treatment) or acetone carrier only (control) lines the inner chamber (Fig. 6). The HITSS has been standardized and used to evaluate the 3 behavioral responses of *Ae. aegypti* against DEET, Bayrepel®, and SS220 (Grieco et al., 2005, 2007). Recently, HITSS has been used to define the behavioral responses among six field populations of *Ae. aegypti* from Thailand against three synthetic pyrethroids (Thanispong et al., 2010). From this study, it was clearly shown that the HITSS assay is an effective and easy to use tool for distinguishing the three actions of chemicals and screening new compounds.

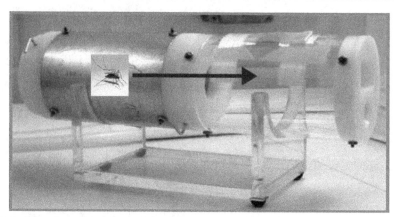

Fig. 4 High throughput screening system (HITSS): (CIA) Contact Irritant Assay (Grieco et al., 2005)

Fig. 5. High throughput screening system (HITSS): (SRA) Spatial Repellency Asay (Grieco et al., 2005)

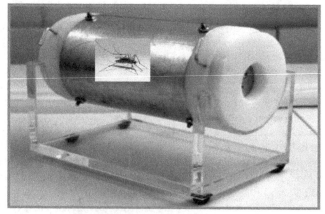

Fig. 6 High throughput screening system (HITSS): (TOX) Toxicity Assay (Grieco et al., 2005)

2.3 Field-based assay: experimental huts

To better understand the behavior of a mosquito exposed to a residual avoidance of under more natural, realistic conditions, field studies should be performed using experimental huts. So-called hut studies provide valuable information that can facilitate vector control operations by helping select the most appropraite and effective tool in combating disease vectors. Since the 1940s, most attention has been directed to the study of *Anopheles* mosquitoes, yet few investigations have been carried out to describe the innate behaviors of mosquitoes entering, resting, biting (blood feeding) and exiting human dwellings (Gahan & Lindquist, 1945; Giglioli, 1948; Smith, 1965; Roberts et al., 1984; Bangs, 1999; Grieco et al., 2000; Pates & Curtis, 2005; Roberts et al., 1997). Another important vector species, *Ae. aegypti*, has received even less attention using of experimental hut assays (Suwannachote et al., 2009). The discovery of using natural pyrethrum extract to prevent human-vector contact inside homes was well known prior to 1945 (Muirhead-Thomson, 1951) and the first report of behavioral responses of malaria vectors to DDT was documented in 1947 (Kennedy, 1947, Brown, 1983). Strong behavioral responses to DDT were progressively reported up until 1975 (Muirhead-Thomson, 1960; Elliott & de Zulueta, 1975; Brown, 1983). For example, de Zulueta and Cullen (1963) observed significant reductions in the numbers of biting mosqutioes on human indoors and resting on walls in houses sprayed with DDT. However, there was no clear explanation on how (mechanism) DDT functioned by either repelling or preventing mosquitoes from locating host stimuli inside sprayed houses. A significant scientific observation on behavioral avoidance of *Anopheles ganbiae* was carried out in the mid 1960s (Smith & Webley, 1968) wherein 2 huts, 1 control and one treated with DDT were equipped with verandah traps and gas chromatography was used to evaluate the subsequent response of mosquitoes. A hut treated with DDT was observed to prevent between 60 and 70% of *An. gambiae* from normally entering indoors as an outflow of DDT was continuously decreasing. Additionally, those females that did enter the sprayed hut became stimulated (irritated) and escaped the hut much quicker than those mosqutioes present in the unsprayed hut. Another study comparing biting patterns of *Anopheles minimus* in Thailand showed a 71.5% decline in attempted blood feeding inside the DDT treated hut and a 42.8% reduction in blood feeding success in a deltamethrin treated hut (Polsomboon et al., 2008). Using the same huts, observing human landing patterns of *Anopheles dirus* complex found that the relative risk (odds) of female mosquitoes entering and attempting to feed were half the number when exposed to DDT compared with the deltamethrin treated hut (Malaithong et al., 2010).

Of the possible responses a mosquto exposed to a chemcial can preform, spatial repellency is prehaps the most interesting and important. Observing true repellency presents problems of accurate messurement in the natural field situation. One of classic methods is to use the experimental huts wherein 2 experimental huts, 1 untreated control and 1 treated with an active ingredient is fitted with entrance (measure of repellency) and exit (irritancy) traps as illustrated in (Figs. 7-9). A landmark study was conducted by Grieco et al., (2007) on the movement patterns of *Ae. aegpyti* into and out of the experimental huts alternatively equipped with entrance (Fig. 8) and exit traps (Fig. 9) to describe the 3 chemical actions and mosquito responses as previously proposed in the mathematical framework for understanding the impact and relevance of repellency, irritability and toxicity on mosquito populations and disease transmission (Roberts et al. 2000). This study clearly showed that the 3 actions described and help to validate the model of actions. It was concluded that contact irritancy is the predominant action of synthetic pyrethroids, whereas spatial repellency is the primary action of DDT. Dieldrin, a once commonly used cyclodiene compound, exhibited primarily a toxic action only.

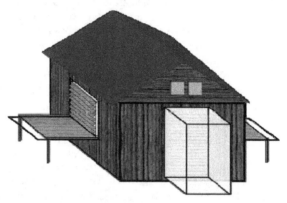

Fig. 7 Experimental hut with traps attached to exterior walls.

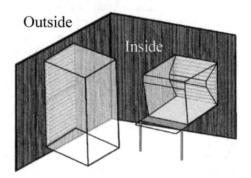

Fig. 8 Experimental hut with entrance traps attached to interior walls.

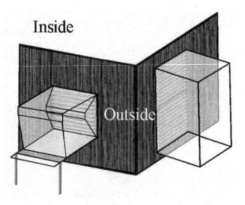

Fig. 9. Experimental hut with exit traps attached to exterior walls.

3. Behavioral responses

To facilitate the study of behavioral responses of mosquitoes to chemical compounds, several test systems have been developed (Roberts et al., 1997; Chareonviriyaphap et al., 2001). The following is a brief review of important historical findings and more recent work from studies in Thialand on behavioral responses of mosquitoes to DDT and various synthetic pyrethroids commonly used in vector control.

3.1 Behaviroral responses to DDT

Some agricultural and medically important insects, including vectors of malaria, demonstrate what has been termed "behavioral resistance" to DDT (Lockwood et al., 1984). However, the term "behavioral avoidance" as an innate response rather than a permanent genetic shift in behavior is preferred since the development of behavioral changes due to insecticide selective pressure in nature has not been adequately documented (Muirhead-Thomson, 1960). The behavioral responses of mosquitoes have been investigated using either specially constructed experimental huts and/or ERB test system. The first study on the irritant effect of DDT residual deposits was conducted on *Anopheles quadrimaculatus* where females were found to be irritated after short contact with treated surfaces with many quickly escaping the DDT treated house without taking a blood meal (Gahan & Lindquist, 1945). Subsequent observations found that *An. quadrimaculatus* had received a lethal dose and perished within 24 hours (Metcalf et al., 1945). Unfortunately, these studies made the observations without having control (untreated houses) for comparison. Moreover, the high mortality seen with *An. quadrimaculatus* may have been caused by further contacts with toxic active ingredients while attempting to leave a treated house through a small outlet (Muirhead-Thomson, 1960). In the studies with *Anopheles albimanus* in Panama, Trapido (1954) concluded that wild-caught mosquitoes lacking re-exposure to DDT for a long period of time, showed the same susceptibility levels to DDT as those from a laboratory colony with no a history of previous exposure. Malaria vectors in some countries (e.g., Brazil, Thailand) have never developed resistance to DDT (Roberts et al., 1984; Chareonviriyaphap et al., 2001), suggesting that the particular mosquito population possibly avoids making direct physical contact with the chemical, thereby precluding any selection for resistance. Table 1 lists *Anopheles* species tested and levels of behavioral responses to DDT and synthetic pyrethroids. DDT was found to be a strong contact irritant and more moderately as a repellent among 3 test populations of *An. albimanus* and that most specimens that quickly escaped DDT exposure survived (Chareonviriyaphap et al., 1997). This finding is in agreement with the results of the field studies by Roberts and Alecrim (1991) who reported a strong repellent action of DDT with *Anopheles darlingi* in a sprayed house. In a similar study, both irritancy and repellency escape responses were observed in 2 populations of *Anopheles minimus* with a major action of DDT being contact excitation (Chareonviriyaphap et al., 2001). Another study was made on the 2 complex species within the Minimus Subgroup; *An. minimus* (species A) and *Anopheles harrisoni* (species C). DDT produced a rapid and striking irritancy response in both species. Additionally, repellency was more pronounced with DDT on *An. minimus* but was seen to be much weaker with *An. harrisoni* (Pothikasikorn et al. 2005). With *Ae. aegypti*, results demonstrated that the higher the degree of physiological resistance to DDT, the greater the apparent suppression of both contact irritant and noncontact repellency responses, yet avoidance behavior was still a significant event (Thanispong et al., 2009).

Species	Strain	Permethrin I	Permethrin R	Deltamethrin I	Deltamethrin R	Cypermethrin I	Cypermethrin R	λ-cyhalothrin I	λ-cyhalothrin R	Bifenthrin I	Bifenthrin R	DDT I	DDT R	References
An. albimanus	Lab	+	-	+	-							+	-	Chareonviriyaphap et al., 1997
An. albimanus	Field	++++	-	++++	-							++++	++	Chareonviriyaphap et al., 1997
An. albimanus	Field	++++	-	++++	-							++++	++	Chareonviriyaphap et al., 1997
An. albimanus	Field	++++	-	++++	-							++++	++	Chareonviriyaphap et al., 1997
An. minimus	Lab			++++	+			+++	-			+++	+	Chareonviriyaphap et al., 2001
An. minimus	Field			++	+			++++	-			+++	+	Chareonviriyaphap et al., 2001
An. minimus	Lab			++++	+++									Chareonviriyaphap et al., 2004
An. minimus	Lab			++++	++									Chareonviriyaphap et al., 2004
An. dirus	Lab			+++	+++									Chareonviriyaphap et al., 2004
An. maculatus	Field			++++	+									Chareonviriyaphap et al., 2004
An swadtwongsporni	Field			++++	++									Chareonviriyaphap et al. 2004
An. dirus	Field			+++	++									Chareonviriyaphap et al. 2004
An. minimus	Field			++++	++			++++	-			++++	+++	Potikasikorn et al., 2005
An. harrisoni	Field			++++	-			++++	-			++++	-	Potikasikorn et al., 2005
An. maculatus	Field	+++	-									+	-	Muenworn et al., 2006
An swadtwongsporni	Field	+++	-									+++	-	Muenworn et al., 2006
An. minimus	Field					++	+							Pothikasikorn et al., 2007
An. minimus	Field									+	-			Tisgratog et al., 2011
An.harrisoni	Field									+++	+			Tisgratog et al., 2011

I: Irritancy (Excitation and movement away from the source following direct contact with chemical stimulant)

R: Repellency (noncontact, spatial detection of chemical that results in movement away from the source)

++++: 81-100% escaped from treated chamber

+++: 61-79% escaped from treated chamber

++: 41-59% escaped from treated chamber

+: 21-40% escaped from treated chamber

-: ≤ 20% escaped from treated chamber but statistically significant ($P < 0.05$) from the matched control

Table 1. Degree of behavioral responses of female *Anopheles* species to synthetic pyrethroids and DDT at an operational field dose (mg/cm²).

3.2 Behavioral response to synthetic pyrethroids

A number of synthetic pyrethroids, i.e. allethrin, deltamethrin, permethrin, cypermethrin, alpha-cypermethrin, cyfluthrin among others, are commonly used by home owners, private business and government sectors to control both household mosquitoes and those as important vectors. These pyrethroids have found to exhibit a moderate to strong repellent effect for many agricultural and medically important insects (Lockwood et al., 1984) and were observed to cause mosquitoes to move away ('avoidance') from sprayed areas (Miller, 1990; Lindsay et al., 1991). The extensive and continuing use of pyrethroids should be a major stimulus to intensify observations on the significance of pyrethroid-induced avoidance behavior in mosquito vectors and other arthropods. Given the role of indoor residual spraying of homes as a means of controlling malaria transmission, the precise role of excitation and repellent actions of pyrethriods should be well defined for specific malaria vectors prior to beginning any large scale control program. Following the refinement of the ERB test system allowing separation of the 2 types of primary behavioral actions (Roberts et al., 1997), a series of important findings on excito-repellency behavior in *Anopheles* mosquitoes have been subsequently reported (Chareonviriyaphap et al., 1997, 2001, 2004; Pothikasikorn et al., 2005; Muenworn et al., 2006; Pothikasikorn et al., 2007). In general, synthetic pyrethroids produce much stronger irritant responses in *Anopheles* compared to repellency action (Table 1). For example, lambda-cyhalothrin and deltamethrin act as strong irritants on test populations of *An. minimus* complex mosquitoes while showing relatively weak repellency action (Chareonviriyaphap et al., 2001). Pothikasikorn et al. (2005) confirmed that Minimus complex species, *An. minimus* and *An. harrisoni* show a rapid irritancy to lambda-cyhalothrin and deltamethrin. Chareonviriyaphap et al., (2004) produced an extensive study to define the excito-repellency action of deltamethrin on 4 *Anopheles* species, all representing important malaria vectors in Thailand. Again, the findings demonstrated that deltamethrin produced a pronounced irritancy action compared to a much weaker repellency effect. Although repellency was less profound than contact excitation, the escape responses were statistically significant compared to the matched controls. A number of *Ae. aegypti* populations have been tested against a series of synthetic pyrethroids (deltamethrin, permethrin, alphacypermethrin, cyphenothrin, d-tetramethrin and tetramethrin) (Table 2). In general, all test populations of *Ae. aegypti* populations exhibit moderate to strong irritancy as compared with repellency (Grieco et al., 2005; Chareonviriyaphap et al., 2006; Paeporn et al., 2007; Thanispong et al., 2009, 2010). In addition, a few populations of *Culex quinqaefacsciatus* have been tested against the 3 principal classes of insecticides used in vector control; pyrethroids (deltamethrin), organophosphates (fenitrothion) and carbamates (propoxur) (Table 3). Striking differences in behavioral responses were seen between populations and active ingredients. Greater contact escape action was observed in a long-established colony exposed to deltamethrin and fenitrothion compared to two recent field populations (Sathantriphop et al., 2006).

To summarize, the behavioral responses to insecticides by mosquitoes are important components of a chemical's overall effectiveness in reducing human-vector contact and transmision of disease. To date, there is no convincing example of behavioral resistance in mosquito species to insecticides, rather all evidence indicates actions on the part of exposed mosquitoes are part of an innate behaviral repertoire. Behavioral response can be split into 2 distinct categories, stimulus-dependent and stimulus-independent actions (Georghiou, 1972). A stimulus-dependent response requires sensory stimulation of the insect in order for an avoidance action to proceed. In general, this form of avoidance enables the insect to

Strains	Deltamethrin		Permethrin		α-cypermethrin		Cyphenothrin		d-tetramethrin		Tetramethrin		DDT		References
	I	R	I	R	I	R	I	R	I	R	I	R	I	R	
Field-R	++	-													Kongmee et al., 2004
Field-R	+	-													Kongmee et al., 2004
Field-R	+	-													Kongmee et al., 2004
Field-R	++++	-													Kongmee et al., 2004
Field-R	++	-													Kongmee et al., 2004
Field-S	+++	-													Kongmee et al., 2004
Lab-S	++	-													Kongmee et al., 2004
Lab-S	++	-													Kongmee et al., 2004
Lab-S	++	-													Kongmee et al., 2004
Field-RR			+	-											Paeporn et al., 2007
Field-RR			-	-											Paeporn et al., 2007
Field					++++	-							++	-	Thanispong et al., 2009
Field-S					+++	-							-	-	Thanispong et al., 2009
Lab-S					++++	++							++	++	Thanispong et al., 2009
Field-S							++++	-	+++	+	++	+			Mongkalagoon et al., 2009
Field-S							++++	+	+++	+	+++	-			Mongkalagoon et al., 2009
Field-S							++++	+	+++	-	+++	+			Mongkalagoon et al., 2009

I: Irritancy (contact excitation), R: Repellency (noncontact/spatial), R: Resistant, S: Susceptible to test compound
++++: 81-100% escaped from treated chamber
+++: 61-79% escaped from treated chamber
++: 41-59% escaped from treated chamber
+: 21-40% escaped from treated chamber
-: ≤ 20% escaped from treated chamber but statistically significant ($P < 0.05$) from the matched control

Table 2. Degree of behavioral responses of female *Aedes aegypti* to synthetic pyrethroids and DDT at an operational field dose (mg/cm²).

detect a chemical on direct contact or spatially before acquiring a lethal dose (Muirhead-Thomson, 1960). On the other hand, a stimulus-independent response does not require direct sensory stimulation of insect for avoidance to occur but rather involves other natural behavioral components such as exophily (outside resting) or zoophily (non-human blood preference) in which an insect avoids exposure to a chemical by preferentially utilizing habitats without active ingredients present (Byford & Sparks, 1987). This type of response has also been included in so-called "phenotypic and genotypic behaviors" (WHO, 1986). Stimulus-dependent behavioral responses include the avoidance behaviors discussed in detail in this chapter. The term *avoidance behavior* is generally used to describe actions that are stimulated by some combination of excitation (irritancy) and repellency, the former taking place following physical contact while spatial repellency results without physical contact with an insecticide (Roberts et al., 1987).

Populations	Deltamethrin		Fenitrothion		Propoxur		References
	I	R	I	R	I	R	
Bangkok*	++	-	++	-	+	-	Sathantriphop et al., 2006
Nontaburi*	+	-	+	-	+	-	Sathantriphop et al., 2006
Mae Sot*	+	-	+	-	+	-	Sathantriphop et al., 2006

*Highly resistance to deltamethrin
I: Irritancy, R: Repellency
++++: 81-100% escaped from treated chamber
+++: 61-79% escaped from treated chamber
++: 41-59% escaped from treated chamber
+: 21-40% escaped from treated chamber
-: $\leq$ 20% escaped from treated chamber but statistically significant ($P < 0.05$) from the matched control

Table 3. Degree of behavioral responses of *Culex quinquefasciatus* to deltamethrin, propoxur and fenitrothion at an operational field dose (mg/cm^2).

4. Conclusion

Any compound used to control (eliminate, reduce or otherwise prevent harm) insect populations have been termed an "insecticide". An insecticide can also be defined as any compound that is used solely to kill insects. This single term of reference involving only one type of action (knockdown/death) is completely inadequate to describe the more meaningful and complete action of many insecticide compounds used against insect populations. Insects can respond to insecticides in at least 2 different ways; behavioral action, namely avoidance and toxicity. In the past, the prevailing practice has been to classify chemicals simply as toxicants for killing insects. As seen in this chapter, we introduce the term "chemical" in place of "insecticide" as it is more appropriate for recognizing the 2 other primary actions on mosquitoes vice toxicity alone.

Chemicals protect humans from the bites of mosquitoes through 3 different actions: excitation, repellency and toxcity (Roberts et al., 2000 & Grieco et al., 2007). Historically the vast majority of chemical studies have focused on the direct toxicological responses (susceptibility and resistance) of chemicals on mosquito populations whereas very little emphasis has been placed on the vector's behavior in response to sub-lethal exposure. Knowledge of the mosquito's behavioral responses to particular chemicals is highly significant in the prioritization and design of appropriate vector prevention and control strategies. Today, the development of insecticide resistance in insect pests and disease vectors occurs worldwide and on a increasing scale. However, resistance has still remained limited in many areas in spite of the long-term use of chemicals for control. This phenomenon suggests that behavioral responses likely play a significant role in how certain chemcials preform to interrupt human-vector contact while also reducing the selection pressure on a target insects for developing resistance (Roberts et al., 2000).

As discussed, at least 2 different types of mosqutio behavioral response outcomes to chemicals are recognized; excitation and repellency. Whether acting from direct contact or from a distance (spatially) both response activities are based on stimulus-response actions that result in clear movement away from an area with the chemical present. There have been numerous attempts to accurately measure the behavioral responses of mosquitoes to chemicals using various types of excito-repellency test systems (e.g., ERB and HITSS). However, no single system has yet been completely accepted as a standardized method of testing or analyzing behavioral responses. Currently, no test system as recommended by the WHO can discriminate between contact irritancy and noncontact repellency. The WHO tests are based on the concept that mosquitoes respond to chemicals only after physical contact and that spatial repellency plays little or no meaningful role in disease control, or may actually be a determental attribute of a chemical. The test systems describe in this chapter have the capacity to differentiate both key behavioral responses and thus assigning each with relative importance in the potential prevention of disease transmission. Both the ERB and HITSS systems have been used to study behavioral responses of several important mosquito vectors and pests (*Anopheles, Aedes,* and *Culex*) to various test chemicals currently used in vector control programs while providing valuable and highly reproducible results. This knowledge will allow better decision making on chemical selection, application method and future product development. From the detailed investigations previously mentioned and elsewhere, we conclude that the behavioral responses of mosquitoes to chemicals are an important, if not critical, components of disease control operations.

5. Acknowledgments

The author would like to thank Dr. Michael J. Bangs for the critical review of this chapter. I am especially grateful to the Thailand Research Fund (TRF) and the Kasetsart University Research and Development Institute (KURDI) for providing the financial support over the many years.

6. References

Bangs, M.J. (1999). The Susceptibility and Behavior of *Anopheles albimanus* Weidemann and *Anopheles vestitipennis* Dyar & Knab (Diptera: Culicidae) to Insecticides in Northern Belize, Central America. Doctoral Dissertation, Uniformed Services University of the Health Sciences, Bethesda, Maryland, pp. 448.

Bondareva, N.L.; Artem'ev, M.M. & Gracheva, G.V. (1986). Susceptibility and Irritability Caused by Insecticides to Malaria Mosquitoes in the USSR. Part 1. *Anopheles pulcherrimus. Meditsinskaia Parazitologha I Parazitarnye Bolezni* (Moskva),Vol. 6, pp. 52-55.

Brown, A.W.A. (1983). Insecticide Resistance as a Factor in the Integrated Control of Culicidae. In Integrated Mosquito Control Methodologies. Volume 1. Laird M & Mile J. Eds. Academic Press: New York, pp. 161-235.

Byford, R.L & Sparks, T.C. (1987). Chemical Approaches to the Management of Resistant Horn Fly, *Haematobia irritans* (L.) Populations. *In* M. 0. Ford, D. W. Holloman, B. P. S. Khambay and R. M. Sawicki (eds.), Combating Resistance to Xenobiotics: Biological and Chemical Approaches, pp. 178-189. Ellis Horwood: Chichester UK.

Chareonviriyaphap, T.; Roberts, D.R., Andre, R. G., Harlan, H. & Bangs, M.J. (1997). Pesticide Avoidance Behavior in *Anopheles albimanus* Wiedemann. *Journal of the American Mosquito Control Association*, Vol. 13, No. 2, pp. 171-183, ISSN 1046-3607.

Chareonviriyaphap, T. & Aum-Aung, B. (2000). An Improved Excito-Repellency Escape Chamber for Behavioral Test in Mosquito Vectors. *Mekong Malaria Forum*, Vol. 5, pp.82-87.

Chareonviriyaphap, T.; Sungvornyothin, S., Ratanatham, S., & Prabaripai, A. (2001). Pesticide-Induced Behavioral Responses of *Anopheles minimus*, a Malaria Vector in Thailand. *Journal of the American Mosquito Control Association*, Vol. 17, No.1, (March 2001), pp, 13-22, ISSN 1046-3607.

Chareonviriyaphap, T.; Prabaripai, A., & Sungvornyothin, S. (2002). An Improved Excito-Repellency for Mosquito Behavioral Test. *Journal of Vector Ecology*, Vol. 27, No.2, (December 2002), pp, 250-252, ISSN 1081-1710.

Chareonviriyaphap, T.; Prabaripai, A. & Bangs, M.J. (2004). Excito-Repellency of Deltamethrin on the Malaria Vectors, *Anopheles minimus*, *Anopheles dirus*, *Anopheles sawadwongporni*, and *Anopheles maculatus*, in Thailand. *Journal of the American Mosquito Control Association*, Vol. 20, No.1, (March 2004), pp. 45-54, ISSN 1046-3607.

Chareonviriyaphap, T.; Parbaripai, A., Bangs, M.J., Kongmee, M., Sathantriphop, S., Muenvorn, V., Suwonkerd, W. & Akratanakul, P. (2006). Influence of Nutritional and Physiological Status on Behavioral Responses of *Aedes aegypti* (Diptera: Culicidae) to Deltamethrin and Cypermethrin. *Journal of Vector Ecology*, Vol. 31, No.1, (June 2006), pp. 89-101, ISSN 1081-1710.

Charlwood, J.D. & Paraluppi, N.D. (1978). The Use of Excito-Repellency Box with *Anopheles darlingi* Root, *An. nuneztovari* Gabaldon and *Culex pipiens quinquefaciatus* Say, Obtained from the Areas near Manaus, Amazonas. *Acta Amazonica*, Vol. 8, No. 4, pp. 605-611, ISSN 0044-5967.

de Zulueta, J. & Cullen, J.R. (1963). Deterrent Effect of Insecticides on Malaria Vectors. *Nature*, Vol.200, (November 1963), pp. 860-61, ISSN 0028-0836.

Evans, R. G. (1993). Laboratory Evaluation of the Irritancy of Bendiocarb, Lambdacyhalothrin, and DDT to *Anopheles gambiae*. *Journal of the American Mosquito Control Association*, Vol. 9, No.3, (September 1993), pp. 285-293, ISSN 1046-3607.

Elliott, R. & de Zulueta, J. (1975). Ethological Resistance in Malaria Vectors. Behavioral Response to Intradomestic residual insecticides. WHO/VBC75.569, pp. 1-15.

Gahan, J.B. & Lindquist, A.W. (1945). DDT Residual Sprays Applied in Buildings to Control *Anopheles quadrimaculatus*. *Journal of Economic Entomology*, Vol.38, No.2, pp. 223-230, ISSN 0022-0493.

Georghiou, G.P. (1972). The Evolution of Resistance to Pesticides. *Annual Review of Ecology and Systematics*, Vol.3, No.1 (November 1972), pp. 133-168, ISSN 0066-4162.

Giglioli, G. (1948). An Investigation of the House-Frequenting Habits of Mosquitoes of the British Guiana Coastland in Relation to the Use of DDT. *American Journal of Tropical Medicine and Hygiene*, Vol. 28, No.1, (January 1948), pp. 43-70, ISSN 0002-9637.

Grieco, J.P.; Achee, N.L., Andre, R.G. & Roberts, D.R. (2000). A Comparison Study of House Entering and Exiting Behavior of *Anopheles vestitipennis* (Diptera: Culicidae) using Experimental Huts Sprayed with DDT or Deltamethrin in the Southern District of Toledo, Belize, CA. *Journal of Vector Ecology*, Vol. 25 No.1, (June 2000), pp. 62-73, ISSN 1081-1710.

Grieco, J.P.; Achee, N.L., Sardelis, M.R., Chauhan, K .R. & Roberts, D.R. (2005). A Novel High Throughput Screening System to Evaluate the Behavioral Response of Adult Mosquitoes to Chemicals. *Journal of the American Mosquito Control Association*, Vol. 21, No.4, (December 2005), pp. 404-411, ISSN 1046-3607.

Grieco, J.P.; Achee, N.L., Chareonviriyaphap, T., Suwonkerd, W., Chauhan, K.R., Sardelis, M. & Roberts, D.R. (2007). A New Classification System for the Actions of IRS Chemicals Traditionally Used for Malaria Control. *PLoS ONE*, Vol.2, Issue. 8, (August 2008), pp. e716, ISSN 1932-6203.

Gubler, D.J. (1997). Dengue and Dengue Haemorrhagic Fever: Its History and Resurgence as a Global Public Health Problem. In: Gubler, DJ and Kuno, G (eds.). Dengue and Dengue Haemorrhagic Fever. New York: CAB International. 22 pp.

Kennedy, J.S. (1947). The Excitant and Repellent Effects on Mosquitoes of Sub-lethal Contacts with DDT. *Bulletin of Entomological Research*, Vol. 37, No. 4, (March 1947), pp. 593-607, ISSN 0007-4853.

Kongmee, M.; Prabaripai, A., Akaratanakul, P., Bangs, M.J. & Chareonviriyaphap, T. (2004). Behavioral Responses of *Aedes aegypti* (Diptera: Culicidae) Exposed to Deltamethrin and Possible Implications for Disease Control. *Journal of Medical Entomology*, Vol. 41, No. 6, (November 2004) pp.1055-1063, ISSN 0022-2585.

Lindsay, S.W.; Adiamah, J.H., Miller, J.E. & Armstrong, J.R.M. (1991). Pyrethroid-Treated Bednet Effects on Mosquitoes of *Anopheles gambiae* Complex in the Gambia. *Medical and Veterinary Entomology*, Vol. 5, No.4, (October 1991), pp. 477-483, ISSN 0269-283X.

Lockwood, J.A.; Sparks, T.C. & Story, R.N. (1984). Evolution of Insect Resistance to Insecticide: a Reevaluation of the Roles of Physiology and Behavior. *Bulletin of the Entomological Society of America*, Vol. 30, No. 4, (November 1984), pp. 41-51, ISSN 0013-8754.

Malaithong, N.; Polsomboon, S., Poolprasert, P., Parbaripai, A., Bangs, M.J., Suwonkerd, W., Pothikasikorn, J., Akratanakul, P. & Chareonviriyaphap, T. (2010). Human-Landing Patterns of *Anopheles dirus* sensu lato (Diptera: Culicidae) in Experimental Huts Treated with DDT or Deltamethrin. *Journal of Medical Entomology*, Vol. 47, No. 4, (September 2010), pp. 823-32, ISSN 0022-2585.

Metcalf, R.L.; Hess, A.D., Smith, G.E., Jeffery, G.M. & Ludwig, G.W. 1945. Observations on the Use of DDT for the Control of *Anopheles quadrimaculatus*. Public Health Report, Vol. 60, No.27, pp. 753-774, ISSN 0033-3549.

Miller, J.E. (1990). Laboratory and Field Studies of Insecticide Impregnated Fibers for Mosquito Control. Ph.D. Dissertation, University of London. 336 pp.

Miller, J. R.; Siegert, P.Y., Amimo, F.A. & Walker, E.D. (2009). Designation of Chemicals in Terms of the Locomotor Responses They Elicit from Insects: an Update of Dethier et al. (1960). *Journal of Economic Entomology*, Vol. 102, No. 6, (December 2009), pp. 2056-2060, ISSN 0022-0493.

Mongkalangoon, P.; Grieco, J.P., Achee, N.L., Suwonkerd, W. & Chareonviriyaphap, T. (2009). Irritability and Repellency of Synthetic Pyrethroids on an *Aedes aegypti* Population from Thailand. *Journal of Vector Ecology*, Vol. 34, No. 1, (June 2009), pp. 217-24, ISSN 1081-1710.

Muenworn, V.; Akaratanakul, P., Bangs, M.J., Parbaripai, A. & Chareonviriyaphap, T. (2006). Insecticide Induced Behavioral Responses in two Populations of *Anopheles maculatus* and *Anopheles swadwongporni*, Malaria Vectors in Thailand. *Journal of the American Mosquito Control Association*, Vol. 22, No. 4, (December 2006), pp. 689-698, ISSN 1046-3607.

Muirhead-Thomson, R.C. (1951). Mosquito Behaviour in Relation to Malaria Transmission and Control in the Topics. pp. viii+219; 16 plates: 22 Text Figure. London: Arnold.

Muirhead-Thomson, R.C. (1960). The Significance of Irritability, Behaviouristic Avoidance and Allied Phenomena in Malaria Eradication. *Bulletin of World Health Organization*, Vol. 22, pp. 721-734, ISSN 0042-9686.

Peaporn, P.; Supaphathom, K., Sathantriphop, S., Chareonviriyaphap, T., Yaicharoen, R. (2007). Behavioral Responses of Deltamethrin and Permethrin Resistant Strains of *Aedes aegypti* When Exposed to Permethrin in an Excito-Repellency Test System. *Dengue Bulletin*, Vol. 31, pp. 153-159, ISSN 0250-8362.

Pates, H. & Curtis, C. (2005). Mosquito Behavior and Vector Control. *Annual Review of Entomology*, Vol. 50, pp.53–70, ISSN 0066-4170.

Polsomboon, S.; Poolprasert, P., Suwonkerd, W., Bangs, M.J., Tanasinchayakul, S., Akratanakul, P. & Chareonviriyaphap, T. (2008). Biting Patterns of *Anopheles minimus* Complex (Diptera: Culicidae) in Experimental Huts Treated with DDT and Deltamethrin. *Journal of Vector Ecology*, Vol. 33, No. 2, (December 2008), pp. 285-292, ISSN 1081-1710.

Pothikasikorn, J.; Chareonviriyaphap, T., Bangs, M.J. & Prabaripai, A. (2005). Behavioral Responses to DDT and Pyrethroids between *Anopheles minimus* Species A and C, Malaria Vectors in Thailand. *American Journal of Tropical Medicine and Hygiene*, Vol. 73, No. 2, (August 2005) pp. 343-349, ISSN 0002-9637.

Pothikasikorn, J.; Overgaard, H., Ketavan, C., Visetson, S., Bangs, M.J. & Chareonviriyaphap, T. (2007). Behavioral Responses of Malaria Vectors, *Anopheles minimus* Complex, to Three Classes of Agrochemicals in Thailand. *Journal of Medical Entomology*, Vol.44, No. 6, (November 2007), pp. 1032-1039, ISSN 0022-2585.

Quinones, M.L. & Suarez, M.F. (1989). Irritability to DDT of Natural Populations of the Primary Malaria Vectors in Colombia. *Journal of the American Mosquito Control Association*, Vol. 5, No. 1, (March 1989), pp. 56–59, ISSN 1046-3607.

Rachou, R.G. ; Moura- Lima, M., Duret, J.P.& Kerr, J.R. (1963). Experiences with the Excito-Repellency Test Box—Model OPS, pp. 442-447. In Proceedings, 50th Annual Meeting of the New Jersey Mosquito, the 19th American Mosquito Control Association, ISSN 1046-3607.

Ree, H.I. & Loong, K.P. (1989). Irritability of *Anopheles farauti*, *Anopheles maculatus*, and *Culex quinquefasciatus* to Permethrin. *Japanese Journal of Sanitary Zoology*, Vol. 40, No. 1, pp. 47–51, ISSN 0424-7086.

Reiter, P. & Gubler, D.J. (1997). Surveillance and Control of Urban Dengue Vectors. In: Gubler, DJ and Kuno, G (eds.). Dengue and Dengue Haemorrhagic Fever. New York: CAB International. pp. 425-462

Roberts, D.R. (1993). Insecticide Repellency in Malaria Vector Control: a Position Paper. VBC Report No. 81131. United States Agency for International Development, Arlington, VA. 72 pp, ISSN 0736-718X.

Roberts, D.R.; Alecrim, W.D., Tavares, A.M. & McNeil, K.M. (1984). Influence of Physiological Condition on the Behavioral Responses of *Anopheles darlingi* to DDT. *Mosquito News*, Vol.44, No.3, pp. 357-361, ISSN 0027-142X.

Roberts, D.R.; Alecrim, W.D., Tavares, A.M. & Radke, M.G. (1987). The House-Frequenting, Host Seeking and Resting Behavior of *Anopheles darlingi* in Southeastern Amazonas, Brazil. *Journal of the American Mosquito Control Association*, Vol.3, No.3, (September 1997), pp. 433-441, ISSN 1046-3607.

Roberts, D.R. & Alecrim, W.D. (1991). Behavioral response of *Anopheles darlingi* to DDT sprayed house walls in Amazonia. *The Bulletin of the Pan American Health Organization*, Vol.25, No. 3, pp. 210–217, ISSN 0042-9686.

Roberts DR, & Andre RG. (1994). Insecticide Resistance Issues in Vector Borne Disease Control. *American Journal of Tropical Medicine and Hygiene*, Vol.50, No.6, pp. 21–34, ISSN 0002-9637.

Roberts, D.R.; Chareonviriyaphap, T., Harlan, H.H. & Hshieh, P. (1997). Methods for Testing and Analyzing Excito-Repellency Responses of Malaria Vectors to Insecticides. *Journal of the American Mosquito Control Association*, Vol.13, No.1, (March 1997), pp. 13-17, ISSN 1046-3607.

Roberts, D.R.; Alecrim, W.D., Hshieh, P., Grieco, J., Bangs, M.J., Andre, R.G. & Chareonviriyaphap, T. (2000). A Probability Model of Vector Behavior: Effects of DDT Repellency, Irritability, and Toxicity in Malaria Control. *Journal of Vector Ecology*, Vol.25, No. 1, (June 200), pp.48-61, ISSN 1081-1710.

Roberts, D.R. & Tren, R. (2010). The Excellent Powder: DDT's Political and Scientific History. Dog Ear Publishing, pp. 432, ISBN: 978-160844-376-5.

Rozendaal, J.A.; Van Hoof, J.P.M., Voorham, J. & Oostburg, B.F.J. (1989). Behavioral Responses of *Anopheles darlingi* in Suriname to DDT Residues on Housewalls. *Journal of the American Mosquito Control Association*, Vol.5, No. 3, (September 1989), pp. 339-350, ISSN 1046-3607.

Sathantriphop, S.; Ketavan, C., Prabripai, A., Vietson, S., Bangs, M.J., Akatanakul, P. & Chareonviriyaphap, T. (2006). Susceptibility and Avoidance Behavior by *Culex quinquefasciatus* Say to three Classes of Residual Insecticides. *Journal of Vector Ecology*, Vol. 31, No. 2, (December 2006), pp. 266-274, ISSN 1081-1710.

Service, M.W. & Townson, H. (2002). The *Anopheles* Vector. pp. 59-84. In: Warrell DA, Gilles HM (eds). Essential malariology (4th ed.). Arnold, London UK, ISBN 9780340740644.

Smith, A. (1965). A Verandah-Trap Hut for Studying the House Frequenting Habits of Mosquitoes and for Assessing Insecticides. II. The Effect of Dichlorvos (DDVP) on Egress and Mortality of *Anopheles gambiae* Giles and *Mansonia uniformis* (Theo.) entering naturally. *Bulletin of Entomological Research*, Vol. 56, pp. 275-286, ISSN 0007-4853.

Smith, A. & Webley, D.J. (1969). A Varandah-Type Hut for Studying the House-Frequenting Habits of Mosquitoes and for Assessing Insecticide. III. The Effect of DDT on Behaviour and Mortality. *Bulletin of Entomological Research*, Vol. 59, No. 3, pp. 33-46, ISSN 0007-4853.

Sinka, M.E.; Bangs, M.J., Manguin, S., Chareonviriyaphap, T., Patil, A.P., Temperley, W.H., Gething, P.W., Elyazar, I.R., Kabaria, C.W., Harbach, R.E. & Hay, S.I. (2011). The Dominant *Anopheles* Vectors of Human Malaria in the Asia-Pacific Region: Occurrence Data, Distribution Maps and Bionomic Précis. *Parasites Vectors* , Vol.4, (May 2011), pp. 89, ISSN 1756-3305.

Sungvornyothin, S.; Chareonviriyaphap, T., Prabaripai, A., Trirakhupt, T., Ratanatham, S. & Bangs, M.J. (2001). Effects of Nutritional and Physiological Status on Behavioral Avoidance of *Anopheles minimus* (Diptera: Culicidae) to DDT, Deltamethrin and Lambdacyhalothrin. *Journal of Vector Ecology*, Vol. 26, No.2, (December 2001), pp. 202-215, ISSN 1081-1710.

Suwannachote, N.; Grieco, J.P., Achee, N.L., Suwonkerd, W., Wongtong, S. & Chareonviriyaphap, T. (2009). Effects of Environmental Conditions on the Movement Patterns of *Aedes aegypti* (Diptera: Culicidae) into and out of Experimental Huts in Thailand. *Journal of Vector Ecology*, Vol. 34, No. 2, (December 2009), pp. 267-275, ISSN 1081-1710.

Tanasinchayakul, S.; Polsomboon, S., Prabaripai, A. & Chareonviriyaphap, T. (2006). An Automated, Field Compatible Device for Excito-Repellency Assay in Mosquitoes. *Journal of Vector Ecology*, Vol.31, No. 1, (June 2006), pp. 210-212, ISSN 1081-1710.

Thanispong, K.; Achee, N.L., Bangs, M.J., Grieco, J.P., Suwonkerd, W., Prabaripai, A., & Chareonviriyaphap, T. (2009). Irritancy and Repellency Behavioral Responses of three strains of *Aedes aegypti* Exposed to DDT and Alpha-cypermethrin. *Journal of Medical Entomology*, Vol. 46, No. 6, (November 2009), pp. 1407-1414, ISSN 0022-2585.

Thanispong, K.; Achee, N.L., Grieco, J.P., Bangs, M.J., Suwonkerd, W., Prabaripai, A., Chauhan, K.R. & Chareonviriyaphap, T. (2010). A High Throughput Screening System for Determining the three Actions of Insecticides against *Aedes aegypti* (Diptera: Culicidae) Populations in Thailand. *Journal of Medical Entomology*,Vol. 47, No. 5, (September 2010), 833-841, ISSN 0022-2585.

Tisgratog, R.; Thananchai, C., Bangs, M.J., Thainchum, K., Juntarajumnong, W., Prabaripai, A., Chauhan, K.R., Pothikasikorn, J.,& Chareonviriyaphap, T. (2011). Chemical Induced Behavioral Responses in *Anopheles minimus* and *Anopheles harrisoni* in Thailand. *Journal of Vector Ecology*, Vol. 36, No.2 (December 2011), pp. xxx-xxx, ISSN 1081-1710.

Trapido, H. (1954). Recent Experiments on Possible Resistance to DDT by *Anopheles albimanus* in Panama. *Bulletin of the World Health Organization*, 11: 885-889, ISSN 0042-9686.

World Health Organization. (1970). Insecticide Resistance and Vector Control. 17th Report of the World Health Organization Expert Committee on Insecticide. Instructions for Determining the Irritability of Adult Mosquitoes to Insecticide. *World Health Organization Technical Report Series*, pp. 433. ISSN 0512-3054.

World Health Organization. (1986). Resistance of Vectors and Reservoirs of Disease to Pesticides: *Tenth Report of the World Health Organization Offset Publication*, No 13, ISSN 0303-7878.

World Health Organization. (1992). Vector Resistance to Pesticides. Technical Report Series No. 818. *World Health Organization*, Geneva, pp. 62, ISSN 0042-9686.

World Health Organization. (1999). Prevention and Control and Dengue and Dengue Haemorrhagic Fever: Comprehensive Guidelines. *World Health Organization Regional Publication*, SEARO No. 29, New Delhi., pp. 134, ISSN 0378-2255.

9

Biological Control of Mosquito Larvae by *Bacillus thuringiensis* subsp. *israelensis*

Mario Ramírez-Lepe[1] and Montserrat Ramírez-Suero[2]
[1]Instituto Tecnológico de Veracruz, UNIDA, Veracruz,
[2]Université de Haufe-Alsace, LVBE EA-3991, Colmar,
[1]Mexico
[2]France

1. Introduction

Chemical insecticides provide many benefits to food production and human health and has proven very effective at increasing agriculture and forestry productivities. However, they also pose some hazards as contamination of water and food sources, poisoning of non-target fauna and flora, concentration in the food chain and selection of insect pest populations resistent to the chemical insecticides (Wojciech & Korsten 2002). It is well documented that chemical pesticides reduced natural-enemy populations and chemical applications can disrupt biological control and may cause outbreaks of secundary pests previously suppressed by natural enemies (Bartlett, 1964) and pest species develop pesticide resistance but natural enemies not (Johnson & Tabashnick, 1999).

The use of synthetic organic pesticides has had serious economic, social and environmental ramifications. Economically, the rapidy increasing cost for development and production of petrochemically derived insecticides, together with the declining effectiveness due to widespread insect resistance. As a result the chemical pesticide industry continues to develop new more expensive compounds and increasing pesticide prices. Socially and ecologically they have caused death and disease in human and damaged the environment. It is estimated that only a minute fraction of the insecticides applied is required for suppression of the target pest. The remainder, more than 99.9%, enters the environment through soil, water and food cycles (Metcalf, 1986).

Alternative methods of insect management offer adequate levels of pest control and pose fewer hazards. One such alternative is the use of microbial insecticides, that contain microorganisms or their by-products. Microbial insecticides are especially valuable because their toxicity to non-target animals and humans is extremely low. Compared to other commonly used insecticides, they are safe for both, the pesticide user and consumers of treated crops. Microbial insecticides also are known as biological pathogens, and biological control agents. Chemical insecticides are far more commonly used in the world than microbial control, however some microbial control agents, at least in part, can be used to replace some hazardous chemical pest control agents. A number of biological control agents formulated with bacteria, fungi, virus, pheromones, and plant extracts have been in use mainly for the control of insects responsible for the destruction of forests and agriculture crops (McDonald & Linde, 2002).

The microbial insecticides most widely used in the world are preparations of *Bacillus thuringiensis* (*Bt*). The insecticidal activity of *Bt* is due to the proteic parasporal inclusions that are produced during sporulation. Insecticides based on the proteinaceous δ-endotoxin of *Bt* constitute part of a more ecologically rational pest control strategy. *Bt* strains have been isolated world wide from many habitats, including soil, insects, stored-product dust, and deciduous and coniferous leaves, all of which have a limited host range, however together span a wide range of insects orders which include: Lepidoptera, Diptera, Coleoptera, Hymenoptera, Homoptera, Phthiraptera, Orthoptera, Acari, and Mallophaga and other organisms such as nematodes, mites, and protozoa (Federici, 1999). *Bacillus thuringiensis* subsp. *israelensis* (*Bti*) or serotype H-14, exhibit acute toxicity towards dipteran insects such as larval mosquitoes and black flies (de Barjac 1978) and is currently used in mosquito control programs world wide (Priest, 1992). The World Health Organization's (WHO) Onchocerciasis Control Program (OCP) in West Africa using *Bti* toxins has been one of the success stories of international co-operation in the control of infectious diseases program (Webb 1992; Drobniewski 1993). Due to the importance of *Bti* to control several tropical diseases such malaria and dengue, our purpose in this chapter is to provide an overview for non-*Bt* specialists of the basic knowledge of *Bti*.

2. *Bacillus thuringiensis* strains

2.1 Origin of some strains of *B. thuringiensis*
A large number of strain of *Bt* have been isolated from which to date. *Bt* as currently recognized is actually a complex of subspecies. They have grouped in 79 serotypes (Zeigler, 1999). The first *Bt* strain was isolated from diseased larvae of the silkworm, *Bombi mori*, in Japan by Ishiwata (1901). Iwabushi (1908) describe the bacillus as *Bacillus sotto*. Aoki & Chigasaki (1915) and Mitani & Wataral (1916) purified a highly toxic substance from sporulated *Bt* cultures. It was not officially described, however, until it was reisolated by Berliner in 1915 from diseased larvae of the Mediterranean flour moth, *Anagasta kuehniella*, in Thuringia, Germany, hence the derivation of species name thuringiensis (Federici, 1999). A *Bt* strain was isolated again by Mattes (1927) and described briefly the inclusion rhomboidal body. The activity of the *Bt* strains against lepidopteran larvae was described by Metalnikov (1930) and by Husz (1931). The association of the inclusion bodies of *B. thuringiensis* with toxicity against insects was established by Steinhaus. In 1951 published a paper which described the morphology of *Bt* and its possible use in the biological control against the alfalfa caterpillar. In France, a product named "Sporeine" was developed and used against *Ephestia kuhniella* (Lepidoptera) (Jacobs, 1950). Hannay, (1953) described a parasporal body in bipyramidal shape produced by the bacterium during sporulation and suggested that the crystal was involved in the toxic activity. The protein nature of the crystals was determined by Hanna and Fitz-James (1955).

2.2 Features of *B. thuringiensis* strains
Bt is a facultative anaerobic, gram-positive bacterium that forms characteristic protein inclusions adjacent to the endospore (Fig.1). The crystalline inclusion are composed of proteins known as ICPs crystal proteins. Cry proteins, or δ-endotoxins is the basis for commercial insecticidal formulations of *Bt*. Insecticides containing *Bt* in pest control programs is now considered as a viable strategy, which has proven to be both safe and reliable over the last 45 years (Chungjatupornchai et al. 1988).

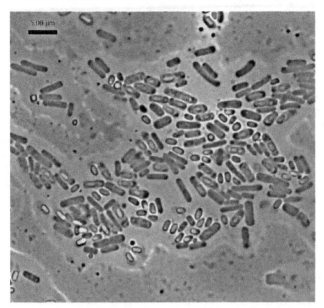

Fig. 1. *Bacillus thuringiensis* subsp. *israelensis*.

2.3 Advantages and disadvantages of *Bt*

According to Federici (1999) the main reasons for the success of *Bt* include (i) the high efficacy of its insecticidal proteins (ii) the existence of a diversity of proteins that are effective against a range of important pests (iii) its relatively safety to nontarget insect predators and parasites (iv) its easy to mass production at a relatively low cost, and (v) its adaptability to conventional formulations and application technology.

Advantages and disadvantages of *Bt* with chemical insecticides were summarized by Rowe and Margaritis (1987).

Advantages of *Bt:*

- High specificity, i.e. no mammalian or nontarget effects; use permitted up to date of harvest.
- No broad insect resistance observed or expected to develop.
- Adaptable to many types of formulations; potential to incorporate feeding stimulants or baits to increase the attractiveness of the formulations to the insects and thereby increase their efficacy.
- Probability of producing more potent formulations and reducing production costs through improved-fermentation technology.
- High probability that strain selection and/or genetic engineering will lead to better control of pest insects by newly found or created strains of Bt having novel host spectra or increased activity

Disadvantages of *Bt:*

- Narrow host spectrum
- Lack of patent protection on new strains
- Proper timing of application require due to slower effect than chemical insecticides

- Activity dependent of ingestion, and feeding activity depends on environmental conditions
- Relatively higher cost compared to chemical insecticides

2.4 Clasification of strains of *Bacillus thuringiensis*

Bt is a crystalliferous spore-forming bacterium close genetic relationship with *B. cereus, B. anthracis* and *B. mycoides* (Höffe & Whiteley, 1989). The classification of *Bt* is difficult because DNA sequencing studies of conserved gene regions of theses species have suggested that they belong to a single specie. *Bt* strains are distinguished from *B. cereus, B. anthracis* and *B. Mycoides* by the ability to produce parasporal crystalline inclusions during sporulation (de Barjac, 1978). Crystal formation is the criterion for distinguishing between *B. cereus* and *Bt*, otherwise they could be considered as the same specie. Research based on a comparative study of 16s rRNA sequences, *Bt* and *B. cereus* var. *mycoides* differed from each other and from *B. anthracis* and *cereus* by less than nine nucleotides (Ash and Collins 1992). Chen & Tsen (2002) amplified 16S rDNA and *gyrB* gene by PCR and they found that the discrimination between *B. cereus* and *Bt* strains, when a large number of Bacillus strains were tested was difficult. They proposed, to distinguish *Bt* from *B. cereus*, a single feature, such as the presence of a parasporal crystal protein or cry gene is reliable.

Several attemps were made to classify *Bt* strains. de Barjac and Frachon (1990) didn´t find correlation between biochemical reactions and 27 H serotypes using 1,600 *B. thuringiensis* isolates. They demonstrated that the current biochemical tests have no value as the sole criteria for differentiating *Bt* strains. Another approach ot these authors for classification was the use of susceptibility to certain bacterial viruses called phages. There are 14 bacteriophages that have been used for *Bt* but phage typing is inconsistent with serotyping and does not permit classification. They found frequent cross-reactions.

One widely clasification system for *Bt* strains is based in the determination of the H-flagellar antigen technique described by Barjac & Bonnefoi (1962). This technique needs very motile bacterial cultures to prepare flagellar suspensions. These suspensions are titrated against antisera directed against *B. thuringiensis* strains of each serotype. Presently *B. thuringiensis* strains are classified within 79 serotypes. Table 1 shows the classification of *Bt* strains by serotypes.

Bt strains that had discovered previously to 1977 were pathogens towards lepidopteran larvae. However, in 1975 was discovered *B. thuringiensis* serovar *israelensis* toxic to mosquito larvae (Goldberg & Margarit 1977) and in 1983 a strain from *Bt* serovar *morrisoni* was found to be pathogenic to Coleoptera larvae (Krieg et al. 1983). According to these findings, serological clasification, although still in use as a basic method to clasify *Bt* strains could not be related with pathogenicity. Subsequently studies showed that, within a serotype, different activity, spectra can be found. For example, some strains of *B. thuringiensis* serotype *morrisoni* in their parasporal inclusión bodies contain different proteins and have activity against Diptera, Coleoptera or Lepidoptera.

With the knowledge of sequence of the genes that encode the proteins cry it was proposed a classification based on the cry toxin genes. Höfte and Whiteley (1989) proposed a nomenclature clasification scheme for *Bt* crystal proteins based in their structural aminoacid sequence, deduced from the DNA and host range. They named *cry* (crystal protein) genes and their related proteins, "Cry proteins". They clasified 42 *Bt* crystal protein genes into 14 distinct genes grouped into 4 major clases. The classes were CryI (Lepidoptera-specific), CryII (Lepidoptera-and Diptera-specific), CryIII (Coleoptera-specific), and CryIV (Dipter-specific).

Serotype	Serovar	Serotype	Serovar
1	thuringiensis	28a, 28c	jeghatensan
2	finitimus	29	amagiensis
3a, 3b, 3c	kurstaki	31	toguchini
3a, 3c	alesti	32	cameroun
3a, 3d	sumiyoshiensis	33	leesis
3a, 3d, 3e	fukuokaensis	34	konkukian
4a, 4b	soto/dendrolimus	35	seoulensis
4a, 4c	kenyae	36	malayensis
5a, 5b	galleriae	37	andralousiensis
5a,5c	canadensis	38	owaldocruzi
6	entomocidus/subtoxicus	39	brasilensis
7	aizawai/pacificus	40	huazhongensis
8a.8b	morrisoni	41	sooncheon
8a, 8c	ostriniae	42	jinghongiensis
8b, 8d	nigeriensis	43	guiyangiensis
9	tolworthi	44	higo
10a, 10b	darmstadiensis	45	roskildensis
10a, 10c	londrina	46	chanpaisis
11a, 11b	toumanoffi	47	wratislaviensis
11a, 11c	kyushuensis	48	balearica
12	thompsoni	49	muju
13	pakistani	50	navarrensis
14	israelensis	51	xiaguangiensis
15	dakota	52	kim
16	indiana	53	asturiensis
17	tohokuensis	54	poloniensis
18a, 18b	kumamtoensis	55	palmanyolensis
18a, 18c	yosoo	56	rongseni
19	tochigiensis	57	pirenaica
20a, 20b	yunnanensis	58	argentinensis
20a, 20c	pondicheriensis	59	iberica
21	colmeri	60	pingluonsis
22	shanongiensis	61	sylvestriensis
23	japonensis	62	zhaodongensis
24a, 24b	neolonensis	64	azorensis
24a, 24c	novosibirsk	65	pulsiensis
25	coreanensis	66	graciosensis
26	silo	67	vazensis
27	mexicanensis	none	wuhanensis
28a, 28b	monterrey		

Table 1. Classification of *B.thuringiensis* strains by serotype (modified after Zeigler, 1999).

The *cryI* genes can be distinguished from the other *cry* genes simply by sequence homology (>50% aminoacid identy) and encode 130 to 140 kDa proteins which accumulate in bypiramidal crystalline inclusions during the sporulation of *Bt*. *cryII* genes encode 65-kDa proteins which form cuboidal inclusions in strains of several species. *cryIII* genes produces romboidal crystals containing one major protein, a 72kDa protein. cryIV class genes (*cryIVA*, cryIVB, cryIVC, and cryIVD) as well as *cytA* were all isolated from the same 72-Mda plasmid present in *Bt* subsp. *israelensis*.

3. *Bacillus thuringiensis* subsp. *israelensis* as an important part of mosquito control

3.1 Properties of *Bti*

In 1975-76 under a World Health Organization sponsored project, a new *Bt* strain was discovered in the Negev desert in Israel by Goldberg and Margalit (1977). The strain was isolated from *Culex* sp. dead larvae mosquito. Later was identified as *Bt israelensis*, serotype H14 according to its fagellar antigenicity by de Barjac (1978).

Bti has all the features taxonomic, morphological, growth, sporulation, of isolation, cultivation of other varieties of *B. thuringiensis* (Fig.1). The larvicidal activity of *Bt israelensis* on mosquito transmitted diseases was the most important feature of the strain. The insecticidal properties of this bacteria are due primarily to insecticidal proteins produced during sporulation. The key proteins are Cyt1A(27.3 kDa), Cry4A (128 kDa), Cry4B (134 kDa) and Cry11A (72 kDa) and in three different inclusion types assembled into a spherical parasporal body held together by lamellar envelope (Ibarra & Federici, 1986). The inclusions are realtively small (0.1 to 0.5μm) and there are usually two to four inclusions per cell which vary in shape from cuboidal to bipyramidal, ovoid or anamorph (Charles & de Barjac 1982; Mikola *et al.* 1982; Yamamoto *et al.* 1983)

3.2 Mosquitoes, important vectors of tropical diseases

Mosquitoes are important vectors of several tropical diseases that suck blood from human and animals. They are vectors of multiple of diseases of man through transmision of pathogenic viruses, bacteria, protozoa and nematodes (Priest, 1992). From the medical point of view, mosquitoes are among the most important insects due their capacity to transmit human diseases such as malaria and dengue.

Vector	Disease
Anopheles	Malaria, lymphatic filariasis
Culex	Lymphatic filariasis, Japanese encephalitis, other viral diseases
Aedes	Yellow fever, dengue, dengue hemorrhagic fever, other viral diseases, lymphatic filariasis
Mansonia	Lymphatic filariasis

Table 2. Some diseases transmitted by mosquito (Rawlins, 1989; Walsh 1986).

There are about 3000 species of mosquito, of which about 100 are vectors of human diseases. Some of the more important of these diseases are listed in Table 2. It is estimated two billion people worldwide living in areas where these are endemic (World Health Organization, 1999). Thus, there is an urgent need for new agents and strategies to control these diseases.

3.3 Susceptibility of mosquito species to *B. thuringiensis* serotype *israelensis*

Bti is highly pathogenic against *Culicidae* (mosquitoes) and *Simuliidae* (blackflies), and has some virulence against certain others Diptera, especially Chironomidae (midges). Mosquito have four distinct stages in their life cycle: egg, larva, pupa and adult. Depending on the specie a female lays between 30 and 300 eggs at a time on the surface of the water, singly (*Anopheles*), in floating rafts (*Culex*) or just above the water line or on wet mud (*Aedes*). Once hatched the larvae grow in four different stages (instars). The first instar measures 1.5 mm in length, the fourth instar about 8-10 mm. The fully grown larvae then changes into a comma shaped pupa. When mature, the pupal skin splits at one end and a fully developed adult emerges. The entire period from egg to adult takes about 7-13 days under good conditions (Wada, 1989). First instar is more suceptible to *Bti* than fourth instar (Mulla *et al.* 1990). Pupa does not feed and therefore is not affected by *Bti*. For almost all species tested, increasing age of the larvae resulted in reduced susceptibility in mosquito (Chen et al. 1984: Mulla *et al.* 1985). *Bti* was found to be specific toxic to larvae of 109 mosquito species (Table 3).

Mosquito genus	Species
Aedes	40
Anopheles	27
Culex	19
Culiseta	5
Mansonia	5
Psorophora	3
Armigeres	3
Toxorhynchites	2
Limatus	2
Trichophospon	1
Uranotaenia	1
Tripteroides	1
	Total 109

Table 3. Larvicidal activity of *Bti* on mosquito species (Glare & O'Callaghan 1998).

Among mosquitoes, different preparations of *Bti* have shown different levels of toxicity to host species. Others factors influencing the susceptibility of mosquito larva to *Bti*. For example, the effect of a given dosage of toxin could produce different results depending on weather the lethal dose is administered all at once or in same doses over a long period (Aly et al. 1988).

Bti has an LC50 in the range of 10–13 ng/ml against the fourth instar of many mosquito species (Federici *et al.* 2003). Generally, *Culex* and *Aedes* are highly susceptible while *Anopheles* are less suceptible, but can be killed with *Bti* (Balaraman *et al.* 1983). Much higher concentrations of *Bti* are required to induce mortality in anopheline larvae than in *Aedes aegypti* larvae (Goldberg & Margalit 1977; Nugud & White 1982)

The exception is *A. franciscanus* that is as susceptible as other genera (García *et al.* 1980, Sun *et al.* 1980). Furhtermore, even within one genus, some species are more suceptible than others (Chui et al. 1993). Sun et al. (1980) suggested that a difference in feeding behavior might account for differences in susceptibilities. Filtering rates vary between genera and species. For example Aly (1988) showed that in the absence of *Bti*, larvae cleared the suspensions with constant relative filtration rates of 632 (*Ae. aegypti*), 515 (*Cx. quinquefasciatus*) or 83.9

μL/Larvae/h (*An. albimanus*). Another factor to be considered for the suceptibility to *Bti* is the behavior of the larvae. *Anopheles* larvae filter-feed on food particles present at the surface of water or a few centimeters below it, where as *Culex* and *Aedes* larvae not only feed faster but are capable of filter-feeding at much deeper water dephts (Aly et al. 1988).

Mosquito species which are not filter feeders do not seem suceptible. For example, against *Culicoides occidentalis* (Colwell 1982) and *Coquillettidia perturbans*, *Bti* larvicides had no effect (Walker et al 1985). Solubilization of the protein reduced the toxicity dramatically (50- to -100 fold) this is attributed to reduced level of toxin ingestion by larvae owing to their filter feeding behaviour (Chungjatupornchai et al. 1988).

3.4 Susceptibility of mosquito species to purified *Bt* proteic crystals

Several efforts have been made to purify the proteins Cry of *Bti* and other *Bt* subps. with the purpose of studying the chemistry of Cry proteins, the synergism between them, and the effect of each crystal on larvae of different mosquito species. It is somewhat difficult to separate the *Bt* spores and crystals because they are of similar size and surface characteristics. For that reason several methods have been used to purify *Bt* crystals proteins. Using NaBr gradients (Chang et al. 1992), sucrose gradients (Debro et al. 1986), renografin gradients (Aronson et al. 1991), and in a separatory funnels (Delafield et al. 1968). Bioassays of purified Cry proteins have been allowed to know that not only *Bti* Cry proteins have activity on mosquito larva. Other purified Cry proteins of other *Bt* strains have also activity on mosquito larvae. Table 4 shows the mosquito toxicity of purified crystals of some *Bt* strains.

Name	Source Strain	Mosquito toxicity
Cry4Aa1	*B.t. israelensis* 4Q2-72	*Aedes agypti, Anopheles stephensi, Culex pipiens* (Diptera: Cuclidae)
Cry4Ba1	*B.t. israelensis* 4Q2-72	*Aedes agypti* (Diptera: Cuclidae)
Cry10Aa1	*B.t. israelensis* ONR60A	*Aedes agypti,* (Diptera: Cuclidae)
Cry11Aa1	*B.t. israelensis* HD-567	*Aedes agypti, Anopheles stephensi, Culex pipiens* (Diptera: Cuclidae)
Cry11Ba1	*B.t. jegathensan* 367	*Aedes agypti, Anopheles stephensi, Culex pipiens* (Diptera: Cuclidae)
Cry11Bb1	*B.t. medellin*	*Aedes agypti, Anopheles albimanus, Culex quinquefasciatus* (Diptera: Cuclidae)
Cry16Aa1	*Clostridium bifermentans malasya* CH18	*Aedes agypti, Anopheles stephensi, Culex pipiens* (Diptera: Cuclidae)
Cry19Aa1	*B.t. jegethesan*	*Anopheles stephensi, Culex pipiens* (Diptera: Cuclidae)
Cry20Aa1	*B.t. fukuokaensis*	*Aedes agypti,* (Diptera: Cuclidae)
Cry21Aa1	*B.t. higo*	*Culex pipiens molestus* (Diptera: Cuclidae)
Cyt1Aa1	*B.t. israelensis* IPS82	*Aedes agypti, Anopheles stephensi, Culex pipiens* (Diptera: Cuclidae)
Cyt1Ab1	*B.t. medellin* 163-131	*Aedes agypti, Anopheles stephensi, Culex pipiens* (Diptera: Cuclidae)
Cyt2Aa1	*B.t. kyushuensis*	*Aedes agypti, Anopheles stephensi, Culex pipiens* (Diptera: Cuclidae)

Table 4. Mosquitocidal activity of Cry and Cyt proteins (Modified after "Zeigler, 1999").

3.5 Resistance

One major problem with insects control via chemical insecticides is the evolution in insects of resistance to those insecticides. The use of *Bti* on biological control of mosquitoes has no resulted in the development of resistence in host populations. Laboratory attemps to induce resistance by continual exposure to *Bti* have generally failed to detect resistence (Lee & Chong 1985; Georghiou & Wirth 1997).

The lack of resistance development to *Bti* could be due to its complex mode of action, involving synergistic interactions between up to four proteins (Becker & Maragrit 1993). Use of a single protein from *Bti* for mosquito control resulted in resistance after only a few generations in the laboratory (Becker and Margalit 1993). Georghiou & Wirth (1997) also showed that resistance could be raised in only a few generations when single *Bti* toxin was used (i.e. Cry 4Aa, 4Ba, 10Aa or 11Aa), and was progressively more difficult to raise in mosquitoes with combinations of two, and three toxins. When all four *Bti* toxins were used, resistance incidence was remarkably low. On the other hand Wirth et al. (2005) have shown that the lack of resistance in *Bti* is due to the presence of the Cyt1Aa protein in the crystal. For example, *Culex quinquefascitus* populations resistant to CryIVA, Cry4B and Cry11A have been obtained in the laboratory but not mosquito larvae resistant to Cry and Cyt1Aa proteins (Georghiou & Wirth 1997)

3.6 Synergism

Bti produces four crystal proteins Cry (4Aa, 4Ba, 10Aa, and 11Aa) and two Cyt (1Aa and 2Ba) (Guerchicoff *et al.* 1997). No single crystal component is as toxic as the intact crystal complex (Chan *et al.* 1993; Wu *et al.* 1994; Chilcott & Ellar, 1998). One possible explanation for this is that two or more proteins act synergisticall, yielding a higher activity than would be expected on the basis of the specific toxicity of the individual protein (Finney, 1971). For example, the toxicity against mosquito larvae of Cyt1Aa is lower compared to each of the four Cry proteins (Crickmore et al. 1995). However, cytA can potentiate the activity of the toxins and synergistic interactions that seems to account for the high toxicity of the *Bti* strains (Delecluse et al. 1993).

Tabashnick (1992) proposed a method to measure synergistic effect. Using the proposed method discuss the data reported by two authors: in bioassays with *A. aegypti* larvae, Wu and Chang (1985) found that mixtures of the 27- and 65-kDa proteins from *B. thuringiensis* subsp. *israelensis* were more toxic than expected on the basis of their individual toxicities, however, Chilcott and Ellar (1988) concluded from their own data that no synergism between these two proteins occurred. With this new interpretation of Tabashnick method, both studies support the same conclusions: positive synergism between the 27-kDa protein (CytA) and either of the CryIV proteins (65 and 130 kDa) and no such synergism between CryIV proteins (65 and 130 kDa). Other studies have been carried out with the aim of increasing the synergistic activity of *Bti*. Ramirez-Suero et al. (2011) evaluated the synergistic effect of *S. griseus* and *Bt aizawai* chitinases with *Bt israelensis* spore-toxin complex against *Aedes aegypti* larvae. The synergistic factor values according to Tabashnik (1992) method were 2 and 1.4, respectively.

3.7 Effect of *Bti* on no-target organisms

Bti has no direct effect on aquatic organisms other than mosquitoes, blackflies and chironomids. Other aquatic organisms, such as shrimps, mites and oysters are generally unaffected (Glare & O'Callaghan, 1998). This large safety margin of preparations of *Bti* for non-target organisms indicate their suitability for mosquito control programs in areas where protection of the natural ecosystem is important (Sinegre et al. 1980)

Several authors have reviewed the non target effects of *Bti* (Becker & Margalit 1993; Lacey & Mulla 1990). Field applications have often been monitored for effects on non-target organisms but no significant non-target effects have been reported (Ali, 1981; Jackson et al. 1994; Hershey et al. 1995)

4. Production of *Bti* by fermentation

4.1 Culture medium for *Bti* production

Commercial production of *Bti* is performed using culture media based on complex nutrients sources. The main purpose of the fermentation is to obtain high quantities of *Bti* crystals. The *Bti* parasporal crystal can account for up to 25% of the sporulated cell dry weight. To optimize the cry production it is necessary to have a suitable culture medium because the toxicity obtained at the end of the fermentation depend on the culture medium and operating conditions. The culture media that have been reported in the literature for high growth and sporulation can be used for any variety of *Bt*. Not always a high cell growth ensures an elevated Cry protein production or an increased insecticidal activity. Various culture mediums have been used for high growth and sporulation of *Bt* in the laboratory: 2XSG medium (Leighton & Doi, 1971), PA medium (Thorne, 1968), G-Tris medium (Aronson et al. 1971), CDGS medium (Nakata, 1964). Other media with inexpensive substrates have been reported by Pearson & Ward (1988), Smith (1982), Foda et al. (1985), Dulmage et al. (1970), Salama et al. (1983), Goldberg et al. (1980).

4.2 Factors affecting Cry production

There are several factors that influence the production of crystals: (1) Carbon source. Glucose is the most appropiate carbon source either for high *Bt* growth and sporulation (Smith, 1982). When glucose has been exhausted in the fermentation, the abscence of this can trigger sporulation. The use of one or other carbon source affects the biological activity and the morphology of the crystals (Dulmage, 1970). (2) Nitrogen source. An appropiate source of aminoacids provides high growth rates and high sporulation of *Bt* strains. Its absence delays sporulation and low yield in Cry proteins (Goldberg et al. 1980) (3) Carbon:Nitrogen ratio. Higher C:N rates glucose do not deplete at the end of fermentation and biomass yield decrease. Several authors have recomended a carbon nitrogen ratio of 7.5:1. (Salama et al. 1983; Foda et al. 1985) (4) Oxygen. High aeration rates are important for high spore and toxin formation. As k_La, increase biomass and Cry protein formation are increased (Rowe & Margaritis, 1987)(5) pH. Optimum pH for *Bt* growth is 6.8-7.2. If pH rises to 9.0 Cry protein can be dissolved (6) Temperature. Optimum temperature of *Bt* is 28-32°C. Higher temperatures favours plasmid losses or *Bt* mutants (Rowe & Margaritis, 1987).

5. Molecular biology of *Bti*

5.1 *Bti cry* and *cytA* genes

All *Bt* strains contain extrachromosomal DNA. *Bt* strains are well known for its numerous plasmids ranging in size from 1.5 MDa to 130 MDa. Plasmids have been found in each variety examined and the plasmid profiles appear to be strain specific. However these plasmid profiles depend on the media type and growth rate of the strain and can be readily gained or lost (Federici, 1999). The mere presence of plasmids in *Bt* does not prove that they are involved in crystal formation, many non-crystalliferous bacteria also contain plasmids The *cry* genes are located on large plasmids although some Cry genes have been reported on the chromosome (Baume & Malvar, 1995). As mentioned earlier, *Bti* produces four different Cry proteins: CryIVA, CryIVB, Cry11A, and the cytolytic CytA protein (Hoffe and Whiteley, 1989). The Cry proteins are codified by *cryIVA*, *cryIVB*, *cry11A*, and *cytA* genes, respectively. These genes responsible for the toxicity of *Bti* have been sequenced by various researchers (Table 5).

Gene name	GenBank Accesion No.	Coding Region	Reference
cryIVAa1	Y00423	1-3540	Ward &Ellar, 1987
cry4Aa2	D00248, E01676	393-3935	Sen *et al.* 1988
cry4Ba1	X07423, X05692	157-3564	Chungjatupornchai *et al.* 1988
cry4Ba2	X07082	151-3558	Tungpradubkul *et al.* 1988
cry4Ba3	M20242	526-3930	Yamamoto *et al.* 1988
cry4Ba4	D00247, E01905	461-3865	Sen *et al.* 1988
cry11Aa1	M31737	41-1969	Donovan *et al.* 1989
cry11Aa2	M22860	1-235	Adams *et al.* 1989
cyt1Aa1	X03182	140-886	Waalwijk *et al.* 1985
cyt1Aa2	X04338	509-1255	Ward & Ellar, 1986
cyt1Aa3	Y00135	36-782	Earp a& Ellar, 1987
cyt1Aa4	M35968	67-813	Galjart *et al.* 1987

Table 5. *cry* and *cytA* genes DNA sequences of *Bti*.

In *Bti* the elements responsible of the toxicity against mosquito larvae are located in a large plasmid of 72 MDa (125 kb) and contribute to the formation of a complex parasporal body (Aronson 1993). Figure 2 shows the partial map of the *Bti* 125 kb plasmid.

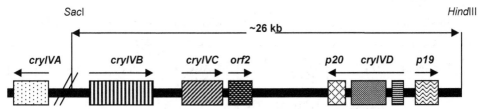

Fig. 2. Map of 26kb DNA fragment carrying the DNA genes responsibles of the toxicity of *Bti* (Modified after Ben-Dov *et al.* 1996).

5.2 *Bti* operon of *cryIVD* gene

The DNA sequence indicated that *cryIVD* gene is the second gene of an operon which includes three genes. A gene that encodes a 19 kDa polypeptide, *cryIVD* gene and a gene that encodes a 20 kDa polypeptide (Dervyn et al. 1995). Transcription of *cryIVD* gene in *Bti* is induced 9 h after the beginning of the sporulation. DNA sequence analysis and potential promoters are recognized by the RNA polimerase associated with the σ^{35} and σ^{28}, specific sigma subunit of the RNA polymerase genes related with control of sporulation of *B*.

thuringiensis. cryIVA and *cryIVB* promotors are activated in the mid-sporulation phase (Ben-Dov et al. 1996), *cryIVA* is regulated by the σ^{35} RNA polymerase gene and *cryIVB* is under control of σ^{35} (Yoshisue et al. 1994). *cytA* gene is transcribed by two promotors, pBtI and pBtII, regulated by the RNA polymerase σ^{35} and σ^{28}, respectively (Brown & Whiteley, 1988, 1990). These results have demostrated that *cryIVD* transcription is subjected to σ^{35} regulation.

5.3 Identication of *Bacillus thuringiensis* pesticidal crystal genes by PCR

The polymerase chain reaction (PCR) is a molecular tool widely used to amplify a single copy or a few copies of a piece of DNA across several orders of magnitude, generating thousands to millions of copies of a particular DNA sequence. The identification of *Bt* toxin genes by PCR can partially predict the insecticidal activity of a given strain. Several studies have reported that the type of *cry* and *cyt* genes present in a *Bt* strain correlates to some extent with its insecticidal activity (Porcar & Juarez-Perez). Thus, the identification of the gene content in a *Bt* strain can be used to predict its insecticidal potential. The PCR-based identication of *B. thuringiensis* cry genes was first developed by Carozzi et al. (1991), who introduced this technique as a tool to predict insecticidal activity. They found correspondence with the toxicity predicted on the basis of the amplification product profiles. Carozzi et al. proposed PCR as an accurate, fast methodology for the identification of novel strains and the prediction of insecticidal activity of new isolates, and they also forecast the possible use of PCR for the discovery of previously unknown cry genes. highly conserved regions and recognizing entire cry gene subfamilies are often used in a preliminary screening prior to performing a second PCR with specific primers. The primers used to amplify *cry4A*, *cry4B*, *cry11A*, and *cytA* genes have been designed by various researchers and are shown in Table 6.

Direct	Sequences (5′⇒3)	Reverse Sequences (5′⇒3)	Amplifies/Prod uct (bp)
EE-4A(d)	GGGTATGGCACTC AACCCCACTT*	Un4(r) GCGTGACATACCC ATTTCCAGGTCC*	cry4A/1529
Dip2A(d)	GGTGCTTCCTATTC TTTGGC**	Dip2B TGACCAGGTCCCT TGATTAC**	*cry4A*/1290
EE-4B(d)	GAGAACACACCTA ATCAACCAACT*	Un4(r) GCGTGACATACCC ATTTCCAGGTCC*	*cry4B*/1951
EE-11A(d)	CCGAACCTACTAT TGCGCCA*	EE-11A(r) CTCCCTGCTAGGA TTCCGTC*	*cry11A*/445
gral cyt1(d)	AACCCCTCAATCA ACAGCAAGG***	gral cyt(r) GGTACACAATACA TAACGCCACC***	*cyt1*/522-525

Table 6. PCR primers pairs and the cry or cyt genes they amplify of *Bti*. Sources: Ben-Dov *et al.* (1997)*, Carozzi et al. (1991)**, Bravo *et al.* (1998)***

5.4 Expression of *Bti* genes in other strains

Expression of *Bti* genes either individually or in combination in crystal-negative *Bt* or other strains have being carried on by several researchers. The genes encoding these proteins have been expressed in *Caulobacter* (Thanabalu et al. 1992), cyanobacteria (Manasherob et al. 2002; Murphy & Stevens 1992), *Escherichia coli* (McLean & Whiteley 1987; Tanapongpipat et al. 2003), *Bacillus subtilis* (Ward et al. 1986), and *Bt* (Crickmore et al. 1990). However, *Bt* toxins have been expressed as active or inactive toxins, especially when expressed in *E. coli* (Ogunjimi et al. 2002).

Quintano et al. (2005) reported the expression of *cry11A* from *Bti* in *S. cerevisiae*. The *cry11A* gene was expressed as fusion proteins with gluthathione *S*-transferase under the control of the *S. cerevisiae HXK1* promoter. The protein was purified by affinity chromatography using gluthathione *S*-transferase–Sepharose beads. Insecticidal activity against third-instar *Aedes aegypti* larvae of the recombinant *S. cerevisiae* cell extracts (LC_{50} = 4.10µg protein/mL) and purified GST-cry11A fusion protein (LC_{50} = 4.10µg protein/mL) was detected in cells grown in ethanol.

Servant et al. (1999) constructed a recombinant *B. sphaericus* strain containing the *cry11A* gene from *Bti*. They found an LC_{50} for the cry11A protein of 1.175 µg/mL against fourth-instar *A. aegypti* larvae. Poncet *et al.* (1997) constructed a recombinant *B. sphaericus* strain with *cry11A* and *p20* genes integrated into the chromosome. In this case, the LC_{50} value reported against third- instar *A. aegypti* larvae was 0.023 µg/mL. Xu *et al.* (2001) studied the expression of cry11A and *cry11A+p20* gene cluster in recombinant *E. coli* and *Pseudomonas putida*. They found that both recombinant bacteria contained higher levels of Cry11A protein when the adjacent *p20* gene was present on the same DNA fragment. Yamagiwa et al. (2004) reported that the solubilized cry11A protein, obtained from a nonrecombinant *Bti*, was less toxic against *Culex pipiens* larvae than the crystal itself (LC50 of 0.267 and 0.008 µg/mL, respectively). These authors obtained 2 GST fusion proteins of 36 and 32 kDa from cry11A. The LC50 against *C. pipiens* larvae obtained using both proteins were 0.818 µg/mL. In other study with the purpose to preserve the toxicity of sunlight-sensitive Cry proteins, Manasherob et al. (2002) constructed a transgenic cyanobacterium *Anabaena* PCC 7120 to express the genes *cry4Aa*, *cry11Aa* and an accesory protein (*p20*) under control of two tandem strong promotors. Cyanobacterium *Anabaena* can multiply in breeding sites of mosquito larvae and serve as their food source. Higher toxicity against *Aedes aegypti* larvae was obtained in this study.

6. Mode of action of *Bti* Cry proteins against mosquito larvae

6.1 Pore-forming-toxins

Bt Cry and Cyt toxins belong to a class of bacterial toxins known as pore-forming toxins (PFT) that are secreted as water-soluble proteins undergoing conformational changes in order to insert into, or to translocate across, cell membranes of their host (Bravo et al. 2007). In most cases, PFT are activated by host proteases after receptor binding inducing the formation of an oligomeric structure that is insertion competent. Finally membrane insertion is triggered, in most cases, by a decrease in pH that induces a molten globule state of the protein (Parker and Feil, 2005). Cry and Cyt proteins are PFT proteins. Both proteins are solubilized in the gut of suceptible diperans insects and proteolytic activated by midgut proteases. For the Cry 11Aa protoxin, proteolytic activation involves amino-terminal processing and intramolecular cleavage leading to two fragments of 36 and 32 kDa that remain associated and retain insect toxicity.

6.2 Mechanism of action of *Bti* toxins

An acepted model for Cry toxin action against mosquito larvae is that it is a multistage process. (i) Ingestion of Cry protein by the larvae (ii) Solubilization of the crystals in the alkaline midgut (iii) Proteolytic activation of the insecticidal solubilised protein (iv) Toxin binds to receptors located on the apical microvellus membrana of ephitelial midgut cell walls (v) Alter the toxin binds the receptor, it is though that there is a change in the toxin conformation allowing toxin insertion into the membrane (vi) Electrophysiological and biochemical evidence suggest that the toxins generate pores in the cell membrane, thus disturbing the osmotic balance, consecuently the cells swell and lyse (vii) The gut becomes paralyzed and the insect stops feeding. Most mosquito larvae die whitin few hours of ingestion, generally cease feeding within 1 hour, show reduced activity by two hours and die six hours after ingestion (Chicott et al. 1990; Marrone & Macintosh, 1993).

Several authors have studied the mechanisms of action of the δ-endotoxin of *Bti* on mosquito larvae. Thomas and Ellar (1983) found that δ-endotoxin active against mosquito larvae was inactivated by prior incubation with lipids extracted from ephitelial midgut *Aedes albopictus* cells. They reported that toxin binds to membrane lipids (phosphatidyl choline, sphingomyelin and phosphatidyl ethanolamine). According to their results, they proposed a mechanism in which the interaction of toxin with membrane lipids causes a detergent-like rearrangement of the lipids and as a consequence cytolisis. Others authors have corroborated these results: Cyt protein, unlike Cry toxins, do not recognize specific binding sites and do not bind to protein receptors, directly interact with membrane lipids inserting into the membrane and forming pores (Knowles *et al.* 1989; Promdonkoy & Ellar, 2000) or destroying the membrane by a detergent-like interaction (Butko, 2003).

The high efficacy of *Bti* is because of the production of multiple toxins with different modes of action. Perez *et al.* (2005) reported that Cyt1Aa protein functions as a receptor for the Cry11Aa toxin and suggest that this interaction explains the synergism between the Cyt1A and Cry11A proteins. Further, the Cyt proteins in *Bti* synergize the toxic effect of Cry11A and Cry4 toxins and, even more, suppresses the resistance to these Cry toxins (Wirth *et al.* 1997).

7. Formulation of *Bti* toxins

7.1 Potency in *Bti* formulations

Formulation is a preparation of an insecticide for a particular application method. Formulation plays an important rol in determining final virulence. The vast majority of the formulations of *Bt* have been developed to control agricultural and forest pests, mainly *Lepidoptera*. However, the feeding habits of *Lepidoptera* are different to feeding habits of *Diptera*. Mosquito larvae feed by filtering water and concentrate organic particles. Product formulations based on *Bti* should consider the mosquito larvae habits and the environmental conditions, promote that Cry proteins retain their toxic activity and promote that the larva have access to them. Products based on strains of *Bti* are given a potency based on bioassays on third or fourth instar mosquito larvae. Bioassays are conducted using 6 to 7 dilutions of the toxin by duplicate in 100 mL cups containing 20 third instar *Aedes aegypti* larvae. Duplicate cups with 20 mosquito larvae in 100 mL of deionized water without test material serves as a control (McLaughlin *et al.* 2004). Concentration-mortality data are obtained, transformed to a log-probit scale, and potency is obtained by comparing the estimated LC_{50} of a test substance with that of a standard with a known potency (de Barjac

1985). The international standard recognized for *Bti* is IPS82. LC_{50} and LC_{90} are the dose require to kill 50 and 90 percent of the mosquito larvae of a tested population after 24h tested duration. Each sample is bioassayed at least 3 times on various days and the results are average values. LC_{50} and LC_{90} are measured in micrograms or milligrams of material per liter, or parts per million (ppm).

For potency calculations, it is used the international recognized standard for mosquito assay, IPS82 (15 000 ITU/mg) provided by Institute Pasteur, Paris, France. Standard vials are kept at -18°C.

Product potency is calculated by the Abbott (1925) formula:

$$\text{Potency (A)} = \text{Potency (std)} \; LC_{50} \, (std) / LC_{50} \, (A)$$

where (a) is the product and (std) is the standard.

The size of the particle could be a factor that influences the potency of the toxin. A product with small particles is more homogenously distributed in the water than a product with larger particles and small particles which sink slower. Changes in LC_{50} are not necesary regarded to reflect changes in amount of toxin, but could be a function of particle size/distribution (Skovmand et al. 1997). Change of particle size also change LC_{50} . Decreasing particle increased LC_{50} and thus decrease the calculated potency. The slope measured between LC_{50} and LC_{90} values should have high value. Higher value will require a smaller quantity to kill a greater number of larvae. The slope of the dosage-mortality curve is in function of the heterogeneity of the product effect. If product availability is in function of particle sizes, particles with broad ranges of particle size distribution will also have low slopes.

7.2 *Bti* formulations

A variety of *Bti* formulations have been studied for mosquito control under laboratory and field conditions. *Bti* fluid formulations are not stable in heat and high humidity and cannot be stored for months under tropical conditions (McGuirre et al. 1996) . In many cases, and especially in areas exposed to the sun, the residual effect is very short (Leong et al. 1980) and the product has to be reapplied. Photoinactivation seems to be one of the major environment factors affecting the stability of *Bti* delta-endotoxin (Morris, 1983). Yu-Tien et al. (1993) reported that *Bti* completely lost its toxicity to mosquito larvae when exposed to irradiation at 253nm. Poszgay et al. (1987) showed that exposure of *B. thuringiensis* toxin to 40 h of ultraviolet light irradiation resulted in lost activity. Cry proteins inactivation by the solar radiation is the result of the destruction of the tryphtophan (Pusztai et al. 1991). Research and development efforts are focusing on formulations to avoid the ultraviolet light effect. Ramirez-Lepe et al. (2003) encapsulated *Bti* spore-toxin within aluminum/ carboxymethylcellulose using green malachite, congo red or ponceau red as photoprotective agents against ultraviolet light in lab conditions. The encapsulated form of the *Bti* spore-toxin complex with photoprotectors avoided the limitation in controlling mosquito larvae caused by ultraviolet light. Yu-Tien et al. (1993) achieved photoprotection of the spore-toxin complex by addition of melanin.

Other *Bti* formulations have been developed. For example, Ramirez-Suero et al. (2005) evaluated maltodextrin, nixtamalized corn flour and corn starch for entraping active materials in *Bti* spore-toxin complexes dried by aspersion. Dried products had water activity values below 0.7 suggesting that the formulations are long shelf-life because keep

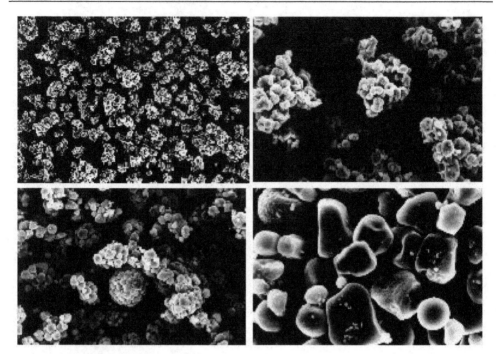

Fig. 3. Scanning electron micrograph of spray/dried formulation of *Bti* spore/toxin complex with Grits a) 379X and b)1500X and Nixtamalized corn flour c) 370X and d) 1500X.

the products without microorganisms for longer periods and increases the larval feeding and as a consequence have higher activity against mosquito larvae. Fig. 3 shows the scanning electron microscope of spray/dried formulations with grits and nixtamalized corn flour. It shows smoth spherical grits and corn particles entrapping the spore/toxin ingredient. Manasherob et al. (1996) encapsulated *Bti* in the protozoan *Tetrahymena pyriformis* and the activity against *A. stephensi* was enhanced 8 times when exposed to protozoan cells filled with *Bti* compared to exposed to the same concentration of *Bti* alone. Elcin (1995) encapsulated *Bti* in alginate microcapsules and increase its stability and its effect against *Culex* sp larvae. Another alternative to protect *Bti* crystals from ultraviolet light is obtaining mutants that protectrs *Bti* crystal. Hoti & Balaraman (1993) obtained a mutant of *Bti* that produced a brown pigment during sporulation, the pigment was identified as melanin. Other *Bti* formulations have been developed. Culigel superabsorbent polymer controlled-release system for the slow release of *Bti* to mosquito larvae (Levy et al. 1990). Combination of chemical and biological agents susch as insect growth regulator s-methoprene + *Bti* (Bassi et al. 1989). Sprayed-dried *Bti* powder as a fizzy tablet (Skovmand & Eriksen 1993). Floating bait formulations designed to improve the effect of bacterial toxins, especially against *Anopheles* spp (Aly et al. 1987).

Highly concentrated liquid formulations are available for control of floodwater mosquitoes while formulations which float for as long possible have been developed for use in fast-flowing or turbulent waters. Formulations which settle and persist at the bottom are required for buttom feeders. Granules which float on the surface are the most effective against surface feeders such as *Anopheles* spp. (Mulla et al. 2004).

Briquettes for mosquitoes with continual successive generations like *Culex* spp (Becker & Margalit 1993; Kase & Branton 1986). Briquetts or pellets, in particular seem to be useful for overcoming lack of persistence, which is one of main limitations of *Bti*. Granules using plant, such corn (maiz) grits or clay carriers are particular useful in aerial application to breeding sites with dense foliage as salt marshes or rice fields. Sustained release formulations such as floating briquettes or semi submersible pellets are designed to provide long-lasting larvicidal activity in containers or small ponds. Ingestion of the toxin depends on the rate of feeding, the rate at which the toxin falls to the bottom of the pool and becomes inaccesible, and competition to ingestion from other suspended organic materials. In turbid and polluted waters the rate of application needs to be at least two-fold greater than in clear water (Mulla et al. 1985). The feeding habits of mosquito larvae influence formulation design. *Anopheles* larvae are surface feeders and ingest particulate material from water surface such as yeast or flour and filter feed poorly. This has led to the development of formulations that present the toxin at, or just below, the water surface and such prepartions are particulary effective against certain *Anopheles* larvae (Aly et al. 1987)

8. Field application

One of the major drawbacks in the use of *Bti* is its rapid inactivation (24-48 h) in the environment (Mulla et al. 1993). Thus larvae populations of stagnant water mosquitoes recover within 5-7 days following treatment. Therefore the use of *Bti* is limited by the low efficacies of current preparations under field conditions (Tyanum & Mulla 1999). Since there is a little persistence of the toxin further applications are necessary to effect continuous control. Formulation and applications techniques can extend the persistence of activity for over one month in some situations, but activity remains sensitive to factors like UV degradation. In areas exposed to the sun, the residual effect is very short (Leong et al. 1980) and the product has to be reapplied. Other factors that affect the toxicity of *Bti* are particle sedimentation (Rushed & Mulla, 1989), protein adsorption onto silt particles, organic matter, elevated temperatures (Ohana et al. 1987), consumption by other organisms to which the toxin is not lethal (Blaustein and Margalit, 1991), dissolved tannins (Lord & Undeen, 1990) and inactivation by sunlight. Other factors that have been observed are that *Bti* does not recycle, under simulated field conditions, is unable to germinate and multiply in mud at the botton of pools although it did remain viable viable for up to 22 days, and higher water depth where applied decrease its activity (Ohana et al. 1987)

Early reports showed that a primary powder formulations of *Bti* had virtually no residual effect against mosquito larvae beyond application (Aguilar-Meza et al. 2010), although the delta-endotoxin remained chemically stable in neutral and acid waters (Sinegre et al 1980). Extend persistence with *Bti* is possible through use of improved formulations. Gunasekaran *et al.* (2002) tested a floating sustained release formulation of *Bti* in polluted water habitats against *Culex quinquefasciatus* larvae. Briquetts may result in more prolonged control than liquid formulations as these products have greater persistence through slow release (Kase & Branton 1986).

On the other hand, Aguilar-Meza et al. (2010) tested the residual insecticidal activity after field exposure of an aluminum-carboxymethylcellulose microencapsulated formulation of *Bti* spore-toxin complex with malachite green as photoprotective agent. The formulation improved the activity against *Aedes aegypti* larvae for 30 days and was comparable to that of the chemical insecticide temephos.

9. Conclusions

Bti is a bacterium that has been applied with success in biological control programs against mosquitoes and flies larvae all over the world. The study in each of its facets addressed in this review will open new perspectives to improve their effectiveness in biological control.

10. Acknowledgments

We thank Dr. Vladimir Sanchez-López of the University of Papaloapan, Tuxtepec, Oaxaca and Tiburcio Laez Aponte of the INECOL, Xalapa, Veracruz for taking optical an electronic microscope microphotographs, respectively.

11. References

Abbott, W.S. (1925). A method of computing the effectiveness of an insecticide. *Journal of Economic Entomology*, Vol. 18, pp. 265-267.

Adams, L.F.; Visick, J.E. & Whiteley H.R. (1989). A 20-kilodalton protein is required for efficient production of the *Bacillus thuringiensis* subsp. *israelensis* 27-kilodalton crystal protein in *Escherichia coli*. *Journal of Bacteriology*, Vol 171, pp. 521-530

Aguilar-Meza, O.; Ramırez-Suero, M.; Bernal, J.S. & Ramırez-Lepe, M. (2010). Field evaluation against *Aedes aegypti* larvae of aluminum-carboxymethylcellulose-encapsulated formulation of *Bacillus thuringiensis* serovar *israelensis*. *Journal of Economic Entomology*, Vol.103, pp. 570–576

Aronson, A.I. Angelo, N. & Holt, S.C. (1971). Organic Nutrients Required For Growth and Sporulation of *Bacillus Cereus*. *Journal of Bacteriology*, Vol.106, p.p.1016-1025.

Ali, A. (1981). *Bacillus thuringiensis* serovar. *israelensis* (ABG-6108) against chironomid midges and some nontarget invertebrates. *Journal of Invertebrate Pathology* Vol 38, pp. 264–272.

Aly, C.; Mulla, M.S. & Schnetter, W. (1987). Floating formulations increase activity of *Bacillus thuringiensis* var. *israelensis* against *Anopheles* larvae. *Journal of the American Mosquito Control Association* Vol 3, pp. 583–588.

Aly, C.; Mulla, M.S.; Xu, B.Z. & Schnetter, W. (1988). Rate of ingestion by mosquito larvae (Diptera:*Culicidae*) as a factor in the effectiveness of a bacterial stomach toxin. *Journal of Medical Entomology*, Vol 25, pp.191–196.

Aoki, K & Chigasaki, Y. (1915) Uber des toxin von sog. Sotto-bacillen Mitt. Med. Fak. Kais. Univ. Tokyo. Vol. 14, pp. 59-80

Aronson, A.I.; Han, E.S; McGaughey, W & Johnson, D. (1991). The solubility of inclusion proteins from *Bacillus thuringiensis* is dependent upon protoxin composition and is a factor in toxicity to insects. *Applied and Environmental Microbiology*, Vol 57, pp.981-986

Ash, C. & Collins, M.D. (1992). Comparative analysis of 23s ribosomal RNA gene sequences of *Bacillus anthracis* and emetic *Bacillus cereus* determined by PCR-direct sequencing. *FEMS Microbiology Letters*, Vol 73, pp. 75-80

Bassi, D.G.; Weathersbee, A.A.; Meisch, M.V. & Inman, A. (1989). Efficacy of Duplex and Vectobac against *Psorophora columbiae* and *Anopheles quadrimaculatus* larvae in Arkansas ricefields. *Journal of the American Mosquito Control Association*, Vol 5, pp. 264-266.

Bauer, L.S. (1995). A threat of the insecticidal crystal protein of *Bacillus thuringiensis*. *Florida Entomologist*, Vol 78, pp. 414-443

Baum, J. A. & T. Malvar, T. (1995). Regulation of insecticidal crystal protein production in *Bacillus thuringiensis*. *Molecular Microbiology*, Vol 18, pp. 1-12.

Bartlet, B.R. (1964). Integration of Chemical and Biological Control. In: *Biological Control of insect pests and weeds*, P. DeBach, (Ed.), 489-511, Chapman & Hall, London.

Becker, N. (2000). Bacterial control of vector-mosquitoes and black flies. In *Entomopathogenic Bacteria: From Laboratory to Field Application*, (Ed.) J.F. Charles, A. Delécluse and C. Nielsen-LaRoux), pp. 383-398. Dordrecht, The Netherlands: Kluwer.

Becker, N. & Margalit, J. (1993). Use of *Bacillus thuringiensis israelensis* against mosquitoes and blackflies. In *Bacillus thuringiensis, an EnvironmentalBiopesticide: Theory and Practice*, P. F. Entwistle, J. S. Cory, M. J. Bailey and S. Higgs (Ed.)), pp. 147-170. John Wiley & Sons, New York.

Ben-Dov, E.; Zaritsky, A.; Dahan, E.; Barak, Z.; Sinai, R.; Manasherob, R.; Khamraev, A.; Troitskaya, E.; Dubitsky, A.; Berezina, N. & Margalith, Y. (1997). Extended screening by PCR for seven *cry*-group genes from field-collected strains of *Bacillus thuringiensis*. *Applied and Environmental Microbiology*, Vol 63, pp. 4883-4890

Bravo, A.; Sarabia, S.; Lopez, L.; Oniveros, H.; Abarca, C.; Ortiz, A.; Ortiz, M.; Lina, L.; Villalobos F.J.; Peña, G.; Nuñez-Valdez, M.E.; Soberón, M. & Quintero. R. (1998). Characterization of *cry* genes in a Mexican *Bacillus thuringiensis* strain collection. *Applied and Environmental Microbiology*, Vol 64, pp. 4965-4972

Bravo, A.; Gill, S.S. & Soberon, M. (2007). Mode of Action of *Bacillus thuringiensis* Cry and Cyt toxins and their potential for insect control. *Toxicon*, Vol.49, No.4 (March 2007). pp.423-435

Brown, K. L. & Whiteley, H.R. (1988). Isolation of a *Bacillus thuringiensis* RNA polymerase capable of transcribing crystal protein genes. *Proceedings of the National Academy of Sciences*, Vol 85, pp. 4166–4170.

Brown, K. L., & H. R. Whiteley. (1990). Isolation of the second *Bacillus thuringiensis* RNA polymerase that transcribes from a crystal protein gene promoter. *Journal of Bacteriology*, Vol 172, pp. 6682–6688.

Butko, P. (2003). Cytolytic toxin Cyt11A and its Mechanism of Membrane Damage: Data and Hypotesis. *Applied and Environmental Microbiology*. Vol69, No.5, (May 2003) pp 2415-2422

Carozzi, N. B.; Kramer, V.C.; Warren, G.W.; Evola, S. & Koziel, M.G. (1991). Prediction of insecticidal activity of *Bacillus thuringiensis* strains by polymerase chain reaction product profiles. *Applied and Environmental Microbiology*, Vol 57, pp. 3057-3061

Crickmore, N.; Zeigler, D.R.; Feitelson, J.; Schnepf, E.; Vanrie, J.; Lereclus, D.; Baum, J. & Dean, D.H. (1998). Revision of the nomenclature for the *Bacillus thuringiensis* pesticidal crystal proteins. *Microbiology and Molecular Biology Reviews*, Vol 62, pp. 807-813

Crickmore, N.; Nicholls, C.; Earp, D.J.; Hodgman, T.C.; & Ellar, D.J. (1990). The construction of *Bacillus thuringiensis* strains expressing novel entomocidal ™-endotoxin combinations. *Journal of Biochemistry*, Vol.270, pp. 133–136.

Chang, C.; Dai, S.M Frutos, R.; Federici, B.A. & Gill, S.S. (1992). Properties of a 72-kilodalton mosquitocidal protein from *Bacillus thuringiensis* subsp. morrisoni PG-14 expressed in *Bacillus thuringiensis* subsp. *kurstaki* by using the shuttle vector pHT3101. *Applied and Environmental Microbiology*, Vol 58, pp. 507-512

Charles, J. F. & deBarjac, H. (1982). Sporulation et crystal- logenese de *Bacillus thuringiensis* var *israelensis* en microscopie electronique.*Annals of Microbiology* (Inst.Pasteur)133A:425-442.

Chen, M.L.& Tsen H.Y. (2002). Discrimination of Bacillus cereus and Bacillus thuringiensis with 16S rRNA and gyrB gene based PCR primers and sequencing of their annealing sites. *Journal of Applied Microbiology*. Vol.92, pp. 912–919

Chilcott, C. N. & Ellar, D.J. (1988). Comparative toxicity of *Bacillus thuringiensis* var. *israelensis* crystal proteins *in vivo* and *in vitro*. *Journal of General Microbiology*, Vol.134, pp.2552-2558

Chilcott, C.N.; Knowles, B.H.; Ellar, D.J. & Drobniewski, F.A. (1990). Mechanism of Action of *Bacillus thuringiensis israelensis* Parasporal Body. In *Bacterial Control of Mosquitoes and Blackfles*, eds de Barjac, H. & Sutherland, D.J. pp. 45-65. New Brunswick : Rutgers University Press.

Chungjatupornchai, W.; Höfte, H.; Seurinck, J.; Angsuthanasombat, C. & Vaeck, M. (1988). Common features of *Bacillus thuringiensis* toxins specific for *Diptera* and *Lepidoptera*. *European Journal of Biochemistry*, Vol.173, pp.9-16

de Barjac, H. & Bonnefoi, A. (1973). Mise au point sur la classification des *Bacillus thuringiensis*. *Entomophaga*. Vol 18, pp. 5-17

de Barjac, H. 1978. A new subspecies of *Bacillus thuringiensis* very toxic for mosquitoes; *Bacillus thuringiensis* serotype H-14. *Compte Rendu Académie Sciences Paris Series D*. Vol. 286 pp.797-800.

de Barjac, H. & Frachon, E. (1990). Classification of *Bacillus thuringiensis* Strains Entomophaga. Vol. 35, pp. 233-240.

de Barjac, H. & Larget-Thiery, I. (1985). Characteristics of IPS82 standard for bioassay of *B. thuringiensis* H-14 preparations. WHO Rep/VBC/84.892.

Dervyn, E.; Poncet, S.; Klier, A. & Rapoport, G. (1995). Transcriptional regulation of the *crylVD* gene operon from *Bacillus thuringiensis* subsp. *israelensis*. *Journal of Bacteriology*, Vol.177, pp.2283-2291

Donovan, W. P.; Dankocsik, C. & Gilbert, M.P. (1988). Molecular characterization of a gene encoding a 72-kilodalton mosquito-toxic crystal protein from *Bacillus thuringiensis* subsp. *israelensis*. *Journal of Bacteriology*, Vol.170, pp.4732-4738

Dulmage, H.T. (1970). Production of the spore- endotoxin complex by variants of *Bacillus thuringiensis* in two fermentation media. *Journal of Invertebrate Pathology*, Vol.16, pp.385-389.

Earp, D.J. & Ellar, D.J. (1987). *Bacillus thuringiensis* var. *morrisoni* strain PG14: nucleotide sequence of a gene encoding a 27 kDa crystal protein. *Nucleic Acids Research*, Vol 15, pp. 3619-3624

Dervyn, E.; Poncet, S.; Klier, A,; Rapoport, A.; Yoshisue, H.; Nishimoto, T.; Sakai, H. & Komano, T. (1994). Identification of a promoter for the crystal protein-encoding gene *crylVB* from *Bacillus thuringiensis* subsp. *israelensis*. *Gene* Vol.137, pp.247-251

Debro, L.; Fitz-James, P.C. &Aronson, A. (1986). Two different parasporal inclusions are produced by *Bacillus thuringiensis subsp. finitimus*. *Journal of Bacteriology*, Vol.58, pp. 507-512

Delafield, F.P.; Somerville, H.J. & Rittenberg S.C. (1968). Immunological homology between crystal and spore protein of *Bacillus thuringiensis*. *Journal of Bacteriology*, Vol.96 pp.713-720

Dunkle, R.L., & B.S. Shasha. 1989. Response of starch encapsulated *Bacillus thuringiensis* containing ultraviolet screens to sunlight. Environ. Entomol. 18: 1035-1041.

Drobniewski, F.A. (1993) Onchocerciasis and the Onchocerciasis Control Programme. *Microbiology Europe* Vol.1, pp. 26-28,

Elcin, Y.M.; Cokmus, C. & Sacilik, C.S. (1995). Aluminum carboximethylcellulose encapsulation of *Bacillus sphaericus* 2362 for control of *Culex* spp. (Diptera: *Culicidae*) larvae. *Journal of Economic Entomology* Vol.88 pp. 830-834

Federici, B.A. (1999). *Bacillus thuringiensis* in Biological Control. . In: *Handbook of Biological Control*. T. Fisher (Ed.)Academic Press (Ed.) 575-593, ISBN 10: 0-12-257305-6

Federici, B.A.; Park, H.W.; Bideshi, D.K.; Wirth M.C. & Johnson, J.J. (2003). Review. Recombinant Bacteria for Mosquito Control. *The Journal of Experimental Biology*, Vol.206, pp.3877-3885

Finney, D.J. (1971). Probit Analysis. Cambridge Univ. Press, Cambridge, United Kingdom.

Foda M.S.; Salama, H.S. & Selim, M. (1985). Factors afecting growth physiology of *Bacillus thuringiensis*. *Applied Microbiology and Biotechnology*, Vol.22, pp.50-52.

Galjart, N. J.; Sivasubramanian, N. & Federici, B.A. (1987). Plasmid location, cloning and sequence analysis of the gene encoding a 23-kilodalton cytolytic protein from *Bacillus thuringiensis* subsp. *morrisoni* (PG-14). Current Microbiology, Vol.16, pp. 171-177

Glare T.R. & O'Callaghan M. (1998). Environmental and Health Impacts of *Bacillus thuringiensis israelensis*. *Report for the Ministry of Health*. Biocontrol & Biodiversity, Grasslands Division, Ag. Research. Lincoln, New Zealand.

Goldberg, L. J. & Margalit, J. (1977). A bacterial spore demonstrating rapid larvicidal activity against *Anopheles sergentii, Uranotaenia unguiculata, Culex univitattus, Aedes aegypti* and *Culex univitattus*. *Mosquto News*, Vol.37, pp. 355–358

Georghiou, G.P. & Wirth, M.C. (1997). Influence of Exposure to Single versus Multiple Toxins of *Bacillus thuringiensis* subsp. *israelensis* on Development of Resistance in the Mosquito *Culex quinquefasciatus* (Diptera: *Culicidae*). *Applied and Environmental Microbiology*, Vol.63, No. 3, pp.1095-1101

Goldberg, I.; Sneh, B.; Battae, E. & Klein, D. (1980). Opt i mi zat i on o f a medium for a high product i on o f spore-crystal preparat i on o f *Bacillus thuringiensis* effective against Egyptian cotton leaf worm *Spodoptera littorallis*. *Biotechnology Letters* Vol.2, pp. 419-426.

Hannay, L.C. (1953). Cristalline inclusion in aerobic spore forming bacteria. Nature. Vol. 172, pp. 1004.

Hannay, C.L. & Fitz-James P.C. (1955). The protein crystals of *Bacillus thuringiensis Berl*. *Canadian Journal of Microbiology*. Vol 1, pp. 694-710.

Hershey, A. E., Shannon, L., Axler, R., Ernst, C. & Mickelson, P. (1995). Effects of methoprene and *Bti* (*Bacillus thuringiensis* var. *israelensis*) on non-target insects. *Hydrobiologia*, Vol. 308, pp.219-227

Höfte, H. & Whiteley, H.R. (1989). Insecticidal crystal proteins of *Bacillus thuringiensis*. *Microbiology Reviews* Vol.53 pp. 242-255.

Hoti, S.L, & Balaraman, K. (1993). Formation of melanin-pigment by a mutant of *Bacillus thuringiensis* H-14. *Journal of General Microbiology* Vol.139, pp. 2365-2369.

Husz, B. (1931). Experiments during 1931 on the use of *Bacillus thuringiensis Berl*. in controlling the corn borer. *Science Reports International of Corn Borer Investigation*, Vol. 4, pp. 22-23.

Ibarra, J. E. & Federici, B. A. (1986). Isolation of a relatively nontoxic 65-kilodalton protein inclusion from the parasporal body of *Bacillus thuringiensis* subsp. *israelensis*. *Journal of Bacteriology*, Vol.165, pp.527-531

Ignofo, C.M., Shasha, B.S. & Shapiro, M (1991). Sunlight ultraviolet protection of the *Heliothis* nuclear polyhedrosis virus through starch encapsulation technology. *Journal of Invertebrate Pathology*, Vol 57: 134-136.

Ishiwata, S. (1901). On a kind of severe flacherie (sotto disease). *Dainihon Sanshi Kaiho* Vol.114, pp.1-5

Iwabushi, H. (1908) Popular silkworm Pathology A.A. Meibundo (Ed.) Tokyo, pp. 428.

Kase, L.E. & Branton, L. (1986). Floating particle for improved control of aquatic insects. U.S. Pat. 4631857

Knowles, B. H.; Blatt, M.R.; Tester, M.; Horsnell, J.M.; Carroll, J.; Menestrina G. & Ellar, D.J. (1989). A cytolytic delta-endotoxin from *Bacillus thuringiensis* var. *israelensis* forms cation-selective channels in planar lipid bilayers. *FEBS Letters*, Vol.244, pp.259–262

Krieg V.A.; Huger A.M.; Langenbruch G.A. & Schnetter W. (1983). *Bacillus thuringiensis* var. *tenebrionis*: Ein neuer, gegenüber lair ven von coleopteren wirksamer pathotyp. *Z. Äng. Ent.* Vol.96, pp. 500 508

Jacobs, S.E. (1950).Bacteriological control of the flour moth *Ephestia kuhniella*. *Proceedings of the Society of Apply Bacteriology,* Vol.13, pp. 83-91.

Jackson, J. K.; Sweeney, B. W.; Bott, T. L.; Newbold, J. D. & Kaplan, L. A. (1994). Transport of *Bacillus thuringiensis* var. *israelensis* and its effect on drift and benthic densities of nontarget macroinvertebrates in the Susquehanna River, northern Pennsylvania. *Canadian Journal of Fisheries and Aquatic Sciences* Vol.51, pp.295-314

Johnson, M.W. & Tabashnik B.E. (1999). Enhanced Biolgical Control Through Pesticide Selectivity. In: *Handbook of Biological Control.* T. Fisher (Ed.)Academic Press (Ed.) 297-317, ISBN 10: 0-12-257305-6

Lacey L.A.; Urbina, M.J. & Heitzzman, C.M.(1989). Sustained release formulations of *Bacillus sphaericus* and *Bacillus thuringiensis* (H-14) for control of container-breeding *Culex quinquefasciatus. Mosquito News* Vol.44, pp.26-32.

Lacey, L. A. & Mulla, M. S. (1990). Safety of *Bacillus thuringiensis* var. *israelensis* and *Bacillus sphaericus* to nontarget organisms in the aquatic environment. In *Safety of Microbial Insecticides.* Laird-M, Lacey-LA and Davidson-EWSO (Ed.), pp. 259 pp. CRC Press Inc.; Boca Raton; USA.

Leighton, T.J. & Doi, R.H.(1971). The Stability of Messenger Kibonucleic Acid during Sporulation in *Bacillus subtilis. The Journal of Biological Chemistry.* Vol.246, pp.3189-3195.

Leong, K.L.; Cano, R.J. & Kubinski, A.M (1980). Factors affecting *Bacillus thuringiensis* total field persistence. *Environmental Entomology,* Vol.9 pp.593-599.

Levy, R.; Nichols, M.A. & Miller, T. W. (1990). Culigel superabsorbent polymer controlled-release system: application to mosquito larvicidal bacilli. *Proceedings and Abstracts, Vth International Colloquium on Invertebrate Pathology and Microbial Control, Adelaide, Australia ,* 107

Lord, J.C. & Undeen A.H. (1990). Inhibition of the *Bacillus thuringiensis* var. *israelensis* toxin by dissolved tannins. Environmental Entomology, Vol.19 pp.1547-1551

Manasherob, R., Ben-Dov, E., Margalit, J., Zaritsky, A. & Barak, Z. (1996). Raising activity of *Bacillus thuringiensis* var. *israelensis* against *Anopheles stephensi* larvae by encapsulation in *Tetrahymena pyriformis* (Hymenostomatidia: Tetrahymenidae). *Journal of the American Mosquito Control Association,* Vol.12, pp. 627-631

Manasherob, R.; Ben-Dov, E.; Xiaoqiang, W.; Boussiba, S., & Zaritsky, A. (2002). Protection from UV-B damage of mosquito larvicidal toxins from *Bacillus thuringiensis* subsp. *israelensis* expressed in *Anabaena* PCC 7120. *Current Microbiology,* Vol 45, pp. 217–220.

Marrone, P.G. & MacIntosh, S.C. (1993). Resistance to *Bacillus thuringiensis* and Resistance Management. *In: Bacillus thuringiensis,* an Environmental Biopesticide: Theory and

Practice, Entwistle, P. F., Cory, J. S., Bailey, M. J., and Higgs, S., Eds., John Wiley and Sons, Chichester, UK, pp.221-235.

McDonald B.A. & Linde C. (2002). Pathogen Population Genetics, Evolutionary Potential, and Durable Resistance. *Annual Review of Phytopathology*, Vol. 40, pp. 349-379

McGuirre, M.R.; Shasha, B.S.; Eastman, C.E. & Sagedchi H.O. (1996). Starch and fluor-based sprayable formulations: Effect on rainfastness and solar stability of *Bacillus thuringiensis*. *Journal of Economic Entomology*, Vol.89, pp. 863-869

McLaughlin, R.E.; Dulmage, H.T.; Alls, R.; Couch, T.L.; Dame, D.A.; Hall, I.M.; Rose, R.I. &.Versol, P.L. (1984). U.S. standard bioassay for the potency assessment of *Bacillus thuringiensis* var. *israelensis* serotype H-14 against mosquito larvae. *Bulletin of the Entomological Society of America*, Vol.30, pp. 26-29

McLean, K.M. & Whiteley, H.R. (1987). Expression in *Escherichia coli* of a cloned crystal protein gene of *Bacillus thuringiensis* subsp. *israelensis*. *Journal of Bacteriology*, Vol.169, pp. 1017-1023

Mattes, O. (1927) Parasitare kronkheiten der mehlmotten larven und versuche uber ihre Verwendborkeit als biologische Bekampfungsmittel. *Sisher Ges. Beforder. Ges Naturw. Marburg.* Vol.62., pp. 381-417

Metalnikov, S.; Hengula, B. & Strail, D.M. (1930). Experiments on the application of bacteria against the corn borer. *Science Reports International Corn Borer Investigation*, Vol3, pp. 148-151.

Mikkola, A. R.; Carlberg, G.A.; Vaara, T. & Gyllenberg, H.G. (1982). Comparison of inclusions in different *Bacillus thuringiensis* strains. An electron microscope study. FEMS Microbiology Letters, Vol.13, pp.401-408

Mitani, K & Watari, J. (1916). A new method to isolate the toxin of *Bacillus sotto* Ishiwata by passing through a bacerial filter and a preliminary report on the toxic action of this toxin to the silkworm larvae. *Archi Gensanshu Serzojo Hokoku.* Vol 3, pp. 33-42.

Metcalf, R.L. (1986). The Ecology of Insecticides and the Chemical Control of Insects. *In Ecological Theory and Integrated Pest Managment in Practice*, M. Kogan (Ed.) 251 297 John Wiley & Sons, New York.

Morris, O.N. (1983). Protection of *Bacillus thuringiensis* from inactivation from sunlight. *Canadian Entomology*, Vol 115, pp.1215-1227.

Mulla, M. S. & Darwazeh, H.A. (1985). Efficacy of formulations of *Bacillus thuringiensis* H-14 against mosquito larvae. *Bulletin fo the Society of Vector Ecology.* Vol.10, pp.14–19

Mulla, M. S.; Darwazeh, H.A.; Ede, L.; Kennedy, B. & Dulmage H.T. (1985). Efficacy and field evaluation of *Bacillus thuringiensis* (H-14) and *B. sphaericus* against floodwater mosquitoes in California. *Journal of American Mosquito Control Association* Vol1, pp.310–315.

Mulla, M. S.; Darwazeh, H.A. & Zgomba, M. (1990). Effect of some environmental factors on the efficacy of *Bacillus sphaericus* 2362 and *Bacillus thuringiensis* (H-14) against mosquitoes. *Bulletin of the Society of Vector Ecology*, Vol.15, pp.166–175.

Mulla, M. S.; Chaney, J.D. & Rodchareon, J. (1993). Elevated dosages of *Bacillus thuringiensis* var. *israelensis* fail to extend control of *Culex* larvae. *Bulletin of the Society of Vector Ecology*, Vol.18, pp.125–132.

Mulla, M. S.; Thavara, U.; Tawatsin, A. & Chompoosri, J. (2004). Procedures for the evaluation of field efficacy of slow-release formulations of larvicides against *Aedes aegypti* in water-storage containers. *Journal of the American Mosquito Control Association* Vol.20, pp.64–73

Nakata, H.M. (1964). Organic Nutrients Required For Growth And Sporulation Of *Bacillus Cereus, Journal of Bacteriology*, Vol.88, pp.1522-1524;

Ohana, B.; Margalit, J. & Barak, Z. (1987). Fate of *Bacillus thuringiensis* subsp *israelensis* under simulated field conditions. *Applied and Environmental Microbiology*, Vol.53, pp.828-831.

Ogunjimi, A.A.; Chandler, J.M.; Gbenle, G.O.; Olukoya, D.K., & Akinrimisi, E.O. (2002). Heterologous expresión of *cry2* gene from a local strain of *Bacillus thuringiensis* isolated in Nigeria. *Biotechnology Applied Biochemistry*, Vol.36, pp.241-246

Pearson, D. & Ward, O.P. (1988). Effect of culture conditions on growth and sporulation of *Bacillus thuringiensis* susp. *israelensis* and development media for production of the protein crystal endotoxin. *Biotechnology Letters*, Vol.10, pp.451-456.

Perez, C.; Fernandez, L.E.; Sun, J.; Folch, J.L.; Gill, S.S. & Soberon M. (2005*). Bacillus thuringiensis* subsp. *israelensis* Cyt1Aa synergizes Cry11Aa Toxin by Functioning as a Membrane-Bound Receptor. *Proceedings of the National Academy of Sciences*. Vol 102, No. 51. pp. 18303-18308

Poncet, S.; Bernard, C.; Dervyn, E.; Cayley, J.; Klier, A.; & Raport, G. (1997). Improvement of *Bacillus sphaericus* toxicity against dipteran larvae by integration, via homologous recombination, of the *cry11A* toxin gene from *Bacillus thuringiensis* subsp. *israelensis*. *Applied and Environmental Microbiology*, Vol.63, pp 4413-4420

Porcar, M. & Juarez-Perez, V. (2003). PCR-based identification of *Bacillus thuringiensis* pesticidal crystal genes. *FEMS Microbiology Reviews*. Vol. 26, pp. 419-432.

Pozsgay, M.; Fast. P.; Kaplan, H. & Carey, P.R. (1987). The effect of sunlight on the protein crystals from *Bacillus thuringiensis* var. *kurstaki* HD1 and NRD12: A raman spectroscopy study. *Journal of Invertebrate Pathology*, Vol.50 pp.620-622

Priest, F.G. (1992). A review. Biological control of mosquitoes and other biting flies by *Bacillus sphaericus* and *Bacillus thuringiensis*. *Journal of Applied Bacteriology*, Vol.72, pp.357-369.

Promdonkoy, B. & Ellar, D.J. (2000). Membrane pore architecture of a cytolytic toxin from *Bacillus thuringiensis*. *Journal of Biochemistry*.Vol. 350, pp.275-282.

Pusztai, M.; Fast, P.; Gringorten, L., Kaplan, H.; Lessard, T. & and Carey, P.R. (1991). The mechanism of sunlight-mediated inactivation of *Bacillus thuringiensis* crystals. *Journal of Biochemistry*, Vol.273, pp.43-47

Quintana-Castro, R.; Ramírez-Suero, M.; Moreno-Sanz, F. & Ramírez-Lepe, M. (2005). Expression of the *cry11A* gene of *Bacillus thuringiensis* ssp. *israelensis* in *Saccharomyces cerevisiae*,.*Canadian Journal of Microbiology*, Vol.51, pp.165-170

Ramırez-Lepe, M.; Aguilar, O.; Ramırez-Suero, M. & Escudero, B. (2003). Protection of the spore-toxin complex of *Bacillus thuringiensis* serovar *israelensis* from ultraviolet irradiation with aluminum-CMC encapsulation and photoprotectors. *Southwestern. Entomology*, Vol.28, pp.137-143.

Ramírez-Suero, M.; Robles-Olvera, V. & Ramírez-Lepe, M. (2005). Spray-dried *Bacillus thuringiensis* serovar *israelensis* formulations for control of *Aedes aegypti* larvae. *Journal of Economic Entomology*, Vol.98, pp.1494-1498

Ramırez-Suero, M.; Valerio-Alfaro, G.; Bernal, J.S. & Ramirez-Lepe M. (2011). Synergisitic effect of chitinases and *Bacillus thuringiensis israelensis* spore-toxin complex against *Aedes aegypti* larvae. *The Canadian Entomologist* Vol,143: pp. 157-164.

Rawlins, S.C. (1989) Biological control of insect pests affecting man and animals in the tropics. *CRC Critical Reviews in Microbiology* Vol.16, pp.235-252

Rowe, G.E. & Margaritis, A.M. (1987). Bioprocess Developments in the Production of Bioinsecticides by *Bacillus thuringiensis*. *CRC Critical Reviews in Biotechnology*. Vol.6, No.4, pp. 87-127.

Rushed, S.S. & Mulla, M.S. (1989). Factors influencing ingestion of particulate materials by mosquito larvae (Diptera: Culicidae). *Journal of Medical Entomology*, Vol.26, pp. 210-216

Salama, H.S.; Foda, M.S. & Dulmage H.T. (1983). Novel Media for Production of δ-endotoxins form *Bacillus thuringiensis*. *Journal of Invertebrate Pathology*, Vol.41, pp.8-19.

Sen, K.; Honda, G.; Koyama, N.; Nishida, M.; Neki, A.; Sakai, H.; Himeno, M. & Komano, T. (1988). Cloning and nucleotide sequences of the two 130 kDa insecticidal protein genes of *Bacillus thuringiensis* var. *israelensis*. *Agriculture Biology Chemistry*, Vol.52, pp.873-878

Servant, P.; Rosso, M.L.; Hamon, S.; Poncet, S.; Delécluse, A. & Rapport, G. (1999). Production of cry11A and cry11B toxins in *Bacillus sphaericus* confers toxicity towards *Aedes aegypti* and resistant *Culex* populations. *Applied and Environmental Microbiology*, Vol.65, pp.3021-3026.

Sinegre, G.; Gaven, B.; Jullien, J. L. & Crespo, O. (1980). Effectiveness of the serotype H-14 of *Bacillus thuringiensis* against the principal species of man-biting mosquitoes on the Mediterranean coast of France. *Parassitologia* Vol.22, pp.223-231.

Skovmand, O. & Eriksen, A. G. (1993). Field trials of a fizzy tablet with *Bacillus thuringiensis* subsp. *israelensis* in forest spring ponds in Denmark. *Bulletin of the Society for Vector Ecology*, Vol. 18, pp.160-163.

Skovmand, O.; Hoeg, D.; Pedersen, H.S., & Rasmussen T. (1997). Parameters Influencing Potency of *Bacillus thuringiensis* var. *israelensis* Products. *Journal of Economic Entomology* Vol 90, No.2. pp. 361-369.

Smith, A.R. (1982). Effect of strain and medium variation on mosquit production by *Bacillus thuringiensis* var. *israelensis*. *Canadian Journal of Microbiology*, Vol.28, pp.1090-1092.

Tabashnik B.E. (1992). Evaluation of Synergism among *Bacillus thuringiensis* Toxins. *Applied and Environmental Microbiology*, Vol.58, pp.3343-3346

Tanapongpipat, S.; Luxananil, P.; Boonhiang, P.; Chewawiwat, N.; Audtho, M. & Panyim, S. (2003). A plasmid encoding a combination of mosquito-larvicidal genes from *Bacillus thuringiensis* subsp. *israelensis* and *Bacillus sphaericus* confers toxicity against broad range of mosquito larvae when expressed in Gram-negative bacteria. FEMS Microbiology Letters, Vol.228, pp.259-263

Thanabalu, T.; Hindley, J.; Brenner, S.; Oei, C. & Berry, C. (1992). Expression of the mosquitocidal toxins of *Bacillus sphaericus* and *Bacillus thuringiensis* subsp. *israelensis* by recombinant *Caulobacter crescentus*, a vehicle for biological control of aquatic insect larvae. *Applied and Environmental Microbiology*, Vol.58, pp. 905- 910

Thomas, W.E. & Ellar, D.J. (1983) *Bacillus thuringiensis* var. *israelensis* crystal delta-endotoxin : effects on insect and mammalian cells *in vitro* and *in vivo*. *Journal of Cell Science* Vol.60, pp.181-197.

Thorne, C.B. (1968). Transducing Bacteriophage for *Bacillus cereus*. *Journal Of Virology*, Vol.2, pp. 657-662.

Tungpradubkul, S.; C. Settasatien, & Panyim, S. (1988). The complete nucleotide sequence of a 130 kDa mosquito-larvicidal delta-endotoxin gene of *Bacillus thuringiensis* var. *israelensis*. *Nucleic Acids Research* Vol.16, pp.1637-1638.

Tyanyun, S. & Mulla, M.S. (1999). Field evaluation of new water dispersible granular formulations of *Bacillus thuringiensis* subp. *israelensis* and *Bacillus sphaericus* against Culex mosquito in microcosms. *Journal of the American Mosquito Control Association* Vol 15 pp. 356-361.

Waalwijk, C.; A.; Dullemans, M.; VanWorkum, M.E.S. & Visser, B. (1985). Molecular cloning and the nucleotide sequence of the Mr28,000 crystal protein gene of *Bacillus thuringiensis* subsp. *israelensis*. *Nucleic Acids Research*, Vol.13, pp. 8207-8217

Walsh, J. (1986) River blindness, a gamble pays off. *Science*, Vol.232, pp.922–925.

Ward, E.S.; Ridley, A.R.; Ellar, D.J. & Todd, J.A. (1986). *Bacillus thuringiensis* var. *israelensis* endotoxin. Cloning and expression of the toxin in sporogenic and asporogenic strains of *Bacillus subtilis*. *Journal of Molecular Biology*, Vol.191, pp. 13–22

Ward, E.S., & Ellar, D.J. (1986). *Bacillus thuringiensis* var. *israelensis* delta-endotoxin: nucleotide sequence and characterization of the transcripts in *Bacillus thuringiensis* and *Escherichia coli*. *Journal of Molecular Biology*, Vol.191, pp.1-11

Ward, E. S., & Ellar, D.J. (1987). Nucleotide sequence of a *Bacillus thuringiensis* var. *israelensis* gene encoding a 130 kDa delta-endotoxin. *Nucleic Acids Research*, Vol 15, pp. 7195

Webb, G. (1992) The Onchocerciasis Control Programme. *Transactions of the Royal Society of Tropical Medicine and Hygiene*, Vol. 86, pp.113-114

Wirth, M. C.; Georghiou, G. P.; Federici, B. A. (1997). CytA enables CryIV endotoxins of *Bacillus thuringiensis* to overcome high levels of CryIV resistance in the mosquito, *Culex quinquefasciatus*. *Proceedings of the National Academy of Sciences*, Vol.94, pp.10536-10540

Wojciech J. J. & Korsten L. (2002). Biological Control of Postharvest Diseases of Fruits. *Annual Review of Phytopathology* Vol. 40, pp. 411-441.

World Health Organization (1999). *World Health Report*. Geneva: World Health Organization.

Wu, D. & Chang, F.N (1985). Synergism in mosquitocidal activity of 26 and 65 kDa proteins from *Bacillus thuringiensis* subsp. *israelensis* crystal. FEBS Letters, Vol.190, pp.232-236

Xu,Y.; Nagai, M.; Bagdasarian, M.; Smith,T.W. & Walker, E.D. (2001). Expression of the *P20* Gene from *Bacillus thuringiensis* H-14 Increases Cry11a Toxin Production and Enhances Mosquito- Larvicidal Activity in Recombinant Gram-Negative Bacteria. *Applied And Environmental Microbiology*, Vol.67, p.p.3010–3015

Wada Y. (1989). Control of Japanese encephalitis vectors. *Southeast Asian journal of tropical medicine and public health*, Vol.20, pp. 623–626

Yamagiwa, M., Sakagawa, K., & Sakai, H. (2004). Functional Analysis of two processed fragments of *Bacillus thuringiensis* cry11A toxin. *Bioscience Biotechnology and Biochemistry*, Vol.68, pp.523–528

Yamamoto, T.; IWatkinson, I.A.; Kim, L.; Sage, M.V.; Stratton, R.; Akande, N.; Li, D.; D.P. & Roe, B.A. (1988). Nucleotide sequence of the gene coding for a 130-kDa mosquitocidal protein of *Bacillus thuringiensis israelensis*. *Gene*, Vol 66, pp.107-120.

Yamamoto, T.; Iizuku, T. & Aronson, J.N. (1983). Mosquitocidal protein of *Bacillus thuringiensis* subsp. *israelensis*: identification and partial isolation of the protein. *Current Microbiology* Vol.9, pp.279-284.

Yu-Tien, L.; Men-Ju, S.; Dar-Der, J.; Wang, W.I.; Chin-Chi, C. & Cheng-Chen, C. (1993). Protection from ultraviolet irradiation by melanin mosquitocidal activity of *Bacillus thuringiensis* var. *israelensis*. *Journal of Invertebrate Pathology*, Vol 62:131-136.

Zeigler, D.R. (1999). *Bacillus thuringiensis Bacillus cereus. Bacillus Genetic Stock Center.* Catalog of strains 7th. ed. Vol.2

10

Essential Plant Oils and Insecticidal Activity in *Culex quinquefasciatus*

Maureen Leyva, Olinka Tiomno, Juan E. Tacoronte,
Maria del Carmen Marquetti and Domingo Montada
Institute Tropical "Pedro Kouri"
Cuba

1. Introduction

Plants have great importance for man because they are one of their sources of food, they provide us through the process of photosynthesis the oxygen we breathe and are essential to maintain the ecological balance (Corbino, 2000).

Essential oils are volatile, usually distillable liquid fractions responsible for the aroma of the plant. The vast majority of them are pleasant smell and its metabolic and evolutionary significance lies in the role they play as attractor of pollinating agents (for its pleasant aroma), constitute elements of defense against the attack of parasites, herbivorous animals and insects, allow the adaptation of the plant when water is scarce and are part of the substances in reserve as the giver of H+ in the processes of electron. The organoleptic characteristics of the essential oils may be given by major components, although in other cases they are substances present in tiny quantities (traces) which define the taste, smell, or therapeutic properties (Scholes, 1995; Worwood, 1992).

These natural substances are known as secondary metabolites, name which refers to substances that are not involved in the basic mechanisms of life of the plant but that comply with specific functions. (Corbino, 2000). They tend to accumulate in large amounts without negative effects and represent a problem in the cell or on the plant. These metabolites have the property to form glycosides and are found soluble in the plant. For many years these metabolites were regarded as final products of metabolic processes without specific function or directly as a waste of plant products (Lopez, 2008).

The study of these substances was initiated by organic chemists of the 19th century and early 20th century who were interested in these substances because of its importance in the medical industry, the manufacture of flavoring, etc. In fact, the study of the secondary metabolites stimulated the development of separating techniques, and spectroscopy for determining their structure and synthesis which constituted the basis of the contemporary organic chemistry (Lopez, 2008).

They can be found in different parts of the plant: leaves (wormwood, basil, buchú, cidrón, eucalyptus, mint, lemongrass, marjoram, mint, patchouli, quenopod, rosemary, Sage, lemon balm, etc.), in the roots (angelic, asaro, saffron, calamus, turmeric, galanga, ginger, sandalwood, Sasafras, Valerian, vetiver, etc.), in the pericarp of the fruit (lemon, Tangerine, Orange, etc.), seeds (anise, cardamom, dill, fennel, cumin, etc.), in the stem (cinnamon, caparrapí, etc.), flowers (arnica, lavender, chamomile, pyrethrum, thyme, clove scent, rose, etc.) and in the fruit (caraway, coriander, bay leaf, nutmeg, parsley, pepper, etc.). (www.buscasalud.com)

2. Essential oils

Essential oils are complex mixtures of up to over 100 secondary metabolites, which can be classified in terpenoid compounds, phenolic and alkaloids based on their biosynthetic origins. Terpenes containing a single unit of isoprene are called monoterpenes, which contain three units named sesquiterpenes (Ikan, 1991; Silva, 2002).

Monoterpenes and sesquiterpenes are terpenes from 10 to 15 atoms of carbons biosynthetically derived from geranilpyrophosfates (SPG) and farnesilpyrophosfate (FPP) respectively. Monoterpenes and in general all the natural terpenoid compounds are synthetized by the route of the acetilCoA through a common intermediate which is the mevalonic acid. However, it has been proposed that some terpenoids are not originated by this route, and instead use an alternative route that may involve pyruvate, glyceraldehyde-3-phosphate (Fig 1) (Adams et al, 1998; Sponsel, 1995).

Essential oils are widely distributed in some 60 families of plants including the Compositae, Labiadas, Lauraceae, Myrtaceae, Rosaceae, Rutaceae, Umbelliferae, etc. Monoterpenoides are mainly found in plants in the family of Ranunculales, Violales, and Primulales, while they are rare in Rutales, Cornales, Lamiales, and Asterales. Moreover, the sesquiterpenoides are commonly found in Rutales, Cornales, Magnoliales and Asterales. Although in both essential oils: mono - sesquiterpenes and phenylpropan are found in free-form, those which are linked to carbohydrates, have been more recently investigated, because it is considered to be the immediate precursor of the essential oil.

From the chemical point of view, despite its complex composition with different types of substances, essential oils can be classified according to the type of substances of which they are the major components. According to this fact, the essential oils rich in monoterpenes are called essential oils sesquiterpenoids, those rich in phenylpropan are the essential oils phenylpropanoids. Although this classification is very general there are more complex classifications that take into account other chemical aspects (Fig 2,3,4) (González, 1984; Ikan, 1991; Judd et al, 2002; Stachenko, 1996).

2.1 Methods of extraction

According to www.chemkeys.com 2000; Gascon et al. 2002; www.herbotecnia. com.ar, 2002; www.losandes.com.ar 2002. essential oils can be extracted from plant samples through various methods samples such as: expression, distillation with steam, extraction with volatile solvents, enfleurage, extraction with supercritical fluids and hydrodistillation.

2.2 Method of isolation

From such oils, it is possible to its isolation through the use of one or several chromatographic methods such as thin layer chromatography and HPLC.

For the chromatographic column and thin layer methods the silica gel are widely used as stationary phase. Polar pure or mixed solvents such as: Toluene-acetate of ethyl benzene, chloroform, dichloromethane, ethyl, chloroform-benzene, chloroform-ethanol - chloroform-benzene acetic acid benzene-acetate are used as mobile phase.

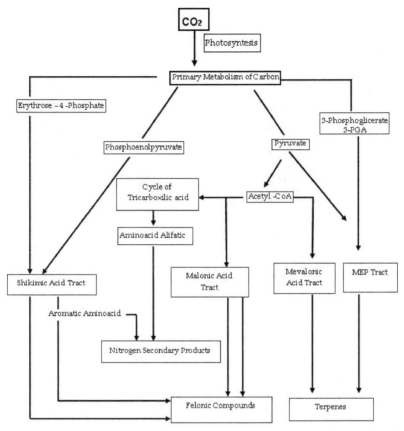

Fig. 1. General pathways of plant secondary metabolism.

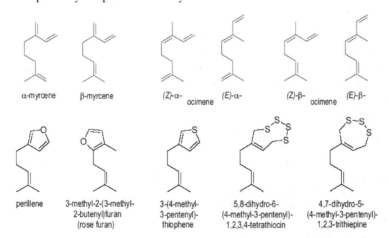

Fig. 2. Natural Monoterpenes.

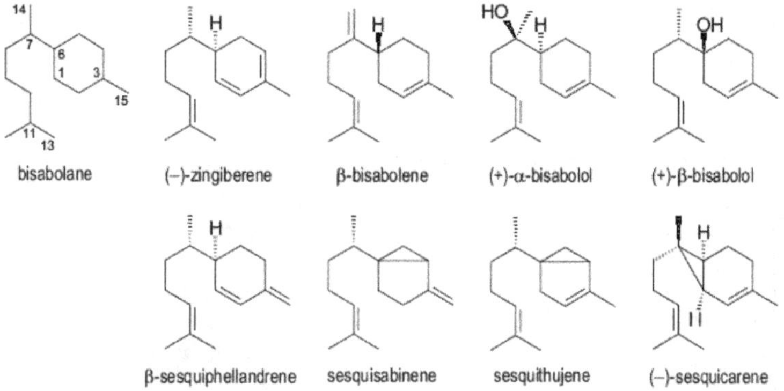

bisabolane   (−)-zingiberene   β-bisabolene   (+)-α-bisabolol   (+)-β-bisabolol

β-sesquiphellandrene   sesquisabinene   sesquithujene   (−)-sesquicarene

Fig. 3. Natural Sesquiterpenes.

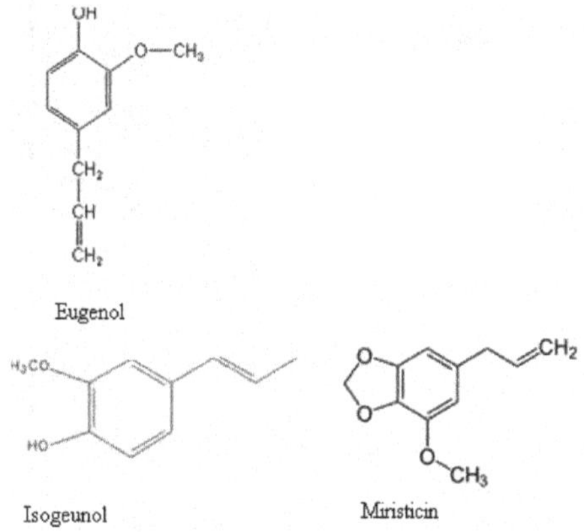

Eugenol

Isogeunol   Miristicin

Fig. 4. Naturals phenylpropanoids.

However, currently more efficient and faster techniques of separation are used such as HPLC high efficiency liquid chromatography and gas (CG) chromatography, as well as also combinations "ON-LINE" HPLC- CG EM. These same methods are used for the analysis of the flower essences. The last technique, thanks to the recent development of capillary columns for high resolution, allows analyzing complex mixtures present in essential oils, and identify the components from the retention times through the so-called Kovats retention indexes (Ik). These values are characteristic for each component and there are databases with indexes of many components of essential oils. Ik values are determined in two chromatographic columns, one polar (e.g. CARBOWAX 20 M) and a non-polar (e.g. OV - 101 also called DB-1).

In addition, the docked technique chromatography of gases - mass spectrometry, (GC-MS) allows obtaining the spectrum of mass of each component which builds the molecular weight and structural information. Likewise there are databases with spectra of masses of many components, something similar goes for the index Kovats (determined in two columns of different polarity) and the spectra of mass are criteria for chemical allocation of many components of essential oils not just monoterpenes but also other types of substances characteristic of such oils. More recently the column chromatographic Chiral for separation of optically active components have been developed, and have developed methods for combined analysis HPLC HPLC-NMR of mixtures of sesquiterpenes and mass spectrometry.

2.3 Test of recognition of monoterpenes and sesquiterpenes (Pino, 1999; Stam, 1970)

Due to the diversity of functional groups that can be present in the components of mono - and sesquiterpenics of an essential oil, there is a specific test for their recognition. However there are few experimental procedures that allow to recognize some of them by their coloration with different reagents, its absorption of 254 nm UV light and its Rf in thin layer chromatography. For example, limonene is recognized in the plates of TLC because it does not absorb UV 254 nm light, adds bromine, does not form a derivative 2, 4-dinitrophenylhydrazone and produces Brown with sulfuric acid.

There is a procedure which combines the TLC with micro reactions (oxidation, reduction, dehydration, hydrolysis, etc.) described in the classic book already today of Ikan. Other reagents useful for revealing monoterpenes and sesquiterpenes are anisaldehid-sulfuric acid, sulfuric acid -vanillin and phosphomolibdic acid.

2.3.1 Spectral characterization

Monoterpenes and sesquiterpenes, in a number, can be characterized chemically from the gas chromatography and mass spectra data as noted above, but when there are doubts of such characterization some spectral methods such as infrared ultraviolet and magnetic resonance spectrometry can be used.

2.3.1.1 Infrared spectroscopy

Infrared Spectroscopy allows detection of the presence of groups hydroxyl, Carbonyl, aromatic rings, bond double C = C cis and trans, etc. To determine the spectrum just put a drop of the component in a cell of NaCl. For example, in the infrared spectrum of the 3-p-menten-7 present in the oil of cumin, the intense band in 1725 cm^{-1} indicates a group not conjugated Carbonyl. The peak at 2710 cm^{-1} is assigned to a proton aldehydic C-H. The bond around on 1375 cm^{-1} indicates a group isopropyl, and the band of 817 cm^{-1} average intensity indicates a double bond tri-substituted.

2.3.1.2 Ultraviolet spectroscopy

UV spectrum ultraviolet of monoterpenes and sesquiterpenes allows the recognition of functional groups and chromophoric. E.g. limonene presents an absorption maximum at 262 nm (ξ = 6400).

Nuclear magnetic resonance (^{1} H - 13 C); due to the developments of the NMR has database of Spectra, especially of 13C -NMR for the monoterpenes and more distributed sesquiterpenes. The NMR spectrometry-13 C has the additional advantage that the carbons of terpenoid chemical displacement (and other natural and synthetic substances) can be calculated by computer programs available in the market as Chemwind and ACD-Lab.

In addition, the recent development of two-dimensional methods homo - and heteronuclear, has allowed the fine structural determination of terpenoids and other natural substances, eliminating the ambiguity in the allocation of the observed signals.

2.3.1.3 Mass Spectrometry

There are databases that contain information and spectra of masses of the components of mono - and sesquiterpenoids of essential oils. These databases are now available in many commercial instruments of analysis as a chromatograph of gases coupled with mass spectrometers.

2.4 Phenylpropan

Phenylpropanes are secondary metabolites derived from phenylalanine, which is first converted to cinnamic acid and after a series of hidroxilations lead to acid cumaric, caffeic acid, pherulic acid etc. The conversion of these acids to their corresponding esters produces some of the volatile components responsible for the fragrance of herbs and flowers (Lopez, 2008).

Phenylpropanes of essential oils are extracted with the same methodology described above for mono - and sesquiterpenes. However, due to its aromatic ring they have advantages in detection by TLC and HPLC because absorb ultraviolet light (254 nm) and do not require to be disclosed with chemical agents, or need to be derivative, and therefore can be isolated and analyzed more easily.

2.4.1 Recognition of phenylpropan

There are test assay for recognition to the aromatic ring as a reaction with formaldehyde and sulfuric acid. Likewise in the case of phenylpropan with phenolic hydroxyl these can be recognized by the test assay of ferric chloride, which produces green and blue coloring with phenolic substances in general.

Substances with aromatic ring, their infrared spectra show signs characteristic of these compounds and give information about the type of replacement of the aromatic ring in addition to the functional groups present in the molecule. For example, the spectrum of eugenol shows, among other, bands in 3500 (wide) due to the hydroxyl group, 1510 characteristic of aromatic, and three bands in 990, 920 and 938 cm^{-1} characteristics of a vinyl mono substituted group. The IR from the cinamaldehyde spectrum displays bands in (weak) 3330, 3050, 2820, 2750, 1660 (intense, due to the carbonyl group), 975, 740 and 695 cm^{-1} among others.

Unlike the majority of mono - and sesquiterpenes, the phenylpropanes absorb UV light with a maximum around 254 nm depending on the groups present in the molecule chromophors. For example, the isoeugenol shows maxima at 260 (15850) and 305 (7000), safrole in 286 nm, the myristicine at 276 nm, the isosafrole in 264 nm, trans- cinnamic acid in 273 nm and cis-cinnamic acid in 264 nm.

Phenylpropan ^{1}H-NMR spectra show signs of aromatic protons around 6-8 ppm whose multiplicities and coupling constants allow a clear structural assignment even with low resolution spectra. Trans-anethole there is a double signal around 1.9 ppm due to the protons from the methyl group, a singlet in 3.9 due to the protons of the methoxy group, a complex signal around 6.1 ppm due to the two protons olefinics provision trans each other, and a double around 6.9 ppm characteristic of 4 protons of an aromatic ring p-based.

2.5 Natural insecticides way-of-action (Silva 2002)

As growth regulators, this effect can manifest itself in several ways. The first is related to the molecules that inhibit the metamorphosis. Other compounds make the insect to have an earlier metamorphosis, occurring at a time which is not favorable. It has been observed that certain molecules can alter the function of the hormones that regulate these mechanisms so that there are insects with malformations, dead or sterile (Gunderson, 1985; Sangykurn, 1999).

Studies carried out from different concentrations of extract of Meliaceas show that this extract inhibits the feeding and negatively affects the development and survival of different species of insects. The anti-alimentary activity of this compound shows that at doses ranging from 5.5 to 27. 6 µg/cm² caused an inhibitory activity of more than 75% the insects. The way of action of these compounds extracted from various species of Meliaceas are taken from a combination between an anti feeding effect and post-digestive toxicity. (www.cannabiscultura.com)

The use of plants as repellents is an ancient knowledge. This practice is basically done with compounds that have odor or irritant effects such as chili and garlic. There are homemade recipes that describe the use of fennel (*Foniculum vulgare*), rude (*Ruta graveolens*) and eucalyptus (*Eucalyptus globules*) among other aromatic plants to repel moths of clothing.

2.5.1 Mode of action of the natural insecticide

Treatments with natural compounds such as essential oils or pure compounds (Awde & Ryan, 1992; Keane & Ryan, 1999;Ryan & Byrne, 1988) may cause symptoms that indicate neurotoxic activity including hyperactivity, seizures and tremors followed by paralysis (knock down), which are very similar to those produced by the insecticides pyrethroids (Kostyukovsky et al., 2002).

Acetyl cholinesterase enzyme catalyzes the hydrolysis of the neurotransmitter excess acetylcholine in the synaptic space between choline and acetic acid. It has been recognized that essential oils could their effect through ACE inhibition.

Many hydrophobic compounds incite the deactivation of protein and enzyme inhibition; the acetilcolinesterase (AchE) is an enzyme that is particularly susceptible to the hydrophobic inhibition (Ryan et al. 1992).

Is described in this context essential oil interfering with AchE are acting as potent of the central nervous system where all synapses cholinergic are virtually located (Bloomquist, 1999) of the cholinesterase inhibitors are known as anticolinesterasic, the chemical products that interfere with the action of this enzyme are potent neurotoxins (López, 2008).

The majority of plant insecticide are extracts made by a group of active compounds of diverse chemical nature, which is hardly found in the same concentrations so the pressure of selection on the plague will not be always the same, i.e. in general, insects take more to develop resistance to a blend of compounds than any of its components separately (Silva, 2003).

2.6 Future prospects

With the beginning of the new millennium, assessment of a large number of essential oils from native or typical flora has increased in many countries of the Americas and Africa due to its potential use as an alternative method of control.

There have been studies with oils in laboratory conditions reporting larvicide action in species of the genus *Culex, Anopheles, Aedes* (Ansari, 2005; Assarn, 2003; Albuquerque et al, 2004; Amer et al, 2006a ; Bassole et al., 2003; Carvalho et to 2003; Cavalcanti et al, 2004; Cetin et al, 2006; Cheng et to 2004; Chung et 2010 b Dharmagda et al, 2005; of Mendonca et al., 2005; Faley et al., 2005; Hafeez et 2011; Khandagle et al, 2001; Mathew & Thoppil 2011;Morais et al, 2006; Prajapati et al., 2005; Pushpanathan et al, 2006; Phasomkusolsil & Soowera 2010; Pitarokili et al, 2011; Raví et al, 2006; Tare et al, 2004; Tomas et al, 2004).

In the same way it has been evaluated the repellent capacity of some oils in the same species of mosquitoes; Amer et al, 2006b;Barnard et al, 2004; Byeoung-Soo, 2005; Caballero-

Gallardo, et-2011; Chang et al, 2006; Jaenson et al, 2006; Das et al., 2003; Kim et al, 2004; Oyedale et al, 2002; Odalo et al., 2005; Pohilt et 2011; Phasomkusolsil & Soowera 2010; Tawatsin et al, 2006; Trongtokit et al. 2005a; 2005b; Yang et al., 2005; Zhu et al, 2006;
Less has been evaluated adulticide effect found only two reports described by Chaiyasit et al 2006 and Miot et al 2004.

3. Studies developing in Cuba with essential oils for controlling of public health pests

In Cuba there is a large plant biodiversity with a lot of endemic species. Myrtaceas, Piperaceas, Zingiberaceae, Pinaceae families possess insecticidal action potential based to their essential oils. To this day, there have been verified in public health pests such as; *Blatella germanica, Musca domestica* and *Aedes aegypti* (Aguilera et al 2003; 2004;Leyva et al, 2007 a, b, 2008a, b; 2009a, b; 2010) and more recently in *Culex quinquefasciatus.*
Evaluated essential oils are essentially from *Pimenta racemosa, Eugenia melanadenia, Psidium rotundatum* and *Melaleuca leucadendron* of the botanical family Myrtaceae, *Piper auritum, Piper aduncum* of the family Piperacea, *Curcuma longa* in the family Zingiberaceae, *Artemisia abrotamum* in the family Asteraceae, *Pinus tropicalis* and *Pinus caribbaea* of the Pinacea family. A derivative of essential oil of pinaceas has also been evaluated; turpentine oil obtained by hidrodestylation of the resin of endemic pine trees, mostly composed of terpenes (α-and β-pinene and a modified turpentine where a part of the original turpentine oil was modified by photoizomerization of the α and β-pinene to compound of the type verbenone and pulegone.)
In all tested oils, there has been found necessary dosages that elicited from 5% to 95% of mortality. After testing CHI2 they have a probability greater than 0.05 by which can raise all the mortalities occurring in each are associated with the used dose.
Studies with these oils in larvae of *Aedes aegypti* show high insecticidal action by the low lethal concentrations obtained and the high slopes of the regression lines; however, in *Blatella germanica* and *Musca domestica* the bioassays were conducted in adults, showing a very low activity given the high lethal concentrations obtained.
Natural essences of plants owe its insecticide and repellent action to the presence in its composition of derivative monoterpenic as d-limonene, α -terpineol, β - myrceno, linolool, 1.8-cineole, 4-terpineol, thymol, carvacrol, α-pinene and β-pinene.) Studied oils possess structures lactones as the 1.8 cineole, 4-terpinol, α and β - pinene, cineole, safrole, turmenon, chavicol, eugenol, beta-phellandrene, longifoleno among others, which account for a 18-68% of its total membership, and are that they can be attributed the high insecticidal action in *Aedes aegypti.*
One aspect assessed for reasons of eco-sustainability was ovicidal and inhibitory action of the development by the oil of turpentine in larvae of *Ae aegypti*
The largest ovicidal effect was shown with the dose diagnosis of oil of turpentine with photochemical treatment. In a test of hypothesis of different proportions for each dose of oil, the diagnostic doses were compared with every CL_{95} finding differences between the doses tested with modified oil of turpentine for p=0.05 while for oil not modified there was only significant difference comparing the CL_{95} and the diagnostic doses. (p=0.05). By comparing the CL_{95} and the diagnostic dose, he found significant difference for a p = 0.05. If we compare both types of essences we may conclude that although modified turpentine oil presents a CL_{95} dose higher than oil without changing this it has a greater effect ovicidal and therefore more protector.

The percentage of hatching inhibition was 36.47% of the total number of larvae. The larvae completing their development to adults corresponded to 60.54% of surviving larvae. As a figure of interest the greatest number of semi emergency in adults occurred at 72 hours after exposure to the oil of turpentine (with a predominance of males). At 96 hour, the largest proportion was in favor of females, who became non-existent for this sex at 120 and 168 hours (5-7 d)

More recently, oil of turpentine and the bicyclic α-pinene obtained by fractional distillation of oil of turpentine have been evaluated in *Culex quinquefasciatus*, important vector of West Nile virus and encephalitis

The purification of the α-pinene was made using column chromatography in system of variable stationary phase (SiO_2, Al_2O_3, and their mixtures in dependence of the polarities of the mobile phases for elution and separation).) Molecular characterization was developed, using techniques of FTIR and NMR, as well as methodologies for its extension to field conditions type TLC (chromatography on thin layer in the presence of plates of glass with silica gel 200-G254 activated with AgNO3 / acetonitril via wet impregnation with methanol). The plates are eluted with mixtures of solvents of media polarity (ethyl-acetone - alcohol isopropyl acetate). Figure 5 shows the spectroscopic characterization of the α pinene.

In table 1 and 2 is shown the concentrations finally used, the obtained mortalities and lethal concentrations that causes 50% to 95% percent of mortality, the slope of the regression line and diagnostic dose.

The lowest lethal concentrations were obtained with oils of turpentine and turpentine modified followed by *Curcuma longa, Psidium rotundatum* and *Chenopodium ambrosoide*. By applying a χ2-square test (p < 0.05) to the results of the bioassays, the results showed that the mortalities are associated with the used dose. The higher slopes of the lines were equal with the oil of turpentine, *Eugenia melanadenia, Psidium rotundatum* and *Chenopodium ambrosiode*, corresponding to the lower value of the slope to the terpene α-pinene. To the slope values obtained were applied a χ2 test finding significant difference between them for p < 0.0001, indicating that these oils have a different response for the same strain.

In our assessed results we find the insecticidal action of oil in *Cx quinquefasciatus* and is important to note that lethal concentrations obtained are higher than concentrations found in *Aedes aegypti* to these same oils (Leyva,2008a; 2009a,b). This response could be that this species has developed physiological resistance and some tolerance to insecticides with which it has active defense before any insecticide action.

Oil of *Curcuma longa* presents in its composition more than 50% of sesquiterpenes, apart from the monoterpenes and α -pinene to which can be attributed the insecticide action. If we compare studies conducted with *Curcuma aromatica* (Choochate et al., 2005) and *Curcuma zedoaria* in *Aedes aegypti* these oils show lower values of LC_{50} and CL_{99} (Champakaew et al, 2007).

Within the family Piperaceae, notably members of the genus Piper have noted insecticidal action, molluscicide (Chansang et al., 2005; Parmar et 1997) this is due to the presence of the secondary metabolites as alkaloids, phenylpropanoids, lignans, neolignanos, terpenes, flavononas, among others (Pino et al, 2004; Smith & Kassim 1979). *Piper aduncum* specifically shows in its composition a high percent of dilapiol, α-pinene and 1.8- cineole (Bottia et 2007). In Mexico studies were not mortalities obtained with a kind of Piper in *Culex quinquefasciatus* in aqueous extracts to 5 and 15% where used in the aerial part of plants (Perez-Pacheco et al 2004).

The larvicidal activity of *Chenopodium ambrosioide* was evaluated in *Ae aegypti*, with extracts metanolics to different concentrations resulting most important inhibition of pupal development and an increase in the time of larval development (Suparvarn et al 1974) has also been evaluated in mosquito repellent activity (Gillij et al 2008). *Ch ambrosiodes* has as one of its components majority the carvacrol reported in the literature as insecticide terpene (Silva et al 2008; Kordali et al 2008). *Pimenta racemosa, Eugenia melanadenia* and *Psidium rotundatum* presented in its composition more than 50% of monoterpenes and sesquiterpenes standing out in his majority composition 1,8 cineole, 4-terpineole and α-pinene terpenes with high insecticidal action (Pino et al, 2004,2005).

Oil of turpentine in *Aedes aegypti* obtained strong insecticide and inhibitory action of development in previous studies (Lucia et al, 2007; Leyva et al, 2009b 2010). In this study isolated terpene α -pinene (one of the major components of turpentine) was higher than the own turpentine lethal concentrations, this may be due to the synergistic effect with several monoterpens (α and β -pinene) at the same time and not isolated and evaluated by themselves in insects.

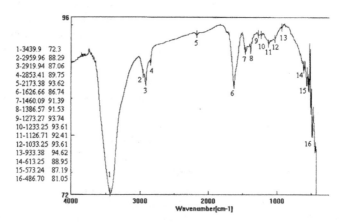

Fig. 5. Spectroscopic characterization of the α pinene.

Conc (%)	Mortality (%)	Lethal Conc		Slope of the Line
				Turpentine
0.0035	5	CL$_{50}$	0.0042	20.9
0.0040	31			Diagnostic
0.0050	94	CL$_{95}$	0.0055	Dose
0.0060	100			0.0108
				(108 mg/L)
				Modified Turpentine
0.0030	5	CL$_{50}$	0.0038	14.4
0.0040	60			Diagnostic
0.0050	93	CL$_{95}$	0.0050	Dose
0.0060	100			0.0112
				(112 mg/L)

α-pinene				
0.0040	35	CL$_{50}$	0.0052	3.7
0.0050	46			Diagnostic
0.0060	56	CL$_{95}$	0.014	Dose
0.0070	67			0.044
0.0080	78			(440 mg/L)

Table 1. Concentrations used and rates obtained with derivatives of Pinaceae in *Cx quinquefasciatus* larvae.

Conc (%)	Mortality (%)	Lethal Conc		Slope of the Line
		Chenopodium ambrosoide		
0.0040	10			11.8
0.0050	42	CL$_{50}$	0.0051	Diagnostic
0.0060	80			Dose :
0.0070	95	CL$_{95}$	0.0071	0.0162
0.0080	98			(162 mg/L)
		Curcuma longa		
0.0040	46			5.22
0.0050	66	CL$_{50}$	0.0041	Diagnostic
0.0060	80			Dose :
0.0070	86	CL$_{95}$	0.0086	0.0232
0.0080	94			(232 mg/L)
		Psidium rotundatum		
0.0040	28			8.16
0.0050	58	CL$_{50}$	0.0047	Diagnostic
0.0060	80			Dose
0.0070	92	CL$_{95}$	0.007	0.018
0.0080	97			(180 mg/L)
		Piper aduncum		
0.0100	36			6.72
0.0120	53	CL$_{50}$	0.0110	Diagnostic
0.0140	73			Dose
0.0160	85	CL$_{95}$	0.0200	0.050
0.0180	90			(500 mg/L)
0.0200	95			
		Pimenta racemosa		
0.0040	10			5.23
0.0050	20	CL$_{50}$	0.0070	Diagnostic
0.0060	44			Dose
0.0070	50	CL$_{95}$	0.0140	0.0380
0.0080	62			(380 mg/L)
0.0100	79			
		Eugenia melanadenia		
0.0240	14			18.5
0.0260	41	CL$_{50}$	0.026	Diagnostic
0.0280	70			Dose
0.0300	78	CL$_{95}$	0.033	0.0718
0.0320	90			(718 mg/L)

Table 2. Concentrations used and rates obtained with essential oils tested in *Cx quinquefasciatus* larvae.

4. Conclusions

An interest in natural products from plants has been increased due to the development of resistance to synthetic insecticides, which are applied in order to reduce the populations of insects.

The biological activity of natural compounds is based on its structure and the doses used for such purposes. Vegetable insecticides have the great advantage of being compatible with other acceptable low-risk options from the ecological point of view on the control of insects. The secondary metabolites produced by plants against the attacks of predators and insects make them natural candidates in the control of species of insects, both vector of diseases and pests of agriculture. It is not logical to come to jump to the idea that they will completely replace the synthetic insecticides. Logical thinking is to have in them a complementary use to optimize and increase the sustainability of current integrated pest control strategies.

There are many publications of lists of plants with insecticidal properties. To use such plants, it is not enough to be regarded as promising or proven insecticidal properties. Analysis of risks to the environment and health should also be made. It is not to recommend the use of plants that are endangered, with limited biomass or that their use involves major alterations to the density in which they are in the nature. An ideal insecticide plant must be perennial, be widely distributed and in large amounts in nature or that can be cultivated, using renewable plant bodies such as leaves, flowers or fruits, not be destroyed every time you need to collect material to (avoid the use of roots and bark), agro-technitian minimum requirements and be eco-sustainability, have additional uses (such as medicines), not having a high economic value, be effective at low doses, possess potential scaling biotechnology (Silva;2002).

Insecticide plants have the advantage of having other uses as medicinal, a rapid degradation which decreases the risk of residues in food and therefore can be more specific for pest insect and less aggressive with natural enemies. They also develop resistance more slowly in comparison with synthetic insecticides. By the other hand, the disadvantages include that they can be degraded more quickly by ultraviolet rays so its residual effect is low, however not all insecticides from plants are less toxic than synthetic and residual is not established.

We can conclude that studies in our country, essential oils and their derivatives from the different families of evaluated plants have a high insecticide activity on larvae of *Aedes aegypti* and *Culex quinquefasciatus* and derived specifically from pinaceas, they have ovicide and inhibitory action of development in *Aedes aegypti*, which are potential candidates for control alternatives in these species of insects.

5. References

Adam, KP; Thiel, R; Zapp, J; Becker, H. (1998). Involment of the mevalonic acid pathway and glyceraldehyde-pyruvate pathway in the terpenoid biosynthesis of the liverworts *Ricciocarpos natansand Conocephalum conicum*. *Archives of Biochemistry and Biophysics*. 354: 181-187.

Aguilera, L; Navarro, A; Tacoronte, JE; Leyva, M ;Marquetti MC. (2003) Efecto letal de Myrtaceas cubanas sobre *Aedes aegypti* (Díptera: Culicidae) *Rev Cubana Med Trop*;55(2):100-4.

Aguilera L, Tacoronte JE, Navarro A, Leyva M, Bello A, Cabreras MT. (2004) Composición química y actividad biológica del aceite esencial de *Eugenia melanadenia* (Myrtales:

Myrtaceae) sobre *Blattella germanica* (Dictyoptera: Blattellidae). *Revista CENIC Ciencias Químicas.* : 35(3).

Alburquerque, MR; Silveira,ER; De Ochoa, DE; Lemus, TL; Souza, EB; Santiago, GM; Pessoa, OD. (2004) Chemical composition and larvicidal activity of the essential oils from *Eupatorium bentonicaea* Baker (Astaraceae) *J Agric Food Chem.*;52 (22): 6708-11.

Amer, A & Mehlhorn, H. (2006) Repellency effect of forty-one essential oils against *Aedes, Anopheles* and *Culex* mosquitoes. *Parassitol Res.*99 (4):478-90.

Amer, A & Methlhorn, H. (2006).Larvicidal effects of various essentials oils against, Aedes, Anopheles and Culex larvae (Diptera: Culicidae). *Parasitol Res.*99(4):466-72.

Ansari, MA; Mittal, PK; Razdan, RK; Sreehari, U. (2005) Larvicidal and mosquito repellent activities of Pine (*Pinus longifolia*, Familia: Pinaceae) oil. *J Vect Borne Dis.* 95-99.

Assarm, AA; El-Sobky, MM. (2003) Biological and histopathological studies of some plant extracts on larvae of *Culex pipiens* (Diptera:Culicidae) *J. Egypt Soc Parasitol l.*33(1):189-200.

Awde, JA; Ryan, MF (1992) Piperazine as a reversible competitive inhibitor of acetylcholinesterase. *Cienc. Biol.Ecol.Syst.* 12: 37-42.

Barnard, DR & Xue, RD. (2004) Laboratory evaluation of mosquito repellent against *Ae albopictus, Culex nigripalpus* (Diptera: Culicidae). *J Med Entomol.* 41 (4): 726-30.

Bassole, IH; Guelbeogo, WM; Nebie, R; Constantini, C; Sagnon, N; Kabore, Z; Traore, SA. (2003) Ovicidal and larvicidal activity against *Ae aegypti* and *An gambiae* complex mosquitoes of essential oils extracted from three spontaneus plants of Burkina Faso. *Parassitologia* 45 (1):23-6.

Bottia, E; Díaz, O; Mendivelso, D; Martínez, J; Stashenko, E. (2007) Comparación de la composición química de los metabolitos secundarios volátiles de cuatro plantas de la familia Piperaceae obtenidos por destilación–extracción simultánea. Scientia et Technica;13(33). ISSN 0122-1701

Bloomquist, J. (1999) Insecticides: Chemistries and Characteristics. Departmentof entomology. Virginia Polytechnic institute and state university. Blacksslomg Virginia. In http://ipmworld.umn.edu/chapters/bloomq.htm.

Byeoung-Soo, P; Won-Sik, C; Jeong-Han, K; Kap-Ho, K; SPNG-Eun, L. (2005). Monoterpenes from Thyme (Thymus vulgaris) as potencial mosquito repellents. *J Am Mosq. Control Association* .21 (1);80-83.

Caballero-Gallardo, K; Olivero-Verbel, J; Stashenko, EE.(2011) Repellent Activity of Essential Oils and Some of Their Individual Constituents against *Tribolium castaneum* Herbst.*J Agric Food Chem.* Feb 3. [Epub ahead of print]

Carvalho, AF; Melo, VM; Carveiro, AA; Machoo, MI; Beantim, MB; Rabelo, EF. (2003) Larvicidal activity of the essential oil fro *Lippia sidioide* against *Aedes aegypti. Mem Oswaldo Cruz.*98 (4): 569-71.

Cavalcanti, ES; Morais, SM; Lima, MA; Santana, EW. (2004) Larvicidal activity of essentials oils from Brazilian plants against *Ae aegypti. Mem Inst Oswaldo Cruz.*99(5):541-4.

Cetin H, Yani A. Study of the larvicidal activity of Origanum (Labiatae) species from Southwest Turkey. *J. Vector Ecol.* 2006; 31 (1): 118-22.

Chaiyasit, D; Choochate, W; Rattanachanpichai, E; Chaithong, U; Chaiwong P et al. (2006) Essential oils as potencial adulticides against two populations of *Aedes aegypti*, the laboratory and natural field strain in Chiang Mai province northern Thailand. *Parasitol Res* 1

Chang, KS; Talk, JH; Kim, SI; Lee, WJ; Ahn, YJ. (2006) Repellency of *Cinnamomum cassia* bark compounds and cream countain cassia oil to de *Ae aegypti* under laboaratory and indoor conditions. *Pest Manag Sci*. 62 (11):1032-1038.

Choochate, W; Chaiyasit, D; Kanjarapotti, D; Rottana, E; Champiochi, E et al (2005). Chemical composition and antimosquito potencial of rhizome extract and volatile oil derivade from *Curcuma aromatica* against *Aedes aegypti* (Diptera: Culicidae). *J Vector Ecol*.30 (2):302-9.

Champakaew, D; Choochote, W; Pongpaibul, Y; Chaithong, U; Jitpakdi, A; Tuetun, B; Pitasawat, B. (2007) Larvicidal efficacy and biological stability of a botanical natural product, zedoary oil-impregnated sand granules, against *Aedes aegypti* (Diptera, Culicidae). *Parasitol Res*.100(4):729-37.

Chansang, U ; Zahiri ,NS ; Bansiddhi, J ; Boonruad, T ; Thongsrirak, P ; Mingmuang, J ; Benjapong , N ; Mulla, MS. (2005) Mosquito larvicidal activity of aqueous extracts of long pepper (*Piper retrofractum* vahl) from Thailand. *J Vector Ecol*.30(2):195-200.

Cheng, SS; Liu, JY; Tasai, KH ; Chen, WJ; Chang, ST. (2004) Chemical composition and mosquito larvicidal activity of essential oils from leaves of different *Cinnamomum osmophloem* provenanaces. *J Agric Food Chem*.52 (14):4395-400.

Chung, IM; Song, HK; Yeo, MA; Moon, HI. (2010) Composition and immunotoxicity activity of major essential oils from stems of *Allium victorialis* L. var. platyphyllum Makino against *Aedes aegypti* L. *Immunopharmacol Immunotoxicol*. Dec 16. [Epub ahead of print]

Chung, IM; Ro, HM; Moon, HI (2010) Major essential oils composition and immunotoxicity activity from leaves of *Foeniculum vulgare* against *Aedes aegypti* L. *Immunopharmacol Immunotoxicol*. Nov 16. [Epub ahead of print]

Corbino, G . (2000) Productos naturales vegetales de interés económico Instituto Nacional de Tecnología Agropecuaria. Estación Experimental San Pedro. Buenos Aires, Argentina.Lab.Prod.InVitro.(11-2-2005)In :   http://www.inta   .gov.ar/sanpedro /1_sala_de_lectura/difusion/novedades.../corbinocondim.ht.

Das, NG; Barauah, I; Talukdar, PK; Das, SC. Evaluation of botanicals as repellent against mosquitoes. *J Vect Borne Dis* ;40 (1-2): 49-53.

De Mendoca, FA; Da Silva, KF; Dos Santos, KK; Ribeneiro Junior, KA; Sant'Ana, AE. (2005) Activities of some brazilian plants against larvae of the mosquito *Ae aegypti.Fitoterapia*;76 (7-8):629-36.

Dharmagada, VS; Naik, SN; Mittal, PK; Vasudevan, P. (2005) Larvicidal activity of *Tagetes patula* essential oil against three mosquito species. *Bioresour Technol* 96 (11);1235-40.

Faley, DH & Frances, SP. (2005) Laboratory evaluation of coconut oil as a larvicide for *Anopheles faraoti* and *Culex annulirostris*. *J Am Mosq.Control Assoc*.;21(4): 477-9.

Gascón, AD; Pelayes, MA. (2002). *Generalidades sobre los procesos extractivos en la obtención de aceites esenciales* Departamento de Tecnología Agroindustrial Facultad de Ciencias Agrarias Chacras de Coria – Mendoza.

Gillij, YG ; Gleiser, RM ; Zygadlo, JA. (2008) Mosquito repellent activity of essential oils of aromatic plants growing in Argentina . *Bioresour Technol*.99(7): 2501-15.

González, P.(1984) Capítulo VI In *"Utilización Terapéutica de Nuestras Plantas Medicinales"*, Universidad de La Salle, Bogotá.

Gunderson, CA; Samuelian, JH; Evans ,CK. (1985) Effects of the mint monoterpene pulegone on *Spodoptera eridinan* (Lepidoptera: Noctuidae). *Environ Entomol*.14:859-63.

Hafeez, F; Akram, W; Shaalan, EA. (2011).Mosquito larvicidal activity of citrus limonoids against *Aedes albopictus*. *Parasitol Res*. Jan 7. [Epub ahead of print]

http://www.buscasalud.com/boletín/analisis/2003_01_1117_09_32.html (12-1-2005)

http://www.chemkeys.com/esp/md/tee_6/ecoqui_2/metqui_5/metqui_5.htm (8-1-2007)

http://www.herbotecnia.com.ar/poscosecha-esencias.html (8-1-2007)

http://www.losandes.com.ar/imprimir.asp?nrc=163985   (8-1-2007)

http://www.cannabiscultura.com/f1/showthread.php?p=84457 (4-3-2011)

Ikan, R.(1991) Natural products. A laboratory guide. Second Edition. Academia Press (Eds).360 pp

Jaenson, TG; Palsson, K; Borg-Karlson, AK. (2006) Evaluation of extracts and oils of mosquito (Diptera: Culicidae) repellent plants from Sweden and Guinea Bissau. *J Med. Entomol*.43 (1):113-9.

Judd, WS; Campell, CS; Kellogg, EA; Stevens, PF; Donoghue, MJ. (2002) Secondary plants compounds. Plant systematic: a phylogenetic approach. Chapter 4, Structural and Biochemical Characters. Second Edition. Sinaeur Axxoc, USA.

Keane, S;Ryan, MF. (1999). Purification, characterization and inhabitation by monoterpenesof acetylcholinesterase from the waxmoth *Galleria mellonella*. *Insect Biochemistry and Molecular Biology* 29:1097-1104.

Khandagle, AJ; Tare, VS; Raut, KD; Morey, RA. (2011).Bioactivity of essential oils of *Zingiber officinalis* and Achyranthes aspera against mosquitoes. *Parasitol Res*. Feb 11. [Epub ahead of print]

Kim, SI; Chang, KS; Yang, YC; Kim, BS; Ahn, YJ. (2004) Repellency of aereosol and cream products containg fennel oil to mosquitoes under laboratory and field conditions. *Pest Manang Sci*.;60 (11):1125-30.

Kordali, S; Cakir, A; Ozer, H; Cakmakci, R; Kesdek, M; Mete, E. (2008)Antifungal, phytotoxic and insecticidal properties of essential oil isolated from Turkish *Origanum acutidens* and its three components, carvacrol, thymol and p-cymene. *Bioresour Technol*. May 28.

Kostyukovsky, M; Rafaeli, A; Gileadi, C; Demchenko, N; Shaaya, E. (2002). Activation of octopaminergic receptors by essential oil constituent isolated from aromatic plants: posible mode of action against insect pest. *Pest Management Science*. 58: 1101-06.

Leyva M , Aguilera L, Tacoronte J, Montada D, Bello A, Marquetti MC. (2007) Estudio de laboratorio del aceite esencial de *Pimenta racemosa* (Myrtales: Myrtaceae) y su posible utilización para el control de *Musca domestica* (Diptera: Muscidae) *Rev CNIC Ciencias Químicas*

Leyva M, Tacoronte J, Marquetti MC.(2007) Composición química y efecto letal de *Pimenta racemosa* (Myrtales: Myrtaceae) en *Blatella germanica* (Dictyoptera: Blattellidae) Rev. Cub. Med. Trop . 59 (2)

Leyva M, Tacoronte JE, Marquetti MC, Scull R, Montada D, Rodríguez Y, Bruzón R.(2008)Actividad insecticida de aceites esenciales de plantas en larvas de *Aedes aegypti* (Diptera: Culicidae) *Rev Cubana Med Trop* 60(1):78-82.

Leyva M, Tacoronte JE,Marquetti MC, Montada D.(2008) Actividad insecticida de tres aceites esenciales en *Musca domestica* (Diptera: Muscidae) Rev Cubana Med Trop 60(3).

Leyva M, Tacoronte JE, Marquetti MC ,Scull R , Tiomno O , Mesa A, Montada D. (2009) Utilización de aceites esenciales de pinaceas endémicas como una alternativa de control en *Aedes Aegypti Rev .Cub . Med.Trop*;61(3):239-43.

Leyva M, Marquetti MC, Tacoronte JE, Scull R , Tiomno O , Mesa A, Montada D. (2009)Actividad larvicida de aceites esenciales de plantas contra *Aedes aegypti* (L) (Díptera: Culicidae) *Biomed* 20: 5-13.

Leyva M, Marquetti MC, Tacoronte JÁ, Tiomno O, Montada D (2010) Efecto inhibidor del aceite de trementina sobre el desarrollo de larvas de *Aedes aegypti* (Diptera: Culicidae) *Rev.Cub.Med.Trop.*;62(3):

López, MD.(2008) Toxicidad volatil de monoterpenoides y mecanismos bioquímicos en insectos plagas del arroz almacenado. Tesis Doctoral. Universidad de Murcia. Facultad de Quimica. España 245 pp

Lucia, A; Gonzalez Audino, P; Saccacini, E; Licastro, S; Zerba, E; Masuh, H (2007).Larvicidal effect of *Eucalyptus grandis* essential oil and turpentine and their major components on *Ae aegypti* larvae *J.Am.Mosq Control Assoc.*23:293-303

Mathew, J; Thoppil, JE.(2011) Chemical composition and mosquito larvicidal activities of Salvia essential oils. *Pharm Biol.* Feb 2. [Epub ahead of print]

Miot, HA; Batistilla, RF; Batista, K; de A Volpato, DE, Augusto, LS; Madeira, NG; Haddad, V; Miot, LD. (2004) Comparative study of the topical effectiveness of the Andiroba oil *(Coparais guianensis)* and DEET 50 % as repellent for Ae sp. *Rev Inst Med. Trop Sao Paaulo.*46 (5): 253-6.

Morais, SM; Cavalvanti, ES; Bertini, LM; Oliveira, CL; Rodríguez, JR; Cardoso, JH. (2006) Larvicidal activity of essential oils from Brazilian Croton sepecies against *Ae aegypti*. *J Am Mosq. Control Assoc.* 22(1):161-4.

Odalo, JO; Omolo, MO; Malebo, H; Angira, J; Njeru, PM; Ndiege, IO; Hassanali ,A. (2005) Repellency of esential oils of some plants from Kenya coast against *An gambiae*. *Acta Tropical.*95 (3):210-8.

Oyedale, AO; Gbolade, AA; Sosan, MB; Adewoyin, FB; Sayelu, OL; Orafidaya, OO. (2002) Formulation of an effective mosquito-repellent topical product from lemon gross. *Phytomedicine.*9(3):259-62.

Parmar, V.S; Jain, S.C; Bisht, K.S; Jain, R; Taneja, P; Jha, A ;et al. (1997) Phytochemistry of the genus *Piper*. *Phytochemistry.* 46:597-673

Pérez-Pacheco, R; Rodríguez –Hernández, C ; Lara-Reyna, J; Montes –Belmont, R; Ramírez – Valverde, G. (2004) Toxicidad de aceites, esencias y extractos vegetales en larvas de mosquito *Culex quinquefasciatus* Say (Diptera: Culicidae) *Acta Zoológica Mexicana.* 20(1): 141-152.

Phasomkusolsil, S; Soonwera, M. (2010) Potential larvicidal and pupacidal activities of herbal essential oils against *Culex quinquefasciatus* say and *Anopheles minimus* (theobald).*Southeast Asian J Trop Med Public Health.* Nov;41(6):1342-51.

Phasomkusolsil, S; Soonwera, M. (2010) Insect repellent activity of medicinal plant oils against *Aedes aegypti* (Linn.), *Anopheles minimus* (Theobald) and *Culex quinquefasciatus* Say based on protection time and biting rate.*Southeast Asian J Trop Med Public Health.* Jul;41(4):831-40.

Pino, J. Borges P. (1999) Los componentes volátiles de las especias. I. Métodos de obtención y de análisis. *Alimentaria.*30 (1) 39-45.

Pino, JA; Marbot, R; Bello, A; Urquiola, A (2004) Essential Oils of *Piper peltata* (L.) Miq. and *Piper aduncum* L. from *Cuba Journal of Essential Oil Research: JEOR*, Mar/Apr 2004

Pino, J A; Marbor, R; Bello, A;Urquiola, A. (2005) Essential oil of *Eugenia melanadenia* Krug et Urb. from Cuba". *Journal of Essential Oil Research* JEOR. FindArticles.com. (09-2-2011) In: . http://findarticles.com/p/articles/mi_qa4091/is_200307/ai_n9287900/

Pitarokili, D; Michaelakis, A; Koliopoulos, G; Giatropoulos, A; Tzakou, O. (2011). Chemical composition, larvicidal evaluation, and adult repellency of endemic Greek Thymus essential oils against the mosquito vector of West Nile virus. *Parasitol Res.* Feb 8.

Pohlit, AM; Lopes, NP; Gama, RA; Tadei, WP; de Andrade-Neto, VF. (2011) Patent Literature on Mosquito Repellent Inventions which Contain Plant Essential Oils - A Review.*Planta Med* Feb 15. [Epub ahead of print]

Prajapati, V; Tripathi, AK; Aggarwall, KK; Kanuja, SP (2005) Insecticidal repellent and oviposition-deterrente activity of selected essential oils against *An stephensi, Ae aegypti* and *Cx quinquefasciatus. Bioresour Technol.* 96 (16): 1749-57.

Pushpanathan, T; Jebanesan, A; Govindarajan, M. (2006) Larvicidal, ovicidal and repellent activities of *Cymbopogan citratus* Stapf (Graminae) essential oil against the filarial mosquito *Culex quinquefasciatus* (Say) (Diptera : Culicidae). *Trop Biomed.* 23(2):208-12.

Ravi, K; Bhavani, K; Sita Devi, P; Rajaswara Rao, BR; Janarahan, RK. (2006) Composition and larvicidal activity of leave and stem essential oil of *Chloroxylon swetenia* against *Ae aegypti* and *An stephensi. Bioresour Technol.* 97(18):24-31.

Ryan, MF; Byrne, O (1988) Plant-insect coevolution and inhibition of acetilcholinesterase. *Journal of Chemical Ecology* 14:1965-75.

Ryan, MF; Awde, J; Moran, S. (1992) Insect pheromones as reversible competitive inhibitors of acetilcholinesterase. *Invertebrate reproduction and development.* 22:31-38.

Scholes, M. (1995). *La Terapia de los Aromas.* Sterling Press, New York.

Stamm, M. D. (1970).Modernos Conocimientos sobre la Química de los Aceites Esenciales. Métodos de Análisis. *C.S.I.C. Manuales de Ciencia Actual 3.* Patronato "Alonso de Herrera", Madrid.

Sangkyun L; Tsao, R; Coats, J. (1999) Influence of dietary applied monoterpenoids and derivates on survival and growth of european corn borer (Lepidoptera: Pyralidae). *J Econ Entomol.*92(1):56-67.

Smith, RM & Kassim, H. (1979) The essential oil of *Piper aduncum* from Fiji. New Zealand. *J Sci* . 22:127-128.

Sponsel VM. (1995). The biosyntesis and metabolism of gibberellins in higher plants. In Plant hormones. Davies PJ (Ed). Kuwer Academic Publishers Boston. 66-97 pp.

Suparvarn, P; Knapp, F; Sigafus, R. (1974) Biologically active plant extracts for control of mosquito larvae. *Mosquito News.*34(4):398-402

Silva, WJ; Dória, GA; Maia, RT; Nunes, RS; Carvalho, GA; Blank, AF; Alves, PB; Marçal, RM; Cavalcanti, SC. (2008) Effects of essential oils on *Aedes aegypti* larvae: Alternatives to environmentally safe insecticides. *Bioresour Technol.*;99(8):3251-5

Silva, G. (2002) *Insecticidas vegetales.* (11-12-2006) In: http://ipmworld.umn.edu/ cancelado /Spchapters/GSilvaSp.htm

Silva, G.(2003) Resistencia a los insecticidas In: Silva, G y Hepp, R (Eds). Bases para el manejo racional de insecticidas. Universidad de Concepción, Fundación para la Innovación Agraria: 237-60pp

Stashenko, E. (1996) In: *Memorias del IV Congreso Nacional de Fitoquímica,* Universidad Industrial de Santander, Escuela de Química, Bucaramanga, febrero de 1996, pp. 29-53.

Tare, V; Deshpade, S; Sharma, RN. (2004) Susceptibility of two different strains of *Ae aegypti* (Diptera: Culicidae) to plants oils. *J Econ. Entomol.* 97 (5):1734-6.

Tawatsin, A; Asavadachanukorn, P; Tahavara, U; Wongsenkongman, P; Bansidhi, J; Boonroad, J. (2006) Repellency of essential oils extracted from plantas in Thailand against four mosquito vectors (Diptera: Culicidae) *Southeast Asian J Med Trop. Med Public Heath.* 37 (5):915-31.

Thomas, TG. (2004) Mosquito larvicidal properties of essential oil of indigenus plant *Ipomoea airica* L. *Jpn J Infect Dis.*57 (4).

Trongotokit, Y; Curtis, CF; Rangsriyum, Y. (2005) Efficacy of repellent products against caged and free flying *An stephensi* mosquitoes Southeast Asian. *J Trop Med. Public Health.* 36 (6): 1423-31.

Trongotokit, Y; Rangsriyum, Y; Komalamisra, N; Apiwathnarson, C. (2005) Comparative repellency of 38 essential oils against mosquito bites. *Phytother Res.*; 19(4):303-9.

Worwood, V. (1992) *The Complete Book of Essential Oils and Aromatherapy.* New World Library, New York.

Yang, P& Ma, Y. (2005) Repellent effects of plants essential oils against *Ae albopictus. J Vector Ecol.* 30 (2): 231-4.

Zhu, J; Zeng, X; Yanma, J; Liu, J; Qian, K; Tucker, B; Schultz, G; Coats, J; Zang, A. (2006) Adults repellency and larvicidal activity of five planta essential oil against mosquitoes. *J Am Mosq Control Assoc* 22(3):515-22.

11

Susceptibility Status of *Aedes aegypti* to Insecticides in Colombia

Ronald Maestre Serrano
Doctoral Student in Tropical Medicine
Cartagena University
Colombia

1. Introduction

Dengue is a disease of public health interest for Colombia due to its impact on morbidity and mortality. This disease mantains an endemic behavior in the departments of the country located below 2200 meters above sea level. Between the years 1978 to 2010 it was recorded approximately a number of 1,011,852 cases of dengue in Colombia. Vector populations of *Aedes aegypti* in the country have been continuously pressed for over five decades with insecticides, being the most widely used tool for interrupting transmission of dengue virus during outbreaks and epidemics in the absence of an available vaccine (Figure 1). The selection pressure excercised with these chemicals has led to the emergence of resistant populations of the vector to organochlorine, organophosphate, pyrethroid and carbamate-like molecules.

Fig. 1. Application of insecticides in the home by fogging durin outbreaks or epidemics.

The records of vector resistance in the country have increased considerably since 2007 when in Colombia was created the national network for the surveillance of susceptibility to insecticides used in public health to *A. aegypti* and major vector of malaria. With this information a base line has been defined for some Colombian departments in the region and basic knowledge has been generated for decision making from the operating standpoint of vector control programs. However, more research on issues related to the enzymatic mechanisms and molecular causes of resistance in vector populations is required as most results to date have been from diagnostic doses through standardized techniques by the Centers for Disease Control and Prevention (CDC) and World Health Organization (WHO). In this chapter these results are socialized and we discuss some ideas related to the topic.

2. Study area

The continental territory of the Republic of Colombia is located in the northwest corner of South America on the equator. Colombia is located in the north to 12°26'46" north latitude at the place called Punta Gallinas in the peninsula of Guajira. To the south, is located at 4° 12' 30" de latitude south, the East is located at 60°50'54" west longitude from Greenwich, by the West is located at 79°02' 33" west longitude from Greenwich.

Colombia territory also includes the archipelago of San Andres and Providencia in the Caribbean sea between 12° and 16°30´north latitude and 78° and 82° west longitude of Greenwich. Colombia has a land area of 1.141.748 km² ranking fourth among the countries of South America.

Colombia is a country with a wide variety of climates, from the coldest to 0°C to temperatures exceeding 28°C especially in the Caribbean regions, Pacific, Amazon and the Orinoco, accounting for nearly 80% of Colombia territory. Colombia is divided administratively into 32 states, these in turn are divided into 1102 municipalities.

3. The situation of dengue in Colombia

In Colombia, dengue is a disease of public health interest for its impact on morbidity and mortality. Its endemic behavior is due to endemic multiple factors including the re-emergence and intense viral transmission with a rising trend, the behavior of epidemic cycles getting shorter, the increased frequency of outbreaks and serious of the disease, the simultaneous circulation of the four serotypes, infestation by *A. aegypti* in over 90% of the national territory located below the 2,200 meters, and other social factors such as uncontrolled increase and lack of basic sanitation in urban centers (Escobar, 2009). The epidemiological pattern of disease in the last decades has been upward, characterized by exponential increase in endemic areas during the different decades. Cyclical behavior has been characterized by epidemic peaks every three or four years, dealing with the reentry of new serotypes to the country.

In Colombia from 1978 to 2010, has officially been recorded a total of 1.011.852 cases of dengue. For this time period there is record history of the following epidemics in the country:

1977: First recorded epidemic of dengue in Colombia.

2002: Epidemic in 76,996 cases, of which 5269 corresponded to Dengue hemorrhagic fever.

2007: Epidemic in 43,227 cases, of which 4665 corresponded to Dengue hemorrhagic fever.

2010: It is recorded in the country the worst epidemic of this disease with 157,152 confirmed cases and 217 deaths.

The departments that have historically had more transmission of dengue in the country are: Atlántico, Santander, Norte de Santander, Valle del Cauca, Antioquia, Tolima, Huila, Cundinamarca, and Casanare.

4. Resistance to insecticides

4.1 Definition
Resistance is defined as the development of the ability to tolerate doses of toxics, which can be lethal to most individuals in a normal population of the same species and is the result of positive selection pressure exerted by the insecticidal over genes initially in low frequency (WHO, 1957).

4.2 Mechanisms of resistance
The two main resistance mechanisms are alterations in the target site and metabolic resistance, also called increase in the rate of detoxification of insecticides. However, there are other less frequent as there are the resistance behavior and decreased penetration. (Miller, 1988; Bisset, 2002; Fonseca and Quiñones, 2005)

4.2.1 Altered target site
This is generated when no silent mutations occur in structural genes that generate a disturbance of amino acids responsible of anchor of the insecticide at a specific site.

4.2.1.1 GABA Receptor

In bugs this is an hetero multi numeric channel acting as a site of action for cyclodiene and avermectins. The resistance is given by *Rdl* gene (Resistance to dieldrin), which encodes for RDL, a GABA receptor subunit. This type of resistance was first identified in *Drosophila* first identified this type of resistance (Fonseca and Quiñones, 2005).

4.2.1.2 Voltage-gated sodium channel

They are the target of action of pyrethroids and DDT. The sodium para-channel protein is a complex of over 2000 amino acids, composed of 4 homologous domains separated by hydrophilic links, each domain contains six segments. The first mutation in the sodium channel that conferred resistance was detected in *Musca domestica*. subsequently mutations in other insects have been identified including Culicidae resistant to pyrethroids (Martínez et al, 1998; Brengues et al, 2003; Anstead, 2005; Saavedra et al, 2007; Brooke, 2008; Chang et al 2009; Yanola et al, 2010).

4.2.1.3 Insensitive acethylcholinesterase

Insecticide resistance, attributed to Ache insensitivity is found in a number of *Anopheles* and *Culex* species. In general, this mechanism produces a broad spectrum resistance to most organophosphates and carbamates, although more pronounced to carbamatos. (Bisset, 2002).

4.2.2 Metabolic resistance
It is conferred by an increase in detoxification insecticide or an inability to metabolize the toxic compound. The most important form of metabolic resistance is given by detoxificating enzymes of the type glutathione S-transferase, mixed function oxidases and esterases (Bisset 2002).

4.2.2.1 Carboxylesterases

They Catalyze the hydrolysis of carboxylic esters and changes in their expression levels, this is the resistance mechanism that occurs most often in insects. In mosquitoes, high levels of these enzymes have been associated with resistance to organophosphate and pyrethroid insecticides (Cui et al, 2007; Rodriguez et al, 2007). Esterases is a family of six proteins

Esterases are a family of six proteins grouped in α and β hydrolases superfamily. In the Diptera these enzymes are encoded by a gene cluster on the same chromosome, each gene may suffer modification that confers resistance to the insecticide (Fonseca and Quiñones, 2005; Santacoloma, 2008).

4.2.2.2 P450 Enzymes or mixed function oxidases

P450 enzymes are encoded by cytochrome P450 genes. P450 enzymes also have other names such as: cytochrome P450 monooxygenases, mixed function oxidases, monooxygenases with polisustrato, microsomal oxidases and heme-thiolate proteins. In insects these enzymes are involved in growth and development through the processing of fatty acids, hormones and pheromones; in the metabolism of secondary plants products and synthetic chemicals such as insecticides. The cytochrome P-450 is implicated as the major factor in many cases of metabolic resistance to carbamates and detoxify organophosphates, pyrethroids and DDT as well. It expresses in tissues of the gut, fat, reproductive tract and Malpighian tubes (Feyereisen, 1999; Poupardin et al, 2008).

4.2.2.3 Glutatione s-transferase

The glutathione s-transferases (GSTs, EC 2.5.1.18) are phase II enzymes involved in xenobiotics detoxification in many organisms. These enzymes metabolize a wide range of toxic hydrophobic compounds such as drugs, insecticides and endogenous toxic substrates, catalyzing the conjugation of glutathione to the hydrophilic center of toxic substances, allowing the increase in compounds solubility. The GSTs are divided into three main groups: the cytosolic, microsomal and mitochondrial. The mitochondrial GSTs have not been found in insects species including mosquitoes. The cytosolic GSTs in insects are grouped into six different kinds: Delta, Epsilon, Omega, Sigma, Theta and Zeta. Most GSTs involved in the metabolism of xenobiotics in insects belong to the class Delta Epsilon. GSTs in insects have been implicated in the insecticide resistance through its direct metabolism, sequestration or protection against secondary toxic effects such as increased lipid peroxidation induced by exposure to insecticides. Glutathione S- transferase mediated detoxification has been reported for insecticides of the organophosphate type and DDT. The main role of GSTs in the resistance to organophosphates is the detoxification of the insecticide by a conjugation reaction. The specific GSTs also catalyze the metabolism of DDT to a non-toxic substance: 1,1-dichloro-2 ,2-bis-(p-chlorophenyl) ethane (DDE), through the process of dehydrochlorination. Epsilon class of GST has been involved in resistance to DDT in *Anopheles gambiae* and *A. aegypti* (Lumjuan et al, 2007).

4.3 Type of resistance
4.3.1 Cross resistance

It occurs when a single gene confers resistance to a number of chemicals in the same group, as it is the case of phosphotriesterases that provides resistance to several organophosphates or *kdr* gene that confers resistance DDT and the pyrethroids (WHO, 1957).

4.3.2 Multiple resistance

It occurs when two or more resistance mechanisms are operating in the same insect. The multiple resistance term does not necessarily involve the term cross-resistance, because a bug may be resistant to 2 or more insecticides and each resistance can be attributed to different mechanisms (Bisset, 2002).

4.4 Techniques used to detect resistance insecticides in mosquitoes
4.4.1 Biological test of the World Health Organization (WHO)
Technique standardized by the World Health Organization (WHO). In adult mosquitoes the technique consists of exposing individuals of the populations we intent to evaluate to papers impregnated with a single and specific dose of an insecticide during a determined time (1 hour with the exception of the organophosphate fenitrothion, to which will be exposed for 2 hours). The mosquitoes are then transfered to a paper that hasn't been impregnated, where they rest under controlled conditions of temperature, relative humidity and an energy source consisting of a sugar solution (Figure 2A). The mortality reading is done after 24 hours of exposure to the insecticide (WHO 1981a).

The susceptibility of larvae is measured by exposing individuals to a diagnostic concentration or several temephos concentrations to determine the lethal concentration (LC_{50} or CL_{95}) to calculate resistance factor using water as solvent (Figure 2B). Individuals remain exposed for 24 hours. (WHO 1981b) (Santacoloma, 2008).

Fig. 2A. Biological test of the World Health Organization (WHO) (mosquitoes).

4.4.2 Test for biological control center and prevention (CDC)
The impregnated bottles technique was standardized by the Center for Diseases Control and Prevention (CDC) (Brogdon and McAllister, 1998). The objective of this technique is to measure the time it takes dose of an insecticide to reach the target site of action in the

Fig. 2B. Biological test of the World Health Organization (WHO) (Larvae).

mosquito. Involves subjecting a sample of mosquitoes between 15 and 25 individuals of the population to be evaluated on a glass surface pre-impregnated with the dose of insecticide. (Santacoloma, 2008) (Figure 3).

Fig. 3. Test for biological control center and prevention (CDC).

4.4.3 Synergists
These are chemicals that specifically inhibit insecticide-metabolizing enzymes, enhancing their action. Among the synergists more used to detect resistance mechanisms in Insects are the S, S, S - tributilfosforotioato (DJF) and esterases inhibitor and of the enzyme glutathione transferase (GST), the triphenyl phosphate (TFF) specific esterase inhibitor, piperonyl butoxide (PB), an inhibitor of monooxygenases, and ethacrynic acid (EA), a specific inhibitor of the enzyme glutathione transferase (GST) (Rodríguez, 2008).

4.4.4 Biochemical test
The biochemical assays are used to define metabolic mechanisms that may be responsible for the physiological resistance in an insect population (WHO, 1992). The metabolic mechanisms include tests to determine the target enzyme decreased sensitivity or the increased enzyme activity. For the first mechanism in particular, measures the change in acetylcholinesterase associated with resistance to carbamates and organophosphates. For the second, evaluates the increased activity of esterases, mixed function oxidases and glutathione s-transferase for kidnapping or increased detoxification of insecticides (Santacoloma, 2008).

4.4.5 Molecular tests
These tests consist in the amplification of specific gene sequences through polymerase chain reaction technique (PCR) to detect mutations.

4.5 Current state of susceptibility to insecticides *Aedes aegypti* in Colombia
Since the late forties, when first reported resistance to DDT in *Aedes tritaeniorhynchus* (Weidemann) and *Aedes solicitans* (Walker) resistance has been recorded in over a hundred species for one or more insecticides of public health use worldwide (Brown, 1986; Fonseca & Quiñones, 2005). For *A. aegypti* in America resistance has been reported to organochlorines, organophosphates, pyrethroids and carbamates in Argentina, Brazil, Mexico, El Salvador, Peru, Panamá, Venezuela, Cuba, Puerto Rico, among other Caribbean countries, whose resistance mechanism in some of these stocks has been associated with altered levels of alpha esters, beta esterases, mixed function oxidases, glutathione s- transferase, as well as mutations in the voltage-gated sodium channel. (Rawlins, 1998; Bisset et al, 2001; Brengues et al, 2003; Macoris et al, 2003; Aparecida et al, 2004; Rodríguez et al, 2004; Chavez et al; 2005, Flores et al, 2005; Pereira da-Cunha et al, 2005, Alvarez et al, 2006, Pereira-Lima et al, 2006; Beserra et al, 2007; Saavedra et al, 2007; Bisset et al, 2009; Martins et al, 2009; Albrieu-Llinas et al, 2010; Polson et al, 2010).
Colombia has applied insecticides for control of vector insects for over five decades. The DDT was the first applied to control malaria and during the campaign for the eradication of *A. aegypti* conducted in the early 1950. This insecticide was banned in the late 60's, due to the findings of resistance worldwide (Brown, 1986). Since 1970, organophosphates including temephos were applied and from the early 90's the use of pyrethroids was started. From that time on, the country has been rotating the application of molecules for mosquito control such as: deltamethrin, lambda-cyhalothrin, malathion, fenitrothion and in the last three years cyfluthrin and pirimiphos-methyl. However, the resistance to these insecticides has been documented gradually, making it difficult to take control actions within programs of Vector Borne Diseases in different regions of the country.

In Colombia until the 1990's there were few works assessing the state of susceptibility in Culicidae populations of interest in public health. Between 1959 and 1987 the first cases of DDT resistance were registered in populations of *Anopheles albimanus* (Wiedemann) in the municipalities of El Carmen (Bolivar); Codazzi, Robles and Valledupar (Cesar); Acandí (Choco) and *An. darlingi* (Root) in some locations of Quibdó municipality (Chocó) (Quiñones et al, 1987). Later Suarez et al, (1996) recorded in the 90's decade the first case of resistance to temephos in the *A. aegypti* species in Cali, Valle del Cauca, and Bisset et al, (1998) evaluated the susceptibility in a strain of *Culex quinquefasciatus* (Say) from Medellin, Antioquia, encountering resistance to organophosphates malathion, primifos-methyl, chlorpyrifos, temephos, fenthion and pyrethroids deltamethrin and permethrin.

In the absence of enough studies in Colombia on susceptibility status of populations of *A. aegypti* to several insecticides of use in public health and in compliance with public policies enshrined in the American continent resolutions CD39.R11 1996, CD43R4 2001 of the Pan American Health Organization, during the years of 2005 and 2007 a national project was conducted, this was funded by Colciencias (Colombian Science and Research Organization) and implemented by the Learning and Control of Tropical Diseases Program (PECET) of the University of Antioquia, the International Centre for Training and Medical Research (CIDEIM), the National University in Colombia, the National Institute of Health (NIH) and 12 departments of health seeking to generate baseline susceptibility of vector populations in Colombia. This multicentered project gave rise to the national surveillance network susceptibility to insecticides for *A. aegypti* and main vectors of malaria led by the National Institute of Health (INS). Since then the record of resistance to *A. aegypti* in Colombia expanded through biological tests by the CDC and WHO as well as the determination of impairment of enzymes involved in resistance.

With these results it has been observed for Colombia widespread resistance to DDT (Figure 4A-4B) and variability in susceptibility to the following insecticides: temephos, lambda-cyhalothrin, deltamethrin, permethrin, cyfluthrin, etofenprox, malathion, fenitrothion, pirimiphos methyl, bendiocarb and propoxur in different regions, with deterioration in some populations in the levels of nonspecific esterases, mixed function oxidases and in smaller proportion to glutathione s-transferesas (Figure 5A, 5B, 5C, 5D) (Rojas et al, 2003; Cadavid et al, 2008; Fonseca et al, 2006; Fonseca et al, 2007; Orjuela et al; 2007, Salazar et al, 2007; Santacoloma et al, 2008; Fonseca et al, 2009; Maestre et al, 2009; Maestre et al, 2010; Ardila and Brochero, 2010, Gomez et al, 2010; Maestre et al, 2010, Fonseca et al, 2011).

For temephos resistance has been observed in Cundinamarca, Guaviare, Meta, Santander, Cauca, Valle del Cauca, Nariño, Huila, Caldas, Sucre, Atlantico, La Guajira. (Anaya et al, 2007; Santacoloma, 2008; Maestre et al, 2009; Ocampo et al, 2011) (Figure 6).

Pyrethroids that despite being used in Colombia more recently compared to organophosphates, have shown higher levels of resistance despite the increased use time of organophosphate (Figures 7A, 7B, 7C, 7D, 7E), (8A, 8B, 8C). Among the pyrethroids lambda is the insecticide wich displays higher frequency of resistance in vector populations in the country. However, there are pyrethroids such as permethrin and etofenprox that despite having no use in public health have resistance generated in populations of *A. aegypti* from Casanare, Antioquia, Chocó and Putumayo (Ardila and Brochero, 2010; Fonseca et al, 2011).

For the carbamate propoxur discordance in susceptibility results was observed between the WHO technique in which resistance and CDC susceptibility is recorded. Further studies are required to determine the state of populations' susceptibility to the insecticide (Fig. 9 AB). Furthermore, few studies in the country have evaluated the susceptibility of the vector to the insecticide Bendiocarb to which there has been resistance registered in the populations in the department of Cauca, Valle del Cauca, Huila and Nariño (Figure 10 AB) (Ocampo et al, 2011).

Currently it has not been registered for the country mutations in voltage-gated sodium channel gene. It is therefore recommended studies to perform studies to explain resistance observed to most of the pyrethroids evaluated in different country populations and may be related to a crossed resistance to DDT. It is also recommended to perform studies to determine cross resistance to other molecules such as organophosphate and carbamate, as well as multi resistance studies.

For Colombia it is recommended to maintain a system of permanent time and space surveillance that allows health authorities to use insecticides with technical criteria to maintain effective control interventions in vector populations.

Resistance

Fig. 4A. Susceptility status of *Aedes aegypti* populations to DDT in Colombia (CDC test).

Fig. 4B. Susceptility status of *Aedes aegypti* populations to DDT in Colombia (OMS test).

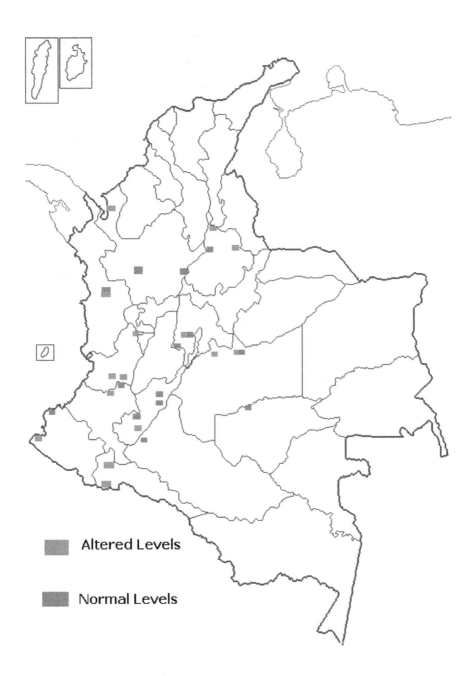

Fig. 5A. Non-specific esterases (NSE).

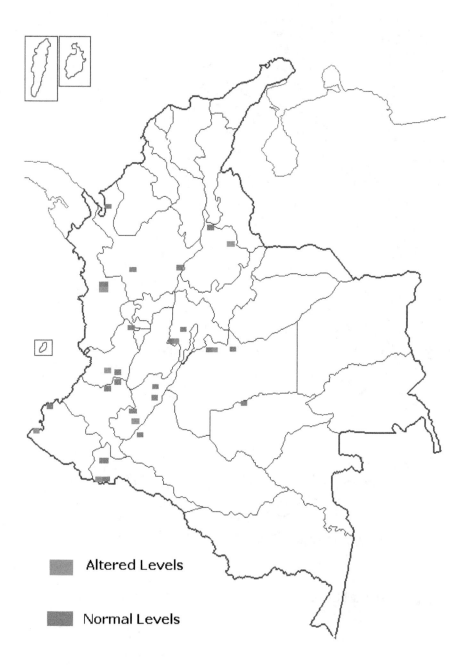

Fig. 5B. Mixed-function oxidases (MFO).

Altered Levels

Normal Levels

Fig. 5C. Glutathione-*S*-transferases (GST).

5D

Fig. 5. Biochemical mechanism of resistance in population of *Aedes aegypti* in Colombia: Non-specific esterases (NSE) (5A); Mixed-function oxidases (MFO) (5B); Glutathione-*S*-transferases (GST) (5C); acethylcholinesterase (AChE) (5D).

Fig. 6. Susceptility status of *Aedes aegypti* populations to Temephos in Colombia (OMS test).

Fig. 7A. Susceptility status of *Aedes aegypti* populations to lambda-cyhalothrin in Colombia (CDC test).

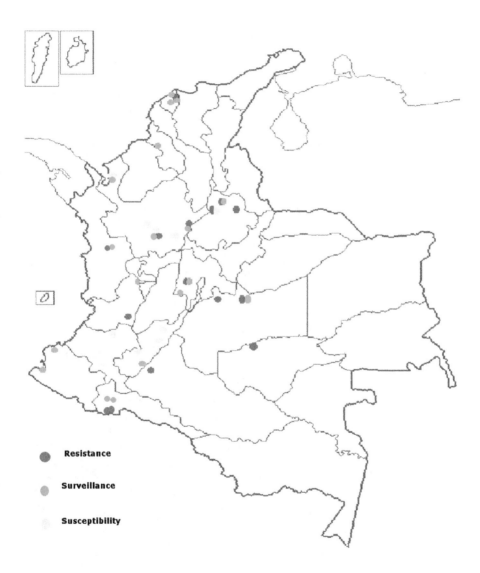

Fig. 7B. Susceptility status of *Aedes aegypti* populations to lambda-cyhalothrin in Colombia (OMS test).

Fig. 7C. Susceptility status of *Aedes aegypti* populations to Deltamethrin in Colombia (CDC test).

Fig. 7D. Susceptility status of *Aedes aegypti* populations to Deltamethrin in Colombia (OMS test).

Fig. 7F. Susceptility status of *Aedes aegypti* populations to cyfluthrin in Colombia (OMS test).

Fig. 7G. Susceptility status of *Aedes aegypti* populations to Permethrin in Colombia (CDC test).

Fig. 7H. Susceptility status of *Aedes aegypti* populations to Permethrin in Colombia (OMS test).

Resistance

Surveillance

Susceptibility

Fig. 7I. Susceptility status of *Aedes aegypti* populations to etofenprox in Colombia (OMS test).

Fig. 8A. Susceptility status of *Aedes aegypti* populations to Malathion in Colombia (CDC test).

Fig. 8B. Susceptility status of *Aedes aegypti* populations to Malathion in Colombia (OMS test).

Fig. 8C. Susceptility status of *Aedes aegypti* populations to Fenitrothion in Colombia (CDC test).

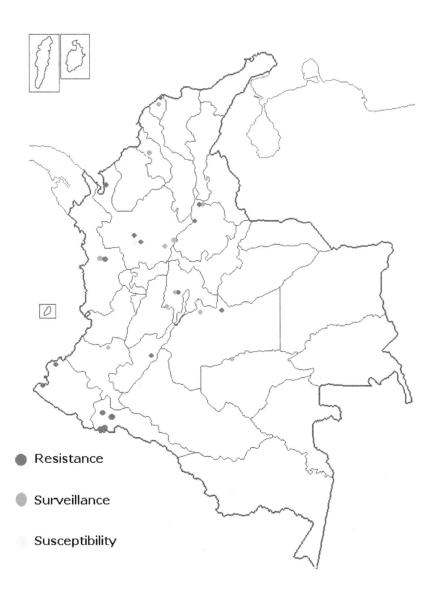

● Resistance

● Surveillance

Susceptibility

Fig. 8D. Susceptility status of *Aedes aegypti* populations to Fenitrothion in Colombia (OMS test).

Fig. 8E. Susceptility status of *Aedes aegypti* populations to pirimiphos methyl in Colombia (CDC test).

Fig. 8F. Susceptility status of *Aedes aegypti* populations to pirimiphos methyl in Colombia (OMS test).

Fig. 9A. Susceptility status of *Aedes aegypti* populations to propoxur in Colombia (CDC test).

Fig. 9B. Susceptility status of *Aedes aegypti* populations to propoxur in Colombia (OMS test).

Fig. 10A. Susceptility status of *Aedes aegypti* populations to bendiocarb in Colombia (CDC test).

Fig. 10B. Susceptility status of *Aedes aegypti* populations to bendiocarb in Colombia (OMS test).

5. Conclusion

The selection pressure exert by insecticides for more than five decades on the populations of *A. aegypti* in Colombia, has generated widespread resistance to DDT and variability in susceptibility to phosphorated, carabamates and pyrethroids insecticides. As a possible cause of metabolic resistance to these insecticides, alteration in nonspecific esterase levels has been registered, mixed function oxidases and s-glutathione transferases. However, it is unknown whether there are alterations in target sites of action, especially in voltage-dependent sodium channel genes that explains the generalized resistance to DDT and most pyrethroids in the country.

6. Acknowledgment

Sponsored by the project Strengthening UNIMOL Group 2011 Vice-Rectory of Research - Res. 4357 (2010) Colombia.

7. References

Albrieu-Llinás, G.; Seccacini, E.; Gardenal, CN. & Licastro, S. (2010). Current resistance status to temephos in *Aedes aegypti* from different regions of Argentina. *Mem Inst Oswaldo Cruz*, Vol.105, No.1, pp.113-116.

Álvarez, L.; Briceño, A. & Oviedo, M. (2006). Resistencia al temefos en poblaciones de *Aedes aegypti* (Diptera: Culicidae) del occidente de Venezuela. *Revista Colombiana de Entomología*, Vol.32, No.2, pp.172-5.

Anaya, Y.; Cochero, S.; Rey, G. & Santacoloma, L. (2007). Evaluación de la Susceptibilidad a insecticidas en *Aedes aegypti* (Diptera: Culicidae) capturados en Sincelejo. *Biomedica*, Vol.27, No.2, pp.257-8.

Anstead, JA.; Williamson, MS. & Denholm, I. (2005). Evidence for multiple origins of identical insecticide resistance mutations in the aphid Myzus persicae. *Insect Biochemistry and Molecular Biology*, Vol.35, pp.249–256.

Aparecida-Braga, I.; Pereira-Lima, J.; Da Silva-Soares, S. & Valle, D. (2004). *Aedes aegypti* resistance to temephos during 2001 in several municipalities in the states of Rio de Janeiro, Sergipe and Alagoas, Brazil. *Memórias do InstitutoOswaldo Cruz*, Vol.99, No.2, pp.199-203.

Ardila-Roldan, S. & Brochero, HL. (2010).Status of insecticide resistance in natural population from *Aedes aegypti* and CAP surveys of dengue vector in Casanare, Colombia. *Journal of American Mosquito Control Association*, Vol.26, No.3, pp.306-20.

Beserra, E.; Fernández, C.; De Quiroga, M. & De Castro, F. (2007). Resistance of *Aedes aegypti* (L.) (Diptera: Culicidae)populations to organophosphates temephos in the Paraíba State, Brazil. *Neotropical Entomology*, Vol.36, No.2, pp.303-7.

Bisset, JA.; Rodríguez, MM.; Molina, D.; Díaz, C. & Soca, LA. (2001). Esterasas elevadas como mecanismo de resistencia a insecticidas organofosforados en cepas de *Aedes aegypti*. *Revista Cubana de Medicina Tropical*, Vol.53, No.1, pp.37-43.

Bisset J. (2002).Uso correcto de insecticidas: control de la resistencia. *Revista Cubana de Medicina Tropical*, Vol.54, No.3, pp.202-219.

Bisset-Lazcano JA.; Rodríguez, MM.; San Martín, JL.; Romero, JE. & Montoya, R. (2009).Evaluación de la resistencia a insecticidas de una cepa de *Aedes aegypti* de El Salvador. *Revista Panamericana de Salud Publica*, Vol.26, No.3, pp.229-34.

Brogdon, W. & Mcallister, J. (1998).Simplification of adult mosquito bioassays through use of time mortality determinations in glass bottles. *Journal of the American Mosquito Control Association*, Vol.14, No2, pp.159-64.

Brengues, C.; Hawkes, NJ.; Chandre, F.; McCarroll, L.; Duchon, S. & Guillet, P. *et al.* (2003). Pyrethroid and DDT cross-resistance in *Aedes aegypti* is correlated with novel mutations in the voltagegated sodium channel gene. *Med Vet Entomol.*, Vol.17, pp.87–94.

Brooke B. (2008). *kdr*: can a single mutation produce an entire insecticide resistance phenotype?. *Transactions of the Royal Society of Tropical Medicine and Hygiene*, Vol.102, pp.524–525.

Brown, A. (1986). Insecticide resistance in mosquitoes: a pragmatic review. *Journal of American Mosquito Control Association*, Vol.2, pp.123-40.

Cadavid, JM.; Valderrama-Hernández, R.; Saenz-Osorio, O.; Quintero-Ruiz, B.; Rodríguez Ríos, C. & Contreras-Samper, A. (2008). Susceptibility of *Aedes aegypti* to insecticides in three high dengue transmission localities of Antioquia, Colombia *Journal of American Mosquito Control Association*, Vol.22, No.4, pp.748.

Chang, Ch.; Shen, WK.; Wang, TT.; Lin, YH.; Hsu, EL. & Dai, SM. (2009). A novel amino acid substitution in a voltage-gated sodium channel is associated with knockdown resistance to permethrin in *Aedes aegypti*. *Insect Biochemistry and Molecular Biology*, Vol.39, pp. 272–278

Chávez, J.; Vargas, J. & Vargas, F. (2005). Resistencia a deltametrina en dos poblaciones de *Aedes aegypti* (Diptera: Culicidae) del Perú. *Revista Peruana de Biología*, Vol.12, No1, pp.161-4.

Cui, F.; Weill, M.; Berthomieu, A.; Raymond, M. & Qiao, Ch. (2007). Characterization of novel esterases in insecticide-resistant mosquitoes. *Insect Biochemistry and Molecular Biology*,Vol.37, pp.1131–37.

Escobar JP. (2009). Políticas y orientaciones técnicas OPS/OMS para prevención y control del dengue y dengue hemorrágico. *Biomédica*, Vol.29, No1, pp.123-5.

Feyereisen R. (1999). Insect P450 enzymes. *Annu. Rev. Entomol.*, Vol.44, pp.507-33.

Flores, A.; Albeldaño-Vásquez, W.; Fernández-Salas, I.; Badii, M.; Loaiza-Becerra, H.;

Ponce-Garcia, G.; Lozano-Fuentes, S.; Brogdon, W.; Black, W.; Beaty, B. (2005). Elevated á-esterase levels associated withpermethrin tolerance in *Aedes aegypti* (L.) from Baja California, México. *Pesticide Biochemistry and Physiology*, Vol.82, pp.66-78.

Fonseca, I. & Quiñones, M. (2005). Resistencia a insecticidas en mosquitos (Diptera: Culicidae): Mecanismos, detección y vigilancia en salud pública. *Revista Colombiana de Entomología*, Vol.31, No2, pp.107-115.

Fonseca I., Brant T.,Quiñones ML., Brogdonm, WG. (2006). Increased glutathione S transferases associated to DDT resistance in *Aedes aegypti* from Colombia. *Journal of American Mosquito Control Association*, Vol.22, No4, pp.738.

Fonseca, I.; Bolaños, D.; Gomez, W. & Quiñones, ML. (2007). Evaluación de la susceptibilidad de larvas de *Aedes aegypti* a insecticidas en el deprtamento del Antioquia. *Biomedica*, Vol.27, No.2, pp.176-7.

Fonseca-González, I. (2008). Estatus de la resistencia a insecticidas de los vectores primarios de malaria y dengue en Antioquia, Chocó, Norte de Santander y Putumayo, Colombia. [Tesis de doctorado]. Medellín: Universidad de Antioquia.

Fonseca, I. & Bolaños, D. (2009).Variación temporal en la susceptibilidad a malatión y Lambdacialotrina en *Aedes aegypti* (L) de Quibdo, Colombia. *Biomédica*, Vol.29, No.1, pp.218-9.

Fonseca-Gonzalez, I.; Quiñones, ML.; Lenhartc, A. & Brogdon, WG. (2011). Insecticide resistance status of *Aedes aegypti* (L.) from Colombia. *Pest Management Science*, Vol.67, No.4, pp.430-7.

Gómez-Camargo DE.; Maestre, RY.; Pisciotti, I.; Malambo, DI. & Gómez-Alegría, CJ. (2010). Insecticide susceptibility of *Aedes aegypti* in Cartagena (Colombia). *The American Journal of Tropical Medicine and Hygiene*, Vol.83, No.5, pp.298-9.

Hemingway, J. & Black IV, WC. (2007). A mutation in the voltage-gated sodium channel gene associated with pyrethroid resistance in Latin American *Aedes aegypti*. *Insect Molecular Biology*, Vol.16, No.6, pp.785-98.

Lumjuan, N.; Stevensona, BJ.; Prapanthadarab, L.; Somboonc, P.; Brophyd, PM.; Loftuse, BJ.; Seversonf, DW. & Ransona, H. (2007). The *Aedes aegypti* glutathione transferase family. *Insect Biochemistry and Molecular Biology*, Vol.37, pp.1026-35.

Macoris, M.; Andrighetti, M.; Takaku, L.; Glasse, C.; Garbeloto, V. & Bracco, J. (2003). Resistance of *Aedes aegypti* from the state of Sao Paulo, Brazil to Organophosphates insecticidas. *Memórias do Instituto Oswaldo Cruz*, Vol.98, No.5, pp.703-8.

Maestre, R.; Rey, G.; De las Salas, J.; Vergara, C.; Santacoloma, L.; Goenaga, S. & Carrasquilla, MC. (2009). Susceptibilidad de *Aedes aegypti* (Diptera: Culicidae) a temefos en Atlántico-Colombia. *Revista Colombiana de Entomología*, Vol.35, No.2, pp.202-205.

Maestre-Serrano, R.; Goenaga, S.; Carrasquilla-Ferro MC.; Rey-Vega G.; Santacoloma-Baron L. & Vergara-Sánchez C. (2009). Estado de la susceptibilidad/resistencia a insecticidas en cuatro poblaciones de *Aedes aegypti* del departamento del Atlántico. *Biomedica*, Vol.29, No.1, pp.231-2.

Maestre, RY.; Florez, Z.; Cabrera, C.; Goenaga, S.; Gómez, D. & Gómez, C. (2010). Susceptibility status of *Aedes aegypti* to insecticides in La Guajira (Colombia). *The American Journal of Tropical Medicine and Hygiene*, Vol.83, No.5, pp.53.

Maestre, R.; Rey, G.; De las Salas, J.; Vergara, C.; Santacoloma, L.; Goenaga, S. & Carrasquilla, MC. (2010). Estado de la susceptibilidad de *Aedes aegypti* a insecticidas en Atlántico (Colombia). *Revista Colombiana de Entomología*,Vol.36, No.2, pp.242-248.

Maestre-Serrano, R.; Goenaga, S.; Flores, Z.; Cabrera C.; Gómez, D. & Gómez, C. (2010). *The Amercican Journal of Tropical Medicine and Hygiene* Vol.83, No.5, pp.53.

Martins, AJ.; Pereira-Lima, JB.; Peixoto, AA. & Valle, D. (2009). Frequency of Val1016Ile mutation in the voltage-gated sodium channel gene of *Aedes aegypti* Brazilian populations. *Tropical Medicine and International Health*, Vol.14, No.11, pp.1351-55.

Martinez-Torres, D.; Chandre, F.; Williamson, MS.; Darriet, F.; Berge, JB.; Devonshire, AL.; Guillet, P.; Pasteur, N. & Pauron, D. (1998). Molecular characterization of pyrethroid knockdown resistance (kdr) in the major malaria vector Anopheles gambiae s.s. *Insect Molecular Biology*, Vol.7, pp.179–184.

Miller TA. 1988. Mechanisms of resistance to pyrethroid insecticides. *Parasitology Today*, Vol.4, pp.8-12.

Ocampo, C.; Salazar-Terreros, M.; Mina, N.; McAllister, J. & Brogdon, W. (2011). Insecticide Resistance status of *Aedes aegypti* in 10 localities of Colombia. *Acta Trop.*, Vol.118 No.1, pp.37-44.

Orjuela, L.; Herrera, M.; Quintana, N. & Quiñones, M. (2007). Estado de la susceptibilidad a insecticidas del mosquito *Aedes aegypti* en lo municipios de Chinchiná y La Dorada, Caldas. *Biomedica*, Vol.27, No.2, pp.176.

Pereira-da-Cunha, M.; Pereira-Lima JB.; Brogdon, WG.; Efrain-Moya G. & Valle D. (2005). Monitoring of resistance to the pyrethroid cypermethrin in Brazilian *Aedes aegypti* (Diptera: Culicidae) populations collected between 2001 and 2003. *Memorias do Instituto Oswaldo Cruz*, Vol.100, No.4, pp.441-44.

Pereira-Lima, E.; Da Oliveira-Filho, A.; Da Oliveira-Lima, J.; Ramos-Junior, A.; Da Goes-Cavalcanti, L. & Soares-Pontes, R. (2006). *Aedes aegypti* resistance to temephos in counties of Ceará State. *Revista da Sociedade Brasileira de Medicina Tropical*, Vol.39, No.3, pp.259-63.

Polson, KA.; Rawlins, SC.; Brogdon, WG. & Chadee, DD. (2010). Biochemical mechanisms involved in DDT and pyrethroid resistance in Trinidad and Tobago strains of *Aedes aegypti*. *The American Journal of tropical medicine and hygiene*, Vol.83, No.5, pp.51.

Poupardin, R.; Reynaud, S.; Strode, C.; Ranson, H.; Vontas, J. & David JP. (2008). Cross-induction of detoxification genes by environmental xenobiotics and insecticides in the mosquito Aedes aegypti: Impact on larval tolerance to chemical insecticides. *Insect Biochemistry and Molecular Biology*, Vol.38, pp.540-51.

Quiñones, ML.; Suarez, MF. & Fleming, GA. (1987). Estado de la susceptibilidad al DDT de los principales vectores de malaria en Colombia y su implicación epidemiológica. *Biomédica*, Vol.7, pp.81-6.

Rawlins S. (1998). Distribución espacial e importancia de la resistencia a insecticidas de poblaciones de *Aedes aegypti* en el Caribe. *Revista Panamericana de Salud Pública*, Vol.4, No.4, pp.243-51.

Rodríguez, M.; Bisset, J.; Fernández, D. & Pérez, O. (2004). Resistencia a insecticidas en larvas y adultos de *Aedes aegypti*: prevalencia de esterasa A4 asociada con la resistencia a temefos. *Revista Cubana de Medicina Tropical*, Vol.56, No.1, pp.54-60.

Rodríguez, MM.; Bisset, JA. & Fernández, D. (2007). Determinación *in vivo* del papel de las enzimas esterasas y glutation transferasa en la resistencia a piretroides en *Aedes aegypti* (Diptera: Culicidae). *Revista Cubana de Medicina Tropical*, Vol.59, No.3, pp.209-12.

Rodríguez, M. (2008). Estudio de la resistencia a insecticidas en *Aedes aegypti* (Diptera: Culicidae). Tesis presentada al grado científico de Doctor en Ciencias de la Salud. Laboratorio de Toxicología y genética. Instituto de Medicina Tropical "Pedro Kourí". 128p.

Rojas, W.; Gonzalez, J.; Amud, MI.; Quiñones, ML. & Velez, ID. (2003). Evaluación de la susceptibilidad de Aedes aegypti del municipio de Barrancabermeja Santander a los insecticidas Malatión Fenitrotion, temefos, Lambdacihalotrina, deltametrina, permetrina, propoxur y DDT. *Biomédica*, Vol.23, No.1, pp.56

Rojas-Alvarez, DP. (2010). Comportamineto epidemiologico del dengue en Colombia, año 2009. *Informe quincenal Epidemiologico Nacional*, Vol.15, No.9, pp.129-37.

Saavedra-Rodriguez, K.; Urdaneta-Marquez, L.; Rajatileka, S.; Moulton, M.; Flores, AE.; Fernandez-Salas, I.; Bisset, J.; Rodriguez, M.; Mccall, PJ.; Donnelly, MJ.; Ranson, H.;

Salazar, M.; Carvajal, A.; Cuellar, ME.; Olaya, A.; Quiñones, J.; Velásquez, OL.; Viveros, A. & Ocampo, C. (2007). Resitencia a insecticidas en poblaciones de *Aedes aegypti* y *Anopheles* spp en los departamentos de Huila, valle Cauca y Nariño. *Biomédica*, Vol.27, No.2, pp.177.

Santacoloma, L. (2008). Estado de la susceptibilidad a insecticidas de poblaciones naturales de *Aedes aegypti* Linnaeus 1762 vector del dengue y *Anopheles darlingi* root 1926 vector primario de malaria (Diptera: Culicidae) en cinco departamentos de Colombia. [Tesis de maestría]. Bogotá: Universidad Nacional de Colombia.

Santacoloma-Varon, L.; Chavez-Cordoba, B. & Brochero, HL. (2010). Susceptibilidad de *Aedes aegypti* a DDT, Deltametrina y Lambdacialotrina en Colombia. *Revista Panamericana de Salud Publica*, Vol.27, No.1, pp.66-73. http://www.todacolombia.com/geografia/ubicacion.html

Valderrama-Eslava, EI.; Gomzalez, R. & Jaramillo, GI. (2008). Evaluación de la susceptibilidad de *Aedes aegypti* (L) (Diptera: Culicidae) a un insecticida Organosfosforado y un piretroide en cuatro poblaciones del Valle del Cauca, mediante dos tipos de bioensayos. *Boletin del Museo de Entomología de la Universidad del Valle*, Vol.9, No.2, pp.1-11.

World Health Organization (WHO). (1957). Seventh report Expert Committee on insecticides *WHO Tech Report Ser*, Vol.125, pp.37.

World Health Organization (WHO). (1981a). Instructions for determining the susceptibility or resistance of adult mosquitoes to organoclorine, organophosphorus and carbamate insecticides. Establishment of the base-line. WHO/ VBC/81.805.

World Health Organization (WHO). (1981b). Instructions for determining the susceptibility or resistance of mosquitoes larvae to insecticide. Unpublished document.WHO/ VBC/81.807.

Yanola, J.; Somboon, P.; Walton, C.; Nachaiwieng, W. & Prapanthadara L. (2010). A novel F1552/C1552 point mutation in the *Aedes aegypti* voltage-gated sodium channel gene associated with permethrin resistance. *Pesticide Biochemistry and hysiology*, Vol.96, pp.127-13.

12

Metabolism of Pyrethroids by Mosquito Cytochrome P450 Enzymes: Impact on Vector Control

Pornpimol Rongnoparut[1], Sirikun Pethuan[1],
Songklod Sarapusit[2] and Panida Lertkiatmongkol[1]
[1]*Department of Biochemistry, Faculty of Science, Mahidol University,*
[2]*Department of Biochemistry, Faculty of Science, Burapha University,*
Thailand

1. Introduction

Cytochrome P450 enzymes (P450s) are heme-containing monooxygenases that catalyze metabolisms of various endogenous and exogenous compounds. These P450s constitute a superfamily of enzymes present in various organisms including mammals, plants, bacteria, and insects. P450 enzymes are diverse and metabolize a wide variety of substrates, but their structures are largely conserved. A universal nomenclature has been assigned to P450 superfamily based on their amino acid sequence homology (Nelson et al., 1996). In eukaryotes, P450 is membrane-bound and in general functions to insert one molecule of oxygen into its substrate, with its heme prosthetic group playing a role in substrate oxidation. This catalytic reaction requires a pair of electrons shuttled from NADPH via the NADPH-cytochrome P450 reductase (CYPOR) enzyme, a P450 redox partner, to target P450s (Ortiz de Montellano, 2005). In contrast in bacteria and mitochondria, ferredoxin reductase and iron-sulfur ferredoxin proteins act as a bridge to transfer reducing equivalent from NAD(P)H to target P450s. In insects, P450s are membrane-bound enzymes that play key roles in endogenous metabolisms (i.e. metabolisms of steroid molting and juvenile hormones, and pheromones) and xenobiotic metabolisms, as well as detoxification of insecticides (Feyereisen, 1999). It becomes evident that P450s are implicated in pyrethroid resistance in insects.

Insecticides form a mainstay for vector control programs of vector-borne diseases. However intensive uses of insecticides have led to development of insecticide resistance in many insects thus compromising success of insect vector control. In particular pyrethroid resistance has been found widespread in many insects such as house flies, cockroaches, and mosquitoes (Acevedo et al., 2009; Awolola et al., 2002; Cochran, 1989; Hargreaves et al., 2000; Jirakanjanakit et al., 2007). Two major mechanisms have been recorded responsible for insecticide resistance, which are alteration of target sites and metabolic resistance (Hemingway et al., 2004). Metabolic resistance is conferred by increased activities of detoxification enzymes such as P450s, non-specific esterases (Hemingway et al., 2004; Price, 1991). Initial approaches to detect involvement of detoxification mechanisms in metabolic resistance are to compare activities of detoxification enzymes between resistant and susceptible insect strains, and by identification of corresponding genes that display higher expression level in resistant insects (Bautista et al., 2007; Chareonviriyaphap et al., 2003; Tomita et al., 1995; Yaicharoen et al., 2005). Examinations in various insects such as house fly, cotton ballworm, and mosquito have implicated involvement of up-regulation of different P450 genes in pyrethroid resistance (Liu & Scott, 1998; Müller et al., 2007; Ranasinghe & Hobbs, 1998; Rodpradit et al., 2005; Tomita et al., 1995). Such P450 overexpression has been assumed constituting a defense mechanism against insecticides and responsible for insecticide resistance, presumably by virtue of enhanced insecticide detoxification.

Recent advanced methods employing microarray-based approach, when genomic sequence information for insects is available, have identified multiple genes involved in pyrethroid resistance in mosquitoes. Genes in CYP6 family, in particular, are reported to have an implication in insecticide resistance. In *Anopheles gambiae* malaria vector, microarray analyses reveal that several CYP6 P450 genes could contribute to pyrethroid resistance, these include CYP6M2, CYP6Z2 and CYP6P3 (Djouaka et al., 2008; Müller et al., 2007). These genes were observed up-regulated in pyrethroid resistant mosquitoes (Müller et al., 2008; Stevenson et al., 2011). CYP6M2 and CYP6P3 have shown ability to bind and metabolize pyrethroids, on the other hand CYP6Z2 is able to bind pyrethroids but does not degrade pyrethroids (Mclauglin et al., 2008). Genetic mapping of genes conferring pyrethroid resistance in *An. gambiae* also supports involvement of CYP6P3 in pyrethroid resistance (Wondji et al., 2007). Up-regulation of CYP6 genes has also been found in other resistant insects, for instance CYP6BQ9 in pyrethroid resistant *Tribolium castaneum* (Zhu et al., 2010), CYP6D1 in *Musca domestica* that is able to metabolize pyrethroids (Zhang & Scott, 1996), and CYP6BG1 in pyrethroid resistant *Plutella xylostella* (Bautista et al., 2007). In *T. castaneum* knockdown of CYP6BQ9 by dsRNA resulted in decreased resistance to deltamethrin (Zhu et al., 2010). Similar finding has been observed for CYP6BG1 in permethrin resistant *P. xylostella*, supporting the role of overexpression of these CYP6 genes in pyrethroid resistance (Bautista et al., 2009). In *An. minimus* mosquito, CYP6AA3 and CYP6P7 are upregulated and possess activities toward pyrethroid degradation (Duangkaew et al., 2011b; Rongnoparut et al., 2003).

2. Cytochrome P450 monooxygenase (P450) and NADPH-cytochrome P450 reductase (CYPOR) enzymes isolated from *An. minimus*

In this chapter, we focus on investigation of the P450s that have been shown overexpressed in a laboratory-selected pyrethroid resistant *An. minimus* mosquito. We describe heterologous expression of the overexpressed P450s in baculovirus-mediated insect cell expression system and characterization of their catalytic roles toward pyrethroid insecticides. Tools utilized in functional investigation of *An. minimus* P450s have been developed and described. In parallel the *An. minimus* CYPOR has been cloned and protein expressed via bacterial expression system. Amino acid sequence of *An. minimus* CYPOR is intrigue in that several important residues that might play role in its functioning as P450 redox partner are different from those of previously reported enzymes from mammals and house fly. The *An. minimus* CYPOR is different in enzymatic properties and kinetic mechanisms from other CYPORs. In this context we speculate that *An. minimus* CYPOR could influence electron delivery to target mosquito P450 enzymes, and could act as a rate-limiting step in P450-mediated metabolisms. These results together could thus gain an insight into pyrethroid metabolisms in this mosquito species and knowledge obtained could contribute to strategies in control of mosquito vectors.

An. minimus is one of malaria vectors in Southeast Asia, including Thailand, Loas, Cambodia and Vietnam. We previously established a deltamethrin-selected mosquito strain of *An. minimus* species A, by exposure of subsequent mosquito generations to LD_{50} and LT_{50} values of deltamethrin (Chareonviriyaphap et al., 2002). Biochemical assays suggested that deltamethrin-resistant *An. minimus* predominantly employ P450s to detoxify pyrethroids (Chareonviriyaphap et al., 2003). We next set out on isolation of P450 genes that have a causal linkage in conferring deltamethrin resistance in this mosquito species. Using reverse-transcribed-polymerase chain reaction (RT-PCR) in combination with degenerate PCR primers whose sequences were based on CYP6 conserved amino acids, we have isolated CYP6AA3, CYP6P7, and CYP6P8 complete cDNAs from deltamethrin-resistant *An. minimus* (Rongnoparut et al., 2003). The three genes showed elevated transcription level in deltamethrin resistant populations compared to the parent susceptible strain. We found that fold of mRNA increase of CYP6AA3 and CYP6P7 is correlated with increase of resistance during deltamethrin selection. However, this correlation was not observed for CYP6P8 (Rodpradit et al., 2005). The three mosquito P450s could thus be used as model enzymes for characterization of their metabolic activities toward insecticides and possibly for future development of tools for mosquito vector control. This can be accomplished by determining whether they possess catalytic activities toward pyrethroid insecticides, thus assuming a causal linkage of overexpression and increased pyrethroid detoxification leading to pyrethroid resistance. Equally important, elucidating properties of the *An. minimus* CYPOR and its influential role in P450 system is beneficial for understanding of P450 metabolisms of this mosquito species.

2.1 *In vitro* insecticide metabolisms

We have heterologously expressed CYP6AA3, CYP6P7, and CYP6P8 in *Spodoptera frugiperda* (*Sf9*) insect cells via baculovirus-mediated expression system. The expression procedure employed full-length CYP6AA3, CYP6P7, and CYP6P8 cDNAs as templates to produce recombinant baculoviruses, and subsequently infected *Sf9* cells for production of P450 proteins. RT-PCR amplification and sodium dodecyl sulfate-polyacrylamide gel electrophoresis (SDS-PAGE) analysis were performed to verify expression of P450 mRNAs and proteins in the infected *Sf9* cells. Expression of CYP6AA3, CYP6P7, and CYP6P8, each is predominantly detected in membrane fractions of infected cells after 72 hours of infection, with expected molecular size of approximately 59 kDa detected on SDS-PAGE (Kaewpa et al., 2007; Duangkaew et al., 2011b). The expressed proteins display CO-reduced difference spectrum of a characteristic peak at 450 nm (Omura & Sato, 1964). Total P450 content obtained from baculovirus-mediated expression of CYP6AA3, CYP6P7, and CYP6P8 ranges from 200 to 360 pmol/mg membrane protein. The expressed CYP6AA3, CYP6P7, and CYP6P8 proteins were used in enzymatic reaction assays testing against pyrethroids and other insecticide groups. Knowledge of the metabolic profile of these P450s could give us insight into functioning of these P450s within mosquitoes towards insecticide metabolisms, i.e. how mosquitoes detoxify against a spectrum of insecticide classes through P450-mediated metabolisms.

In enzymatic assay, each P450 in the reaction was performed in the presence of NADPH-regenerating system and was reconstituted with *An.minimus* CYPOR (Kaewpa et al., 2007), as CYPOR is required to supply electrons to P450 in the reaction cycle. Insecticide metabolism was determined by detection of disappearance of insecticide substrate at different times compared with that present at time zero as previously described (Booseupsakul et al., 2008). This time course degradation was detected through HPLC analysis. Table 1 summarizes enzyme activities of CYP6AA3 and CYP6P7 toward insecticides and metabolites detected. Insecticides that were tested by enzyme assays were type I pyrethroids (permethrin and bioallethrin), type II pyrethroids (deltamethrin, cypermethrin, and λ-cyhalothrin), organophosphate (chlorpyrifos), and carbamate (propoxur). Additional insecticides (bifenthrin, dichlorvos, fenitrothion, temephos, and thiodicarb) belonging to these four insecticide classes were tested by cytotoxicity assays (see Section 2.3). Chemical structures of these insecticides are shown in Fig. 1.

Insecticides	CYP6AA3 Activity (metabolites)	CYP6P7 Activity
Type I pyrethroids		
Bioallethrin	-	-
Permethrin	+ (1 major unknown product)	+, ND
Type II pyrethroids		
Cypermethrin	+ (3-phenoxybenzaldehyde and 2 unknown products)	+, ND
Deltamethrin	+ (3-phenoxybenzaldehyde and 2 unknown products)	+, ND
λ - Cyhalothrin	+, ND	-
Organophosphate		
Chlorpyrifos	-	-
Carbamate		
Propoxur	-	-

Table 1. Presence (+) and absence (-) of P450 activities in insecticide degradation and metabolites obtained. ND, products not determined

The results shown in Table 1 demonstrate that CYP6AA3 and CYP6P7 share overlapping metabolic profile against both type I and II pyrethroids, while no detectable activity was observed toward chlorpyrifos and propoxur (Duangkaew et al., 2011b), nor in the presence of piperonyl butoxide (a P450 inhibitor). Differences in activities of both enzymes could be noted, for CPY6AA3 could metabolize λ-cyhalothrin while CYP6P7 did not display activity against λ-cyhalothrin. For CYP6P8 we initially detected absence of pyrethroid degradation activity, further tests using cytotoxicity assays described in Section 2.3 suggest that CYP6P8 is not capable of degradation of pyrethroids, organophosphates and carbamates.

Determination of products obtained from CYP6AA3-mediated pyrethroid degradations using GC-MS analysis reveal multiple products for type II pyrethroid cypermethrin degradation and for earlier described deltamethrin metabolism (Boonseupsakul et al., 2008). These products were 3-phenoxybenzyaldehyde and two unknown products with chloride and bromide isotope distribution derived from cypermethrin and deltamethrin metabolisms, respectively. In contrast there was only one unknown product that was predominantly detected from CYP6AA3-mediated permethrin (type I pyrethroid) degradation, with mass spectrum profile showing characteristic chloride isotope distribution of permethrin-derived compound. Unlike cypermethrin and deltamethrin metabolisms, we did not obtain 3-phenoxybenzaldehyde from permethrin degradation (Boonseupsakul, 2008).

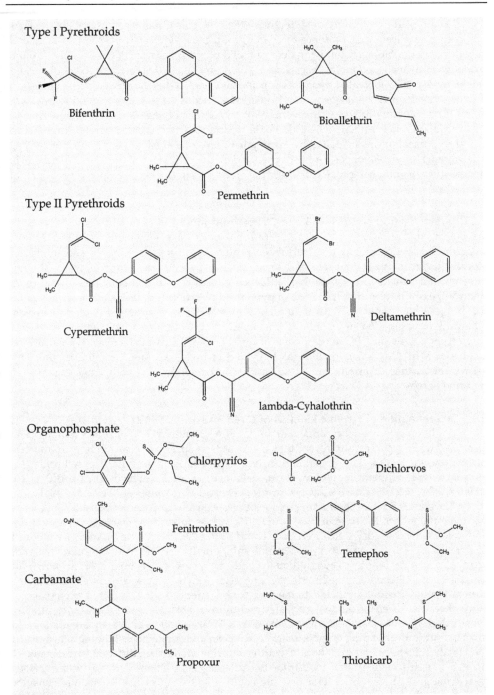

Fig. 1. Chemical structures of insecticides used in the study.

Type I and type II pyrethroids are different by the presence of cyano group (see Fig. 1). Thus our results implicate that presence of cyano group may play role in CYP6AA3-mediated pyrethroid degradations resulting in detection of 3-phenoxybenzaldehyde, possibly through oxidative cleavage reaction. In *An. gambiae* CYP6M2-mediated deltamethrin metabolism and house fly CYP6D1-mediated cypermethrin metabolism, 4'-hydroxylation of deltamethrin and cypermethrin is the major route of their metabolisms since 4'-hydroxylation products were predominantly detected (Stevenson et al., 2011; Zhang & Scott, 1996). The 4'-hydroxylation and 3-phenoxybenzaldehyde products have been observed in *in vitro* pyrethroid metabolisms mediated by mammalian microsomal enzymes (Shono et al., 1979). The absence of detection of 3-phenoxybenzaldehyde in CYP6AA3-mediated permethrin degradation could be predicted that the reaction underwent monooxygenation of permethrin.

2.2 Characterization of CYP6AA3 and CYP6P7 enzymes

As described, both CYP6AA3 and CYP6P7 enzymes have enzymatic activities against pyrethroid insecticides and their metabolic profiles are different. Kinetics and inhibition studies further support their abilities to metabolize pyrethroids, however with different enzyme and kinetic properties that influence substrate and inhibitor selectivity. Such knowledge could have an implication in pyrethroid detoxification in *An. minimus* mosquito, for example how the two P450s redundantly metabolize overlapping sets of pyrethroids. Alongside investigation of pyrethroid metabolisms, we examined their activities toward fluorescent compounds for development of rapid enzymatic assays. Finally we performed cell-based MTT (3-[4,5-dimethylthiazol-2-yl]-2,5-diphenyltetrazolium bromide) cytotoxicity assays for further determination of substrates and inhibitors of both P450 enzymes as reported herein.

2.2.1 Determination of enzyme kinetics of CYP6AA3 and CYP6P7 enzymes

We recently reported kenetic paremeters for CYP6AA3 and CYP6P7 enzymes (Duangkaew et al., 2011b). Kinetic results reveal that CYP6AA3 has preference in binding to and has higher rate in degradation of permethrin type I pyrethroid than type II pyrethroids (K_m values toward permethrin, cypermethrin, deltamethrin, and λ-cyhalothrin of 41.0 ± 8.5, 70.0 ± 7.1, 80.2 ± 2.0, and 78.3 ± 7.0 μM, respectively and V_{max} values of 124.2 ± 1.2, 40.0 ± 7.1, 60.2 ± 3.6, and 60.7 ± 1.1 pmol/min/pmol P450, respectively). In contradictory CYP6P7 does not have preference for type of pyrethroids (K_m values toward permethrin, cypermethrin, and deltamethrin of 69.7 ± 10.5 , 97.3 ± 6.4, and 73.3 ± 2.9, respectively and V_{max} values of 65.7 ± 1.6, 83.3 ± 7.6, and 55.3 ± 5.7 pmol/min/pmol P450, respectively) and does not metabolize λ-cyhalothrin. Thus although both enzymes are comparable in terms of capability to metabolize pyrethroids *in vitro*, their kinetic values are different. Enzyme structure could account for their differences in kinetic properties and substrate preference.

Since there has been no known crystal structure available for insect P450s, we initially constructed homology models for CYP6AA3, CYP6P7, and CYP6P8 in an attempt to increase our understanding of molecular mechanisms underlying their binding sites toward insecticide substrates and inhibitors. The three enzyme models are different in geometry of their active-site cavities and substrate access channels. Upon docking with various insecticide groups, results of its active site could predict and explain metabolic behavior toward pyrethroid, organophosphate, and carbamate insecticides (Lertkiatmongkol et al., 2011). These results suggest that differences in metabolic activities among P450 enzymes in insects could be attributed to structural differences resulting in selectivity and different enzymatic activities against insecticides.

In human, CYP2C8, CYP2C9, CYP2C19, and CYP3A4 have been reported abilities to metabolize both type I and II pyrethroids (Godin et al., 2007; Scollon et al., 2009). The preference for type I pyrethroid in CYP6AA3 is similar to human CYP2C9 and CYP2C19, while similar metabolic activity toward both types of pyrethroids found for CYP6P7 is similar to that of human CYP2C8 enzyme (Scollon et al., 2009). Nevertheless efficiency of CYP6AA3 and CYP6P7 in deltamethrin degradation is 5- to 10-fold less effective than human CYP2C8 and CYP2C19. It is noteworthy that more than one P450s residing within an organism can metabolize pyrethroids as described for human and mosquito, multiple rat P450s are also found capable of pyrethroid metabolisms (Scollon et al., 2009). When comparing to *An. gambiae* CYP6P3, both CYP6AA3 and CYP6P7 possess at least 10 fold higher K_m than CYP6P3, but V_{max} values of both *An. minimus* CYP6AA3 and CYP6P7 are at least 20 fold higher (Müller et al., 2008). Higher values of K_m and V_{max} of CYP6AA3 and CYP6P7 than those values of *An. gambiae* CYP6M2 (Stevenson et al., 2011) are also observed.

2.2.2 CYP6AA3 and CYP6P7 are inhibited differently by different compounds

To obtain a potential fluorogenic substrate probe for fluorescent-based assays of CYP6AA3 and CYP6P7, we previously screened four resorufin fluorogenic substrates containing different alkyl groups (Duangkaew et al., 2011b) and results in Table 2 suggest that among test compounds, benzyloxyresorufin could be used as a fluorescent substrate probe since both CYP6P7 and CYP6AA3 could bind and metabolize benzyloxyresorufin with lowest K_m (values of 1.92 for CYP6AA3 and 0.49 for CYP6P7) and with highest specific activities (Duangkaew et al., 2011b). The assays of benzyloxyresorufin-O-debenzylation activity were further used for inhibition studies of both mosquito enzymes.

| | Specific activity (pmole resorufin/min/pmole P450) | |
Compounds	CYP6AA3	CYP6P7
Benzyloxyresorufin	6.81 ± 0.65	4.99 ± 0.74
Ethyloxyresorufin	2.88 ± 0.21	3.61 ± 0.17
Methyloxyresorufin	0.02 ± 0.01	-
Penthyloxyresorufin	0.01 ± 0.01	-

Table 2. Specific activities of CYP6AA3 and CYP6P7 toward resorufin derivatives.

Using fluorescence-based assays, we could initially determine what compound types that give mechanism-based inhibition pattern by pre-incubation of enzyme with various concentrations of test inhibitors in the presence or absence of NADPH for 30 min before addition of substrates and IC_{50} values have been determined as described (Duangkaew et al., 2011b). As known, mechanism-based inactivation inhibits enzyme irreversibly, rendering this mechanism of inhibition more efficient than reversible inhibition. Nevertheless information on mode of inhibition for inhibitors is potential for understanding of catalytic nature of enzymes. We thus determined mode of inhibition for all compounds tested. As shown in Table 2, the compounds we have tested are phenolic compounds and their chemical structures are shown in Fig. 2.

It is apparent that none of test flavonoids and furanocoumarins shows mechanism-based inhibition pattern, but piperonyl butoxide (PBO) and piperine that are methylenedioxyphenyl compounds show NADPH-dependent mechanism-based inhibition activities against both enzymes. Piperine has been commonly found in *Piper sp.* plant extracts, it possesses acute toxicity to mammals (Daware et al., 2000). Inhibition results shown in Table 3 also elucidate that α-naphthoflavone displayed strongest inhibitory effect. Its inhibition pattern suggests that α-naphthoflavone uncompetitively inhibit both enzymes by binding to CYP6AA3– and CYP6P7-benzyloxyresorufin complex. Moreover, a difference was noted for xanthotoxin as it uncompetitively inhibits CYP6AA3 but mixed-type inhibited CYP6P7. Thus inhibition results together with different metabolic profiles thus confirm that CYP6AA3 and CYP6P7 have different enzyme properties. We thus also tested crude extracts of two plants (*Citrus reticulate* and *Stemona spp.*) that were reported containing phenolic compounds (Kaltenegger et al., 2003; Jayaprakasha et al., 1997) and are found in Thailand. Initial results suggest that compounds within both plants may not possess mechanism-based activities against CYP6AA3 and CYP6P7, and both extracts did not inhibit both enzymes as efficient as flavonoids and methylenedioxyphenyl compounds.

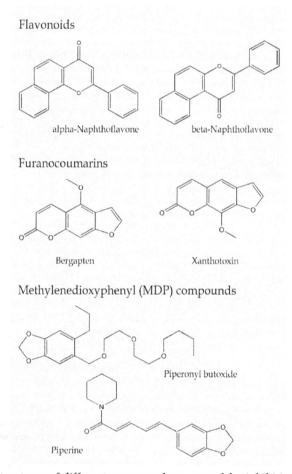

Fig. 2. Chemical structures of different compound types used for inhibition assays of mosquito P450s.

| Inhibitor | Inhibition type (K_i, µM) | | IC_{50} (µM) | | | |
| | CYP6AA3 | CYP6P7 | CYP6AA3 pre-incubation | | CYP6P7 pre-incubation | |
			w/o NADPH	w/ NADPH	w/o NADPH	w/ NADPH
Flavonoids						
α-Naphthoflavone	Uncompetitive (0.84)	Uncompetitive (2.02)	0.37 ± 0.06	0.38 ± 0.06	2.90 ± 0.27	3.03 ± 0.45
-Naphthoflavone	ND	ND	19.22 ± 3.13[a]	34.44 ± 5.95[a]	17.25 ± 3.67	33.35 ± 9.90
Furanocoumarins						
Xanthotoxin	Uncompetitive (52.45)	Mixed-type (47.14)	51.04 ± 2.15	52.17 ± 2.86	33.77 ± 3.54[a]	78.93 ± 10.04[a]
Bergapten	Mixed-type (93.27)	Mixed-type (65.59)	93.77±10.87[a]	170.3 ± 16.88[a]	52.76 ± 6.77[a]	114.0 ± 11.81[a]
Methylenedioxyphenyl (MDP) compounds						
Piperine	Mechanism-based (ND)	Mechanism-based (ND)	15.26± 1.21[a]	4.86 ± 0.79[a]	52.86 ± 6.92[a]	3.48 ± 0.36[a]
Piperonyl butoxide (PBO)	Mechanism-based (ND)	Mechanism-based (ND)	9.91 ± 0.81[a]	4.04 ± 0.31[a]	31.77 ± 3.21[a]	16.22 ± 1.81[a]
Crude Extracts						
Citrus reticulata	ND	ND	236.1± 32.6	234.9 ± 9.54	116.4 ± 16.54	141.1 ± 15.1
Stemona spp.	ND	ND	56.11 ± 7.05	63.91 ± 5.2	71.77 ± 5.73[a]	105.7 ±10.18[a]

Table 3. Mode of inhibition and inhibition constants of CYP6P7- or CYP6AA3-benzyloxyresorufin-O-debenzylation activities of flavonoids, furanocoumarins, and MDP compounds (Duangkaew et al., 2011b). Crude plant extracts reported herein are ethanolic extracts. Values marked with 'a' are significantly different between reactions with (w/) and without (w/o) NADPH. ND, not determined.

2.3 Use of cell-based MTT cytotoxicity assays to determine insecticide substrates and inhibitiors of *An. minimus* P450 enzymes

Since *in vitro* reconstitution system demonstrated CYP6AA3 and CYP6P7 enzymatic activities against pyrethroids, further investigation of the ability of CYP6AA3 and CYP6P7 enzymes to eliminate pyrethroid toxicity from cells was assessed in P450-infected *Sf9* cells. This can be accomplished because other than targeting on sodium channels of nervous system, pyrethroids possess toxic effects on cells such as inhibition of cell mitochondrial complex I or causing DNA damage and cell death (Gassner et al., 1997; Patel et al., 2007; Naravaneni & Jamil, 2005). Similar cell death and cytotoxic to cells caused by organophosphates and carbamate insecticides have also been reported (Maran et al., 2010; Schmuck & Mihail, 2004). This is supported by that we previously observed cytotoxic effects of deltamthrin on insect *Sf9* cells. When using *Sf9* cells that express CYP6AA3 in MTT assays, cell mortality was drastically decreased in the presence of insecticides due to degradation of deltamethrin by CYP6AA3 and thus posing cytoprotective role on *Sf9* cells (Boonseupsakul et al., 2008). Use of insect cells to test for toxicity effects of compounds such as fungal metabolites (Fornelli et al., 2004) and pyridalyl insecticide (Saito et al., 2005) has been previously reported. Moreover, insect cells expressing P450 have also been successfully used to test detoxification capability of enzyme against cytotoxic xenochemicals (Grant et al., 1996; Greene et al., 2000). In this context, we used MTT assays to determine insecticide detoxification by P450 expressed in *Sf9* cells. Insecticides tested were pyrethroids (deltamethrin, permethrin, cypermethrin, bifenthrin, bioallethrin and λ-cyhalothrin), organophosphates (chlorpyrifos, dichlorvos, fenitrothion and temephos), carbamates (thiodicarb and propoxur). Various concentrations (1-500 μM) of insecticides were used for determination of cytotoxic effect of insecticides toward CYP6AA3-, CYP6P7-, and CYP6P8-expressing cells and compared to the control parent *Sf9* cells. Cell viability of insecticide treated cells was measured by MTT assay as previously described (Boonseupsakul et al., 2008) and plotted against insecticide concentrations. The LC_{50} value of each insecticide was subsequently evaluated and obtained from each plot. Table 4 summarizes LC_{50} values of insecticides against *Sf9* cells and cells with expression of P450s.

We observed that pyrethroids, organophosphates and carbamates are toxic to *Sf9* parent cells. Since LC_{50} values of permethrin, bifenthrin, cypermethrin, and deltamethrin against CYP6AA3- and CYP6P7-expressing cells were approximately 4- to 19-folds significantly greater than those from parent *Sf9* cells, these values imply that CYP6AA3 and CYP6P7 enzymes could cytoprotect *Sf9* cells from pyrethroid toxicity. Conversely there was no significant difference of IC_{50} values between cells treated with organophosphate (chlorpyrifos, fenitrothion and temephos), carbamates (thiodicarb and propoxur) and bioallethrin pyrethroid insecticide, suggesting that expression of P450s did not cytoprotect cells from these insecticides. In addition CYP6P8 did not cytoprotect *Sf9* cells against insecticides tested. It should be noted that LC_{50} value of λ-cyhalothrin in CYP6AA3-expressing cells was significantly greater than *Sf9* parent cells, but not in CYP6P7-expressing cells. These results from MTT cytotoxicity assays are thus in agreement with *in vitro* enzymatic assays as described in Section 2.1. Thus abilities to cytoprotect against insecticide toxicity in infected *Sf9* cells are due to P450-mediated enzymatic activity toward insecticides of CYP6AA3 and CYP6P7 (Duangkaew et al., 2011a). Together with *in vitro* enzymatic and cytotoxicity assays, we can conclude that both CYP6AA3 and CYP6P7 share metabolic activities against pyrethroids, but both enzymes play no role in degradations of organophosphates and carbamates. The results suggest that CYP6P8 plays no role in degradation of insecticides tested in this report. Moreover, such cytotoxicity results implicate that the method could also be applied for primary screening of compounds that have an inhibitory effect towards CYP6AA3 and CYP6P7, as well as P450 enzymes that possess enzymatic activities against these insecticides.

Insecticides	LC_{50} (μM)			
	*Sf*9	CYP6AA3	CYP6P7	CYP6P8
Pyrethroids				
Bioallethrin[b]	30.6 ± 2.1	32.7 ± 2.4	23.3 ± 3.9	29
Permethrin[b]	42.7 ± 1.8	406.7 ± 21.5[a]	214.7 ± 48.8[a]	78
Bifenthrin	45 ± 7.6	210 ± 12.4[a]	135 ± 51[a]	45
Cypermethrin[b]	21.8 ± 0.5	192.7 ± 30.4[a]	216.7 ± 21.4[a]	25
Deltamethrin[b]	27.5 ± 9.2	285.0 ± 27.8[a]	379.5 ± 21.9[a]	10
λ-Cyhalothrin[b]	38.4 ± 4.3	133.3 ± 37.5[a]	42.0 ± 1.8	ND
Organophosphates				
Chlorpyrifos[b]	40.3 ± 6.5	56.3 ± 8.5	41.7 ± 2.8	60
Fenitrothion	25.0 ± 5.3	30.0 ± 6.4	ND	25
Temephos	11.0 ± 3.9	19.0 ± 7.5	ND	ND
Dichlorvos	32.0 ± 8.9	39.0 ± 6.4	ND	ND
Carbamates				
Propoxur[b]	4.0 ± 6.6	4.7 ± 0.3	3.6 ± 0.2	ND
Thiodicarb	28.6 ± 2.3	29.2 ± 4.7	ND	ND

Table 4. Cytotoxicity effects by insecticides on P450-infected cells and the parent SF9 cells using MTT assays. Values reported for CYP6P8 were average obtained from experiments performed in duplicate. Those marked with 'a' were significantly different from parent *Sf*9 cells and those marked with 'b' were reported in Duangkaew et al, 2011a. ND, not determined.

To test whether inhibitors can be screened, MTT assays were performed with P450-expressing cells treated with 100 μM deltamethin in the presence or absence of each test inhibitor. Concentrations of test inhibitory compounds were those of approximately LC_{20} values pre-determined by MTT assays on *Sf*9 cells. In cell-based inhibition assays, cell viability was determined upon co-incubation of test compound and deltamethrin, and normalized with viability of cells treated with test compound without deltamethrin. Inhibition experiments were performed with control *Sf*9 cells in the same manner as CYP6AA3-expressing cells and percent cell viability was plotted against test inhibitor concentrations, and results demonstrated that cell viability of parent *Sf*9 cells was not affected by test compounds (data not shown).
The results shown in Fig. 3 indicate that cell viability of CYP6AA3-expressing cells was decreased upon increasing concentration of test inhibitors. Piperine, piperonyl butoxide, and α-naphthoflavone could inhibit cytoprotective activity of CYP6AA3 more than xanthotoxin. This is thus in compliance with *in vitro* enzymatic inhibition assays, although piperonyl butoxide showed more potential than α-naphthoflavone in inhibiting cytoprotective activity of CYP6AA3. Cell permeability of test compounds could be accounted for differences of cell-based MTT and *in vitro* enzymatic assays. The results however indicate usefulness of cells expressing P450 enzymes to primarily screen for P450 substrates and inhibitors. Our results indicated that PBO and piperine could inhibit P450s

and possess synergistic actions against deltamethrin cytotoxicity in Sf9 cells expressing P450. PBO has been used as pyrethroid synergist to enhance pyrethroid toxicity, as it can bind to P450s thereby inhibiting P450 activity (Fakoorziba et al., 2009; Kumar et al., 2002; Vijayan et al., 2007). Unfortunately PBO has been reported acutely toxic to mammals (Cox, 2002; Okamiya et al., 1998).

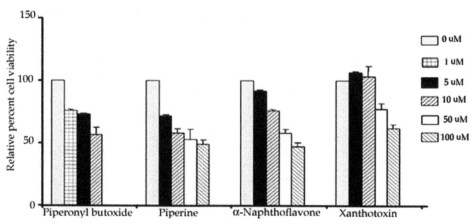

Fig. 3. Inhibition effect of test compounds against cell-based CYP6AA3-mediated deltamethrin detoxification measured by MTT assays.

2.4 *An. minimus* CYPOR and its possible role in regulation of P450-reaction cycle

The NADPH-Cytochrome P450 oxidoreductase (CYPOR) enzyme is a member of di-flavin enzymes that transfers electrons, one by one, from NADPH through FAD and FMN to target enzymes to fulfill functioning of various cytochrome P450 enzymes as well as other enzymes (Murataliev et al., 2004). Other members of this class are those containing a flavoprotein subunit, such as nitric oxide synthase, sulfite reductase, methionine synthase reductase and protein NR 1. Detailed biochemical and structural studies of rat CYPOR reveal several conserved structural domains existed in this enzyme class, these are membrane-bound, FMN-binding, connecting, and FAD/NADPH binding domains (Wang et al., 1997).

The *An. minimus* CYPOR has been cloned and expressed in *E. coli*, and CYPOR could support CYP6AA3- and CYP6P7-mediated pyrethroid metabolisms *in vitro* (Duangkaew et al., 2011b; Kaewpa et al., 2007). However its expression has been of poor yield as a result of inclusion bodies formation. An attempt to obtain soluble protein by deletion of the first 55 amino acid residues comprising of membrane binding region (*Δ55AnCYPOR*) has been successful (Sarapusit et al., 2008). However the protein could not be purified by 2'5'-ADP affinity column, indicating that NADPH binding capacity of mosquito CYPOR is low and this is different from CYPORs of other organisms such as rat and human (Sarapusit, 2009). Low binding affinity to 2'5'-ADP affinity column has also been recently reported in *An. gambiae* CYPOR (Lian et al., 2011). Only under specific condition was *Δ55AnCYPOR* successfully expressed and purified to homogeneity by a combination of Ni²⁺NTA-affinity chromatography and G200-gel filtration chromatography (Sarapusit et al., 2008). Moreover both purified full-length (*flAnCYPOR*) and membrane-deleted *Δ55AnCYPOR* proteins readily lose FAD and FMN cofactors, they undergo aggregation and are unstable compared to rat and human CYPORs (Sarapusit et al., 2008, 2010). While supplementation of FAD could increase activity of both full-length and membrane-deleted forms, FMN supplementation could increase activity of full-length form only (Sarapusit et al., 2008, 2010). This behavior is different from membrane-deleted soluble CYPORs of rat and human in which exogenous FMN is readily incorporated into its FMN-binding site (Döhr et al., 2001; Shen et al., 1989). Due to loss of flavin cofactors and instability of *An. minimus* CYPOR, we have identified two key amino acids (Leu86 and Leu219 in FMN binding domain) by amino sequence alignment and shown that mutations of the two leucine residues into conserved phenylalanine residues that are found conserved among other CYPORs could rescue loss of FAD cofactor and increase protein stability of mosquito CYPOR (Sarapusit et al., 2008, 2010). These mutations do not affect kinetic mechanism and constants of enzyme. Double mutations of leucine to the conserved phenylalanine (L86F/L219F) in full-length *flAnCYPOR*, but not in *Δ55AnCYPOR*, could increase binding of FMN and increase CYP6AA3-mediated pyrethroid metabolism (Sarapusit et al., 2010), indicating that membrane-bound region of *An. minimus* CYPOR could influence both structural folding of FMN domain and mediation of P450 catalysis (Murataliev et al., 2004; Wang et al., 1997).

The enzyme activity and kinetic mechanism of *flAnCYPOR* using cytochrome c as substrate are ionic strength dependent, with its mechanism following random Bi-Bi mechanism at low ionic strength and non-classical two-side Ping-Pong at high ionic strength. These mechanisms are different from rat, human, and house fly CYPORs (Murataliev et al., 2004; Sem & Kasper, 1994, 1995). In addition, *flAnCYPOR* could use extra flavins as additional substrates in which FAD binds at FAD/NADPH domain and FMN binds at FMN domain (as depicted in Fig. 4), resulting in an increase in its rate of electron transfer in CYP6AA3-mediated pyrethroid degradation (Sarapusit et al., 2010).

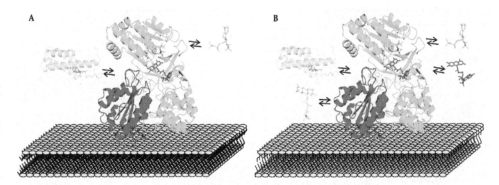

Fig. 4. Schematic representation of enzymatic reactions of CYPOR enzymes. CYPOR enzymes are represented in cartoon model of which FMN domain is in red color and FAD/NADPH domain is in green. Cofactors are represented in the stick mode; FMN is yellow colored, FAD is red, and NADP+ is cyan (rat CYOR: pdb code 1AMO). The cytochrome c substrate (cytychrome c: pdb code 1BBH) is in cyan cartoon model with an orange heme group residing at the center.

In Figure 4, panel A illustrates common CYPOR (such as rat, human CYPORs) to which NADPH and cytochrome c substrate separately binds FAD/NADPH and FMN domains, while in panel B, *flAnCYPOR* could use extra flavins as additional substrates to which FAD cofactor binds FAD/NADPH domain and FMN cofactor binds FMN domain. We thus speculate that *An. minimus* mosquito uses CYPOR in regulation of P450-mediated metabolisms, since it supplies electrons to a collection of P450s within the cell. Although structural basis for loose binding of flavin cofactors in *An. minimus* CYPOR is not known, it is conceivable that its distinct property that adopt extra flavins as substrates may render the enzyme ability to regulate electron transfer to target mosquito enzymes.

3. Conclusion

The results of this study on CYP6AA3 and CYP6P7 could lay groundwork into an understanding of the mechanisms that control substrates and reaction selectivity of both P450 enzymes, thereby increase an understanding of P450-mediated resistance mechanisms to various pesticides. The kinetic values, metabolic profile of pyrethroid insecticide metabolisms and inhibition patterns by different inhibitors of CYP6AA3 are different from CYP6P7. Future approach could aim at the strategy involving finding a collection of substrates together with structural models and mutation analyses of CYP6AA3 and CYP6P7 that affect specific P450 catalysis. Moreover, characterizing inhibitors and inhibition mechanisms of large collection of compounds with known chemical structures against CYP6AA3 and CYP6P7 enzymes could give insight into an understanding of mechanisms of cytochrome P450s that metabolize pyrethroids. It is conceivable that CYP6P8 does not play role in detoxification of pyrethroid, organophosphate, and carbamate insecticides. Further substrate search for CYP6P8 may help to learn about its overexpression in pyrethroid-resistant mosquito. Together with knowledge obtained from enzymatic properties of *An. minimus* CYPOR, this could improve our understanding of P450-mediated detoxification of insecticides, as well as provide a foundation for rational design of P450 synergists specific for P450-mediated pesticide resistance and thus resistant management in mosquito vector control program.

4. Acknowledgment

This work is supported by Grant BRG5380002 from Thailand Research Fund and Mahidol University; grant BT-B01-XG-14-4803 from BIOTEC, National Science and Technology Development Agency.

5. References

Acevedo, G.R.; Zapater, M. & Toloza, A.C. (2009). Insecticide resistance of house fly, *Musca domestica* (L.) from Argentina. *Parasitology Research*, Vol. 105, No. 2, pp. 489-493, ISSN 0932-0113

Awolola, T.S.; Brooke, B.D.; Hunt, R.H. & Coetzee, M. (2002). Resistance of the malaria vector *Anopheles gambiae s.s.* to pyrethroid insecticides, in south-western Nigeria. *Annals of Tropical Medicine and Parasitology*, Vol. 96, No. 8, pp. 849-852, ISSN 003-4983

Bautista, Ma.A.M.; Miyata, T.; Miura, K. & Tanaka, T. (2009). RNA interference-mediated knockdown of a cytochrome P450, CYP6BG1, from the diamondback moth *Plutella xylostella*, reduces larval resistance to permethrin. *Insect Biochemistry and Molecular Biology*, Vol. 39, No. 1, pp. 38-46, ISSN 0965-1748

Bautista, Ma.A.M; Tanaka, T. & Miyata, T. (2007). Identification of permethrin-inducible cytochrome P450s from the diamondback moth, *Plutella xylostella* (L.) and the possibility of involvement in permethrin resistance. *Pesticide Biochemistry and Physiology*, Vol. 87, No. 1, pp. 85-93, ISSN 0048-3575

Boonsuepsakul, S. (2008). Characterization of CYP6AA3 in baculovirus expression system. Ph.D. Thesis, Mahidol University, Bangkok, Thailand

Boonsuepsakul, S.; Luepromchai, E. & Rongnoparut, P. (2008). Characterization of *Anopheles minimus* CYP6AA3 expressed in a recombinant baculovirus system. *Archives of Insect Biochemistry and Physiology*, Vol. 63, No. 1, pp. 13-21, ISSN 0739-4462

Chareonviriyaphap, T.; Rongnoparut, P. & Juntarumporn, P. (2002). Selection for pyrethroid resistance in a colony of *Anopheles minimus* species A, a malaria vector in Thailand. *Journal of Vector Ecology*, Vol. 27, No. 2, pp. 222-229, ISSN 1081-1710

Chareonviriyaphap, T.; Rongnoparut, P.; Chantarumporn, P. & Bangs, M.J. (2003). Biochemical detection of pyrethroid resistance mechanisms in *Anopheles minimus* in Thailand. *Journal of Vector Ecology*, Vol. 28, No. 1, pp. 108-116, ISSN 1081-1710

Cochran, D.G. (1989). Monitoring for insecticide resistance in field-collected strains of the German cockroach (Dictyoptera: Blattellidae). *Journal of Economic Entomology*, Vol. 82, No. 2, pp. 336-341, ISSN 0022-0493

Cox, C. (2002). Piperonyl butoxide. *Journal of Pesticide Reform*, Vol. 22, pp. 12–20, ISSN 0893-357X

Daware, M.B.; Mujumdar, A.M. & Ghaskadbi, S. (2000). Reproductive toxicity of piperine in Swiss Albino mice. *Planta Medica*, Vol. 66, No. 3, pp. 231-236, ISSN 0032-0943

Djouaka, R.F.; Bakare, A.A.; Coulibaly, O.N.; Akogbeto, M.C.; Ranson, H.; Hemingway, J. & Strode, C. (2008). Expression of the cytochrome P450s, CYP6P3 and CYP6M2 are significantly elevated in multiple pyrethroid resistant populations of *Anopheles gambiae s.s.* from Southern Benin and Nigeria. *BMC genomics*, Vol. 9, p. 538, ISSN 1471-2164

Döhr, O.; Paine, M.J., Friedberg, T.; Robert, G.C.K. & Wolf, R. (2001). Engineering of a functional human NADH-dependent cytochrome P450 system *Proceedings of the National Academy of Sciences of the United States of America*, Vol. 98, No.1, pp. 81-86, ISSN 0027-8424

Duangkaew, P.; Kaewpa, D. & Rongnoparut, P. (2011a). Protective efficacy of *Anopheles minimus* CYP6P7 and CYP6AA3 against cytotoxicity of pyrethroid insecticides in *Spodoptera frugiperda* (Sf9) insect cells. *Tropical Biomedicine*, Vol. 28, No. 2, pp. 293-301, ISSN 0127-5720

Duangkaew, P.; Pethuan, S.; Kaewpa, D.; Boonseupsakul, S.; Sarapusit, S. & Rongnoparut, P. (2011b). Characterization of mosquito CYP6P7 and CYP6AA3: Differences in substrate preferences and kinetic properties. *Archives of Insect Biochemistry and Physiology*, Vol. 76, No. 4, pp. 236-248, ISSN 0739-4462

Fakoorziba, M.R.; Eghbal, F. & Vijayan, V.A. (2009). Synergist efficacy of piperonyl butoxide with deltamethrin as pyrethroid insecticide on *Culex tritaeniorhynchus* (Diptera: Culicidae) and other mosquitoe species. *Environmental Toxicology*, Vol. 24, No. 1, pp. 19-24, ISSN 1520-4081

Feyereisen, R. (1999). Insect P450 enzymes. *Annual Review of Entomology*, Vol. 44, pp. 507-533, ISSN 0066-4170

Fornelli, F.; Minervini, F. & Logrieco, A. (2004). Cytotoxicity of fungal metabolites to lepidopteran (*Spodoptera frugiperda*) cell line (*Sf-9*). *Journal of Invertebrate Pathology*, Vol. 85, No. 2, pp. 74-79, ISSN 0022-2011

Gassner, B., Wüthrich, A., Scholtysik, G. & Solioz, M. (1997). The pyrethroids permethrin and cyhalothrin are potent inhibitors of the mitochondrial complex I. *Journal of Pharmacology and Experimental Therapeutics*, Vol. 281, No. 2, pp. 855–860, ISSN 0022-3565

Godin, S.J.; Crow, J.A.; Scollon, E.J.; Hughes, M.F.; DeVito, M.J. & Ross, M.K. (2007). Identification of rat and human cytochrome P450 isoforms and a rat serum esterase that metabolize the pyrethroid insecticides deltamethrin and esfenvalerate. *Drug Metabolism and Disposition*, Vol. 35, No. 9, pp. 1664–1671, ISSN 0090-9556

Grant, D.F.; Greene, J.F.; Pinot, F.; Borhan, B.; Moghaddam, M.F.; Hummock, B.D.; McCutchen, B.; Ohkawa, H.; Luo, G. & Guenthner, T.M. (1996). Development of an *in situ* toxicity assay system using recombinant baculoviruses. *Biochemical Pharmacology*, Vol. 51, No. 4, pp. 503–515, ISSN 0006-2952

Greene, J.F.; Zheng, J.; Grant, D.F. & Hammock, B.D. (2000). Cytotoxicity of 1,2-epoxynaphthalene is correlated with protein binding and *in situ* glutathione depletion in cytochrome P4501A1 expressing *Sf-21* cells. *Toxicological Sciences*, Vol. 53, No. 2, pp. 352-360, ISSN 1096-6080

Hargreaves, K.; Koekemoer, L.L.; Brooke, B.D.; Hunt, R.H.; Mthembu, J. & Coetzee, M. (2000). *Anopheles funestus* resistant to pyrethroid insecticides in South Africa. *Medical and Veterinary Entomology*, Vol. 14, No. 2, pp. 181-189, ISSN 0269-283X

Hemingway, J.; Hawkes, N.J.; McCarroll, L. & Ranson, H. (2004). The molecular basis of insecticide resistance in mosquitoes. *Insect Biochemistry and Molecular Biology*, Vol. 34, No. 7, pp. 653-665, ISSN 0965-1748

Kaltenegger, E.; Brema, B.; Mereiter, K.; Kalchhauser, H.; Kählig, H.; Hofer, O.; Vajrodaya, S. & Greger, H. (2003). Insecticidal pyrido[1,2-a]azepine alkaloids and related derivatives from *Stemona* species. *Phytochemistry*, Vol. 63, No. 7, pp. 803-816, ISSN 0031-9422

Jayaprakasha, G.K.; Singh, R.P.; Pereira, J. & Sakariah, K.K. (1997). Limonoids from *Citrus reticulata* and their moult inhibiting activity in mosquito *Culex quinquefasciatus* larvae. *Phytochemistry*, Vol. 44, No. 5, pp. 843-846, ISSN 0031-9422

Jirakanjanakit, N.; Rongnoparut, P.; Saengtharatip, S.; Chareonviriyaphap, T.; Duchon, S.; Bellec, C. & Yoksan, S. (2007). Insecticide susceptible/resistance status in *Aedes (Stegomyia) aegypti* and *Aedes (Stegomyia) albopictus* (Diptera: Culicidae) in Thailand during 2003-2005. *Journal of Economic Entomology*, Vol. 100, No. 2, pp. 545-550, ISSN 0022-0493

Kaewpa, D.; Boonsuepsakul, S. & Rongnoparut, P. (2007). Functional expression of mosquito NADPH-cytochrome P450 reductase in *Escherichia coli. Journal of Economic Entomology*, Vol. 100, No. 3, pp. 946-953, ISSN 0022-0493

Kumar, S.; Thomas, A.; Sahgal, A.; Verma, A.; Samuel, T. & Pillai, M.K.K. (2002). Effect of the synergist, piperonyl butoxide, on the development of deltamethrin resistance in yellow fever mosquito, *Aedes aegypti* L. (Diptera: Culicidae). *Archives of Insect Biochemistry and Physiology*, Vol. 50, No. 1, pp. 1-8, ISSN 0739-4462

Lertkiatmongkol, P.; Jenwitheesuk, E. & Rongnoparut, P. (2011). Homology modeling of mosquito cytochrome P450 enzymes involved in pyrethroid metabolism: insights into differences in substrate selectivity. *BMC Research Notes*, Vol. 4, p. 321, ISSN 1756-0500

Lian, L.Y.; Widdowson, P.; McLaughlin, L.A. & Paine, M.J. (2011). Biochemical comparison of *Anopheles gambiae* and Human NADPH P450 reductases reveals different 2'-5'-ADP and FMN binding traits. *Plos one*, Vol. 6, No. 5, pp. e20574, ISSN 1932-6203

Liu, N. & Scott, J.G. (1998). Increased transcription of CYP6D1 causes cytochrome P450-mediated insecticide resistance in house fly. *Insect Biochemistry and Molecular Biology*, Vol. 28, No.8, pp. 531-535, ISSN 0965-1748

Maran, E.; Fernández-Franzón, M.; Font, G. & Ruiz, M.J. (2010). Effects of aldicarb and propoxur on cytotoxicity and lipid peroxidation in CHO-K1 cells. *Food and Chemical Toxicology*, Vol. 48, No. 6, pp. 1592–1596, ISSN 0278-6915

Mclaughlin, L.A.; Niazi, U.; Bibby, J.; David, J.-P.; Vontas, J.; Hemingway, J.; Ranson, H.; Sutcliffe, M.J. & Paine M.J.I. (2008). Characterization of inhibitors and substrates of *Anopheles gambiae* CYP6Z2. *Insect Molecular Biology*, Vol. 17, No. 2, pp. 125-135, ISSN 0962-1075

Müller, P.; Donnelly, M.J. & Ranson H. (2007). Transcription profiling of a recently colonized pyrethroid resistant *Anopheles gambiae* strain from Ghana. *BMC Genomics*, Vol. 8, p. 36, ISSN 1471-2164

Müller, P.; Warr, E.; Stevenson, B.J.; Pignatelli, P.M.; Morgan, J.C.; Steven, A.; Yawson, A.E.; Mitchell, S.N.; Ranson, H.; Hemingway, J.; Paine, M.J.I. & Donnelly, M.J. (2008). Field-caught permethrin-resistant *Anopheles gambiae* overexpress CYP6P3, a P450 that metabolises pyrethroids. *PLoS Genetics*, Vol. 4, No. 11, pp. e1000286, ISSN 1553-7390

Murataliev, M..B.; Feyereisen, R. & Walker, F.A. (2004). Electron transfer by diflavin reductases. *Biochimica et Biophysica Acta*, Vol. 1698, No. 1, pp. 1 – 26, ISSN 0006-3002

Naravaneni, R. & Jamil, K. (2005). Evaluation of cytogenetic effects of lambda-cyhalothrin on human lymphocytes. *Journal of Biochemical and Molecular Toxicology*, Vol. 19, No. 5, pp. 304-310, ISSN 1095-6670

Nelson, D.R.; Koymans, L.; Kamataki, T.; Stegeman, J.J.; Feyereisen, R.; Waxman, D.J.; Waterman, M.R.; Gotoh, O.; Coon, M.J; Estabrook, R.W.; Gunsalus, I.C. & Nebert, D.W. (1996). P450 superfamily: update on new sequences, gene mapping, accession numbers and nomenclature. *Pharmacogenetics*, Vol. 6, No. 1, pp. 1-42, ISSN 0960-314X

Okamiya, H.; Mitsumori, K.; Onodera, H.; Ito, S.; Imazawa, T.; Yasuhara, K. & Takahashi, M. (1998). Mechanistic study on liver tumor promoting effects of piperonyl butoxide in rats. *Archives of Toxicology*, Vol. 72, No. 11, pp. 744–750, ISSN 0340-5761

Omura, T. & Sato, R. (1964). The carbon monoxide-biding pigment of liver microsome I. Evidence for its hemoprotein nature. *The Journal of Biological Chemistry*, Vol. 239, pp. 2370-2378, ISSN 0021-9258

Ortiz de Montellano, P.R. (2005). *Cytochrome P450: structure, mechanism, and biochemistry*, Kluwer Academic/Plenum Piblishers, ISBN 0-306-48324-6, New York, USA.

Patel, S.; Bajpayee, M.; Pandey, A.K.; Parmar, D. & Dhawan, A. (2007). *In vitro* induction of cytotoxicity and DNA strand breaks in CHO cells exposed to cypermethrin, pendimethalin and dichlorvos. *Toxicology In Vitro*, Vol. 21, No. 8, pp. 1409–1418, ISSN 0887-2333

Price, N.R. (1991). Insect resistance to insecticides: mechanisms and diagnosis. *Comparative Biochemistry and Physiology - C Pharmacology Toxicology and Endocrinology*, Vol. 100, No. 3, pp. 319-326, ISSN 0742-8413

Ranasinghe, C. & Hobbs, A.A. (1998). Isolation and characterization of two cytochrome P450 cDNA clones for CYP6B6 and CYP6B7 from *Helicoverpa armigera* (Hubner): possible involvement of CYP6B7 in pyrethroid resistance. *Insect Biochemistry and Molecular Biology*, Vol. 28, No. 8, pp. 571-580, ISSN 0965-1748

Rodpradit, P.; Boonsuepsakul, S.; Chareonviriyaphap, T.; Bangs, M.J. & Rongnoparut, P. (2005). Cytochrome P450 genes: molecular cloning and overexpression in a pyrethroid-resistant strain of *Anopheles minimus* mosquito. *Journal of the American Mosquito Control Association*, Vol. 21, No. 1, pp. 71-79, ISSN 8756-971X

Rongnoparut, P.; Boonsuepsakul, S.; Chareonviriyaphap, T. & Thanomsing, N. (2003). Cloning of cytochrome P450, CYP6P5, and CYP6AA2 from *Anopheles minimus* resistant to deltamethrin. *Journal of Vector Ecology*, Vol. 28, No. 2, pp. 150-158, ISSN 1081-1710

Saito, S.; Sakamoto, N. & Umeda, K. (2005). Effects of pyridalyl, a novel insecticidal agent, on cultured *Sf*9 cells. *Journal of Pesticide Science*, Vol. 30, No. 1, pp. 17-21, ISSN 0385-1559

Sarapusit, S. (2009). Study of NADPH-cytochrome P450 oxidoreductase from *Anopheles minimus* mosquito. Ph.D Thesis, Mahidol University, Bangkok, Thailand

Sarapusit, S.; Xia, C.; Misra, I.; Rongnoparut, P. & Kim, J-JP. (2008). NADPH-Cytochrome P450 oxidoreductase from *Anopheles minimus* mosquito: Kinetic studies and the influence of Leu86 and Leu219 for cofactors binding and protein stability. *Archives of Biochemistry and Biophysics*, Vol. 477, No. 1, pp. 53-59, ISSN 0739-4462

Sarapusit, S.; Pethuan, S. & Rongnoparut, P. (2010). Mosquito NADPH-cytochrome P450 oxidoreductase: kinetics and role of phenylalanine amino acid substitutions at Leu86 and Leu219 in CYP6AA3-mediated deltamethrin metabolism. *Archives of Insect Biochemistry and Physiology*, Vol. 73, No. 4, pp. 232-244, ISSN 0739-4462

Schmuck, G. & Mihail, F. (2004). Effects of the carbamates fenoxycarb, propamocarb and propoxur on energy supply, glucose utilization and SH-groups in neurons. *Archives of Toxicology*, Vol. 78, No. 6, pp. 330–337, Epub 2004 Feb 19, ISSN 0340-5761

Scollon, E.J.; Starr, J.M.; Godin, S.J.; DeVito, M.J. & Hughes, M.F. (2009). *In vitro* metabolism of pyrethroid pesticides by rat and human hepatic microsomes and cytochrome P450 isoforms. *Drug Metabolism and Disposition*, Vol. 37, No. 1, pp. 221–228, ISSN 0090-9556

Sem, D.S & Kasper, C.B. (1994). Kinetic mechanism for the model reaction of NADPH-cytochrome P450 oxidoreductase with cytochrome c. *Biochemistry*, Vol. 33, No. 4, pp. 12012–12021, ISSN 0006-2960

Sem, D.S. & Kasper, C.B. (1995). Effect of ionic strength on the kinetic mechanism and relative rate limitation of steps in the model NADPH-cytochrome P450 oxidoreductase reaction with cytochrome c. *Biochemistry*, Vol. 34, No. 39, pp.12768–12774, ISSN 0006-2960

Shen, A.L., Porter, T.D., Wilson, T.E., Kasper, C.B. (1989) Structural analysis of the FMN binding domain of NADPH-cytochrome P-450 oxidoreductase by site-directed mutagenesis. *Journal of Biological Chemistry*, Vol. 254, No. 13, pp. 7584–7589, ISSN 0021-9258

Shono, T.; Ohsawa, K. & Casida, J.E. (1979). Metabolism of *trans*- and *cis*-permethrin, *trans*- and *cis*-cypermethrin and decamethrin by microsomal enzymes. *Journal of Agriculture and Food Chemistry*, Vol. 27, No. 2, pp. 316-325, ISSN 0021-8561

Stevenson, B.J.; Bibby, J.; Pignatelli, P.; Muangnoicharoen, S.; O'Neill, P.M.; Lian, L-Y; Müller, P.; Nikou, D.; Steven, A.; Hemingway, J.; Sutcliffe, M.J. & Paine, M.J.I. (2011). Cytochrome P450 6M2 from the malaria vector *Anopheles gambiae* metabolizes pyrethroids: Sequential metabolism of deltamethrin revealed. *Insect Biochemistry and Molecular Biology*, Vol. 41, No. 7, pp. 492-502, ISSN 0965-1748

Tomita, T.; Liu, N.; Smith, F.F.; Sridhar, P. & Scott, J.G. (1995). Molecular mechanisms involved in increased expression of a cytochrome P450 responsible for pyrethroid resistance in the housefly, *Musca domestica. Insect Molocular Biology*, Vol. 4, No. 3, pp. 135-140, ISSN 0962-1075

Vijayan, V.A.; Sathish Kumar, B.Y.; Ganesh, K.N.; Urmila, J.; Fakoorziba, M.R. & Makkapati, A.K. (2007). Efficacy of piperonyl butoxide (PBO) as a synergist with deltamethrin on five species of mosquitoes. *The Journal of Communicable Diseases*, Vol. 39, No. 3, pp. 159-163, ISSN 0019-5638

Wang, M.; Roberts, D.L.; Paschke, R.; Shea, T.M.; Masters, B.S. & Kim, J.J. (1997). Three-dimensional structure of NADPH-cytochrome P450 reductase: prototype for FMN- and FAD-containing enzymes. *Proceedings of the National Academy of Sciences of the United States of America*, Vol. 94, No. 16, pp. 8411-8416, ISSN 0027-8424

Wondji, C.S.; Morgan, J.; Coetzee, M.; Hunt, R.H.; Steen, K.; Black IV, W.C.; Hemingway, J. & Ranson, H. (2007). Mapping a Quantitative Trait Locus (QTL) conferring pyrethroid resistance in the African malaria vector *Anopheles funetus. BMC Genomics*, Vol. 8, p. 34, ISSN 1471-2164

Yaicharoen, R.; Kiatfuengfoo, R; Chareonviriyaphap, T. & Rongnoparut, P. (2005). Characterization of deltamethrin resistance in field populations of *Aedes aegypti* in Thailand. *Journal of Vector Ecology*, Vol. 30, No. 1, pp. 144-150, ISSN 1081-1710

Zhang, M. & Scott, J.G. (1996). Cytochrome b5 is essential for cytochrome P450 6D1-mediated cypermethrin resistance in LPR house flies. *Pesticide Biochemistry and Physiology*, Vol. 55, No. 2, pp. 150-156, ISSN 0048-3575

Zhu, F.; Parthasarathy, R.; Bai, H.; Woithe, K.; Kaussmann, M.; Nauen, R.; Harrison, D.A. & Palli S.R. (2010). A brain-specific cytochrome P450 responsible for the majority of deltamethrin resistance in the QTC279 strain of *Tribolium castaneum*. *Proceedings of the National Academy of Sciences of the United States of America*, Vol. 107, No. 19, pp. 8557-8562, ISSN 0027-8424

Permissions

The contributors of this book come from diverse backgrounds, making this book a truly international effort. This book will bring forth new frontiers with its revolutionizing research information and detailed analysis of the nascent developments around the world.

We would like to thank Farzana Perveen, for lending their expertise to make the book truly unique. They have played a crucial role in the development of this book. Without their invaluable contribution this book wouldn't have been possible. They have made vital efforts to compile up to date information on the varied aspects of this subject to make this book a valuable addition to the collection of many professionals and students.

This book was conceptualized with the vision of imparting up-to-date information and advanced data in this field. To ensure the same, a matchless editorial board was set up. Every individual on the board went through rigorous rounds of assessment to prove their worth. After which they invested a large part of their time researching and compiling the most relevant data for our readers. Conferences and sessions were held from time to time between the editorial board and the contributing authors to present the data in the most comprehensible form. The editorial team has worked tirelessly to provide valuable and valid information to help people across the globe.

Every chapter published in this book has been scrutinized by our experts. Their significance has been extensively debated. The topics covered herein carry significant findings which will fuel the growth of the discipline. They may even be implemented as practical applications or may be referred to as a beginning point for another development. Chapters in this book were first published by InTech; hereby published with permission under the Creative Commons Attribution License or equivalent.

The editorial board has been involved in producing this book since its inception. They have spent rigorous hours researching and exploring the diverse topics which have resulted in the successful publishing of this book. They have passed on their knowledge of decades through this book. To expedite this challenging task, the publisher supported the team at every step. A small team of assistant editors was also appointed to further simplify the editing procedure and attain best results for the readers.

Our editorial team has been hand-picked from every corner of the world. Their multi-ethnicity adds dynamic inputs to the discussions which result in innovative outcomes. These outcomes are then further discussed with the researchers and contributors who give their valuable feedback and opinion regarding the same. The feedback is then collaborated with the researches and they are edited in a comprehensive manner to aid the understanding of the subject.

Apart from the editorial board, the designing team has also invested a significant amount of their time in understanding the subject and creating the most relevant covers. They scrutinized every image to scout for the most suitable representation of the subject and create an appropriate cover for the book.

The publishing team has been involved in this book since its early stages. They were actively engaged in every process, be it collecting the data, connecting with the contributors or procuring relevant information. The team has been an ardent support to the editorial, designing and production team. Their endless efforts to recruit the best for this project, has resulted in the accomplishment of this book. They are a veteran in the field of academics and their pool of knowledge is as vast as their experience in printing. Their expertise and guidance has proved useful at every step. Their uncompromising quality standards have made this book an exceptional effort. Their encouragement from time to time has been an inspiration for everyone.

The publisher and the editorial board hope that this book will prove to be a valuable piece of knowledge for researchers, students, practitioners and scholars across the globe.

List of Contributors

A. C. Achudume
Institute of Ecology and Environmental Studies, Obafemi Awolowo UniversitY, Ile-Ife, Nigeria

Dong Wang, Hisao Naito and Tamie Nakajima
Nagoya University Graduate School of Medicine, Japan

Farzana Perveen
Chairperson, Department of Zoology, Hazara University, Garden Campus, Mansehra, Pakistan

Apurva Kumar R. Joshi and P.S. Rajini
Food Protectants and Infestation Control Department, Central Food Technological, Research Institute (CSIR lab), Mysore, India

Iván Nelinho Pérez Maldonado, Jorge Alejandro Alegría-Torres, Octavio Gaspar-Ramírez, Francisco Javier Pérez Vázquez, Sandra Teresa Orta-Garcia and Lucia Guadalupe Pruneda Álvarez
Laboratorio de Toxicología, Facultad de Medicina, Universidad Autónoma de San Luis Potosí, San Luis Potosí, SLP, México

Goran Gajski
Institute for Medical Research and Occupational Health

Marko Gerić
Institute for Medical Research and Occupational Health

Sanda Ravlić
Ruđer Bošković Institute, Center for Marine Research

Željka Capuder
Dr. Andrija Štampar Institute of Public Health, Croatia

Vera Garaj-Vrhovac
Institute for Medical Research and Occupational Health

Alhaji Aliyu
Department of Community Medicine, Ahmadu Bello University, Zaria, Nigeria

Theeraphap Chareonviriyaphap
Department of Entomology, Faculty of Agriculture, Kasetsart University, Bangkok, Thailand

Mario Ramírez-Lepe
Instituto Tecnológico de Veracruz, UNIDA, Veracruz, Mexico

Montserrat Ramírez-Suero
Université de Haufe-Alsace, LVBE EA-3991, Colmar, France

Maureen Leyva, Olinka Tiomno, Juan E. Tacoronte, Maria del Carmen Marquetti and Domingo Montada
Institute Tropical "Pedro Kouri", Cuba

Ronald Maestre Serrano
Doctoral Student in Tropical Medicine, Cartagena University, Colombia

Pornpimol Rongnoparut
Department of Biochemistry, Faculty of Science, Mahidol University, Thailand

Sirikun Pethuan
Department of Biochemistry, Faculty of Science, Mahidol University, Thailand

Songklod Sarapusit
Department of Biochemistry, Faculty of Science, Burapha University, Thailand

Panida Lertkiatmongkol
Department of Biochemistry, Faculty of Science, Mahidol University, Thailand

Printed in the USA
CPSIA information can be obtained
at www.ICGtesting.com
JSHW011459221024
72173JS00005B/1132